Fuel System and Emission Control

Classroom Manual

Fifth Edition

Chek-Chart

Warren M. Farnell
Revision Author

James D. Halderman
Series Advisor

PEARSON

Prentice Hall

Upper Saddle River, New Jersey
Columbus, Ohio

Library of Congress Cataloging-in-Publication Data

Farnell, Warren.
  Fuel system and emission control. Classroom manual / Warren M. Farnell, revision author;
  James D. Halderman, series advisor.--5th ed.
    p. cm.
  Includes index.
  ISBN 0-13-140784-8
  1. Automobiles--Fuel systems--Maintenance and repair--Handbooks, manuals, etc.
2. Automobiles--Pollution control devices--Maintenance and repair--Handbooks, manuals, etc.
I. Title.
  TL214.F8F368   2006
  629.25'2--dc22   2005020688

Acquisitions Editor: Tim Peyton
Assistant Editor: Linda Cupp
Editorial Assistant: Nancy Kesterson
Production Coordination: Carlisle Publishers Services
Production Editor: Holly Shufeldt
Design Coordinator: Diane Ernsberger
Cover Designer: Jeff Vanik
Cover art: Corel
Production Manager: Deidra Schwartz
Marketing Manager: Ben Leonard
Senior Marketing Coordinator: Liz Farrell
Marketing Assistant: Les Roberts

This book was set in Times by Carlisle Communications, Ltd. It was printed and bound by Bind-Rite Graphics. The cover was printed by The Lehigh Press, Inc.

Portions of materials contained herein have been reprinted with permission of General Motors Corporation, Service and Parts Operations. License Agreement 0510867.

Pearson Education Ltd.   Pearson Education Australia Pty. Limited
Pearson Education Singapore Pte. Ltd.   Pearson Education North Asia Ltd.
Pearson Education Canada, Ltd.   Pearson Educación de Mexico, S.A. de C.V.
Pearson Education—Japan   Pearson Education Malaysia Pte. Ltd.

10 9 8 7 6 5 4 3 2 1
0-13-140784-8

Introduction

Fuel System and Emission Control is part of the Chek-Chart automotive series. The entire series is job-oriented and designed especially for students who intend to work in the automotive service profession. The package for each course consists of two volumes, a *Classroom Manual* and a *Shop Manual.*

The fifth edition of *Fuel System and Emission Control* has been completely revised to include in-depth coverage of the latest developments in automotive emission controls and fuel systems. Students will be able to use the knowledge gained from these books and from their instructor to diagnose and repair automotive emission controls and fuel systems used on today's automobiles.

This package retains the traditional thoroughness and readability of the Chek-Chart automotive series. Furthermore, both the *Classroom Manual* and the *Shop Manual,* as well as the *Instructor's Manual,* have been greatly enhanced.

CLASSROOM MANUAL

New features in the *Classroom Manual* include:

- New chapters on computer input devices, computer output devices, and emissions.
- Three new chapters covering ignition systems.
- Objectives in each chapter that alert students to the important themes and learning goals.
- Over 65 new illustrations.
- An added appendix that includes OBD II diagnostic trouble codes, major elements of operating I/M programs, vehicle manufacturer service information websites, and links to state emission programs.
- Obsolete material has been deleted.

SHOP MANUAL

Each chapter of the completely revised *Shop Manual* correlates with the *Classroom Manual.* Like the *Classroom Manual,* the *Shop Manual* features an overhauled illustration program. It includes over 135 new or revised figures and extensive photo sequences showing step-by-step repair procedures.

INSTRUCTOR'S MANUAL

The *Instructor's Manual* includes task sheets that cover many of the NATEF tasks for A8 Engine Performance. Instructors may reproduce these task sheets for use by the students in the lab or during an internship. The *Instructor's Manual* also includes a test bank and answers to end-of-chapter questions in the *Classroom Manual.*

The *Instructor's Resource* CD that accompanies the *Instructor's Manual* includes Microsoft® PowerPoint® presentations and photographs that appear in the *Classroom Manual* and the *Shop Manual.* Each photograph included on the *Instructor's Resource* CD cross-references the figure number in either the *Classroom Manual* or the *Shop Manual.* These high-resolution photographs are suitable for projection or reproduction.

Because of the comprehensive material, hundreds of high-quality illustrations, and inclusion of the latest technology, these books will keep their value over the years. In fact, *Fuel System and Emission Control* will form the core of the master technician's professional library.

How to Use This Book

WHY ARE THERE TWO MANUALS?

This two-volume text—*Fuel System and Emission Control*—is not like most other textbooks. It is actually two books, a *Classroom Manual* and *Shop Manual* that should be used together. The *Classroom Manual* teaches you what you need to know about fuel system and emission control theory, systems, and components. The *Shop Manual* will show you how to repair and adjust complete systems, as well as their individual components.

WHAT IS IN THESE MANUALS?

There are several aids in the *Classroom Manual* that will help you learn more.

- Each chapter is based on detailed learning objectives, which are listed in the beginning of each chapter.
- Each chapter is divided into self-contained sections for easier understanding and review. This organization clearly shows which parts make up which systems, and how various parts or systems that perform the same task differ or are the same.
- Most parts and processes are fully illustrated with drawings and photographs.
- A list of Key Terms is located at the beginning of each chapter. These are printed in **boldface type** in the text and are defined in a glossary at the end of the manual. Use these words to build the vocabulary needed to understand the text.
- Review Questions follow each chapter. Use them to test your knowledge of the material covered.
- A brief summary at the end of each chapter helps you review for exams.

The *Shop Manual* has detailed instructions on the test, service, and overhaul of automotive fuel and emission control systems and their components. These are easy to understand and often include step-by-step explanations of the procedure. Key features of the *Shop Manual* include:

- Each chapter is based upon ASE/NATEF tasks, which are listed in the beginning of each chapter.

- Helpful information on the use and maintenance of shop tools and test equipment.
- Detailed safety precautions.
- Clear illustrations and diagrams to help you locate trouble spots while learning to read the service literature.
- Test procedures and troubleshooting hints that help you work better and faster.
- Repair tips used by professionals, presented clearly and accurately.

WHERE SHOULD I BEGIN?

If you already know something about automotive fuel and emission control systems and how to repair them, you will find that this book is a helpful review. If you are just starting in automotive repair, then the book will give you a solid foundation on which to develop professional-level skills.

Your instructor will design a course to take advantage of what you already know, and what facilities and equipment are available to work with. You may be asked to read certain chapters of this manual out of order. That is fine; the important thing is to fully understand each subject before you move on to the next. Study the vocabulary words, and use the review questions to help you comprehend the material.

While reading the *Classroom Manual,* refer to your *Shop Manual* and relate the descriptive text to the service procedures. When working on actual automotive fuel and emission control systems, look back to the *Classroom Manual* to keep basic information fresh in your mind. Working on such complicated modern fuel and emission systems isn't always easy. Take advantage of the information in the *Classroom Manual,* the procedures in the *Shop Manual,* and the knowledge of your instructor to help you.

Remember that the *Shop Manual* is a good book for work, not just a good workbook. Keep it on hand while you're working on a fuel or emission control system. For ease of use, the *Shop Manual* will fold flat on the workbench or under the car, and it can withstand quite a bit of rough handling.

When you perform actual test and repair procedures, you need a complete and accurate source of manufacturer specifications and procedures for the specific vehicle. As the source for these specifications, most automotive repair shops have the annual service information (on paper, CD, or Internet formats) from the vehicle manufacturer or an independent guide.

Acknowledgments

The publisher sincerely thanks the following vehicle manufacturers, industry suppliers, and individuals for supplying information and illustrations used in the Chek-Chart Series in Automotive Technology.

Allen Testproducts
American Isuzu Motors, Inc.
Automotive Electronic Services
Bear Manufacturing Company
Borg-Warner Corporation
Champion Spark Plug Company
DaimlerChrysler Corporation
DeAnza College, Cupertino, CA
Fluke Corporation
Ford Motor Company
Fram Corporation
General Motors Corporation
  Delco-Remy Division
  Rochester Products Division
  Saginaw Steering Gear Division
  Buick Motor Division
  Cadillac Motor Car Division
  Chevrolet Motor Division
  Oldsmobile Division
  Pontiac-GMC Division

Honda Motor Company, LTD
Jaguar Cars, Inc.
Marquette Manufacturing Company
Mazda Motor Corporation
Mercedes-Benz USA, Inc.
Mitsubishi Motor Sales of America, Inc.
Nissan North America, Inc.
The Prestolite Company
Robert Bosch Corporation
Saab Cars USA Inc.
Snap-on Tools Corporation
Toyota Motor Sales, U.S.A., Inc.
Vetronix Corporation
Volkswagen of America
Volvo Cars of North America

The publisher gratefully acknowledges the reviewers of this edition:

Kenneth Mays, Central Oregon Community College
Katherine Pfau, Montana State University, Billings

The publisher also thanks Series Advisor James D. Halderman.

Contents

Chapter 1 — Introduction to Fuel System and Emission Control 1
    Objectives 1
    Key Terms 1
    Introduction 1
    Air Pollution—A Perspective 2
    Major Pollutants 2
    Pollution and the Automobile 4
    Smog-Climatic Reaction with Air Pollutants 5
    Air Pollution Legislation and Regulatory Agencies 5
    Automotive Emission Controls 11
    Summary 14
    Review Questions 15

Chapter 2 — Engine Operating Principles 17
    Objectives 17
    Key Terms 17
    Introduction 17
    Engine Operation 17
    Major Engine Components 20
    Cylinder Arrangement 27
    Engine Displacement and Compression Ratio 28
    Engine Cooling System 31
    Engine Lubrication System 33
    The Ignition System 33
    Engine-Ignition Synchronization 37
    Initial Timing 38
    Other Engine Types 39
    Summary 43
    Review Questions 44

Chapter 3 — Engine Air-Fuel Requirements 47
    Objectives 47
    Key Terms 47
    Introduction 47
    Air Pressure—High and Low 48
    Airflow Requirements 48
    Air-Fuel Ratios 49

    Introduction to Electronic Engine Controls 51
    The Intake System 51
    Fuel Composition 58
    Summary 64
    Review Questions 65

Chapter 4 — Fuel Delivery Systems 67
    Objectives 67
    Key Terms 67
    Introduction 67
    Fuel Tanks and Fillers 68
    Fuel Lines 71
    Fuel Line Layout 77
    Evaporative Emission Control Systems 80
    Pump Operation Overview 85
    Pump Types 86
    Mechanical Fuel Pumps 86
    Electric Fuel Pumps 88
    Fuel Filters 93
    Summary 97
    Review Questions 98

Chapter 5 — Engine Control Systems 101
    Objectives 101
    Key Terms 101
    Introduction 101
    Electronic Control Systems 102
    Electrical Review 102
    Computer Control 107
    Parts of a Computer 112
    Data Lines 116
    Fuel Control System Operating Modes 121
    Summary 123
    Review Questions 124

Chapter 6 — Engine Management Input Devices 127
    Objectives 127
    Key Terms 127
    Introduction 127
    Electrical Operation of Input Devices 128
    Critical Sensor Inputs 133

Contents

Engine Coolant Temperature Sensor
(ECT) 133
Throttle Position Sensor 134
Manifold Absolute Pressure Sensor 135
Ignition Reference Signal 137
Oxygen Sensors 138
Summary 143
Review Questions 144

Chapter 7 — Engine Management Output Devices 147
Objectives 147
Key Terms 147
Introduction 147
Low-Side Drivers 149
High-Side Drivers 149
Pulse Width Modulation 151
Critical Outputs 152
Fuel Injectors 152
Ignition Control 152
Electronic Throttle Control 153
Idle Speed Control 154
Fuel Pump 155
Serial Data 156
Malfunction Indicator Lamp 156
Exhaust Gas Recirculation 157
Other Outputs 158
Evaporative and Purge Solenoids 158
Summary 159
Review Questions 160

Chapter 8 — Electronic Engine Control Systems 161
Objectives 161
Key Terms 161
Introduction 161
Computer Functions—A Review 162
Basic Engine Operating Modes 163
Air-Fuel Ratio, Timing, and EGR Effects on
Operation 164
Full-Function Control Systems 170
Control System Development 171
Common Components 171
System Actuators 178
History of Engine Control Systems 180
Onboard Diagnostics (OBD) Systems 185
Summary 195
Review Questions 196

Chapter 9 — Gasoline Fuel-Injection Systems 199
Objectives 199
Key Terms 199
Introduction 199
Fuel-Injection Operating Requirements 200

Advantages of Fuel Injection 201
Air-Fuel Mixture Control 201
Types of Fuel-Injection Systems 202
Common Subsystems and
Components 207
Specific Systems 220
Trends 229
Summary 232
Review Questions 234

Chapter 10 — Supercharging and Turbocharging 237
Objectives 237
Key Terms 237
Introduction 237
Engine Compression 237
The Benefits of Air-Fuel Mixture
Compression 238
Supercharging 238
Turbochargers 246
Turbocharger Controls 251
Summary 254
Review Questions 255

Chapter 11 — Variable, Flexible, and Bi-Fuel Systems 257
Objectives 257
Key Terms 257
Introduction 257
Variable or Flexible Fuel Vehicles 258
M85 and Methanol-Blended Fuels 259
Methanol and Automotive Engines 260
Common Subsystems and
Components 260
General Motors VFV Fuel System 261
General Motors VFV Emission Control
Systems 266
Ford Flexible Fuel System 266
Ford FF Emission Control Systems 267
E-85 Vehicles 267
Bi-Fuel Vehicles 269
Summary 271
Review Questions 273

Chapter 12 — Emissions, Five Gas Theory, and I/M Programs 275
Objectives 275
Key Terms 275
Introduction 275
The Combustion Process 276
The Five Gases 277
Inspection/Maintenance Programs
(I/M) 282
Summary 288
Review Questions 289

Chapter 13 — Positive Crankcase Ventilation, Air-Injection Systems, Catalytic Converters, and EGR Systems 291
Objectives 291
Key Terms 291
Introduction 291
Crankcase Ventilation 292
Air Injection 295
Catalytic Converters 300
NO_x Formation 305
EGR System History 306
Computer-Controlled EGR 310
Summary 316
Review Questions 318

Chapter 14 — The Ignition Primary Circuit and Components 321
Objectives 321
Key Terms 321
Introduction 321
Need for High Voltage 321
High Voltage Through Induction 322
Basic Circuits and Current 323
Primary Circuit Components 324
Monitoring Ignition Primary Circuit
    Voltages 334
Summary 339
Review Questions 340

Chapter 15 — The Ignition Secondary Circuit and Components 343
Objectives 343
Key Terms 343
Introduction 343
Ignition Coils 344
Distributor Cap and Rotor 353
Ignition Cables 356
Spark Plugs 356
Spark Plug Construction 358
Summary 362
Review Questions 363

Chapter 16 — Electronic Ignition Systems 365
Objectives 365
Key Terms 365
Introduction 365
Basic Electronic Ignition Systems 366
Control Modules and Primary Circuitry 367
Triggering Devices and Ignition Timing 368
Electronic Ignition Dwell, Timing, and
    Advance 377
Original-Equipment Electronic Distributor
    Ignitions 385
Original-Equipment Distributorless
    Ignitions 389
Summary 413
Review Questions 414

Glossary of Technical Terms 417

Appendix A — OBD II Diagnostic Trouble Codes 424

Appendix B — Evaporative System Pressure Conversion Chart 427

Appendix C — Major Elements of Operating I/M Programs 428

Appendix D — Vehicle Manufacturer's Service Information 434

Appendix E — Automotive Fuel and Emission's Related Websites 435

Appendix F — Links to State Emission Programs 436

Index 437

1

Introduction to Fuel System and Emission Control

OBJECTIVES

Upon completion and review of this chapter, you will be able to:

- Have knowledge of the causes of air pollution.
- Have knowledge of the four major pollutants emitted from mobile sources.
- Explain the photochemical reaction.
- Have knowledge of the major air pollution legislation and regulatory agencies.
- Describe the regulations for OBD I and OBD II systems.

KEY TERMS

carbon monoxide (CO)
hydrocarbons (HC)
onboard diagnostic (OBD) system
oxides of nitrogen (NO_X)
ozone

particulate matter (PM10)
photochemical smog
sulfur oxides (SO_X)
temperature inversion

INTRODUCTION

The combustion process in automotive engines produces harmful by-products that are discharged from the engine and become air pollutants. Emission control systems are necessary to minimize the formation and discharge of these pollutants.

When emission control requirements were first introduced, manufacturers and car owners were able to comply by installing add-on or "hang-on" devices that were not an integral part of engine and vehicle design. As regulations became more strict, manufacturers had to include emission controls in basic engine design.

The first emission control regulation was adopted in California in 1961. Today, almost four decades later, emission control regulations are still being tightened and new control systems developed. Sophisticated computer-controlled systems appear on most cars, and emission control requirements are important considerations in the design and operation of all parts of the fuel system. The ignition system, which provides the spark for combustion, plays an equally important role in emission control.

How did these great changes in automotive emission controls come about? What exactly is air pollution, and how does the automobile contribute to it? This chapter examines air pollution and automotive emissions, including the legislation controlling emissions and the ways in which manufacturers have met the regulations.

AIR POLLUTION— A PERSPECTIVE

We can define air pollution as the introduction of contamination into the atmosphere in an amount large enough to injure human, animal, or plant life. There are many types and causes of air pollution, but they all fall into two general groups: natural and man-made. Natural pollution is caused by such things as the organic plant life cycle, forest fires, volcanic eruptions, and dust storms. Although pollution from such sources is often beyond our control, we can control man-made pollution from industrial plants and automobiles.

Most urban and large industrial areas around the world suffer periodic air pollution. During the late 1940s, a unique form of air pollution was identified in the Los Angeles area. When certain pollutants are exposed to sunlight, irritating chemical compounds form a pollutant called photochemical smog. As this phenomenon increased both in intensity and frequency, it posed more of a problem. California took the lead in combating it by becoming the first state to place controls on motor vehicle emissions, figure 1-1. As smog gradually began to appear in other parts of the country, the federal government moved into the area of regulation. To understand why, we must look at the automobile-produced elements that form air pollution and smog.

MAJOR POLLUTANTS

An internal combustion engine emits three major gaseous pollutants into the air: hydrocarbons (HC), carbon monoxide (CO), and oxides of nitrogen (NO_x), figure 1-2. In addition, an automobile engine gives off many small liquid or solid particles, such as lead, carbon, sulfur, and other particulate matter (PM10), which contribute to pollution. By themselves, all these emissions are not smog, but simply air pollutants.

Figure 1-1. During the 1960s, high levels of airborne pollutants in the Los Angeles basin prompted the state of California to enact emission control regulations.

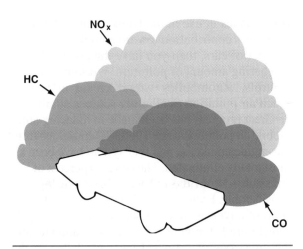

Figure 1-2. Hydrocarbons (HC), carbon monoxide (CO), and oxides of nitrogen (NO$_X$) are the three major automotive pollutants.

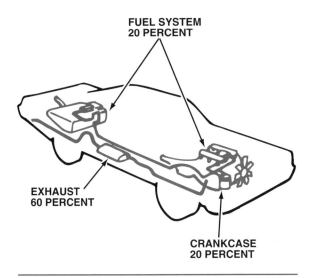

Figure 1-3. Sources of hydrocarbon emissions.

Hydrocarbons

Gasoline is a HC compound. Unburned HCs given off by an automobile are mostly unburned fuel. Over 200 different varieties of HC pollutants come from automotive sources. While most come from the fuel system and the engine exhaust, others are oil and gasoline fumes from the crankcase. Even a car's tires, paint, and upholstery emit tiny amounts of HCs. Figure 1-3 shows the three major sources of HC emissions from an automobile:

- Fuel system evaporation—20 percent
- Crankcase vapors—20 percent
- Engine exhaust—60 percent

HCs are the only major automotive air pollutant that comes from sources other than engine exhaust. HC molecules of all types are changed into other compounds by combustion. If an automobile engine burned gasoline completely, there would be no HCs in the exhaust, only water and carbon dioxide (CO$_2$). When the vaporized and compressed air-fuel mixture is ignited, combustion occurs so rapidly that gasoline near the sides of the combustion chamber may not get burned. This unburned fuel then passes out with the exhaust gases. The problem is worse with engines that misfire or are not properly tuned.

Carbon Monoxide

Although not part of photochemical smog, CO is also found in automobile exhaust in large amounts. A deadly poison, CO is both odorless and colorless. CO is absorbed by the red corpuscles in the body, displacing the oxygen (O$_2$). In a small quantity, it causes headaches, vision difficulties, and delayed reaction times. In larger quantities, it causes vomiting, coma, and death.

Because it is a product of incomplete combustion, the amount of CO produced depends on the way in which HCs burn. When the air-fuel mixture burns, its HCs combine with O$_2$. If the air-fuel mixture contains too much fuel, there is not enough O$_2$ to complete this process, so CO forms. Using an air-fuel mixture with less fuel makes combustion more complete. The leaner mixture increases the ratio of O$_2$, which reduces the formation of CO by producing CO$_2$ instead.

CO$_2$, although not considered a pollutant affecting public health, does contribute to global warming. Currently, scientists are studying the relationship between CO$_2$ levels and global warming.

Oxides of Nitrogen

Air is made up of about 78 percent nitrogen, 21 percent O$_2$, and 1 percent other gases. When the combustion chamber temperature reaches 2500°F (1370°C) or greater, the nitrogen and O$_2$ in the air-fuel mixture combine to form large quantities of NO$_X$. NO$_X$ also is formed at lower temperatures, but in far smaller amounts. One component of NO$_X$, nitrogen dioxide, is extremely toxic to humans. In addition, NO$_X$ combines with other elements in the

Figure 1-6. Smog engulfs the Los Angeles Civic Center in the 1960s. When the base of the temperature inversion is only 1,500 feet (457 meters) above the ground, the inversion layer—a layer of warm air above a layer of cool air—prevents the natural dispersion of air contaminants into the upper atmosphere.

with a blowby device that virtually eliminated crankcase emissions on all cars.

California followed by requiring that 1966 and later new cars sold within its boundaries have exhaust emission controls. The use of exhaust emission control systems was extended nationwide during the 1967–68 model years.

The first federal air pollution research program began in 1955. In 1963, Congress passed the Clean Air Act, providing the states with money to develop air pollution control programs. This law was amended in 1965 to give the federal government authority to set emission standards for new cars, and was amended again in 1977–78. Under this law, emission standards were first applied nationwide to 1968 models.

In addition to the Clean Air Act, the federal government took a new approach to air pollution in

CALIFORNIA PASSENGER CAR NEW VEHICLE STANDARDS

YEAR	HC	CO	NOₓ
1966-69	275 ppm	1.5%	-
1970	2.2 g/m	23 g/m	-
1971	2.2 g/m	23 g/m	4.0 g/m
1972	3.2 g/m	39 g/m	3.2 g/m
1974	3.2 g/m	39 g/m	3.0 g/m
1975-76	0.9 g/m	9 g/m	2.0 g/m
1977-79	0.41 g/m	9 g/m	1.5 g/m
1980	0.41 g/m	9 g/m	1.0 g/m
1981-92	0.41 g/m	7 g/m	0.7 g/m
1993	0.39 g/m	7.0 g/m	0.4 g/m
1994-95	0.25 g/m	3.4 g/m	0.4 g/m
(Tier 1) 96-2004	0.25 g/m	3.4 g/m	0.4 g/m
TLEV	0.125 g/m	3.4 g/m	0.4 g/m
LEV	0.075 g/m	3.4 g/m	0.2 g/m
ULEV	0.040 g/m	1.7 g/m	0.2 g/m

Figure 1-7A. This chart of California exhaust emission limits for new cars shows how standards became more stringent in the wake of the Clean Air Act. California standards are stricter than federal standards. (Courtesy of the California Air Resources Board)

Vehicle Type	Durability Vehicle Basis (mi.)	Vehicle Emission Category	NMOG (g/mi.)	Carbon Monoxide (g/mi)	Oxides of Nitrogen (g/mi)	Formaldehyde (mg/mi)	Particulate from diesel vehicles** (g/mi)
All PCs; LDTs (0-3750 lbs. LVW)	50,000	Tier 1	0.25*	3.4	0.4	n/a	0.08
		TLEV	0.125	3.4	0.4	15	n/a
		LEV	0.075	3.4	0.2	15	n/a
		ULEV	0.040	1.7	0.2	8	n/a
	100,000	Tier 1	0.31	4.2	0.6	n/a	n/a
		Tier 1 diesel option	0.31	4.2	1.0	n/a	n/a
		TLEV	0.156	4.2	0.6	18	0.08
		LEV	0.090	4.2	0.3	18	0.08
		ULEV	0.055	2.1	0.3	11	0.04

Figure 1-7B. Exhaust Mass Emission Standards for New 2001-2003 Model Year Tier 1 Vehicles and TLEV Passenger Cars and Light-Duty Trucks; 2001-2006 Model Year LEV I LEV and ULEV Passenger Cars and Light-Duty Trucks; 2001-2003 Model Year Tier 1 Medium-Duty Vehicles; and 2001-2006 Model Year LEV I LEV, ULEV and SULEV Medium-Duty Vehicles.

1967 with the Air Quality Act. This act and its major amendments of 1970, 1974, and 1977 instituted changes designed to turn piecemeal programs into a unified attack on pollution of all kinds. Canada attacked its own smog problem with vehicle emission requirements established by the Ministry of Transport, which took effect with 1971 models.

The 1990 Clean Air Act

In 1990, the U.S. Congress amended and updated the Clean Air Act for the first time in 13 years. Besides making tailpipe standards more stringent and expanding vehicle Inspection and Maintenance (I/M) Programs, the 1990 law focused on fuel itself, in addition to vehicle technology. Some key aspects of the 1990 Clean Air Act include

- Tighter tailpipe standards
- CO control
- Ozone control
- Reformulated gasoline
- Other controls

Figure 1-8 shows a timetable for implementing certain provisions of the 1990 Clean Air Act.

Tighter Tailpipe Standards

Tailpipe standards for 1990–93 were 0.41 gram per mile (g/m) HC, 3.4 g/m CO, and 1.0 g/m NO$_X$. The 1990 Clean Air Act requires phasing in lower-limited standards for HC and NO$_X$ between

TIMETABLE FOR IMPLEMENTATION OF CERTAIN PROVISIONS OF THE 1990 CLEAN AIR ACT

1992   Limits on maximum gasoline vapor pressure go into effect nationwide.

      Regulations for minimum oxygen content of gasoline go into effect in 39 areas.

1993   Production of vehicles requiring leaded gasoline becomes illegal.

1994   Phase-in of tighter tailpipe standards and cold-temperature CO standards for light-duty vehicles begins.

      Expansion of I/M programs begins in certain cities.

      Requirement for new cars to be equipped with onboard diagnostics systems takes effect.

1995   Reformulated gasoline must be sold in the nine smoggiest cities in the U.S.

      New warranty provisions on emission control systems take effect.

1996   Phase-in of California Pilot Program begins.

      Lead banned from use in motor vehicle fuel.

      All new vehicles must meet tighter tailpipe standards and cold-temperature CO standards.

1997   Federal "Clean Fleet" program begins in 19 states in areas with excessive ozone and CO.

2001   Second phase of California Pilot Program and Federal "Clean Fleet" program begins.

Figure 1-8. Various parts of the Clean Air Act Amendments of 1990 take effect through the 1990s and in the early 21st century.

1994–96—0.25 g/m of nonmethane HC and 0.4 g/m NO$_X$. The law also requires the EPA to study whether even stricter standards are necessary, feasible, and economical. In 1999 the EPA implemented new engine and gasoline standards known as Tier II.

These standards went into effect in 2004. California introduced new amendments known as LEV II standards, which will apply from 2004 through 2010.

Tier II and LEV II standards are fleet averaging programs. The federal Tier II standards allow the manufacturer to produce dirtier and cleaner vehicles; however, the mix of vehicles that the manufacturer sells must have average NO_X emissions below a specified value. California's LEV II standards are also a fleet averaging program; however, the standards are based on hydrocarbon emissions, not NO_X emissions as the federal program is.

Carbon Monoxide Control

The EPA places primary blame for CO pollution on mobile sources (including cars and trucks, as well as other vehicles, such as bulldozers and construction equipment) in 39 U.S. cities.

CO emissions from cars are particularly high during cold weather, when vehicles operate less efficiently. While previous CO standards applied only at 75°F (24°C), the 1990 Clean Air Act establishes an additional CO standard of 10 g/m at 20°F (−7°C). If CO levels are still excessive in six or more cities by 1997, the standard will be tightened to 3.4 g/m, to be met beginning with the 2002 model year.

The 1990 law also requires a higher O_2 content in the gasoline sold during the winter in the 39 CO-saturated cities, a requirement that took effect in 1992. O_2 leans the air-fuel mixture, reducing CO emissions, lowering fuel economy, and increasing CO_2 emissions.

Ozone Control

Ozone—the combination of HC and NO_X, the primary component of smog—is the most widespread air quality problem in the United States. The U.S. EPA rates the following areas as having a "severe" ozone problem:

- New York City and Long Island in New York State, and north New Jersey
- Baltimore, Maryland
- Muskegon, Michigan
- Chicago, Illinois
- Gary, Indiana
- Lake County and Milwaukee, Wisconsin
- Houston, Galveston, and Brazoria, Texas
- San Diego, Southeast desert, and Ventura County, California

Los Angeles, California, receives its very own ozone rating: "extreme." Approximately 15 other cities and urban areas across the United States are rated as having "serious" ozone levels, and about 30 more have "moderate" ozone problems, figure 1-9.

Gasoline vapors are a major source of HC in ozone, so the 1990 Clean Air Act calls for reducing evaporative emissions by such means as improved engine and fuel system vapor traps and wider use of systems to capture vapors during refueling in smoggy cities. In addition, the Clean Air Act places a cap on fuel volatility—reducing its tendency to evaporate.

Reformulated Gasoline (RFG)

The 1990 Clean Air Act established standards for the content of gasoline sold in the nine worst ozone areas. This "cleaner" gasoline must meet or exceed a certain minimum O_2 content, and it must not exceed a certain maximum level of benzene. Also, the gasoline must reduce toxic and smog-forming emissions 15 percent by 1995, and 20 to 25 percent by 2000, without increasing NO_X emissions. Other cities may choose to use this clean fuel, too, and the U.S. EPA expects many to do so.

■ **How New Cars Are Emission-Certified**

Whether you can buy a particular new car each year—and where you can buy it—depend on the manufacturer's success in the emission certification process. The U.S. Environmental Protection Agency (U.S. EPA) performs a constant-volume sampling test for each car, called the Federal Test Procedure (FTP), or the Federal Vehicle Certification Standard Procedure. Some vehicles are also tested by the California Air Resources Board.

As each new model year approaches, manufacturers build prototype, or emission-data, cars for U.S. EPA use. The manufacturers are responsible, for conducting a 50,000-mile (80,000-kilometer) durability test of their emission control systems. Before the U.S. EPA test, the manufacturers drive the test cars for 4,000 miles (6,500 kilometers) to stabilize the emission systems.

The manufacturer preconditions the car before the U.S. EPA test, then it stands for 12 hours at an air temperature of 73°F (22°C) to simulate a cold start. The actual test is done on a chassis dynamometer, using a driving cycle that represents urban driving conditions. The car's exhaust is mixed with air to a constant volume and analyzed for harmful pollutants.

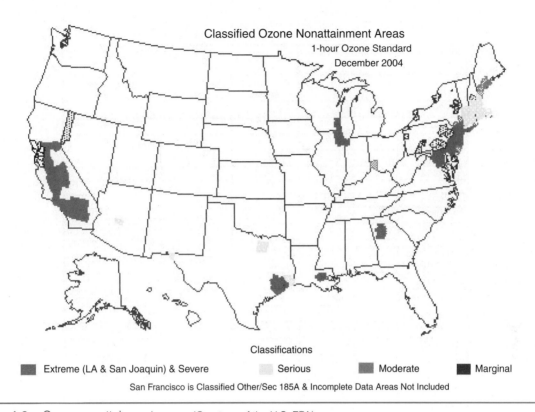

Classified Ozone Nonattainment Areas
1-hour Ozone Standard
December 2004

Classifications

■ Extreme (LA & San Joaquin) & Severe   Serious   ■ Moderate   ■ Marginal

San Francisco is Classified Other/Sec 185A & Incomplete Data Areas Not Included

Figure 1-9. Ozone nonattainment areas. (Courtesy of the U.S. EPA)

The entire test requires about 41 minutes. The first 23 minutes are a cold-start driving test. The next 10 minutes are a waiting, or hot-soak period. The final eight minutes are a hot-start test, representing a short trip in which the car is stopped and started several times while hot. If the emissions test results for all data cars are equal to or lower than the HC, CO, and NO_x standards, the EPA grants certification for the engine "family," and the manufacturer can sell the car to the public.

This certification process explains why some engines disappeared in the wake of clean-air legislation—they were too "dirty" and could not be "cleaned up." It also explains why some powertrain combinations may not be available in California—which has different requirements—when they can be purchased in other states.

Other Controls

The 1990 law calls for the EPA to monitor toxic emissions such as benzene and formaldehyde, regulating them as needed. Both California and the federal government are mandating clean car pro-

grams. The California Pilot Program requires that, starting in 1996, car manufacturers supply at least 150,000 "clean" cars for sale in California. The qualifications of a California Pilot Program clean car are that it emits no more than 0.125 g/m HC, 3.4 g/m CO, and 0.4 g/m NO_x. In 1999, the number of clean cars required increased to 300,000.

Meanwhile, starting in 1998, the federal government requires fleets in certain very polluted cities to have 30 percent of their new cars be clean cars. For these areas, a clean car is one that can use clean fuel and can meet extra-low emission standards. By 2000, the proportion rises to 70 percent of the fleet. This federal program affects 22 metropolitan areas in 19 states.

Corporate Average Fuel Economy (CAFE) Standards

The 1973 energy crisis focused national attention on fuel economy and resulted in the establishment of Corporate Average Fuel Economy (CAFE) standards, figure 1-10, as a part of the Federal Energy Act of 1975. The EPA and the Department of

Fuel Economy Standards for Passenger Cars and Light Trucks
Model Years 1978 through 2004 (in MPG)

Model Year	Passenger Cars	Light Trucks [1]		
		Two-wheel Drive	Four-wheel Drive	Combined [2], [3]
1978	18.0 [4]	...	...	...
1979	19.0 [4]	17.2	15.8	...
1980	20.0 [4]	16.0	14.0	...[5]
1981	22.0	16.7[6]	15.0	...[5]
1982	24.0	18.0	16.0	17.5
1983	26.0	19.5	17.5	19.0
1984	27.0	20.3	18.5	20.0
1985	27.5[4]	19.7[7]	18.9[7]	19.5[7]
1986	26.0[8]	20.5	19.5	20.0
1987	26.0[9]	21.0	19.5	20.5
1988	26.0[9]	21.0	19.5	20.5
1989	26.5[10]	21.5	19.0	20.5
1990	27.5[4]	20.5	19.0	20.0
1991	27.5[4]	20.7	19.1	20.2
1992	27.5[4]	...	...	20.2
1993	27.5[4]	...	...	20.4
1994	27.5[4]	...	...	20.5
1995	27.5[4]	...	...	20.6
1996	27.5[4]	...	...	20.7
1997	27.5[4]	...	...	20.7
1998	27.5[4]	...	...	20.7
1999	27.5[4]	...	...	20.7
2000	27.5[4]	...	...	20.7
2001	27.5[4]	...	...	20.7
2002	27.5[4]	...	...	20.7
2003	27.5[4]	...	...	20.7
2004	27.5[4]	...	...	20.7

Figure 1-10. This chart shows how CAFE standards have progressed from the late 1970s through 2004. (Courtesy of the U.S. EPA)

Transportation (DOT) are responsible for administering the CAFE standards.

The combination of CAFE and clean air laws gives automotive engineers conflicting goals—reducing emissions while improving fuel economy. If the CAFE standards are not reached each year, manufacturers pay a penalty on each vehicle sold—the so-called "gas guzzler" tax. However, manufacturers who exceed the yearly average on a corporate basis gain a credit that can be applied to later years.

By downsizing vehicles, using smaller engines, and paying particular attention to reducing weight and improving aerodynamic efficiency, the domestic automotive industry transformed itself in the early 1980s. During this period, there was a concerted national effort to conserve fuel because of high prices at the pump and a desire to reduce the nation's dependence on foreign oil.

This national effort worked so well that oil prices, which had stabilized by 1983, collapsed in 1985–86. At the same time, car owners indicated a desire to return to larger cars by ignoring the fuel-efficient small cars and purchasing the less-efficient cars with bigger engines and more room. This led General Motors (GM) and Ford to petition the EPA for a "rollback" of CAFE standards to avoid massive fines resulting from meeting customer demand. Over the objections of Chrysler and other small manufacturers who met the standards on time, a temporary return to the 1983 level was enacted for 1986, with the 1987 standard returning to that of 1985. However, in the early 1990s, as gas prices climbed again, CAFE standards returned to the original levels established by Congress. On April 1, 2003, the National Highway Traffic Safety Administration announced it would increase the fuel economy of SUVs by 1.5 mpg. The current standard of 20.7 will increase to 21 mpg for model year 2005, 21.6 mpg for model year 2006, and 22.2 mpg for model year 2007.

Onboard Diagnostic (OBD) Systems Regulation

In the late 1980s, the California Air Resources Board (CARB) mandated that all cars sold in the state of California must have an **onboard diagnostic (OBD) system.** These initial regulations, known as OBD I, were minimal, and most of the major vehicle manufacturers had systems already in production that complied to the standards. In general, three elements were required by OBD I:

1. An instrument panel warning lamp to alert the driver of certain control system failures
2. The ability of the system to record and transmit diagnostic fault codes (DTC) for emission-related failures
3. Electronic system monitoring of the exhaust gas recirculation (EGR) valve, fuel system, and some other emission-related components.

The purpose of the regulations was twofold: provide the driver an early warning of an emissions failure, and to make the technician's task of locating the source of the failure easier. The ultimate goal was a reduction of tailpipe pollutants, which would result in improved air quality.

OBD I may have been founded with good intent, but implementation left much to be desired. The manufacturers were free to interpret the rules as they saw fit and the result was a vast array of systems. Rather than simplify the job of locating and repairing a failure, the technician was now faced with a tangled network of procedures that often required the use of expensive special test equipment and information available only to the dealers.

The overall reduction of tailpipe emissions was marginal. It soon became apparent that more-stringent measures were needed if the ultimate goal, breathable air in southern California, was to be achieved. This led CARB to develop OBD II, a second generation of more-stringent legal requirements.

OBD-II Development

For their second set of regulations, CARB desired more precise control and monitoring of emission-related components, as well as standard service procedures without the use of dedicated special tools. Because the technology required to develop this new program was outside its field of expertise, CARB enlisted the aid of the Society of Automotive Engineers (SAE). CARB would establish the guidelines, and SAE would develop the technology.

The new regulations, OBD II, were gradually phased into production beginning with the 1994 model year. Since 1996, all vehicles sold in California are required to have an OBD II–compliant system.

On a national level, the EPA has adapted the system implemented by California. As it stands, the federal government requires all vehicles to comply with OBD II standards established by CARB through the 1997 model year. In 1998, new federal standards established by the EPA took effect. Later chapters detail OBD II–compliant engine control systems.

AUTOMOTIVE EMISSION CONTROLS

Early researchers dealing with automotive pollution and smog began work with the idea that all pollutants were carried into the atmosphere by the car's exhaust pipe. But auto manufacturers doing their own research discovered that the fuel tank and engine crankcase also give off pollutants. The total automotive emission system, figure 1-11, contains three different types of controls. The emission controls on a modern automobile are not a separate system, but an integral part of an engine's fuel, ignition, and exhaust systems.

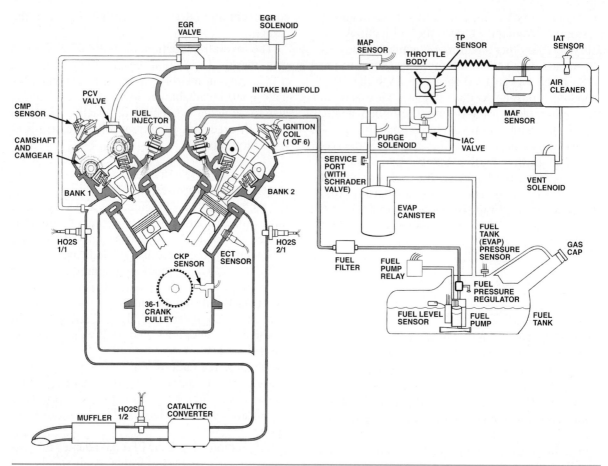

Figure 1-11. The complex emission controls on a modern automobile engine are an integral part of the fuel, ignition, and exhaust systems. (Courtesy of DaimlerChrysler Corporation)

In order to service a car's fuel system and emission controls, you must have a basic understanding of the internal combustion engine and how it works. The chapters of this book will cover engine operating principles and air-fuel requirements, emission controls as they relate to different parts of the fuel system, and major emission controls that can be studied separately from the fuel system.

Automotive emission controls can be grouped into major families:

- Crankcase emission controls
- Evaporative emission controls
- Exhaust emission controls

Positive crankcase ventilation (PCV) systems control HC emissions from the engine crankcase. Evaporative emission control (EEC or EVAP) systems control the evaporation of HC vapors from the fuel tank, pump, and fuel injection system. Various systems and devices control HC, CO, and NO_X emissions from engine exhaust. We can divide exhaust emission controls into the following categories:

- Air injection systems
- Catalytic converters
- Engine modifications
- Spark timing controls
- Exhaust gas recirculation

Air injection systems add air to the exhaust to help burn HC and CO and to aid catalytic conversion. The first catalytic converters installed in the exhaust systems of 1975–76 cars helped the chemical oxidation of the exhaust—that is, burning HC and CO. Later catalytic converters also promote the chemical reduction of NO_X emissions.

■ **OBD II and Vacuum Testing**

In 1996, the Federal government mandated a new set of guidelines concerning emission controls and the internal combustion engine. These guidelines dictated that the self-diagnostic capability of engines be greatly improved. An OBD II computer can monitor the air induction system with far more accuracy and sensitivity than a vacuum gauge. These computers can detect conditions that would lead to vacuum fluctuations but would not register on a vacuum gauge. Although the monitoring is more accurate, the diagnosis of the cause of that condition is not always accurate. Trouble codes have been defined based on what an engineer or group of engineers feels is the most likely cause of the symptom that set the code.

One example is very common on older OBD II applications. There is a group of codes on these applications that implies problems with the ignition system or injection system are causing a misfire. Much of the documentation for the troubleshooting process on these codes refers to testing these areas. In reality, the code could just as easily be caused by a vacuum leak or engine-breathing problem. In spite of the sophistication of ODB II electronic controls, basic tools like the vacuum gauge remain the most effective method of ruling out air induction problems.

Manufacturers have made a variety of changes in engine design and fuel and ignition system operation to help eliminate all three major pollutants. Various systems to delay or retard ignition spark timing help control HC and NO_X emissions. Early spark-timing controls modified the distributor vacuum advance, but later-model cars use electronic engine control systems that eliminate the need for mechanical or vacuum timing devices.

Recirculating a small amount of exhaust gas back to the intake manifold to dilute the incoming air-fuel mixture is an effective way to control NO_X emissions. This is called an exhaust gas recirculation (EGR) system.

Emission Trends

As can be seen in figure 1-12, much progress has been made in the reduction of emissions since 1970. Since 1970, total VOC (volatile organic compounds, mainly HCs) have been reduced by 59 percent from on-road vehicles. CO emissions have decreased by nearly 50 percent from on-road vehicles. The only increase has been a 16 percent increase in NO_X emissions, mainly due to the increase in heavy-duty diesel engines. Another factor limiting further reduction in emissions from on-road vehicles is the fact that there are more vehicles on the road today and these vehicles are driven more miles than in the past.

Percent of Change in Emissions (1970–99)

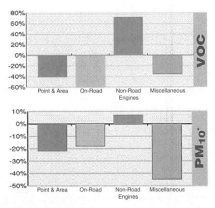

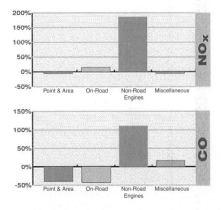

*Note: 1985–99.

Figure 1-12. Americans have made great progress in cleaning the air. For nearly three decades, national emission trends have been declining. A great deal of credit for the improvements goes to cleaner cars and trucks and reformulated fuels. (Courtesy of U.S. Environmental Protection Agency. Office of Air Quality Planning and Standards. *National Air Quality Emissions Trends, 1990–1999.* March 2001. Tables A-2, A-4, A-5 and A-6)

So while significant strides have been made in reducing emissions from mobile sources, many areas of the country still exhibit high levels of pollution due to the increase in vehicle traffic and miles driven. According, in 1999 the EPA determined that more stringent Tier II standards were needed to meet the NAAQS (North American Air Quality Standards).

SUMMARY

The automobile is a major source of air pollution, resulting from gasoline burned in the engine and vapors escaping from the crankcase, fuel tank, and the rest of the fuel system. The major automobile-produced pollutants are unburned HCs, CO, and NO_X. The use of emission controls in recent decades has reduced automotive pollutants by 65 to 98 percent.

Emission controls began as separate "add-on" components and systems but are now integrated into engine and vehicle design. The major emission control systems are PCV systems, evaporative control systems, air injection, spark timing controls, exhaust gas recirculation, catalytic converters, and electronic engine control systems.

Increasingly, stringent emission standards, combined with the CAFE regulations imposed beginning in the late 1970s, and throughout the 1980s and 1990s, have reshaped the domestic automotive industry. The end result has been smaller, more fuel-efficient vehicles that produce less pollution. Manufacturers are still working to improve emission control and gas mileage.

Review Questions

Choose the letter that represents the best possible answer to the following questions:

1. Smog:
   a. Is a natural pollutant
   b. Cannot be controlled
   c. Is created by a photochemical reaction
   d. Was first identified in New York City

2. The three major pollutants in automobile exhaust are:
   a. Sulfates, particulates, carbon dioxide
   b. Sulfates, carbon monoxide, nitrous oxide
   c. Carbon monoxide, oxides of nitrogen, hydrocarbons
   d. Hydrocarbons, carbon dioxide, nitrous oxide

3. Fuel evaporation accounts for what percentage of total HC emissions?
   a. 10 percent
   b. 60 percent
   c. 33 percent
   d. 20 percent

4. Carbon monoxide is a result of:
   a. Incomplete combustion
   b. A lean mixture
   c. Excess oxygen
   d. Impurities in the fuel

5. Which of the following is *not* true?
   a. High engine temperatures increase NO_X emissions.
   b. High engine temperatures reduce HC and CO emissions.
   c. Low engine temperatures reduce HC and CO emissions.
   d. Low engine temperatures reduce NO_X emissions.

6. Particulate matter is:
   a. A by-product of photochemical smog
   b. Created only by diesel engines
   c. Caused by chemical reaction of CO and NO_X
   d. Microscopic particles suspended in the atmosphere

7. Sulfur oxides are harmful because they:
   a. Damage three-way catalytic converters
   b. Combine with water to form sulfuric acid
   c. Are primary automotive pollutants
   d. Are visible in bright sunlight

8. Ozone is created by a combination of sunlight, still air, and:
   a. High levels of CO and HC
   b. High levels of CO and NO_X
   c. High levels of HC and NO_X
   d. High levels of HC and sulfur oxides

9. A temperature inversion increases air pollution by:
   a. Pushing cool air up
   b. Forming a "lid" over stagnant air
   c. Pushing warm air down
   d. Decreasing wind force

10. U.S. federal emission limits are established by the:
    a. California Air Resources Board
    b. Ministry of Transport
    c. Environmental Protection Agency
    d. Department of Transportation

11. Corporate Average Fuel Economy (CAFE) standards were rolled back in the:
    a. 1960s
    b. 1970s
    c. 1980s
    d. 1990s

12. Sulfur by-products of combustion can combine with water to form:
    a. Sulfates
    b. Particulates
    c. Oxides of sulfur
    d. Sulfuric acid

13. The only major type of pollutant that comes from a vehicle source *other than the exhaust* is:
    a. HC
    b. CO
    c. CO_2
    d. NO_X

14. OBD-I seeks to decrease levels of automotive emissions by:
    a. Alerting the driver to a potential problem
    b. Monitoring the efficiency with which certain systems are operating
    c. Measuring the levels of HC and CO coming from the tailpipe
    d. Both a and b

2

Engine Operating Principles

OBJECTIVES

Upon completion and review of this chapter, you will be able to:

- Describe the four-stroke cycle.
- List and describe the purpose of the major engine components.
- Calculate the displacement of an engine given the bore and stroke.
- Be familiar with the different firing orders for various configured engines.

KEY TERMS

bore	inertia
bottom dead center or BDC	injection pump
clearance volume	internal combustion engine
compression ratio	poppet valves
displacement	preignition
eccentric	reciprocating engine
external combustion engines	stroke
firing interval	top dead center or TDC
firing order	two-stroke engine
four-stroke engine	vaporizes
ignition interval	water jackets

INTRODUCTION

To know enough about engine fuel systems and emission controls to service them, you must understand engine design, construction, and operation.

ALIGHM BOLD TEXT

...ERATION

...f fuel to produce me-
...ngine" known to man
...els" the muscle push-
...that the muscle alone
...r way, the automotive
engine uses fuel to perform work, figure 2-1.

It is common to think of the automotive engine as a gasoline engine. Most of the automotive engines in the world are fueled by gasoline. But the correct name for the automotive engine is **internal combustion engine.** It can be designed to run on any fuel that **vaporizes** easily or on any flammable gas.

The automotive engine is called an internal combustion engine because the fuel it uses is

17

Figure 2-1. Today's automotive engines produce high power with low fuel consumption and low exhaust emissions—not an easy design task. (Courtesy of General Motors Corporation)

burned inside the engine. **External combustion engines** burn the fuel outside the engine. A common example of an external combustion engine is the steam engine. Fuel is burned to produce heat to make steam, but this burning takes place anywhere from a few feet to several miles away from the engine. Figure 2-2 shows the basic differences between internal and external combustion engines.

The internal combustion engine burns its fuel inside a combustion chamber. One side of this chamber is open to a piston. When the fuel burns, the hot gases expand very rapidly and push the piston away from the combustion chamber. This basic action of heated gases expanding and pushing is the source of power for all internal combustion engines. This includes piston, rotary, and turbine engines.

Compression and Combustion

Gasoline by itself will not burn; it must be mixed with oxygen in the air. If fuel burns in the open air,

it produces no power because it is not confined. If the same amount of fuel is enclosed and burned, it will expand with some force. To get the most force from the burning of a liquid fuel, it must be vaporized, mixed with air, and compressed to a small volume before it is burned. This compression and combustion is the most efficient way of releasing the energy stored in the air-fuel mixture.

In a piston engine, a piston moving in a cylinder provides compression. An example of this type of compression can be found in a two-section mailing tube with metal ends, figure 2-3. Push the inside tube in very quickly, and it will compress the air inside. Release the inside tube quickly and it will fly out. In a similar way, the piston compresses the air-fuel mixture in the cylinder.

Internal combustion engines are designed to compress and burn the vaporized air-fuel mixture in a sealed chamber, figure 2-4. Here, the combustion energy can work on the movable piston to produce mechanical energy. When the heat from the burning fuel causes the fuel vapor, air, and ex-

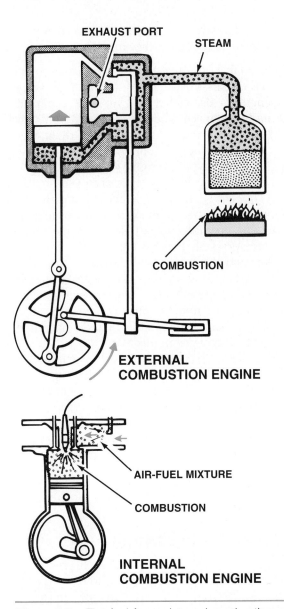

Figure 2-2. The fuel for an internal combustion engine is burned inside the engine. The fuel for an external combustion engine is burned outside the engine.

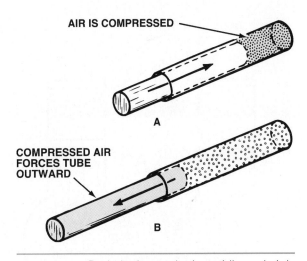

Figure 2-3. Push the inner tube in rapidly, and air is compressed (A). Release the inner tube quickly, and the compressed air forces it out (B).

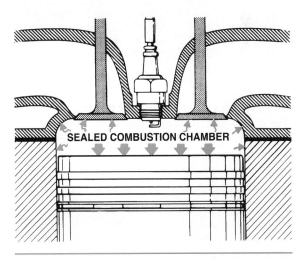

Figure 2-4. For combustion to produce power in an engine, the combustion chamber must be sealed.

haust gases inside the cylinder to expand, it produces much more power than was required to compress it. The burning, expanding gases push the piston to the other end of the cylinder.

Vacuum

The air-fuel mixture enters the combustion chamber past an intake valve. The suction or vacuum that pulls the mixture into the cylinder is created by the descending piston. You can create this same suction in the example of the two-piece mailing tube shown in figure 2-3. Draw the assembled

tubes apart quickly and let go. The suction tends to pull the tubes back together. If an open intake valve were in the end of the outer tube, pulling the inside tube would draw air past the valve. This suction is known as engine vacuum. Suction exists because of a difference in air pressure between the two areas. We will study vacuum and air pressure in more detail in later chapters.

The Four-Stroke Cycle

The piston creates a vacuum by moving down through the cylinder bore. The movement of the piston from one end of the cylinder to the other is called a stroke, figure 2-5. After the piston reaches the end of the cylinder, it will move back to the other end. As

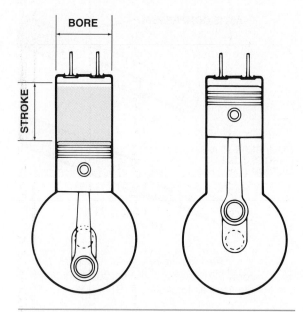

Figure 2-5. One top-to-bottom or bottom-to-top movement of the piston is called a stroke ⌐ piston stroke performs 180 degrees of cranks⌐   n; two strokes perform 360 degrees of ⌐   ⌐.

ALIGN BULK TEXT (handwritten)

long as the en⌐
move, or stroke⌐

An internal ⌐
through four sepa⌐
erating sequence o⌐   ⌐ type
of reciprocating eng⌐   ⌐perating cycle may require eithe⌐   ⌐ur strokes. In the
four-stroke engine, fc⌐ ⌐rokes of the piston in
the cylinder are needed to complete one full operating cycle. Each stroke is named after the action it performs—intake, compression, power,
and exhaust—in that order, figure 2-6.

1. Intake stroke: As the piston moves down,
    the mixture of vaporized fuel and air is
    drawn into the cylinder past the open intake
    valve.
2. Compression stroke: The intake valve closes,
    the piston moves up, and the mixture is com-
    pressed within the combustion chamber.
3. Power stroke: The mixture is ignited by a
    spark, and the expanding gases of combus-
    tion force the piston down in the cylinder.
    The exhaust valve opens near the bottom of
    the stroke.
4. Exhaust stroke: The piston moves up with the
    exhaust valve open, and the burned gases are
    pushed out to prepare for the next intake
    stroke. The intake valve usually opens just
    before the top of the exhaust stroke. This four-

stroke cycle is continuously repeated in every
cylinder as long as the engine is running.

Engines that use the four-stroke sequence are
known as four-stroke engines. This four-stroke cycle engine is also called the Otto-cycle engine after its inventor, Dr. Nicolaus Otto who built the
first successful four-stroke engine in 1876. Most
automobile engines are four-stroke, spark-ignition
engines. Other types of engines include two-stroke
and compression-ignition (diesel) engines.

A two-stroke engine also goes through intake,
compression, power, and exhaust actions to complete one operating cycle. However, the intake
and compression actions are combined in one
stroke, and the power and exhaust actions are
combined in the other stroke. Two-stroke engines
are used in motorcycles, lawn mowers, heavy
trucks, construction equipment, as well as ships.
Beyond a basic explanation of their operation,
they are not covered in this text.

Diesel engines do not use a spark to ignite the air-
fuel mixture. The heat from their high compression
ignites the fuel. Diesel engine operation is explained
later in this chapter. Aside from the differences in ignition and fuel systems, diesel and gasoline engines
are physically quite similar. Although this text does
not feature diesel engine repair and rebuilding,
much of the information on gasoline engines is typical of light-duty diesel engine service.

Reciprocating Engine

Except for the Wankel rotary engine, all production
automotive engines are reciprocating, or piston
type. Reciprocating means "up and down" or
"back and forth." It is the up and down action of a
piston in a cylinder that gives the reciprocating
engine its name. Power is produced by the inline
motion of a piston in a cylinder. However, this linear motion must be changed to rotating motion to
turn the wheels of a car or truck. In the following
paragraphs we will show how the parts of an engine produce reciprocating motion and change it to
rotating motion.

MAJOR ENGINE COMPONENTS

Of the major parts of an automobile engine, so far
we have mentioned only the pistons and cylinders. It takes many more parts, however, to build

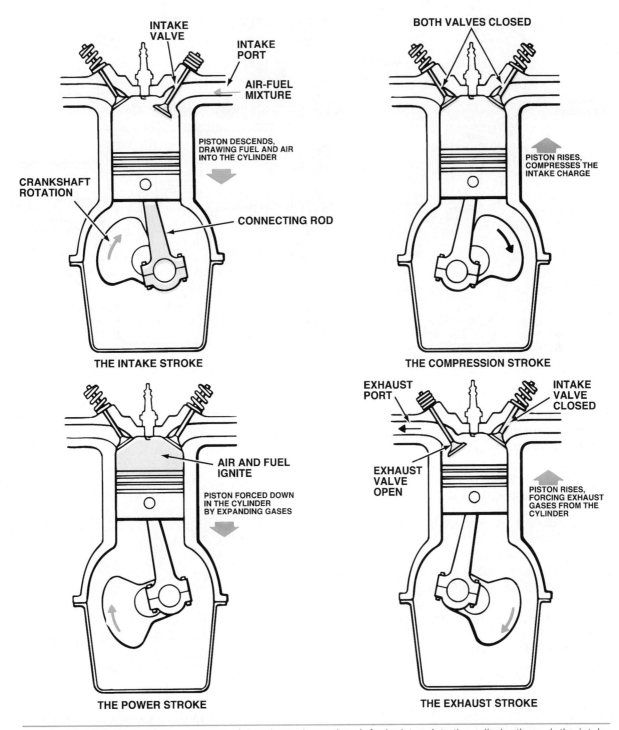

INTAKE VALVE

INTAKE PORT

AIR-FUEL MIXTURE

PISTON DESCENDS, DRAWING FUEL AND AIR INTO THE CYLINDER

CRANKSHAFT ROTATION

CONNECTING ROD

THE INTAKE STROKE

BOTH VALVES CLOSED

PISTON RISES, COMPRESSES THE INTAKE CHARGE

THE COMPRESSION STROKE

AIR AND FUEL IGNITE

PISTON FORCED DOWN IN THE CYLINDER BY EXPANDING GASES

THE POWER STROKE

EXHAUST PORT

INTAKE VALVE CLOSED

EXHAUST VALVE OPEN

PISTON RISES, FORCING EXHAUST GASES FROM THE CYLINDER

THE EXHAUST STROKE

Figure 2-6. The downward movement of the piston draws the air-fuel mixture into the cylinder through the intake valve on the intake stroke. On the compression stroke, the mixture is compressed by the upward movement of the piston with both valves closed. Ignition occurs at the beginning of the power stroke, and combustion drives the piston downward to produce power. On the exhaust stroke, the upward-moving piston forces the burned gases out the open exhaust valve.

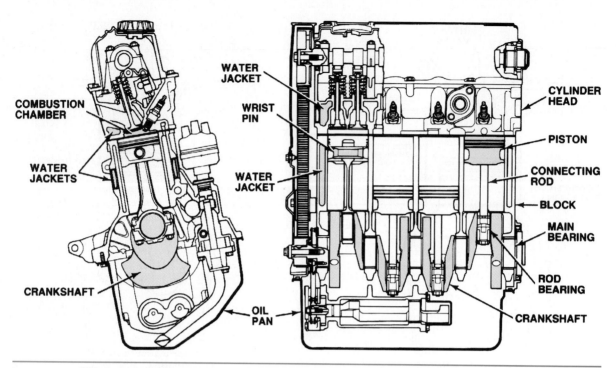

Figure 2-7. This four-cylinder engine has a typical engine block and cylinder head assembly.

a complete engine that will do useful work. The following parts are common to typical four-stroke internal combustion engines.

Cylinder Block and Head

Most automobile engines are built upon a cylinder block, or engine block, figure 2-7. The block is usually an iron or aluminum casting that contains the engine cylinders as well as passages for coolant and oil circulation. The top of the block is covered by the cylinder head, which has more coolant passages and forms most of the combustion chamber. The bottom of the block is covered with an oil pan, or an oil sump.

Crankcase

All piston engines have a crankcase, figure 2-8. It is a housing that supports or encloses the crankshaft. Early automotive designs used a separate crankcase bolted to the cylinders. Most modern automotive engines use a crankcase that is cast in one piece with the cylinder block. The entire casting of block and crankcase is known as the cylinder block, or simply the engine block. The term *crankcase* is still used to describe the open space around the crankshaft, which also usually includes the oil pan.

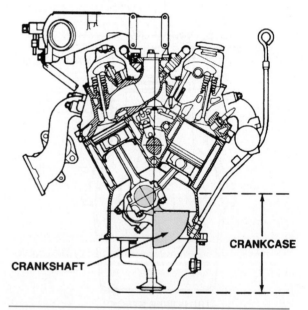

Figure 2-8. The crankcase of this Chrysler V6 engine, like most modern engines, is the lower portion of the cylinder block casting and the oil pan.

Crankshaft and Connecting Rod

The crankshaft revolves inside the crankcase portion of the engine block, figures 2-7 and 2-8. Main bearing caps bolt to the block and hold the crank-

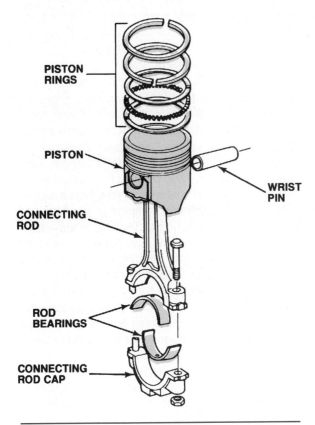

Figure 2-9. This piston is attached to the connecting rod by a wrist pin, which lets it pivot as the engine runs.

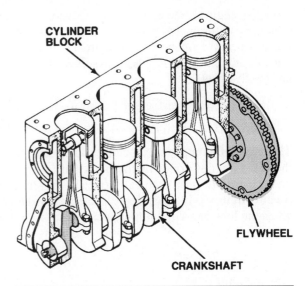

Figure 2-10. Flywheel inertia helps smooth out the impulses of the firing strokes.

shaft in place. Large shell bearings, called main bearings, are used between the crankshaft and caps.

The piston is attached to one end of a connecting rod by a pin called a piston pin or a wrist pin, figure 2-9. The other end of the rod is attached to the crankshaft. Rod bearings, similar to the main bearings, are used between the connecti... the crankshaft.

As the piston stro... the crankshaft. The ... shaft can be used to ... mobile, rotate the bla... the propeller of an ai...

It is important to ... between the revolv... stroking piston. The ...ways makes two strokes for each revolution of the crankshaft. The complete four-stroke cycle requires two crankshaft revolutions.

Because only one of the four strokes is a power stroke, the crankshaft must coast for one and one-half revolutions in a single-cylinder engine. It does this because of the **inertia** of its rotating parts, particularly the flywheel.

Flywheel

Because the power in a piston or rotary engine is applied in impulses, the engine tends to jerk or pulse. This tendency is reduced by the flywheel. The flywheel is a large, heavy disc of metal attached to the end of the crankshaft, figure 2-10. The flywheel works on the principle of inertia. The inertia of the flywheel resists any change in speed. When there is a power impulse, the heavy flywheel resists a rapid increase in engine speed, and during the coasting period (the exhaust, intake, and compression strokes), the flywheel keeps the engine turning because it also resists a decrease in speed.

When an automatic transmission is used, a torque converter is bolted to the flywheel. Because the torque converter is heavy, the flywheel can be much lighter. The flywheel used with a torque converter is often called a "flexplate." The total weight of torque converter and flexplate equals the weight of the flywheel and clutch on a manual transmission engine.

Engine Rotation

The front of the engine is commonly considered to be the end opposite the flywheel. When looking at the front of an engine as just defined, the crankshaft and flywheel of most engines rotate clockwise.

This was the routine definition of an engine's physical features and direction of rotation in the simpler days when nearly all cars had front-mounted engines and rear-wheel-drive. Now that many cars with front-wheel-drive have transversely mounted engines, there is some confusion over which end of the engine is the front. In this book, no matter where the engine is situated, front or rear, in-line or transverse, the front is always the end opposite the flywheel.

Camshaft

The camshaft controls the opening and closing of the valves, and is driven by the crankshaft. Lobes on the camshaft push each valve open as the shaft rotates, figure 2-11. A spring closes each valve when the lobe is not holding it open.

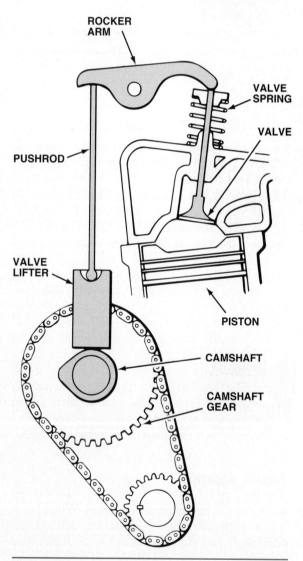

Figure 2-11. Each valve is opened by a lobe on the camshaft and closed by a spring. The camshaft sprocket has twice as many teeth as the crankshaft sprocket, causing the camshaft to rotate at one-half crankshaft speed.

Nicolaus Otto was born in 1832, in a small hamlet in Germany near the Rhine river. Poor economic conditions forced him to drop out of secondary school to become a grocery clerk, and he ended up as a salesman of tea, sugar, and kitchenwares.

Otto was intrigued by the Lenoir internal combustion engine, in 1860 the first internal combustion engine commercially available. Handicapped by his lack of education, he spent three years and all his own money (and much of the money of his friends) trying to improve it. In 1864, Otto formed a company with Eugen Langen, a technologically minded speculator who provided much-needed capital. Their company, today called Klöckner-Humboldt-Deutz AG, is the first and oldest internal combustion engine manufacturing company in the world. The company and its refined engines were immensely successful.

The first four-stroke engine was not built until 1876. All previous internal combustion engines—including Otto's—were noncompression, meaning fuel and air were drawn into the cylinder during part of a piston's downward stroke and then ignited. The expanding gases then pushed the piston down the remainder of its stroke. Many inventors used this design to make the pistons double-acting, with a power stroke each way. Otto's new engine used a downward stroke of the piston to draw in an intake charge, and a second upward stroke to compress it. It also required two more strokes to extract power and push out exhaust gases—the four-stroke engine cycle. Otto's competitors were dubious, believing that to waste three piston strokes for a single power stroke must surely outweigh any advantages of compressing the mixture.

They were quite wrong. Comparing Otto's new engine with his earlier best seller showed that the four-stroke weighed one-third as much, could run almost twice as fast, and needed only 7 percent of the cylinder displacement to produce the same horsepower, with almost identical fuel consumption. Within ten years, a four-stroke engine powered the first motorcycle, and soon after that the engine appeared in what would be called the horseless carriage.

Once during each revolution of the camshaft, the lobe will push the valve open. The timing of the valve opening is critical to the operation of the engine. Intake valves must be opened just before the beginning of the intake stroke. Exhaust valves must be opened just before the beginning of the exhaust stroke. Because the intake and exhaust valves open only once during every two revolutions of the crankshaft, the camshaft must run at half the crankshaft speed.

Turning the camshaft at half the crankshaft speed is accomplished by using a gear or sprocket on the camshaft that is twice the diameter of the crankshaft gear or sprocket, figure 2-11. If you count the teeth on each gear or sprocket, you will find exactly twice as many on the camshaft as on the crankshaft.

Valves

All modern automotive piston engines use poppet valves, figure 2-12. These are valves that work by linear motion. Most water faucet valves operate with a circular motion. The poppet valve is opened simply by pushing on it. The poppet valve must have a seat on which to rest and from which it closes off a passageway. There also must be a spring to hold the valve against the seat. In operation, a push on the end of its stem opens the valve, and when the force is removed, the spring closes the valve.

Valve Arrangement

Intake and exhaust valves on modern engines are located in the cylinder head. Because the valves are "in the head," this basic arrangement is called an I-head design. In the past, poppet valves have been arranged in three different ways, figure 2-13.

- The L-head design positions both valves side-by-side in the engine block. Because the cylinder head is rather flat and contains only the combustion chamber, water jacket, and spark plugs, L-head engines also are called "flat-heads." Still very common on lawn mowers, this valve arrangement has not been used in a domestic automotive engine since the mid-1960s.
- The F-head design positions the intake valve in the cylinder head and the exhaust valve in the engine block. A compromise between the L-head and I-head designs, the F-head was last used in the 1971 Jeep.

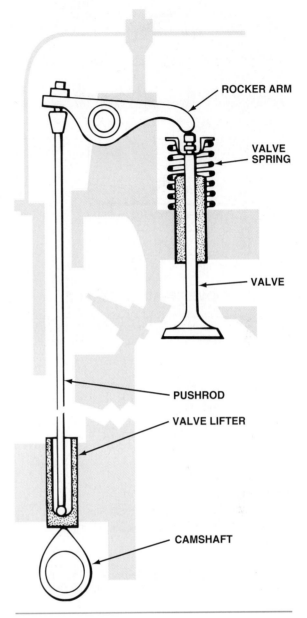

Figure 2-12. Modern automotive engines use poppet valves.

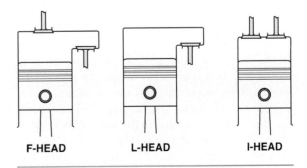

Figure 2-13. Historically, designers have arranged valves for four-stroke engines in these three ways.

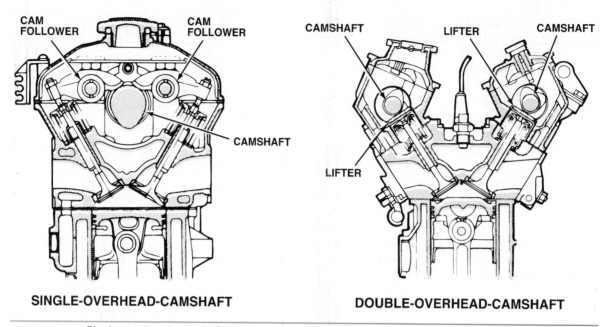

SINGLE-OVERHEAD-CAMSHAFT   **DOUBLE-OVERHEAD-CAMSHAFT**

Figure 2-14. Single-overhead-camshafts usually require an additional component, such as a rocker arm, to operate all the valves. Double-overhead-camshaft engines actuate the valves directly.

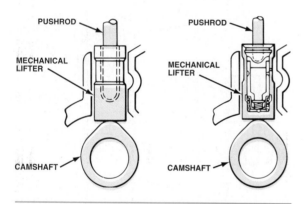

Figure 2-15. Mechanical lifters are solid metal. Hydraulic lifters use engine oil to take up clearance and transmit motion.

- The I-head design, in both overhead-valve or overhead-camshaft form, positions both the intake and the exhaust valves in the cylinder head. All modern automotive engines use this design.

In the overhead-valve engine, the camshaft is in the engine block and the valves are opened by valve lifters, pushrods, and rocker arms, figures 2-11 and 2-12. In the overhead-camshaft engine, the camshaft is mounted in the head, either above or to one side of the valves, figure 2-14. This improves valve action at higher engine speeds. The valves may open directly by means of valve lifters or camshaft followers, or through rocker arms. The double overhead camshaft engine has two camshafts, one on each side of the valves. One

camshaft operates the intake valves; the other operates the exhaust valves.

Valve Lifters
Valve lifters can be mechanical or hydraulic, as shown in figure 2-15. A mechanical valve lifter is solid metal. A hydraulic lifter is a metal cylinder containing a plunger that rides on oil. A chamber below the plunger fills with engine oil through a feed hole and, as the camshaft lobe lifts the lifter, the chamber is sealed by a check valve. The trapped oil transmits the lifting motion of the camshaft lobe to the valve pushrod. Hydraulic lifters are generally quieter than mechanical lifters and do not normally need to be adjusted, because the amount of oil in the chamber varies to keep the valve adjustment correct.

Number of Valves
As you saw in figure 2-6, most automobile engines have one intake and one exhaust valve per cylinder. This means that a four-cylinder engine has 8 valves, a six-cylinder engine has 12 valves, and a V-8 has 16 valves.

Many engines have been built, however, with more than two valves per cylinder. Engines with three valves or four per cylinder are currently used in several Japanese production cars. Engines with four valves per cylinder have been used in racing engines since the 1912 Peugeot Grand Prix cars. Street motorcycle engines with five valves

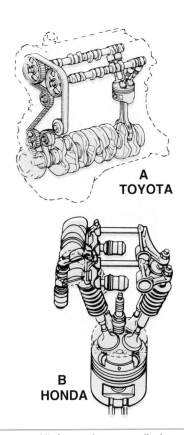

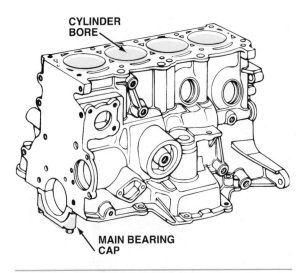

Figure 2-17. Most inline engines position the cylinders vertically, like this Chrysler 4-cylinder.

Figure 2-16. All four-valve-per-cylinder production engines use overhead camshafts. Most have separate intake and exhaust camshafts as in the Toyota example (A). Some have a single camshaft, such as the Honda Acura V-6 design (B) that operates the exhaust valves through short pushrods.

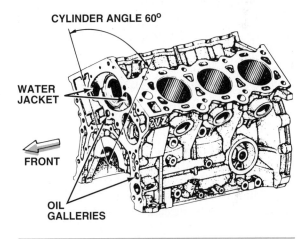

Figure 2-18. This Lexus V-type cylinder block has its cylinders inclined at 60 degrees.

per cylinder have been built in recent years. In spite of performance advantages, the higher costs and greater complexity of engines with more than two valves per cylinder kept such designs from being common in production engines until the mid-1980s, figure 2-16.

CYLINDER ARRANGEMENT

While single-cylinder engines are common in motorcycles, outboard motors, and small agricultural machines, automotive engines have more than one cylinder. Most car engines have 4, 6, or 8 cylinders, although engines with 3, 5, 10, and 12 cylinders are also being produced. Within the engine block, the cylinders are arranged in one of three ways:

- Inline engines have a single bank of cylinders arranged in a straight line, figure 2-17. The cylinders do not have to be vertical. They can be inclined to either side. Most inline engines

have 4 or 6 cylinders, but many inline engines with 3, 5, and 8 cylinders have been built.

- V-type engines, figure 2-18, have two banks of cylinders, usually inclined either 60 degrees or 90 degrees from each other. Most V-type engines have 6 or 8 cylinders, but V-2 (or V-twin), V-4, V-12, and V-16 engines have been built.
- Horizontally opposed, "flat," or "boxer" engines have two banks of cylinders 180 degrees apart, figure 2-19. These engine designs are often air-cooled, and are found in the original VW Beetle and some Ferrari, Porsche, and Subaru models. Ferrari and Subaru designs are liquid cooled. Volkswagen vans use a version of the traditional air-cooled VW horizontally opposed engine with liquid-cooled cylinder heads. Most opposed engines

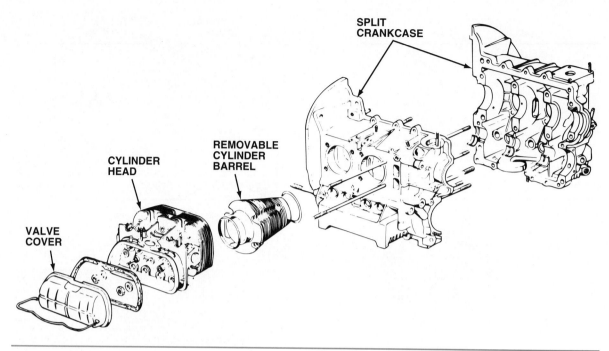

Figure 2-19. An horizontally opposed engine.

have 2, 4, or 6 cylinders, but flat engines with 8, 12, and 16 cylinders have been built.

Engine Balance

When an engine has more than one cylinder, the crankshaft is usually made so that the firing impulses are evenly spaced. Engine speed is measured in revolutions per minute, or in degrees of crankshaft rotation. It takes two crankshaft revolutions or 720 degrees of crankshaft rotation to complete the four-stroke sequence. If a four-stroke engine has two cylinders, the firing impulses and crankshaft throws can be spaced so that there is a power impulse every 360 degrees. On an inline 4-cylinder engine, the crankshaft is designed to provide firing impulses every 180 degrees. An inline 6-cylinder crankshaft is built to fire every 120 degrees. In an 8-cylinder engine, either inline or V-type, the firing impulses occur every 90 degrees of crankshaft rotation.

As you can see, the more cylinders an engine has, the closer the firing impulses will be. On 6- and 8-cylinder engines, for example, the firing impulses are close enough that power strokes overlap slightly. In other words, a new power stroke begins before the power stroke that preceded it ends. This provides a smooth transition from one firing pulse to the next. A 4-cylinder en-

gine has no overlap of its power strokes, which makes it a relatively rough-running engine compared to engines with more cylinders. Therefore, an 8-cylinder engine runs smoother than a 4-cylinder engine.

Figure 2-20 shows common crankshaft arrangements and firing impulse frequencies. Other arrangements are possible, such as V-4, opposed 4-cylinder, and several combinations of 2-cylinder layouts. The cylinder arrangement, cylinder numbering order, and crankshaft design all determine the firing order of an engine, which we will study at the end of this chapter.

ENGINE DISPLACEMENT AND COMPRESSION RATIO

In any discussion of engines, the term engine size comes up often. This does not refer to the outside dimension of an engine, but to its displacement. As the piston strokes in the cylinder, it moves through or displaces a specific volume. Another important engine measurement term is **compression ratio.** Displacement and compression ratio are related to each other, as you will learn in the following paragraphs.

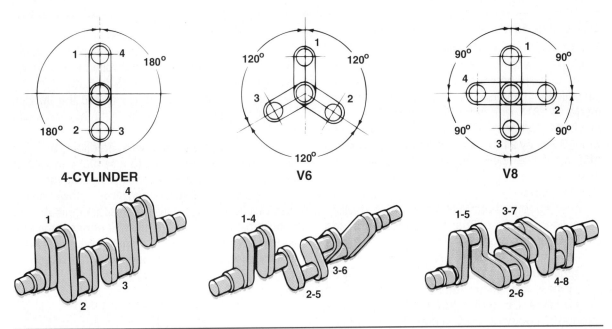

Figure 2-20. The more cylinders an engine has, the closer together the firing impulses are. Here are common crankshaft designs for 4-, 6-, and 8-cylinder engines.

Engine Displacement

Engine displacement is a measurement of engine volume. The number of cylinders is a factor in determining displacement, but the arrangement of cylinders is not. Engine displacement is calculated by multiplying the piston displacement of one cylinder by the number of cylinders. The total engine displacement is the volume displaced by all the pistons.

We learned earlier that the bore and stroke are important engine dimensions. Both are necessary measurements for calculating engine displacement. The displacement of one cylinder is the volume through which the piston's top surface moves as it travels from the bottom of its stroke (**bottom dead center or BDC**) to the top of its stroke (**top dead center or TDC**), figure 2-21. Piston displacement is computed as follows:

$$\text{Piston displacement} = \frac{(\text{Bore})^2}{2} \times 3.1416 \times \text{Stroke}$$

1. Divide the bore (cylinder diameter) by two. This will give you the radius of the bore.
2. Square the radius (multiply it by itself).
3. Multiply the square of the radius by 3.1416 to find the area of the cylinder cross section.

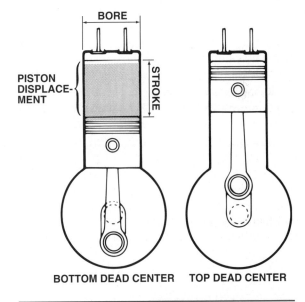

Figure 2-21. The bore and stroke of a piston are used to calculate an engine's displacement.

4. Multiply the area of the cylinder cross section by the length of the stroke.
5. You now know the piston displacement for one cylinder. Multiply this by the number of cylinders to determine the total engine displacement.

■ **Isaac de Rivaz and His Self-Propelled Carriage**

The first self-propelled vehicles using an internal combustion engine were those built by Isaac de Rivaz, a Swiss engineer and government official. He built the first version in 1805, but his improved model of 1813 was more impressive. This "great mechanical chariot" was 17 feet long by 7 feet wide (5.2 m by 21 m), and weighed 2,100 pounds (950 kg). Its top speed was about 3 miles per hour (5 km/hr) on a level road, and it could climb a 12 percent grade.

The engine that drove this marvel used a single cylinder, open at the top, with a bore and stroke of 36.5 cm by 150 cm (14.4 inches by 59 inches). A long rod attached to the piston was connected by a chain to a drum outside the cylinder. When the piston descended, the chain would rotate the drum, turning the drive axle through a rope and pulley. When the piston rose, a ratchet let the rope and pulley freewheel.

Coal gas (a burnable mixture of hydrogen and methane) was stored in a collapsible leather bladder, pumped into a mixing chamber in a carburetor by a bellows for each stroke of the piston. To start an engine cycle, the driver pulled on a lever, which first dropped the floor of the cylinder a few inches to draw in an intake charge, and next closed an electric circuit to cause a spark to jump a gap between two wires in the combustion chamber. The burning, expanding gases shot the piston up to the top of its stroke, as the ratchet mechanism freewheeled. After combustion stopped, the gases in the cylinder cooled and contracted, and

atmospheric pressure pushed the piston down. This engaged the ratchet and rolled the carriage forward about 16 to 20 feet (5 to 6 m). Exhaust was expelled when the driver raised the floor of the cylinder again in preparation for the next stoke of the engine.

His carriage was not very practical, as trips were limited to the 2-mile (3 km) fuel capacity of the leather bag, and the driver had to pull and push the lever for each piston stroke. However, his use of a fuel-mixing carburetor, spark ignition, and a portable fuel tank were all engineering feats that would eventually become common on the internal combustion engines in modern-day "carriages."

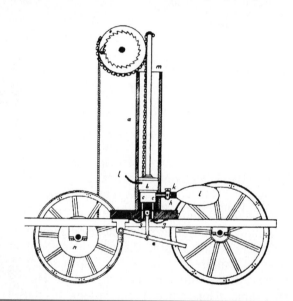

For example, to find the displacement of a 6-cylinder engine with a 3.80-inch bore and a 3.40-inch stroke:

1. $\dfrac{3.80}{2} = 1.9$
2. $1.9 \; [\times] \; 1.9 = 3.61$
3. $3.61 \; [\times] \; 3.1416 = 11.3412$
4. $11.3412 \; [\times] \; 3.40 = 38.56$
5. $38.56 \; [\times] \; 6 = 231.36$

The displacement is 231 cubic inches. Fractions of an inch are usually not included.

Metric Displacement Specifications

When stated in U.S. customary values, displacement is given in cubic inches. The engine's cubic inch displacement is abbreviated cid. When stated in metric values, displacement is given in cubic centimeters (cc) or in liters (one liter equals 1,000 cc). To convert engine displacement specifications from one value to another, use the following formulas:

- To change cubic centimeters to cubic inches, multiply by 0.061 (cc × 0.061 = cu in.).
- To change cubic inches to cubic centimeters, multiply by 16.39 (cid × 16.39 = cc).
- To change liters to cubic inches, multiply by 61.02 (liters × 61.02 = cid).

Our 231-cid engine from the previous example is also a 3,786-cc engine (231 × 16.39 = 3,786). When expressed in liters, this figure is rounded up to 3.8 liters.

Metric displacement in cc can be calculated directly with the displacement formula, using centimeter measurements instead of inches. Here is how it works for the same engine with a bore equaling 96.52 mm (9.652 cm) and a stroke equaling 86.36 mm (8.636 cm):

1. $\dfrac{9.652}{2} = 4.826$
2. $4.826 \; [\times] \; 4.826 = 23.29$
3. $23.29 \; [\times] \; 3.1416 = 73.16$
4. $73.16 \; [\times] \; 8.636 = 631.81$
5. $631.81 \; [\times] \; 6 = 3791 \; cc$

This figure is a few cubic centimeters different from the 3,786-cc displacement we got by converting 231 cubic inches directly to cubic centimeters. This is due to rounding. Again, the engine displacement can be rounded up to 3.8 liters.

Compression Ratio

The compression ratio compares the total cylinder volume when the piston is at BDC to the volume of the combustion chamber when the piston is at TDC, figure 2-22. Total cylinder volume may seem to be the same as piston displacement, but it is not. Total cylinder volume is the piston displacement plus the combustion chamber volume. The combustion chamber volume with the piston at TDC is sometimes called the clearance volume.

Compression ratio is the total volume of a cylinder divided by the clearance volume. If the clearance volume is 1/8 of the total cylinder volume, the compression ratio is 8 to 1. The formula is as follows:

$$\frac{\text{Total volume}}{\text{Clearance volume}} = \text{Compression ratio}$$

To determine the compression ratio of an engine in which each piston displaces 510 cc and which has a clearance volume of 65 cc:

$$510 + 64 = \text{cc (total cylinder volume)}$$
$$\frac{574}{64} = 8.968$$

The compression ratio is 8.986 to 1. This would be rounded and expressed as a compression ratio of 9 to 1. This can also be written 9:1.

A higher compression ratio is desirable because it increases the efficiency of the engine, making the engine develop more power from a given quantity of fuel. A higher compression ratio increases cylinder pressure, which packs the fuel molecules more

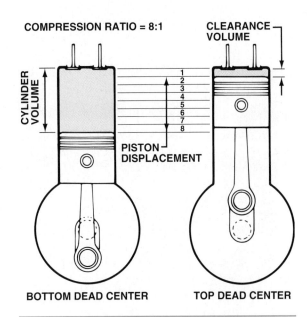

Figure 2-22. Compression ratio is the ratio of the total cylinder volume to the clearance volume.

tightly together. The flame of combustion then travels more rapidly and across a shorter distance.

ENGINE COOLING SYSTEM

The two jobs of the engine cooling system are to:

• Carry excess heat away from the engine
• Maintain uniform temperature throughout the engine

These requirements are critical for modern engines with electronic controls that must maintain precise air-fuel ratios for economy and emission control. Since the average combustion chamber temperature is approximately 1500°F (800°C) and peak combustion chamber temperature can reach 6000°F (3300°C), it is easy to see why engine cooling is necessary.

The engine changes about one-third of combustion-created heat into energy to drive the car, but it needs to get rid of, or dissipate, the remaining two-thirds. In some engines, radiation into the air dissipates the heat, but most engines use a liquid cooling system and the exhaust system to remove heat. We have already mentioned that automotive engine blocks and cylinder heads contain passages, called water jackets, for coolant circulation. Figure 2-23 shows the components and operation of a typical liquid-cooling system.

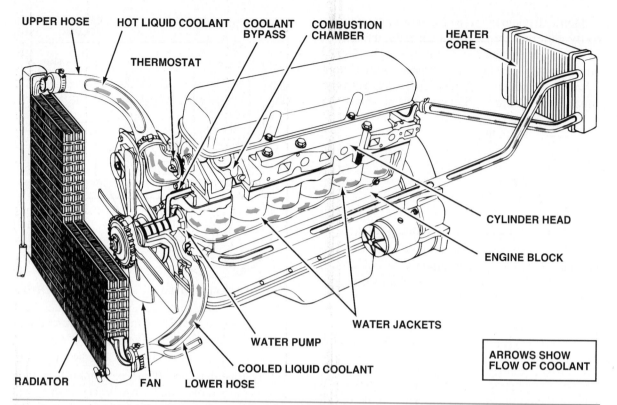

UPPER HOSE   HOT LIQUID COOLANT   COOLANT BYPASS   COMBUSTION CHAMBER   HEATER CORE

THERMOSTAT

CYLINDER HEAD

ENGINE BLOCK

WATER JACKETS

WATER PUMP

COOLED LIQUID COOLANT

RADIATOR   FAN   LOWER HOSE

ARROWS SHOW FLOW OF COOLANT

Figure 2-23. In a typical cooling system, coolant flows from the radiator into the engine, where it absorbs heat. It dissipates some heat in the heater core, then flows into the cylinder head. From there, it goes into the radiator to cool off before starting the cycle again.

■ William Cecil's Internal Combustion Engine

Internal combustion engines are by no means new. The first internal combustion engine that would operate continuously by itself was built in 1820 by William Cecil.

Cecil's engine was constructed with three cylinders arranged in a "T," and connected at their intersection by a rotating valve. The vertical cylinder contained a piston that descended to draw in a charge of air mixed with hydrogen—the two horizontal cylinders remained empty during the intake stroke. At the bottom of the intake stroke, the central valve rotated to briefly open a passage connecting the cylinder to a small flame, igniting the hydrogen. The valve then rotated again, permitting the burning, expanding gases to fill up the two cylinders forming the crosspiece of the "T," pushing out the air they contained through flapper exhaust valves. As the burned gases cooled and contacted, they pulled the flapper valves shut. Atmospheric pressure then pushed the piston back up into the cylinder for the power stroke, and the cycle repeated.

Cecil's engine had a cylinder capacity of about 30 cid (500 cc), and used a flywheel that weighed 50 pounds (23 kg). It would run evenly on a fuel-air mixture of 1 to 4, although best power was obtained with a richer mixture of 1.25 to 1. Because of the flame-type ignition, top speed was limited to about 60 rpm—above this speed the flame could not light the hydrogen reliably.

Cecil demonstrated a working model of his engine to the Cambridge Philosophical Society in 1820, but chose not to pursue its development further. He was ordained two years after inventing the engine, and spent the remainder of his life as a clergyman with the Church of England.

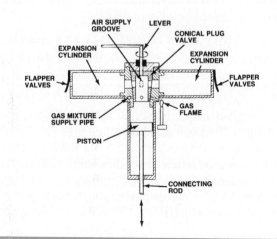

AIR SUPPLY GROOVE   LEVER

CONICAL PLUG VALVE

EXPANSION CYLINDER

EXPANSION CYLINDER

FLAPPER VALVES

FLAPPER VALVES

GAS MIXTURE SUPPLY PIPE

GAS FLAME

PISTON

CONNECTING ROD

A water pump circulates liquid coolant, a mixture of ethylene glycol antifreeze and water, through the engine to absorb heat. The coolant passes out of the engine water jackets into a radiator, where air flowing across the radiator tubes and fins absorbs heat from the coolant. The coolant then returns to the engine to absorb more heat in a continuous process.

The coolant passages in the block and cylinder heads are designed so the coolant flows uniformly through the engine without collecting in pockets. If coolant flow is not uniform, there is an uneven transfer of heat from the engine to the coolant. Some parts of the engine are undercooled and others overcooled.

If an engine runs too cold, the fuel does not vaporize completely. Liquid fuel in the cylinders reduces lubrication by washing the oil from the cylinder walls and diluting the engine oil. This causes a loss of performance, an increase in HC emissions, and premature engine wear.

A too-hot engine may cause preignition, a condition in which the high combustion-chamber temperature ignites the air-fuel charge before the spark plug fires. Oil circulating in an overheated engine loses viscosity and forms varnish and carbon deposits. The low-viscosity oil may be drawn into the combustion chamber, where it increases HC emission and causes poor performance and premature wear, and even engine damage.

The engine cooling system actually is a temperature-regulation system. It must maintain a high-enough temperature for efficient combustion, but not so high as to allow engine damage.

ENGINE LUBRICATION SYSTEM

An automotive engine cannot run without proper lubrication. Engine components work under conditions of extreme heat and must maintain close tolerances. Engine oil performs these important tasks:

- Reduces friction
- Removes heat from engine parts
- Helps keep the engine clean
- Seals the combustion chambers
- Cushions engine parts
- Prevents corrosion and rust

To do these jobs, engine oil is formulated to flow easily without losing its primary characteristic of being a lubricating film.

An automotive engine lubrication system consists of a reservoir (called the oil pan, sump, or crankcase) to hold the oil supply, a pump to develop pressure, a filter for cleaning, and valves to control flow and pressure.

There are two ways to circulate oil through an engine—pressure and splash—that are often combined, figure 2-24. The oil pump pressurizes the oil and delivers it to the filter for cleaning before sending it to the main oil gallery and on to the crankshaft, camshaft, and valvetrain components. Other components are lubricated by splashing oil and by a network of passages, or galleries.

Oil circulates through the passages drilled or cast in the block, head, and crankshaft. These oil galleries serve as a network to carry the oil where it is needed. All engines have at least one main gallery, but larger V-type engines may have two. Smaller galleries lead away from the main gallery to carry oil to the crankshaft and camshaft bearings. Holes, grooves, nozzles, and orifices ensure that proper lubrication reaches all parts of the engine at all times. A bypass valve presents oil starvation in case the filter clogs up.

As the oil carries out its primary job of lubrication, it absorbs engine heat and carries it back to the crankshaft, or sump, where the heat dissipates. Hot oil cannot do its job properly and quickly loses the qualities required for proper lubrication, so many engines have an oil cooler, especially diesel, turbocharged, and air-cooled engines. Depending on cooler design, either airflow or coolant circulation removes heat from the oil.

THE IGNITION SYSTEM

After the air-fuel mixture is drawn into the cylinder and compressed by the piston, it must be ignited. The ignition system creates a high electrical potential or voltage. This voltage jumps a gap between two electrodes in the combustion chamber. The arc (spark) between the electrodes ignites the compressed mixture. The two electrodes are part of the spark plug, which is a major part of the ignition system.

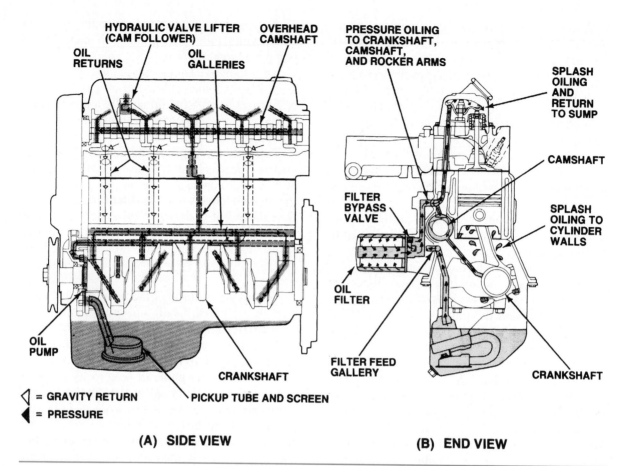

Figure 2-24. The oil pump sends pressurized oil to some engine parts; other parts get lubricated by oil splashing on them when the engine is running.

Ignition Interval

The timing of the spark is critical to proper engine operation. It must occur near or before the start of the power stroke. It the spark occurs too early or too late, full power will not be obtained from the burning air-fuel mixture.

As we have seen, every two strokes of a piston rotate the crankshaft 360 degrees, and there are 720 degrees of rotation in a complete four-stroke cycle. During the four strokes of the cycle, the spark plug for each cylinder fires only once. In a single-cylinder engine, there would be only one spark every 720 degrees. These 720 degrees are called the ignition interval, or firing interval: the number of degrees of crankshaft rotation that occur between ignition sparks.

Common Ignition Intervals

Since a 4-cylinder engine has four power strokes during 720 degrees of crankshaft rotation, one power stroke must occur every 180 de-grees (720 ÷ 4 = 180). The ignition system must produce a spark for every power stroke, so it produces a spark every 180 degrees of crankshaft rotation. This means that a 4-cylinder engine has an ignition interval of 180 degrees.

An inline 6-cylinder engine has six power strokes during every 720 degrees of crankshaft rotation, for an ignition interval of 120 degrees (720 ÷ 6 = 120). An 8-cylinder engine has an ignition interval of 90 degrees (720 ÷ 8 = 90).

Unusual Ignition Intervals

Most automotive engines have 4, 6, or 8 cylinders, but other engines are in use today. Some companies, such as Jaguar, Ferrari, and BMW, produce 12-cylinder engines, with a 60-degree firing interval. Audi and Mercedes vehicles with 5-cylinder engines have a 144-degree firing interval; Suzuki produces a 3-cylinder engine with a 240-degree firing interval used in the Chevrolet Metro.

Other firing intervals result from unusual engine designs. General Motors has produced two

Figure 2-25. Most engines have a pulley bolted to the front end of the crankshaft. Notice the timing mark on the crankshaft pulley.

different V-6 engines from V-8 engine blocks. The Buick version, developed in the 1960s, has alternating 90- and 150-degree firing intervals, figure 2-25. The uneven firing intervals resulted from building a V-6 with a 90-degree crankshaft and block. This engine was modified in mid-1977 by redesigning the crankshaft to provide uniform 120-degree firing intervals, as in an inline six. In 1978, Chevrolet introduced a V-6 engine that fires at alternating 108- and 132-degree intervals.

Spark Frequency

In a spark-ignition engine, each power stroke begun by a spark igniting the air f
Each power stroke need
8-cylinder engine
sparks per engine r
are two 360-degre
720-degree operatin,
running at 1,000 rp
deliver 4,000 sparks
about 4,000 rpm, the i
16,000 sparks per mi ̣   ̣ ̣ ̣ ̣tion system
must perform precisely ̣ ̣ meet these demands.

Firing Order

The order in which the air-fuel mixture is ignited within the cylinders is called the **firing order,**

and it varies with different engine designs. Firing orders are designed to reduce the vibration and imbalance created by the power strokes of the pistons. Engine designers number the cylinders for identification. However, the cylinders seldom fire in the order in which they are numbered.

The ignition system must deliver a spark to the correct cylinder at the correct time. To get the correct firing order, the spark plug cables must be attached to the distributor cap, or to the coil on distributorless ignition systems, in the proper sequence, figure 2-26.

Inline Engines

Straight or inline engines are numbered from front to rear, figure 2-27. The most common firing order for both domestic and imported 4-cylinder engines is 1-3-4-2. That is, the number one cylinder power stroke is followed by the number three cylinder power stroke, then the number four power stroke and, finally, the number two cylinder power stroke. Then, the next number one power stroke occurs again. Some inline 4-cylinder engines have been built with firing orders of 1-2-4-3. One of these two firing orders is necessary due to the geometry of the engine.

The 5-cylinder, spark-ignited engine built by Audi and the 5-cylinder diesel engine built by Mercedes are also numbered front to rear. They share the same firing order of 1-2-4-5-3.

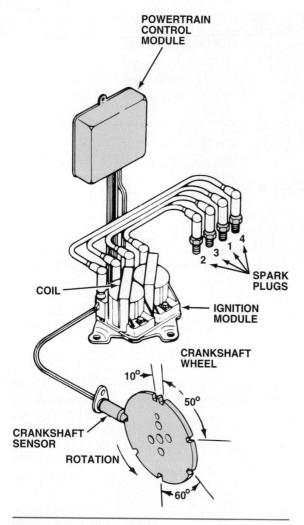

Figure 2-26. The spark plug wires must be connected in the proper sequence, or the engine will run poorly or not at all.

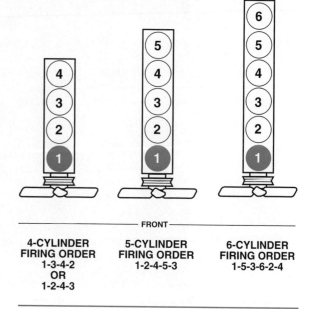

4-CYLINDER FIRING ORDER
1-3-4-2
OR
1-2-4-3

5-CYLINDER FIRING ORDER
1-2-4-5-3

6-CYLINDER FIRING ORDER
1-5-3-6-2-4

Figure 2-27. These are customary cylinder numbering and possible firing orders of inline engines.

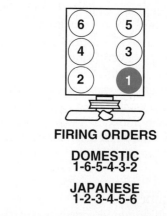

FIRING ORDERS

DOMESTIC
1-6-5-4-3-2

JAPANESE
1-2-3-4-5-6

Figure 2-28. The domestic 90-degree V-6 engines and several Japanese 60-degree engines are numbered this way. However, they use different firing orders.

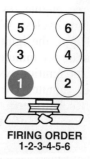

FIRING ORDER
1-2-3-4-5-6

Figure 2-29. General Motors 60-degree V-6 engines and many Japanese V-6 engines are numbered and fired this way.

The cylinders of an inline 6-cylinder engine are numbered from front to rear. The firing order for all inline 6-cylinder engines, both domestic and imported, is 1-5-3-6-2-4.

V-engines

The V-type engine structure allows designers greater freedom in selecting a firing order and still producing a smooth-running powerplant. Consequently, there is great variety of cylinder numbering and firing orders for V-type engines, too many to cover completely. Figures 2-28 through 2-32 show a representative sampling of common cylinder numbering styles and firing orders for these engines.

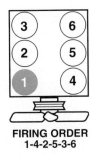

**FIRING ORDER
1-4-2-5-3-6**

Figure 2-30. Ford and Acura V-6 engines are numbered and fired this way.

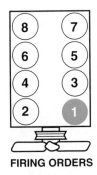

FIRING ORDERS

**DOMESTIC AND LEXUS
1-8-4-3-6-5-7-2**

**INFINITI
1-8-7-3-6-5-4-2**

Figure 2-31. Most Chrysler and General Motors V–8s as well as the Lexus V-8s are numbered and fired this way. The Infiniti V-8 is numbered this way, but fires differently.

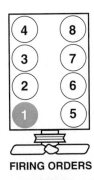

FIRING ORDERS

**FORD
1-3-7-2-6-5-4-8
OR
1-5-4-2-6-3-7-8**

**AUDI AND MERCEDES-BENZ
1-5-4-8-6-3-7-2**

Figure 2-32. Ford V-8 engines are numbered this way, but use two different firing orders, depending on the engine. Audi and Mercedes-Benz V-8 engines are numbered like the Fords, but use a third firing order.

ENGINE-IGNITION SYNCHRONIZATION

During the engine operating cycle, the intake and exhaust values open and close at specific times. The ignition system delivers a spark when the piston is near the top of the compression stroke and both values are closed. These actions must all be coordinated, or engine damage can occur.

Distributor Drive

The distributor must supply one spark to each cylinder during each cylinder's operating cycle. The distributor cam has as many lobes as the engine has cylinders, or in an electronic ignition system, the trigger wheel has as many teeth as the engine has cylinders. One revolution of the distributor shaft will deliver one spark to each cylinder. Since each cylinder needs only one spark for each two crankshaft revolutions, the distributor shaft must turn at only one-half engine crankshaft speed. Therefore, the distributor is driven by the camshaft, which also turns at one-half crankshaft speed.

On cars equipped with a distributorless ignition system (DIS), the spark plugs are fired using a multiple coil pack containing two or three separate ignition coils, according to the number of cylinders. A computer called the ignition control module (ICM) discharges each coil separately in sequence, with each coil serving two cylinders 360 degrees apart in the firing order.

Crankshaft Position

As you learned earlier, the exact bottom of the piston stroke is called bottom dead center (BDC). The exact top of the piston stroke is called top dead center (TDC). The ignition spark occurs near TDC, as the compression stroke is ending. As the piston approaches the top of its stroke, it is said to be before top dead center (BTDC). A spark that occurs BTDC is called an advanced spark, figure 2-33. As the piston passes TDC and starts down, it is said to be after top dead center (ATDC). A spark that occurs ATDC is called a retarded spark.

Burn Time

The instant the air-fuel mixture ignites until its combustion is complete is called the burn time.

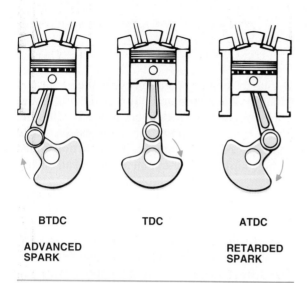

BTDC   TDC   ATDC

ADVANCED   RETARDED
SPARK   SPARK

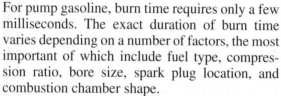

Figure 2-33. Piston position is identified in terms of crankshaft position.

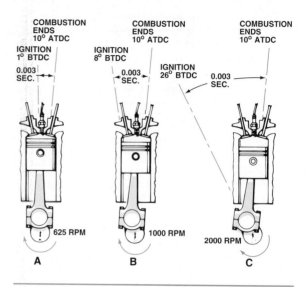

Figure 2-34. As engine speed increases, ignition timing must be advanced.

For pump gasoline, burn time requires only a few milliseconds. The exact duration of burn time varies depending on a number of factors, the most important of which include fuel type, compression ratio, bore size, spark plug location, and combustion chamber shape.

As its name implies, burn time is a function of time and not of piston travel or crankshaft degrees. The ignition spark must occur early enough so that the combustion pressure reaches its maximum just when the piston is beginning its downward power stroke. Combustion should be completed by about 10 degrees ATDC. If the spark occurs too soon BTDC, the rising piston will be opposed by combustion pressure. If the spark occurs too late, the force on the piston will be reduced. In either case, power will be lost. In extreme cases, the engine could be damaged. Ignition must start at the proper instant for maximum power and efficiency. Engineers perform many tests with running engines to determine the proper time for ignition to begin.

Engine Speed

As engine speed increases, piston speed increases. If the air-fuel ratio remains relatively constant, the fuel burning time will remain relatively constant. However, at greater engine speed, the piston will travel farther during this burning time. The spark must occur earlier to ensure that maximum combustion pressure occurs at the proper piston position. Making the

spark occur earlier is called spark advance or ignition advance.

For example, consider an engine, figure 2-34, that requires 0.003 second for the fuel charge to burn and that achieves maximum power if the burning is completed at 10 degrees ATDC.

- At an idle speed of 625 rpm, position A, the crankshaft rotates about 11 degrees in 0.003 second. Therefore, timing must be set at 1 degree BTDC to allow ample burning time.
- At 1,000 rpm, position B, the crankshaft rotates 18 degrees in 0.003 second. Ignition should begin at 8 degrees BTDC.
- At 2,000 rpm, position C, the crankshaft rotates 36 degrees in 0.003 second. Spark timing must be advanced to 26 degrees BTDC.

INITIAL TIMING

As we have seen, ignition timing must be set correctly for the engine to run at all. This is called the engine's initial, or base, timing. Initial timing is the correct setting at a specified engine speed. In figure 2-34, initial timing was 1 degree BTDC. Initial timing is usually within a few degrees of top dead center. For many years, most engines were timed at the specified slow or curb-idle speed for the engine. However, some engines built since 1974 require timing at speeds either above or below the curb-idle speed.

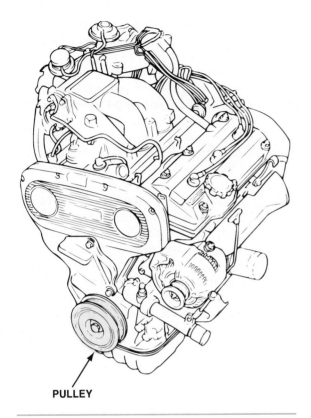

Figure 2-35. Most engines have a pulley bolted to the front end of the crankshaft.

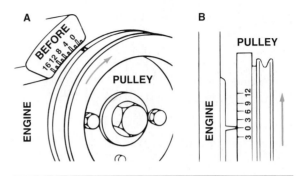

Figure 2-36. These are common types of timing marks.

Timing Marks

We have seen that base timing is related to crankshaft position. To properly time the engine, we must be able to determine crankshaft position. The crankshaft is completely enclosed in the engine block, but most cars have a pulley and vibration damper bolted to the front of the crankshaft, figure 2-35. This pulley rotates with the crankshaft and can be considered an extension of the shaft.

Marks on the pulley show crankshaft position. For example, when a mark on the pulley is aligned with a mark on the engine block, the number 1 piston is at TDC.

Timing marks vary widely, even within a manufacturer's product line. There are two common types of timing marks, figure 2-36:

- A mark on the crankshaft pulley, and marks representing degrees of crankshaft position, on the engine block, position A.
- Marks on the pulley, representing degrees of crankshaft position, and a pointer on the engine block, position B.

Some cars may also have a notch on the engine flywheel and a scale on the transmission cover or bell-housing. In addition to a conventional timing mark, some cars have a special test socket for electromagnetic engine timing. Technicians time these cars with a conventional timing light, or a special test probe that fits into the socket that was also used on the assembly line in the manufacture of the car.

OTHER ENGINE TYPES

Other engine types besides the four-stroke engine have been installed in automobiles over the years, but only three have been used with any real success—the two-stroke, the diesel, and the rotary engines.

The Diesel Engine

In 1892, a German engineer named Rudolf Diesel perfected the compression-ignition engine that bears his name. The diesel engine uses heat created by compression to ignite the fuel, so it requires no spark ignition system.

The diesel engine requires compression ratios of 16 to 1 and higher. Incoming air is compressed until its temperature reaches about 1,000°F (540°C). As the piston reaches the top of its compression stroke, fuel is injected into the cylinder, where it is ignited by the hot air, figure 2-37. As the fuel burns, it expands and produces power.

Diesel engines differ from gasoline-burning engines in other ways. Instead of a carburetor to mix the fuel with air, a diesel uses a precision injection pump and individual fuel injectors. The pump delivers fuel to the injectors at a high pressure and at timed intervals. Each injector measures the fuel exactly, spraying it into the combustion chamber

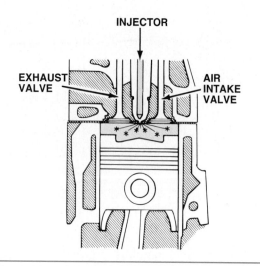

Figure 2-37. Diesel combustion occurs when fuel is injected into the hot, highly compressed air in the cylinder.

at the precise moment required for efficient combustion, figure 2-38. The injection pump and injector system thus perform the tasks of the carburetor and distributor in a gasoline engine.

The air-fuel mixture of a gasoline engine remains nearly constant—changing only within a narrow range—regardless of engine load or speed. But in a diesel engine, air remains constant

and the amount of fuel injected is varied to control power and speed. The air-fuel mixture of a diesel can vary from as lean as 85 to 1 at idle, to as rich as 20:1 at full load. This higher air-fuel ratio and the increased compression pressures make the diesel more fuel efficient than a gasoline engine.

Like gasoline engines, diesel engines are built in both two-stroke and four-stroke versions. The most common two-stroke diesels are the truck and industrial engines made by the Detroit Diesel. In these engines, air intake is through ports in the cylinder wall. Exhaust is through poppet valves in the head. A blower box blows air through the intake port to supply air for combustion and to blow the exhaust gases out of the exhaust valves. Crankcase fuel induction cannot be used in a two-stroke diesel.

For many years, diesel engines were used primarily in trucks and heavy equipment. Mercedes-Benz, however, has built diesel cars since 1936, and the energy crises of 1973 and 1979 focused attention on the diesel as a substitute for gasoline engines in automobiles.

In the late 1970s, General Motors and Volkswagen developed diesel engines for cars. They were followed quickly by most major vehicle manufacturers in offering optional diesel engines for their vehicles. By the early 1980s, many vehi-

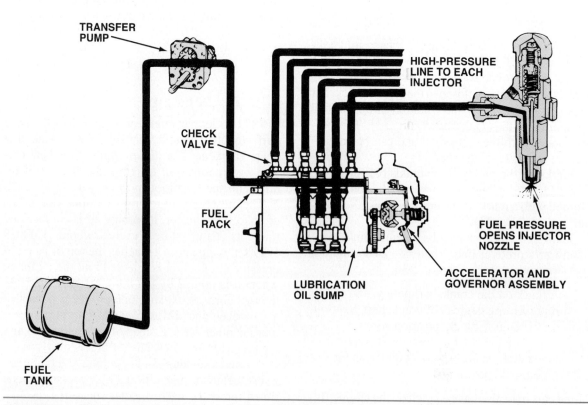

Figure 2-38.   This is a typical automotive diesel fuel-injection system.

cle manufacturers were predicting diesel power for more than 30 percent of the domestic auto population. These predictions did not come true.

In the mid 1980s, increasing gasoline supplies and lower prices in combination with the disadvantages of noise and higher purchasing costs reduced the incentives for customers to buy diesel automobiles. Stringent diesel emission regulations added even more to manufacturing costs and made the engines harder to certify for sale. A few automobile diesel engines, such as the Oldsmobile V8, were derived from gasoline powerplants and suffered reliability problems. In the late 1980s, diesel engine use in the United States and Canada was limited mostly to truck and industrial applications, although late-model diesel automobiles, such as those made by Volkswagen, continue to sell in the United States, and in Europe, where gasoline costs remain high.

The Rotary (Wankel) Engine

The reciprocating motion of a piston engine is both complicated and inefficient. For these reasons, engine designers have spent decades attempting to devise engines in which the working parts would all rotate on an axis. The major problem with this rotary concept has been the sealing of the combustion chamber. Of the various solutions proposed, only the rotary design of Felix Wankel—as later adapted by NSU, Curtiss Wright, and Toyo Kogyo (Mazda)—has proven practical.

Although the same sequence of events occurs in both a rotary and a reciprocating engine, the rotary is quite different in design and operation. A curved triangular rotor moves on an eccentric, or off-center, geared portion of a shaft within an elliptical chamber, figure 2-39.

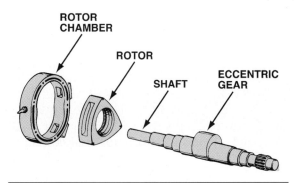

ROTOR CHAMBER

ROTOR

SHAFT

ECCENTRIC GEAR

Figure 2-39. The main parts of a Wankel rotary engine are the rotor chamber, the three-sided rotor, and the shaft with an eccentric gear.

As it turns, seals on the rotor's corners follow the housing shape. The rotor thus forms three separate chambers whose size and shape change constantly during rotation. The intake, compression, power, and exhaust functions occur within these chambers, figure 2-40. Wankel engines can be built with more than one rotor. Mazda production engines, for example, were two-rotor engines.

One revolution of the rotor produces three power strokes or pulses, one for each face of the rotor. In fact, each rotor face can be considered the same as one piston. Each pulse lasts for about three-quarters of a rotor revolution. The combination of rotary motion and longer, overlapping power pulses results in a smooth-running engine.

About equivalent in power output to that of a 6-cylinder piston engine, a two-rotor engine is only one-third to one-half the size and weight. With no pistons, connecting rods, valves, lifters, and other reciprocating parts, the rotary engine has 40 percent fewer parts than a piston engine.

■ Rudolf Diesel

The theory of a diesel engine was first set down on paper in 1893 when Rudolf Christian Karl Diesel wrote a technical paper, "Theory and Construction of a Rational Heat Engine." Diesel was born in Paris in 1858. After graduating from a German technical school, he went to work for the refrigeration pioneer, Carl von Linde. Diesel was a success in the refrigeration business, but at the same time he was developing his theory on the compression ignition engine. The theory is really very simple. Air, when it is compressed, gets very hot. If you compress it enough, say on the order of 20 to 1, and squirt fuel into the compressed air, the air-fuel mixture will ignite.

In its first uses, the diesel engine was used to power machinery in shops and plants. By 1910, it was used in ships and locomotives and a 4-cylinder engine was even used in a delivery van. The engine proved too heavy at that time to be practical for automobile use. It was not until 1927, when Robert Bosch invented a small fuel-injection mechanism, that the use of the diesel engine in trucks and cars became practical.

Unlike many of the early inventors, Diesel did make a great deal of money from his invention, but in 1913 he disappeared from a ferry crossing the English Channel and was presumed to have committed suicide.

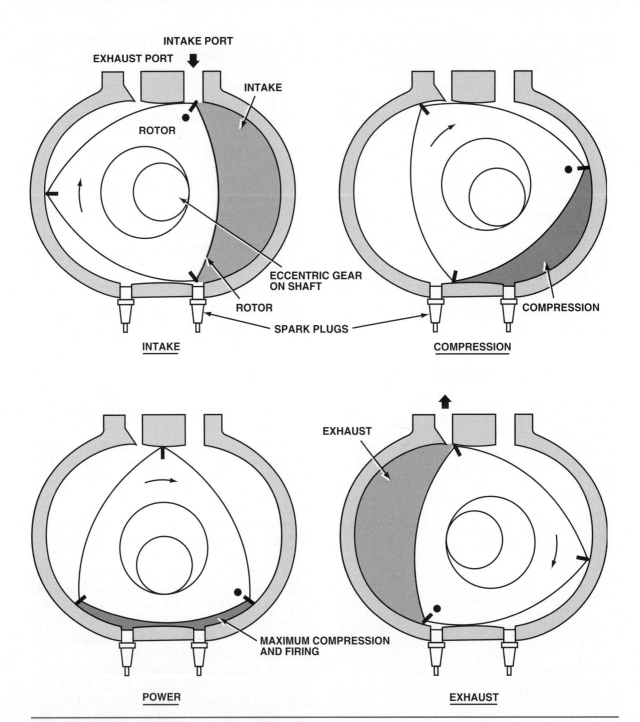

Figure 2-40. These are the four stages of rotary operation. They correspond to the intake, compression, power, and exhaust strokes of a four-stroke reciprocating engine. The sequence is shown for only one rotor face, but each face of the rotor goes through all four stages during each rotor revolution.

While the rotary overcomes many of the disadvantages of the piston engine, it has its own disadvantages. It is basically a very "dirty" engine. In other words, it gives off a high level of emissions and requires additional external devices to clean up the exhaust.

SUMMARY

Most automobile engines are internal combustion, reciprocating four-stroke engines. An air-fuel mixture is drawn into sealed combustion chambers by a vacuum created by the downward stroke of a piston. The mixture is ignited by a spark.

Valves at the top of the cylinder open and close to admit the air-fuel mixture and release the exhaust. These valves are driven by a camshaft and synchronized with engine rotation. The sequence in which the cylinders fire is the firing order. Several valve designs have been used, including I-head, F-head, and L-head designs. Most engine cylinders have two valves, but some engines have three, four, or five valves per cylinder.

Displacement and compression ratio are two frequently used engine specifications. Displacement indicates engine size, and compression ratio compares total cylinder volume to compression chamber volume.

The ignition interval is the number of degrees between ignition sparks. A 4-cylinder engine commonly has an ignition interval of 180 degrees, a V-8 90 degrees, and an inline 6-cylinder 120 degrees (although many intervals have been used over the years). Still, each cylinder fires once every 720 degrees of crankshaft rotation. The ignition spark is provided by the distributor or electronic ignition, and is synchronized with the crankshaft rotation.

Correct ignition timing is essential for the engine to operate. Timing marks on the front of the engine block or the pulley indicate piston position, and can be used to alter the timing from the engine's base or initial timing.

The most successful automobile engine types used besides four-stroke gasoline engines are the two-stroke, the diesel, and the rotary (Wankel).

Review Questions

Choose the letter that represents the best possible answer to the following questions:

1. An automotive internal combustion engine:
   a. Uses energy released when a compressed air-fuel mixture is ignited
   b. Has pistons that are driven downward by explosions in the combustion chambers
   c. Is better than an external combustion engine
   d. Burns gasoline

2. A spark plug fires near the end of the:
   a. Intake stroke
   b. Compression stroke
   c. Power stroke
   d. Exhaust stroke

3. The flywheel:
   a. Keeps the engine turning during the "coasting" or nonpower stroke periods of the engine
   b. Provides a mounting plate for the clutch or torque converter
   c. Smoothes out increases and decreases in engine power output, providing continuous thrust
   d. All of the above

4. The camshaft rotates at:
   a. The same speed as the crankshaft
   b. Double the speed of the crankshaft
   c. Half the speed of the crankshaft
   d. It depends on the engine design

5. Which of the following is never in direct contact with the engine valves?
   a. Pushrods
   b. Valve springs
   c. Rocker arms
   d. Valve seats

6. The bore is the diameter of the:
   a. Connecting rod
   b. Cylinder
   c. Crankshaft
   d. Combustion chamber

7. The four-stroke cycle operates in which order?
   a. Intake, exhaust, power, compression
   b. Intake, power, exhaust, compression
   c. Compression, power, intake, exhaust
   d. Intake, compression, power, exhaust

8. Diesel engines:
   a. Have no valves
   b. Produce ignition by heat of compression
   c. Have low compression
   d. Use special carburetors

9. How many strokes of a piston are required to turn the crankshaft through 360 degrees?
   a. One
   b. Two
   c. Three
   d. Four

10. A "retarded spark" is one that occurs:
    a. At top dead center
    b. Before top dead center
    c. After top dead center
    d. At bottom dead center

11. An internal combustion engine is an efficient use of available energy because:
    a. The combustion of the fuel produces more energy than is required to compress and fire it
    b. Gasoline is so easy to burn
    c. Internal combustion engines turn with so little friction
    d. Compression ratios are high enough to vaporize gasoline

12. Which of the following shows the typical ignition interval of an inline 6-cylinder engine?
    a. 600 degrees ÷ 6 = 100 degrees
    b. 360 degrees ÷ 3 = 120 degrees
    c. 90 degrees ÷ 6 = 540 degrees
    d. 720 degrees ÷ 6 = 120 degrees

13. How many sparks (spark plug firings) per crankshaft revolution are required by a typical 8-cylinder engine?
    a. 8
    b. 6
    c. 4
    d. 2

14. The firing order of an engine is:
    a. The same on all V-8s
    b. Is stated in reference books and on the cylinder block
    c. Can be deduced by common sense
    d. None of the above

15. Ignition timing:
    a. Requires a basic setting specified in degrees BTDC for a particular engine
    b. Varies while engine is running, synchronizing combustion with proper piston position
    c. Is indicated by the alignment of markings on the crankshaft pulley or flywheel with other markings or a pointer fixed to the cylinder block
    d. All of the above

16. In a four-stroke engine, the piston is driven down in the cylinder by expanding gases during the _____ stroke.
    a. Compression
    b. Exhaust
    c. Power
    d. Intake

17. The ignition interval of an engine is the number of degrees of crankshaft rotation that:
    a. Occur between ignition sparks
    b. Are required to complete one full stroke
    c. Take place in a four-stroke engine
    d. All of the above

18. The rotary engine:
    a. Operates without pistons, connecting rods, or poppet valves
    b. Is not a reciprocating engine
    c. Produces three power strokes per revolution of its rotor
    d. All of the above

19. Which of the following is *not* used in calculating engine displacement?
    a. Stroke
    b. Bore
    c. Number of cylinders
    d. Valve arrangement

20. To change cubic centimeters to cubic inches, multiply by:
    a. 0.061
    b. 16.39
    c. 61.02
    d. 1000

21. Compression ratio is:
    a. Piston displacement plus clearance volume
    b. Total volume times number of cylinders
    c. Total volume divided by clearance volume
    d. Stroke divided by bore

3

Engine Air-Fuel Requirements

OBJECTIVES

Upon completion and review of this chapter, you will be able to:

- Understand air-fuel ratios and define stoichiometric ratio.
- List the advantages of throttle body and port type fuel-injection systems.
- Define octane rating.
- Have knowledge of reformulated and oxygenated fuels and their advantages and benefits.
- List the oxygenates that are currently used in reformulated and oxygenated fuels.

KEY TERMS

air-fuel ratio	
antiknock value	port fuel injection
atmospheric pressure	pressure differential
atomization	stoichiometric ratio
catalytic cracking	tetraethyl lead (TEL)
detonation	thermal cracking
electronic fuel	throttle body injection
injection (EFI)	(TBI)
ethanol	vacuum
gasohol	vaporization
methanol	volatility
octane rating	volumetric efficiency

INTRODUCTION

Automobile engines run on a mixture of gasoline and air. Gasoline has several advantages as a fuel, including that it:

- Vaporizes (evaporates) easily
- Burns quickly but under control when mixed with air and ignited
- Has a high heat value and produces a large amount of heat energy
- Is easy to store, handle, and transport

Gasoline also has certain disadvantages as a fuel. In particular, burning gasoline creates harmful pollutants, which enter the atmosphere with the engine exhaust. For the time being, however, no better alternative than gasoline is available to fuel automotive engines.

To understand how the fuel system works in an engine, we must understand the engine's air-fuel requirements. This chapter discusses those requirements, and describes how the fuel gets from the fuel tank to the combustion chamber.

AIR PRESSURE— HIGH AND LOW

You can think of an internal combustion engine as a big air pump. As the pistons move up and down in the cylinders, they pump in air and fuel for combustion and pump out exhaust gases. They do this by creating difference in air pressure. The air outside an engine has weight and exerts pressure, as does the air inside an engine.

As a piston moves down on an intake stroke with the intake valve open, it creates a larger area inside the cylinder for the air to fill. This lowers the air pressure within the engine. Because the pressure inside the engine is lower than the pressure outside, air flows into the engine to fill the low-pressure area and equalize the pressure.

The low pressure within the engine is called **vacuum.** You can think of the vacuum as sucking air into the engine, but it is really the higher pressure on the outside that forces air into the low-pressure area inside. The difference in pressure between the two areas is called a **pressure differential.** The pressure differential principle has many applications in automotive fuel and emissions systems.

An engine pumps exhaust out of its cylinders by creating pressure as a piston moves upward on the exhaust stroke. This creates high pressure in the cylinder, which forces the exhaust toward the lower-pressure area outside the engine.

Pressure differential applies to liquids as well as to air. Fuel pumps work on this principle. The pump creates a low-pressure area in the fuel system that allows the higher pressure of the air and fuel in the tank to force the fuel through the lines to the carburetor or the injection system.

AIRFLOW REQUIREMENTS

All gasoline automobile engines share certain air-fuel requirements. For example, a four-stroke engine can take in only so much air at any time, and how much fuel it consumes depends on how much air it takes in. Engineers calculate engine airflow requirement using these three factors:

- Engine displacement
- Engine revolutions per minute (rpm)
- Volumetric efficiency

The airflow number represents cubic feet per minute (cfm) or cubic meters per minute (cmm). The designer sizes the carburetor or fuel-injection intake airflow capacity to match the engine's maximum requirement. Volumetric efficiency and how it relates to engine airflow is described in the following paragraphs.

Volumetric Efficiency

Volumetric efficiency is a comparison of the actual volume of air-fuel mixture drawn into an engine to the theoretical maximum volume that could be drawn in. Volumetric efficiency is expressed as a percentage, and changes with engine speed. For example, an engine might have 75 percent volumetric efficiency at 1,000 rpm. The same engine might be rated at 85 percent at 2,000 rpm and 60 percent at 3,000 rpm.

If the engine takes in the airflow volume slowly, a cylinder might fill to capacity. It takes a definite amount of time for the airflow to pass through all the curves of the intake manifold and valve port. Therefore, manifold and port design directly relate to engine "breathing," or volumetric efficiency. Camshaft timing and exhaust turning also are important.

If the engine is running fast, the intake valve does not stay open long enough for full volume to enter the cylinder. At 1,000 rpm, the intake valve might be open for one-tenth of a second. As engine speed increases, this time decreases to a point where only a small airflow volume enters the cylinder. Therefore, volumetric efficiency decreases as engine speed increases. At high speed, it may drop to as low as 50 percent.

The average street engine never reaches 100 percent volumetric efficiency. With a street engine, you can expect a volumetric efficiency of about 75 percent at maximum speed, or 80 percent at the torque peak. A high-performance street engine is about 85 percent efficient, or a bit more efficient at peak torque. A race engine usually has 95 percent or better volumetric efficiency. These figures apply only to naturally aspirated engines, however. Turbocharged and supercharged en-

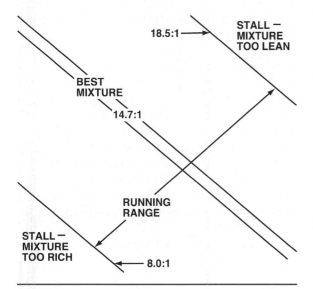

Figure 3-1. An engine can run without stalling on an air-fuel mixture with a ratio between 8 to 1 and 18.5 to 1.

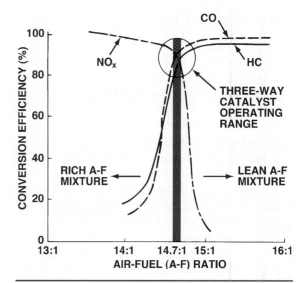

Figure 3-2. With a three-way catalytic converter, emission control is optimum with an air-fuel ratio between 14.65 to 1 and 14.75 to 1.

gines easily achieve more than 100 percent volumetric efficiency.

AIR-FUEL RATIOS

Fuel burns best when the intake system turns it into a fine spray and mixes it with air before sending it into the cylinders. In carbureted engines, the fuel becomes a spray and mixes with air in the carburetor. In fuel-injected engines, the fuel becomes a fine spray as it leaves the tip of the injectors and mixes with air in the intake manifold. In both cases, there is a direct relationship between engine airflow and fuel requirements. This relationship is called the **air-fuel ratio.**

The air-fuel ratio is the proportion by weight of air and gasoline that the carburetor or injection system mixes as needed for engine combustion. This ratio is important, since there are limits to how rich (more fuel) or lean (less fuel) it can be and remain combustible. The mixtures with which an engine can operate without stalling range from 8 to 1 to 18.5 to 1, figure 3-1. These ratios are usually stated this way: 8 parts of air by weight combined with 1 part of gasoline by weight (8:1), which is the richest mixture that an engine can tolerate and still fire regularly; 18.5 part of air mixed with 1 part of gasline (18.5:1), which is the leanest. Richer or leaner air-fuel ratios cause the engine to misfire badly or not run at all.

Air-fuel ratios are calculated by weight rather than volume. If it requires 14.7 pounds or kilograms of air to burn one pound or kilogram of gasoline, the air-fuel ratio is 14.7 to 1.

Stoichiometric Air-Fuel Ratio

The ideal mixture or ratio at which all the fuel combines with all of the oxygen in the air and burns completely is called the **stoichiometric ratio**—a chemically perfect combination. In theory, this ratio is an air-fuel mixture of 14.7 to 1. In reality, the exact ratio at which perfect mixture and combustion occurs depends on the molecular structure of gasoline, which varies somewhat. The stoichiometric ratio is a compromise between maximum power and maximum economy. Late-model computerized vehicles are designed to regulate the air-fuel ratio at stoichiometric.

Emission control is also optimum at this ratio, if the exhaust system uses a three-way oxidation-reduction catalytic converter. As the mixture gets richer, hydrocarbon (HC) and carbon monoxide (CO) conversion efficiency falls off. With leaner mixtures, oxides of nitrogen (NO_X) conversion efficiency also falls off. As figure 3-2 shows, the conversion efficiency range is very narrow—between 14.65 to 1 and 14.75 to 1. A fuel system without feedback control cannot maintain this narrow range.

Engine Air-Fuel Requirements

An automobile engine works with the air-fuel mixture ranging from 8 to 1 to 18.5 to 1. The ideal ratio provides the most power and the most economy, while producing the least emissions. Such a ratio does not exist because engine fuel requirements vary widely, depending on temperature, load, and speed conditions.

Research proves that a 15 to 1 to 16 to 1 ratio provides the best fuel economy, while a 12.5 to 1 to 13.5 to 1 ratio gives maximum power output. An engine needs a rich mixture for idle, heavy load, and high-speed conditions and a leaner mixture for normal cruising and light-load conditions. No single air-fuel ratio provides the best fuel economy and the maximum power output at the same time.

Just as outside conditions such as speed, load, temperature, and atmospheric pressure change the engine fuel requirements, other forces at work inside the engine cause additional variations. Here are two examples:

- The mixture may be imperfect because the fuel may not vaporize completely.
- The mixture from a carburetor or a throttle body does not distribute equally through the intake manifold to each cylinder, so some cylinders get a richer or leaner mixture than others.

For an engine to run well under a variety of outside and inside conditions, the carburetor or injection system must vary the air-fuel mixture quickly to give the best mixture possible for engine requirements at any given moment.

Power Versus Economy

If the goal is to get the most power from an engine, all the oxygen in the mixture must burn because the power output of any engine is limited by the amount of air it can pull in. For the oxygen to combine completely with the available fuel, there should be extra fuel available. This makes the air-fuel ratio richer, with the result that some fuel does not burn.

To get the best fuel economy and the lowest emissions, the gasoline must burn as completely as possible in the combustion chamber. This means combustion with the least amount of leftover waste material, or emissions, provides the greatest economy. The intake system must provide more air to make sure that enough oxygen is available to combine with the gasoline. This results in a leaner air-fuel mixture than the ideal.

The air-fuel ratio required to provide maximum power changes very little, except at low speeds, figure 3-3. Reducing speed reduces the airflow into the engine, resulting in poorer mixing of the air and fuel and less efficiency in distributing it to the cylinders. The mixture must be slightly richer at low speeds to make up for this.

The same is true for maximum fuel economy—the leaner air-fuel ratio remains virtually the same throughout most of the operating range, figure 3-4. However, the mixture must be richer during idle and low speeds, and during higher speeds and under load—two conditions that require more power.

Sometimes the air-fuel mixture gets richer when it is not required or wanted, as with high-altitude

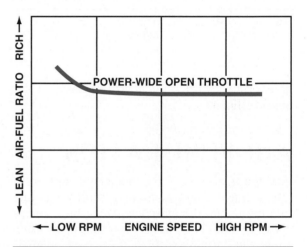

Figure 3-3. The air-fuel ratio that provides the most power is rich overall, and slightly richer at low engine speeds.

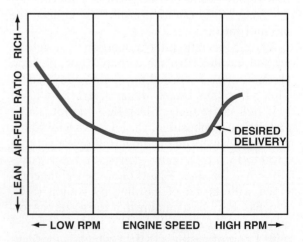

Figure 3-4. The air-fuel ratio that provides the best economy is lean overall, but slightly richer at both higher and lower engine speeds.

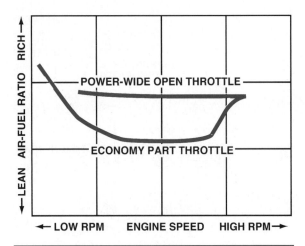

Figure 3-5. The carburetor or fuel-injection system must determine the best air-fuel ratio to balance power and economy.

driving. As altitude increases, **atmospheric pressure** drops and the air becomes thinner. The same volume of air weighs less and contains less oxygen at higher altitudes, so an engine takes in fewer pounds or kilograms of air and less oxygen. The result is a richer air-fuel ratio, which must be corrected to achieve efficient high-altitude engine operation. Altitude-compensating carburetors and fuel-injection airflow and pressure sensors solve this problem. The corrected air-fuel mixture lets the engine burn fuel efficiently, but total horsepower is less, corrected or not. With the intake system providing less fuel to the cylinder to mix with the lower oxygen content of the high-altitude air, fewer calories of energy are available to do work.

For these reasons, the carburetor or injection system must deliver fuel so that the best mileage is provided during normal crusing, with maximum power available when the engine is under load, accelerating, or at high speed, figure 3-5.

INTRODUCTION TO ELECTRONIC ENGINE CONTROLS

Electronic engine controls first appeared on vehicles manufactured for the 1977 model year. Early control systems regulated only a single function, either ignition timing or fuel metering. However, manufacturers rapidly expanded them to include control over both systems, as well as other engine functions.

As illustrated in figure 3-6, the basic parts of the first electronically controlled fuel management systems included:

- A feedback carburetor
- An ignition control module (ICM), powertrain control module (PCM), or both
- An exhaust gas oxygen sensor (O2S)
- A catalytic converter.

Carbureted engines used either direct or indirect actuators to control the fuel-metering rods and air bleeds. With the direct method, a solenoid or stepper motor mounted on or in the carburetor operated the fuel-metering rods or the air bleeds, or both. With indirect control, a remote-mounted, solenoid-actuated vacuum valve regulated the carburetor vacuum diaphragms that operated the fuel-metering rods and air bleeds.

The PCM constantly monitored the oxygen content of the exhaust gas through signals received from the O2S sensor. The PCM sent a pulsed voltage signal to the control device, varying the ratio of on-time to off-time according to the signals received from the O2S sensor. As the percentage of on-time increased or decreased, the mixture became leaner or richer.

With fuel-injection systems, the PCM controls the air-fuel mixture by switching one or more injectors on and off, at a rate based on engine speed. The PCM also varies the length of time the injectors remain open to establish the air-fuel ratio. As the microprocessor receives data from system sensors, it lengthens or shortens the pulse width (on-time) according to engine operating and load conditions. We study more about fuel-injection systems and how they work in later chapters of this *Classroom Manual*.

THE INTAKE SYSTEM

Primary parts of the intake system are a manifold and a fuel delivery system, figure 3-7. The manifold is a casting with a series of enclosed passages that route the air-fuel mixture from the carburetor, or throttle body, to the intake ports in the engine cylinder head. With port fuel-injection systems, the intake manifold routes only air, and the fuel injectors deliver the fuel directly to the intake port.

Airflow

Remember, you can think of an automobile engine as a large air pump. As the pistons move up and

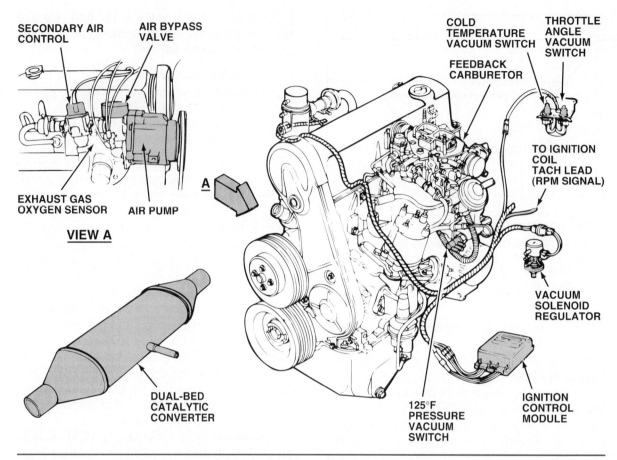

Figure 3-6. The first electronically controlled fuel management systems included a feedback carburetor, an injection control module, an oxygen sensor, and a catalytic converter.

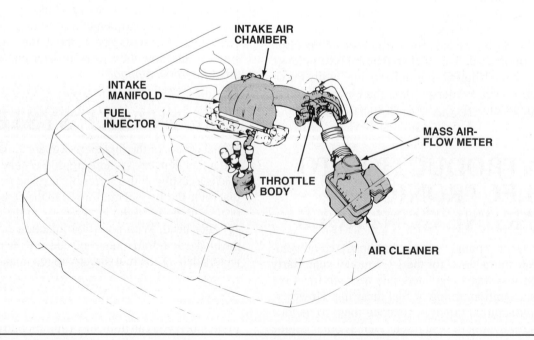

Figure 3-7. The two basic parts of an engine intake system are the intake manifold and the fuel delivery system.

down in the cylinders, they draw in air and fuel for combustion and pump out burned exhaust. They do this by creating a difference in air pressure.

You recall that the downward movement of the piston intake stroke creates a partial vacuum, lowering the air pressure in the combustion chamber and intake manifold. Opening the carburetor or fuel-injection throttle valve, figure 3-8, lets air move from the higher-pressure area outside the engine, through the intake, to the lower-pressure area of the manifold. The throttle valve opening determines how much and how fast the air travels. When the intake valve in the cylinder head opens, the vacuum draws the air-fuel mixture from the intake manifold into the combustion chamber, figure 3-9.

Air-fuel Mixture Control

Before gasoline can do its job as a fuel, it must be metered, atomized, vaporized, and distributed to each cylinder in the form of a burnable mixture. To do this, a metering device mixes the gasoline with air in the correct ratio and distributes the mixture as required by engine load, speed, throttle valve position, and operating temperature. On most late-model engines, a fuel-injection system does the job of air-fuel mixing done by a carburetor on older engines.

As of 1991 carburetors were limited to a few import trucks, but they were phased out by 1994.

However, we study carburetors because there are still vehicles in service using them. Understanding how carburetors work improves your understanding of fuel injection and its advantages.

Carburetors

Carburetors consist of lightweight metal bodies that have one, two, three, or four bores or barrels, figure 3-10. Airflow through the barrels draws fuel out of a float bowl. As air rushes through the barrel, pressure inside the barrel drops. A passage or jet connects the barrell to the fuel-filled float bowl. The higher atmospheric pressure in the float bowl forces the fuel through the jet and into

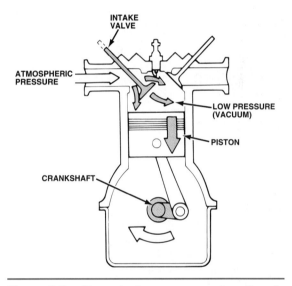

Figure 3-9. Atmospheric pressure pushes the air-fuel mixture into the combustion chamber.

Figure 3-10. The carburetor is made of lightweight metal and draws fuel from a float bowl through the barrels, atomizing and metering the gasoline.

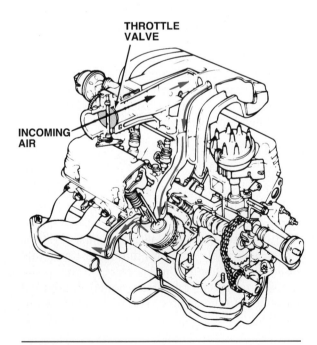

Figure 3-8. Atmospheric pressure pushes air into the intake system.

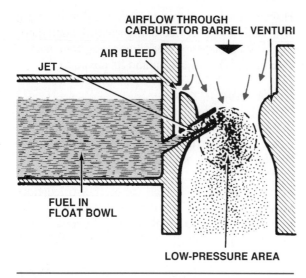

Figure 3-11. Low pressure in the carburetor barrel draws fuel from the float bowl.

the low-pressure stream of air rushing through the barrel, figure 3-11. The airflow speed through the carburetor affects how much fuel comes through the jet. The higher the airflow velocity, the lower the pressure in the carburetor barrel and the more fuel flows through the jet from the float bowl.

Carburetor barrels do not have straight sides because they work better with a construction, or venturi. When air flows through the venturi constriction, it speeds up. The increase in speed lowers the air pressure inside the venturi and draws more fuel into the airflow.

Besides mixing liquid fuel with air, the carburetor must also atomize the liquid as much as possible. A small opening, called an "air bleed," in the fuel inlet passage helps break up the liquid fuel for better atomization. The air bleed mixes air with the liquid fuel as it is drawn into the main airflow. The air bleed opening is above the venturi, where high pressure causes the air to flow in.

The carburetor must also change the air-fuel mixture automatically. It has circuits that deliver a richer mixture for starting, idle, and acceleration, and a leaner mixture for part-throttle operation. Even with all its various circuits, a carburetor is a mechanical device that is neither totally accurate nor particularly fast in responding to changing engine requirements. Adding electronic feedback mixture control improves a carburetor's fuel-metering capabilities under some circumstances, but many mechanical jets, passages, and air bleeds still do most of the work. Adding feedback controls and other emission-related devices result in very complex carburetors that are expensive to repair or replace.

The intake manifold is also a device that, when teamed with a carburetor, results in less than ideal air-fuel control. If there is only one carburetor for several cylinders, it is difficult to position the carburetor an equal distance from all cylinders and remain within the space limitations under the hood. In the longer intake manifold passages, the fuel tends to fall out of the airstream. Bends in the passageways slow the flow or cause puddles of fuel to collect. The manifold runners have to be kept as short as possible to minimize fuel delivery lag, and there cannot be any low points where fuel might puddle. These restrictions severely limit the amount of manifold tuning possible, and even the best designs still have problems with fuel condensing on cold manifold walls.

Multiple carburetors helped eliminate some compromises in carburetor and manifold design by improving mixture distribution. The first fuel-injection systems were mechanical, and they relied on an injection pump, mechanically driven by the crankshaft and timed to engine rpm, to allow the injectors to fire at the proper time.

Electronic Fuel Injection

The development of reliable solid-state components during the 1970s made the best type of fuel-injection system, **electronic fuel injection (EFI)**, practical. EFI provides precise mixture control over all speed ranges and under all operating conditions. Its fuel delivery components are simpler and often less expensive than a feedback carburetor. Most designs allow a wider range of manifold designs. Equally important is that EFI offers the potential of highly reliable and very precise electronic control. With few exceptions, modern engines use one of two types of electronically controlled fuel injection:

- Port injection
- Throttle body injection

With **port fuel injection** (also called multipoint or multipoint injection), individual injection nozzles for each cylinder are located in the ends of the intake manifold passages, near the intake valves, figure 3-12. Only air flows through the manifold up to the location of the injectors. At that point, the system injects a precise amount of atomized gasoline into the airflow. The fuel vaporizes immediately before it passes around the valve and into the combustion chamber, figure 3-13.

With **throttle body injection (TBI)**, a throttle body on top of the intake manifold houses one or two injection nozzles, figure 3-14. The throttle

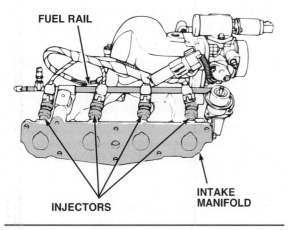

Figure 3-12. Port fuel injection places an injector near each intake port in the engine.

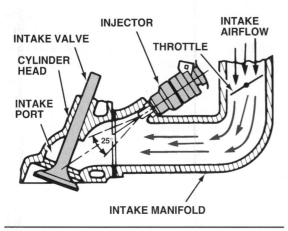

Figure 3-13. The injector sprays atomized fuel into the intake airflow.

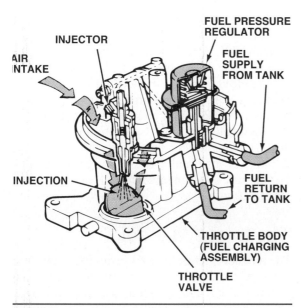

Figure 3-14. Throttle body injection meters fuel better than a carburetor, but provides less-precise delivery than port injection.

body looks like the lower part of a carburetor, but does not have a fuel bowl or any of the fuel-metering circuits of a carburetor. The system injects fuel under pressure into the airflow as it passes through the throttle body.

Fuel Atomization and Vaporization

Proper fuel distribution depends on six factors:

- Correct fuel **volatility**
- Proper fuel **atomization**
- Complete fuel **vaporization**
- Intake manifold passage design
- Intake throttle valve angle
- Carburetor, throttle body, or fuel injector location on the intake manifold

Volatility is a measure of gasoline's ability to change from a liquid to a vapor and is affected by temperature and altitude. The more volatile it is, the more efficiently the gasoline vaporizes. Efficient vaporization promotes even fuel distribution to all engine cylinders for complete combustion.

Refiners control volatility by blending different hydrocarbons that have different boiling points. In this way, it is possible to produce a fuel with a high boiling point for use in warm weather, and one with a lower boiling point for cold-weather driving. Such blending involves some guesswork about weather conditions in various regions, so severe and unexpected temperature changes may cause a number of temperature-related problems ranging from hard starting to vapor lock.

Several factors contribute to changing gasoline from a liquid into a combustible vapor. Gasoline must atomize, or break up into a fine mist, before it properly vaporizes. In a carbureted engine, the liquid fuel first enters the carburetor where it is sprayed into the incoming air and atomized, figure 3-15. The resulting atomized air-fuel mixture then moves into the intake manifold where manifold heat vaporizes the many fine droplets of the atomized fuel.

Rapid vaporization occurs only when the fuel is hot enough to boil. The boiling point is related to pressure: the higher the pressure, the higher the boiling point; the lower the pressure, the lower the boiling point. Because intake manifold pressure is usually quite a bit less than atmospheric pressure, the boiling point of gasoline drops when it enters the manifold, and the fuel quickly begins to vaporize. Fuel-injected engines spray the fuel into the incoming air under much greater pressure than do carburetors, so the fuel atomizes more thoroughly and vaporizes more quickly.

Heat from the intake manifold floor combines with heat absorbed from air particles surrounding

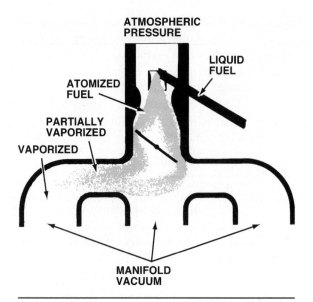

Figure 3-15. In a carburetor, the fuel atomizes as it enters the barrel.

the fuel particles to speed vaporization. The higher the temperature, the more complete the vaporization, so raising the intake manifold temperature helps vaporization. Several things can cause poor vaporization:

- Low mixture velocity or low fuel-injection pressure
- Insufficient fuel volatility
- A cold manifold
- Poor manifold design
- Cold incoming air
- Low manifold vacuum

When poor vaporization occurs, too much liquid fuel reaches the cylinders. Some of this additional fuel escapes through the exhaust as unburned hydrocarbons, and some washes oil from the cylinder walls, causing engine wear. Blowby gases carry the rest past the piston rings.

Carburetors may be precise and consistent, but fuel injection always provides the *correct* amount of fuel for all conditions—cold start, wide open throttle, etc. Carburetion is a compromise. Also, injecting the fuel under pressure provides good vaporization under all loads and speeds. Of the six causes of poor vaporization listed above all apply to carburetors, but only the first four have any effect on throttle body fuel-injection systems, and only the first two on port fuel-injection systems. Fuel injection, accurate metering, precise fuel control, and improved vaporization provide more power across the speed range of the engine, better fuel economy, and reduced exhaust emissions.

■ **Birth of the Internal Combustion Engine**

Like most inventions, the internal combustion engine did not spring full-grown from one person's mind. It developed gradually, as various inventors built on each other's work from 1690 to 1876. Historians generally credit Denis Papin's atmospheric engine concept as the first step toward today's modern gasoline burner. (Papin also invented the pressure cooker.) The principles Papin put forth for an external combustion engine led directly to James Watt and Richard Trevithick's basic steam engine.

Meanwhile, Nicolas Leonard Sadi Carnot, a French physicist, postulated theories in 1824 that advanced the science of heat-exchange thermodynamics. During the 1850s, companies refined volatile fuels from petroleum, and by 1860 a French engineer named J. J. E. Lenoir modified a steam engine that used illuminating gas as fuel. Alphonse Beau de Rochas published a treatise on four-stroke-engine theory in 1862, but never built an engine.

Things really started moving when the firm of Otto and Langen produced an engine in 1867. Their design used a rack-and-gear device to transmit power from a free-moving piston to a shaft and flywheel. A freewheeling clutch in the gear allowed it to rotate freely in one direction and transmit power in the other. The same firm built the first modern internal combustion engine, the Otto Silent Engine, in 1876. That four-stroke design was manufactured in the United States after 1878, and is said to have inspired the early research of Henry Ford and other automobile pioneers.

With TBI, manifold temperature still affects vaporization. As the mixture travels the length of the manifold, some of the air-fuel mixture separates and fuel condenses on the walls of the manifold. Since TBI more consistently atomizes the fuel under all conditions, separation is less likely than with a carburetor. However, with port fuel injection, the manifold carries only air, so the air-fuel mixture does not separate as it travels through the manifold.

The Intake Manifold

The air-fuel mixture flowing from the carburetor or throttle body must be evenly distributed to each cylinder. The intake manifold does this with a series of carefully designed passages that connect the carburetor or throttle body with the engine intake valve ports. To do its job, the intake manifold must provide efficient vaporization and air-fuel delivery. The manifolds for port fuel-injected engines carry only air, and designers size, or tune, them to do this efficiently.

Intake manifolds on older engines are usually cast iron, but modern engines often have aluminum manifolds because aluminum conducts heat better and is lighter. Better heat conductivity helps transfer heat to the air-fuel mixture faster and more uniformly.

The intake manifold on a V-type engine is in the valley between the cylinder banks, figure 3-16, and it is on the side of an inline engine. The average inline engine has the intake and exhaust manifolds on the same side, figure 3-17. High-performance inline engines have two manifolds mounted on opposite sides of the cylinder head. This is called a crossflow design, figure 3-18.

Most intake manifolds are separate pieces that unbolt from the cylinder head, figure 3-19. However, Ford 3.3- and 4.1-liter (200- and 250-cu in.) engines had the intake manifold cast into the head to reduce cost and simplify overall engine manufacturing. The last of the inline Chevrolet 6-cylinder engines in the late 1970s had intake manifolds that were integral with the heads. That design was an attempt to improve mixture temperature and distribution, which are hard to control uniformly for an inline six.

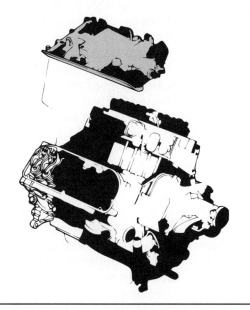

Figure 3-16. The intake manifold is located between the cylinder banks of a V-type engine.

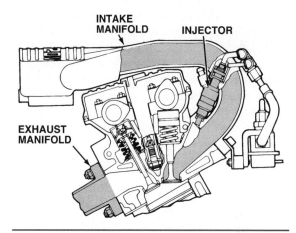

Figure 3-18. High-performance inline engines use a crossflow design.

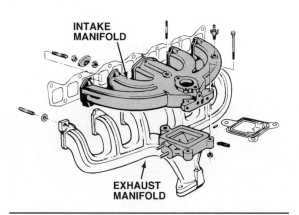

Figure 3-17. Most inline engines have the intake and exhaust manifolds on the same side.

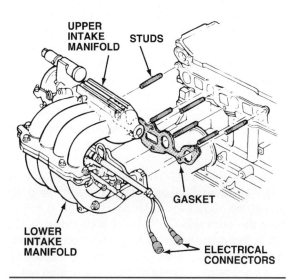

Figure 3-19. Most intake manifolds can be unbolted from the engine.

FUEL COMPOSITION

Gasoline is a clear, colorless liquid—a complex blend of various basic hydrocarbons (hydrogen and carbon). As a fuel, it has good vaporization qualities, and is capable of producing tremendous power when combined with oxygen and ignited. Yet, it is impossible to predict accurately how a certain blend of gasoline will perform in a particular engine, since no two engines are identical. Even mass-produced engines have variations that affect fuel efficiency.

■ Sonic Tuning of Intake and Exhaust Systems

What exactly is a "tuned exhaust"? Why do racers spend so much time and money cutting and welding different header systems for their cars? And what about "tuned intakes"? Why are the velocity stacks on a racing fuel-injection induction system sometimes short, sometimes long, and sometimes even different lengths for different venturis in the same system? There is obviously horsepower to be found in fiddling with the intake and exhaust system, but where does it come from, and how does it work?

Answers to these questions lie in understanding resonance, or sonic tuning. Sonic tuning in the exhaust is a way to scavenge the exhaust gases from a cylinder more efficiently than by piston action alone. It also helps the induction system, like a sort of natural supercharging that gets more mixture into the cylinder than engine vacuum does. Most street engines get around 80 to 85 percent volumetric efficiency at peak torque. With a tuned intake and exhaust system, on the other hand, a race engine gets better than 100 percent volumetric efficiency over a narrow portion of its power band.

Here is how it works in the exhaust system. As the exhaust valve opens, the pressurized gases burst out of the cylinder into the exhaust port, forming a high-pressure pulse in the exhaust gases already in the header primary pipe. This pressure pulse runs through the header pipe at the speed of sound— about 1,700 feet per second at exhaust temperature. The speed of the pulse does *not* depend on the speed of the gases. In fact, the pulse reaches the end of the header pipe much sooner than the gases do. When it reaches the end, it inverts—becoming a negative pressure pulse, or vacuum—and rushes back up the pipe toward the engine.

During this time, the rising piston has been pushing exhaust gases through the exhaust port. When the negative pulse reaches the exhaust port, the extra vacuum helps scavenge the gases left in the cylinder, actually pulling them out past the valve. Because the intake valve is also open (the valves are at overlap), the vacuum also helps pull fresh mixture into the cylinder before the piston even starts its downstroke.

The end result is more horsepower, because the cylinder charge is both denser and less diluted with residual exhaust gases. Obviously, the way to optimize the system is to adjust the length of the header pipes so that the negative pressure pulse arrives at the exhaust valve at just the right time. And equally obviously, a tuned exhaust is only going to work at its best over a narrow range of engine rpm, perhaps several hundred rpm. Making the pipes shorter lets the pressure pulse reach the exhaust port sooner, tuning the pipes for some higher rpm. Making the pipes longer does the opposite, tuning the pipes for some lower rpm. Racers find noticeable differences in horsepower when they change the length of their pipes by as little as three inches.

Pipes with megaphones at the ends make the pressure pulse longer and less distinct, spreading the effect over a wider rpm than a straight, cut-off pipe, which decreases the effect. A reverse-cone megaphone (one that increases and then decreases in diameter) produces an additional power boost by sending a *positive* pressure pulse up the exhaust pipe, chasing the vacuum pulse. At some engine speeds, this pressure pulse catches the fresh mixture escaping from the exhaust port and "stuffs" it back into the combustion chamber.

Pressure pulses resonate back and forth in the induction tract, too. They are caused by the repeated opening and closing of the intake valve in

the path of the moving column of air-fuel mixture. Racers sometimes use sonic tuning in the induction system, but this is much trickier than in the exhaust. Ideally, a positive pulse should arrive late in the intake stroke to reduce reverse pumping, and a negative pulse should arrive just as the intake valve closes to ease the disruption the closing valve has on the moving column of mixture. A third positive pulse can pressurize the port just ahead of the valve, letting the intake charge burst into the cylinder the instant the valve opens.

In practice, however, sonic tuning in the intake is difficult to control. The temperature of the intake charge is hard to predict, because it is simultaneously heated by the hot manifold and cooled by the evaporating gasoline. In addition, opening and closing the throttle valves raises and lowers the pressure in the intake manifold and port, which has a strong effect on pulse speed. And finally, a standard log-type or common plenum chamber manifold gives a designer little room for turning, because the pulses mix together and cancel or reinforce each other wherever the passages meet.

Port fuel-injection systems *do* allow some intake tuning, though, because the long runners are separate between the intake plenum and the intake valve. The manufacturers can adjust the length of the runners to move the power boost up and down the rpm range of the engine, perhaps to fill a "hole" in the power band, or to add a little more low- or high-rpm torque, depending on the expected use of the engine.

Aftermarket induction systems with a separate carburetor or injection nozzle for each cylinder can also be tuned by making the velocity stacks longer or shorter. Some tuners use different-length stacks on the same engine to widen the power band by making different cylinders resonate at different rpm. In general, long runners seem to help low-rpm horsepower, and short runners seem to help high-rpm horsepower, but in practice tuners usually find the best length for a given engine and application by trial and error.

Sonic tuning has the most effect on piston-port two-stroke engines, which require it for cylinder scavenging. Four-strokes with poppet valves gain much less from sonic tuning, because they have a camshaft to force the intake and exhaust gases to move at certain times. Nonetheless, resonance still has a profound effect on racing four-stroke engines. With an intake and exhaust system working in harmony, the designer can specify a camshaft with much more duration than the engine could otherwise tolerate. This means that higher valve lifts are possible while still keeping valve acceleration low for reliability. The result is greater volumetric efficiency from the camshaft, made possible by a tuned intake and exhaust system.

If you are contemplating modifying a street engine, the benefits of sonic tuning are less than you might hope. Big power gains are only possible over a narrow range of engine speeds—useful at the track, but not in traffic. A better approach is to plan a system concentrating on low restriction and high velocity in both the intake and exhaust. Using small-diameter intake runners and header pipes helps horsepower with small-displacement engines and at low rpm, keeping gas velocities high enough so that induction and scavenging are more efficient. For high-rpm use or larger engines, larger intakes and header diameters keep the passages from restricting gas flow.

In laboratory tests, oil refiners calculate and measure the most important gasoline characteristics for a specific job. They blend fuels for particular temperature and altitude conditions. The gasoline you use during the summer is not the same blend available in the winter, nor is the gasoline sold in Denver the same as that sold in Death Valley. Besides temperature and altitude, refiners must consider several other things during the blending process:

- Volatility
- Chemical impurities
- Octane rating
- Additives

Volatility

Volatility is a measure of gasoline's ability to change from a liquid to a vapor and is related to temperature and altitude. The more volatile it is, the more efficiently the gasoline vaporizes. As we have seen, efficient vaporization promotes even fuel distribution to the engine cylinders, and helps provide complete combustion.

Chemical Impurities

Gasoline is refined from crude oil and contains a number of impurities that harm engines and fuel systems. For example, if gasoline has a high sulfur content, some sulfur may reach the engine crankcase, where it combines with water and forms sulfuric acid. Sulfuric acid corrodes engine parts, although proper crankcase ventilation helps avoid damage. Another impurity is gum, which forms sticky deposits that eventually clog carburetor and injector passages and cause sticking piston rings and valves.

To a large extent, the amount of chemical impurities in gasoline depends on the type of crude oil used, the refining process, and the oil refiner's desire to keep production costs low. The more-expensive process of **catalytic cracking** usually produces lower-sulfur gasoline than the less-expensive and more-common **thermal cracking** method.

Octane Rating

When engine compression pressure reaches a certain level, the pressurized air-fuel mixture generates a great deal of heat. This may cause a secondary air-fuel explosion known as **detonation,** figure 3-20, popularly called "knocking" or "pinging." Detonation causes a loss of power and overheats valves, pistons, and spark plugs. Overheating in turn causes more detonation and eventual engine damage.

To prevent detonation, gasoline must have a certain **antiknock value.** This characteristic de-

rives from the type of crude oil and the refining processes used to extract the gasoline. An **octane rating** indicates the antiknock value. Gasoline with a high octane rating resists detonation, while one with a low octane value does not.

Additives

Certain chemicals not normally present in gasoline are added during refining to improve its performance:

- Anti-icers are specially treated alcohols that prevent moisture in the air from freezing in the carburetor or throttle body at low temperatures.
- Antioxidant inhibitors prevent gum formation.
- Phosphorus compunds prevent spark plug misfiring and preignition.
- Metal deactivators prevent gasoline from reacting chemically with metal storage containers in which it is stored and transported.
- Cleaners and detergents prevent the formation or accumulation of compounds that could clog the small passages or orifices in a carburetor or fuel injector.

Major gasoline refiners use different propriety chemicals as detergents or cleaners. Aftermarket cleaners and detergents may be a good idea when fuel quality is below standard.

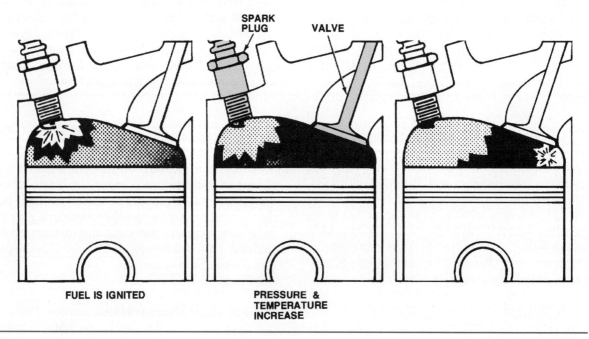

Figure 3-20.   Detonation occurs if gasoline octane is too low.

Tetraethyl lead (TEL) was once a gasoline additive that prevented detonation and provided lubrication for valve seats and many other moving parts. However, TEL is a highly toxic substance, and it remains as a particulate in the exhaust. Lead particulate emissions create a health hazard if a large amount collect in a small geographic area. Lead also destroys catalytic converter effectiveness. For these reasons, the U.S. EPA phased out lead use in gasoline. Refiners ceased producing leaded gasoline on January 1, 1988.

■ Vapor Lock

Vapor lock occurs when gasoline "gas" bubbles form in the fuel lines, reducing fuel flow to the engine. Partial vapor lock makes the air-fuel mixture leaner and reduces engine power. Complete vapor lock causes an engine to stall, and makes restarting impossible until the bubbles in the fuel system disperse—usually when the fuel system cools. Any one or combination of these four factors can cause vapor lock:

- High gasoline temperature due to engine overheating or hot weather
- Low fuel system pressure
- Gasoline with too-high volatility
- Low air pressure due to driving at high altitude

Vapor lock is an uncommon problem today for several reasons. Oil companies are good at reducing the vapor-locking tendencies of gasoline by adjusting its volatility to weather and geographic requirements, generally blending lower-volatility fuels for summer or high-altitude use and higher-volatility fuels in winter. In fact, the trend in toward year-round use of lower-volatility fuel because it evaporates more slowly, putting fewer hydrocarbons into the air.

Vapor lock is theoretically possible in fuel-injected engines, but the high fuel-line pressures normal with fuel injection make vapor lock unlikely. The higher the pressure, the higher the boiling point of the fuel, making vapor difficult to form. on modern vehicles, an electric fuel pump is usually mounted at the fuel tank, pressurizing 98 percent of the fuel system. Older models with mechanical fuel pumps mounted on the engine actually created a low-pressure condition in the entire fuel line from the tank to the pump, increasing the likelihood of vapor lock.

Octane boosters and lead substitutes sold in the automotive aftermarket may provide some octane increase. Refiners must select appropriate additives with care because many are only alcohol solutions, a common but not desirable octane booster. Octane boosters and lead substitutes are expensive and impractical for everyday use. Because of U.S. EPA and health regulations, no aftermarket fuel additive contains actual TEL.

Alcohol Additives and Fuel Quality

Gasoline blended with alcohol is widely available, although many states do not legally require labelling it as such. A mixture of 10 percent ethanol (ethyl alcohol) and 90 percent unleaded gasoline is called "gasohol" or oxygenated fuel. **Gasohol** is a generic term, however, and there are no set standards for the type and amount of alcohol it contains. Several companies sell premium fuels that use ethanol as the octane booster.

Alcohol improperly blended with gasoline causes numerous and serious problems with an automotive fuel system, including:

- Corrosion inside fuel tanks, steel fuel lines, fuel pumps, carburetors, and fuel injectors
- Deterioration of the plastic liner used in some fuel tanks, eventually plugging the in-tank filter
- Deterioration and premature failure of fuel line hoses and synthetic rubber or plastic materials such as O-ring seals, diaphragms, inlet needle tips, accelerator pump cups, and gaskets
- Hard starting, poor fuel economy, lean surge, vapor lock, and other driveability problems.

Fuels with an alcohol content tend to absorb moisture from the air. Once the moisture content of the fuel reaches approximately 1 percent, it combines with the alcohol and separates from the fuel. This water-alcohol mixture then settles at the bottom of the fuel tank. The fuel pickup carries it into the fuel line, to the carburetor or fuel injectors, causing lean surge.

All alcohols are solvents. While **ethanol** is relatively mild, **methanol** (methyl alcohol) is highly corrosive. It attacks fuel system components unless properly mixed with corrosion inhibitors and appropriate suspension agents or cosolvents to keep the water-alcohol combination from separating from the gasoline.

Gasohol has a cleaning effect on service station storage tanks, as well as the vehicle fuel tank. Because of this cleaning action, a combination of rust, sludge, and metallic particles pass into the automotive fuel system. These substances reduce fuel flow through the filter and eventually plug the carburetor or injector passageways.

Fuel economy and driveability are other areas of concern with alcohol-gasoline blends. Alcohols contain fewer British thermal units of energy per gallon or liter than gasoline, which results in reduced fuel mileage. Besides their lower energy content, alcohols are less volatile than gasoline; they require higher temperatures before they ignite and burn. Since the stoichiometric ratio for alcohol is 6.5 to 1 rather than the 14.7 to 1 of gasoline, an alcohol-gasoline blend creates a lean mixture. In turn, this creates or worsens lean surge in some driving conditions and increases the probability of vapor lock.

The problem of alcohol improperly blended with gasoline was so common in the United States during the mid-1980s that automotive tool manufacturers offered alcohol detection kits to determine fuel quality. The detection procedure required water as a reacting agent, but if the alcohol-blending process had used cosolvents as a suspension agent, the test would not show the presence of alcohol unless ethylene glycol (antifreeze) was used instead of water. Consequently, technicians often tested gasoline samples twice, first with water and then with ethylene glycol. The procedure could not differentiate between types of alcohol (ethanol or methanol), nor was it absolutely accurate. However, it was accurate enough to detect whether there was enough alcohol in the fuel for the user to take precautions. Since the mid-1980s, refiners have solved most of the alcohol-blending problems that created a demand for these kits. Today, it is not generally necessary to test for harmful amounts of alcohol in gasoline, although the kits are still on the market.

Reformulated Fuels

Reformulated fuels are fuels that are blended to reduce smog-forming and toxic pollutants typically produced by non-reformulated fuels. The Clean Air Act requires the use of reformulated fuels in areas of the country with the worst pollution problems. The federal program was started in 1995 and is called Phase I RFG. This first phase was designed to reduce the air pollution that causes smog by 64,000 tons per year. This reduction was estimated to achieve the equivalent smog-forming emissions from 16 million vehicles. Reformulated fuel is currently used in 17 states and the District of Columbia. It has been estimated that between 1995 and 1999 this program has reduced smog-forming pollutant levels by 17 percent, figure 3-21.

Despite this reduction achieved by the Phase I RFG, overall air quality in the large metropolitan areas has not been significantly reduced. From 1970 till 1997 the total number of miles traveled has increased from one trillion miles to 2.5 trillion miles. In addition almost half of the vehicles sold today are larger and higher polluting trucks and SUVs. Due to the increase in the number of vehicle miles driven and higher sales of larger more polluting vehicles, the EPA implemented Phase II RFG in January 1, 2000. Phase II RFG will be used in many metropolitan areas, figure 3-22. Phase II RFG is estimated to cut total VOC emissions by 27 percent from 1995 levels, figure 3-23.

Reformulated fuels are required to have 2 percent oxygen by weight. In addition RFG does not evaporate as easily as conventional fuels leading to lower evaporative emissions. RFG will also have lower levels of other compounds that contribute to air pollution such as benzene and sulfur. Benzene is limited to 1 per-

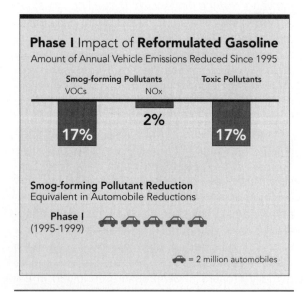

Figure 3-21. It is estimated that RFG has reduced smog-forming pollutants by 17 percent between 1995–1999. (Courtesy of the U.S. EPA)

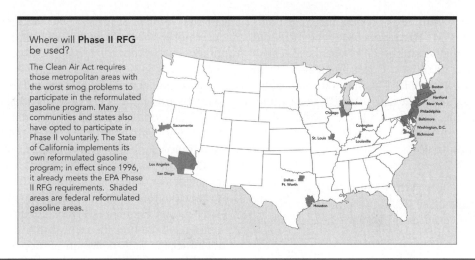

Figure 3-22. Areas of the country where Phase II RFG is required. (Courtesy of the U.S. EPA)

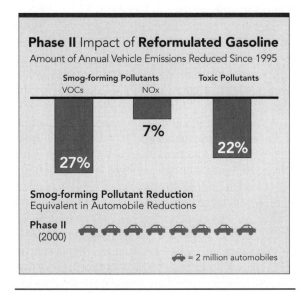

Figure 3-23. Phase II RFG is estimated to cut total VOC emissions by 27 percent from 1995 levels. (Courtesy of the U.S. EPA)

cent in reformulated fuels. Benzene is known to produce cancer in humans so the reduction is a positive health benefit for humans. Sulfur reduces catalytic converter efficiency as well as contributes to acid rain. By 2005 the average sulfur content must be 30 parts per million (ppm) as opposed to levels as high as 300 ppm in some fuels.

Oxygenates

There are currently four commonly used oxygenates in RFG. These are ethanol, methyl tertiary

butyl ether (MTBE), tertiary amyl methyl ether (TAME), and ethyl tertiary butyl ether (ETBE).

Ethanol

Ethanol is an alcohol made from fermenting agricultural products such as corn or sugar cane. It has been the most widely used oxygenate. It is used as an octane enhancer. Ten percent volume of ethanol will raise the octane number by 2.5. Many refiners use ethanol in their mid grades and high test grades. Ethanol is about 35 percent oxygen so a 10 percent blend of ethanol will add about 3.5 percent oxygen by weight to fuel. Since ethanol is a renewable resource it decreases dependency on nonrenewable resources. However, ethanol is more expensive than petroleum products, so blending ethanol increases the cost of the fuel. Ten percent volume is the maximum amount recommended by the automobile manufacturers for their vehicles unless the vehicle is equipped with a flexible fuel system designed to run on various levels of blends.

MTBE (Methyl Tertiary Butyl Ether)

A 6 percent to 8 percent blend of MTBE has been used as an octane enhancer in fuels. In oxygenated fuels it is blended at 15 percent volume. MTBE is derived from methanol, which eliminates the undesirable properties associated with methanol. MTBE is not as sensitive to water as is ethanol and methanol and does not increase the volatility of the gasoline. Controversy has arisen over the use of MTBE in gasoline due to the fact that it is highly soluble in water. In areas where leaks have occurred, it has contaminated ground-water supplies. MTBE travels faster and further in water than other gasoline components. Because of its strong odor and taste, even small amounts in water can be detectable. This has led some states and localities to ban the use of MTBE in gasoline. The EPA is considering the reduction of MTBE as an oxygenate in gasoline due to the groundwater contamination issues.

TAME (Tertiary Amyl Methyl Ether)

The performance of TAME is similar to that of MTBE when used as an octane enhancer or an oxygenate in fuel. One difference is that TAME actually lowers the vapor pressure of the gasoline. It can be used at levels of up to 17.2 percent volume, which can increase the octane number by 3.0 and the oxygen level 2.7 percent weight. As with MTBE, TAME is manufactured from methanol and isoamylene.

ETBE (Ethyl Tertiary Butyl Ether)

The performance characteristics of ETBE are similar to that of MTBE. The maximum level is 17.2 percent volume weight, which can increase the octane number by 3 octane numbers. A 17.2 percent blend would contain 2.7 percent weight oxygen. ETBE also lowers the fuel vapor pressure. This makes it a good choice for RFG.

SUMMARY

We measure engine size by displacement, or the total volume of all the cylinders. Displacement is a determining factor in the engine airflow requirement. Stock engines operate below volumetric efficiency, or the maximum volume of air they could take in. Engines operate efficiently on an air-fuel ratio of 8 to 18.5 to 1, but the exact ratio at which perfect mixture and complete combustion occur, or the stoichiometric ratio, is closer to 14.7 to 1. The stoichiometric ratio is also the best compromise between power and economy, as well as producing the least emissions. Electronic fuel management systems are designed to provide the engine with the proper air-fuel ratio according to engine speed and load demands.

Gasoline needs to be metered, atomized, vaporized, and distributed to each cylinder by a metering device. The metering device on most later-model engines is a fuel injector. Older engines used the fuel-metering circuits in a carburetor. Proper fuel distribution depends on fuel volatility, atomization, and vaporization; intake manifold passage design, intake throttle valve angle; and on carburetor, throttle body, or fuel injector location on the intake manifold.

Gasoline needs proper refining to remove chemical impurities and blending with additives to prevent preignition, detonation, carburetor icing, varnish formation, and misfiring. Tetraethyl lead was once the major octane booster, but the U.S. EPA severely limited its use.

A fuel blend made up of alcohol and gasoline is called oxygenated fuel. If this fuel is misblended, as sometimes happened in the mid-1980s, it permanently damages a fuel system. However, blending problems are less common than when these fuels first appeared. Ford and GM have developed flexible-fuel vehicles for fleet use, which operate on varying percentages of methanol and gasoline. Engines must run on oxygenated fuel, containing a higher proportion of methanol, in certain areas that have severe pollution to reduce CO emissions. Methanol has less energy than gasoline, so the engine needs more of it to run. Using methanol requires different fuel system components and materials.

Review Questions

Choose the letter that represents the best possible answer to the following questions:

1. A disadvantage of gasoline is that it:
   a. Vaporizes easily
   b. Burns quickly
   c. Produces pollutants upon combustion
   d. Has a high heat value

2. Which is *not* a factor in determining airflow requirement?
   a. Engine displacement
   b. Maximum rpm
   c. Carburetor size
   d. Volumetric efficiency

3. Volumetric efficiency:
   a. Is the theoretical maximum air volume that can be drawn into an engine
   b. Increases as engine speed increases to maximum efficiency at maximum rpm
   c. Is expressed in cubic feet per minute or cubic meters per minute
   d. Is the ratio of air entering the engine to engine displacement

4. At maximum speed, the volumetric efficiency of a stock engine is approximately:
   a. 75 percent
   b. 50 percent
   c. 100 percent
   d. Better than 100 percent

5. The richest air-fuel ratio that an internal combustion engine can tolerate is about:
   a. 4:1
   b. 2.5:1
   c. 8:1
   d. 18.5:1

6. To burn one pound of gasoline with maximum efficiency requires about:
   a. 8 pounds of air
   b. 15 pounds of air
   c. 18 pounds of air
   d. 17 pounds of air

7. A rich air-fuel mixture is needed for:
   a. Cruising speed
   b. Light load
   c. Better driveability
   d. Acceleration

8. An internal engine condition affecting fuel requirements is:
   a. Engine load
   b. Mixture distribution
   c. Atmospheric pressure
   d. Engine speed

9. Obtaining maximum power results in:
   a. No change in air-fuel ratios
   b. Leaner mixtures
   c. Unburned oxygen
   d. Excess unburned fuel

10. Maximum fuel economy requires:
    a. Less air
    b. Leaner air-fuel mixtures
    c. Richer air-fuel mixtures
    d. Higher temperatures

11. For maximum power, the air-fuel ratio becomes:
    a. Leaner at low speeds
    b. Richer at low speeds
    c. Richer at high speeds
    d. Leaner during acceleration

12. For maximum fuel economy, the air-fuel ratio becomes:
    a. Leaner during acceleration
    b. Leaner at high speeds
    c. Leaner at low speeds
    d. Leaner for middle speeds

13. Different fuel-blending techniques are used by refiners to:
    a. Increase fuel volatility
    b. Decrease the octane rating
    c. Replace impurities with additives
    d. Add sulfuric acid

14. Technician A says carburetors use two, three, or four barrels. Technician B says carburetors are *not* found on later-model cars.

    Who is right?
    a. A only
    b. B only
    c. Both A and B
    d. Neither A nor B

15. Technician A says a TBI injection system uses one injector at each cylinder port.

    Technician B says the TBI unit is installed on the intake manifold where a carburetor would be.

    Who is right?
    a. A only
    b. B only
    c. Both A and B
    d. Neither A nor B

16. Technician A says port fuel injection is also called multipoint injection.

    Technician B says a port fuel-injection unit looks like the lower part of a carburetor.

    Who is right?
    a. A only
    b. B only
    c. Both A and B
    d. Neither A nor B

17. Which of the following is *not* a quality of tetraethyl (TEL) lead?
    a. Prevents detonation
    b. Lubricates valve seats
    c. Can destroy a catalytic converter
    d. Is a nontoxic substance

18. Gasohol is generally regarded as a blend of:
    a. 10 percent ethanol and 90 percent gasoline
    b. 90 percent ethanol and 10 percent gasoline
    c. 25 percent ethanol and 75 percent gasoline
    d. 75 percent ethanol and 25 percent gasoline

4

Fuel Delivery Systems

OBJECTIVES

Upon completion and review of this chapter, you will be able to:

- Identify and describe the purpose of the components of a typical fuel delivery system.
- Describe the different types of fuel lines used on vehicles.
- Describe the differences between a return and returnless fuel system.
- Describe the purpose of the evaporative emission control system.
- Have knowledge of a non-enhanced and enhanced evaporative system.
- Have knowledge of the various types of fuel pumps used on fuel-injected vehicles.
- Describe the function of the electric fuel pump used on fuel-injection vehicles.

KEY TERMS

adsorption
armature
baffle
check valve
delivery system
evaporative emission
  controls (EVAP)
float valve
impeller
liquid-vapor separator
microns

orifice
percolation
pressure drop
pulsating
residual or rest
  pressure
returnless
vacuum lock
vapor lock
volatile organic
  compounds (VOCs)

INTRODUCTION

Chapters 1–3 introduced the basic parts, principles, and problems involved in the automobile fuel system and emission controls. Chapter 4 covers the major components in the fuel system in more detail. This helps you to develop a working knowledge of the relationships between the fuel system and emission controls. This chapter examines basic fuel delivery and return system operation, and its effects on engine performance, economy, and emissions. The next chapter, Chapter 5, examines engine management systems. Chapter 6 covers engine management impact devices. Chapter 7 looks at engine management output devices.

Creating and maintaining a correct air-fuel mixture requires a properly functioning fuel and air **delivery system.** Fuel delivery (and return) systems use many if not all of the following

67

components to make certain that fuel is available under the right conditions to the carburetor or fuel injection system:

- Fuel storage tank, filler neck, and gas cap
- Fuel tank pressure sensor
- Fuel pump
- Fuel filter(s)
- Fuel delivery lines and fuel rail
- Fuel pressure regulator
- Evaporative emission controls: canister, purge valve, thermostatic vacuum valves
- Fuel return line
- Air cleaners and filters
- Thermostatic air intake controls

FUEL TANKS AND FILLERS

The automobile fuel tank, figure 4-1, is made of two corrosion-resistant steel halves that are ribbed for additional strength and welded together. Exposed sections of the tank may be made of heavier steel for protection against road damage and corrosion.

Some models, such as sport utility vehicles (SUVs) and light trucks, may have an auxiliary fuel tank. Manufacturers make some of the auxiliary tanks for these vehicles of polyethylene plastic and of composites.

Tank design and capacity are a compromise between available space, filler location, fuel expansion room, and fuel movement. Some later-model tanks deliberately limit tank capacity by extending the filler tube neck into the tank low enough to prevent complete filling, or by providing for expansion room, figure 4-2. A vertical **baffle** in this same tank limits fuel sloshing as the vehicle moves.

Regardless of size and shape, all fuel tanks incorporate most if not all of the following features:

- Inlet or filler tube through which fuel enters the tank
- Filler cap with pressure holding and relief features
- An outlet to the fuel line leading to the fuel pump or fuel injector
- Fuel pump mounted within the tank
- Tank vent system
- Fuel pickup tube and fuel level sending unit

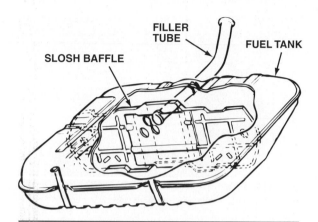

Figure 4-1. The filler tube is located in this tank so that the tank cannot be filled completely. The air space at the top of the tank allows room for fuel expansion.

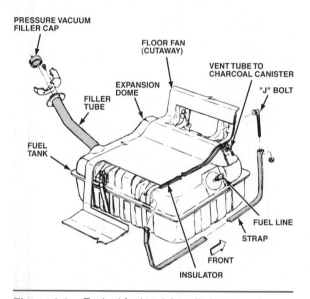

Figure 4-2. Typical fuel tank installation.

Tank Location and Mounting

Most domestic vehicles use a horizontally suspended fuel tank, usually mounted below the rear of the floor pan, just ahead of or behind the rear axle. Fuel tanks are located there so that frame rails and body components protect the tank in the event of a crash. To prevent squeaks, some models have insulated strips cemented on the top or sides of the tank wherever it contacts the underbody.

Fuel inlet location depends on the tank design and filler tube placement. It is located behind a filler cap and often a hinged door in the outer side of either rear fender panel. Some older models

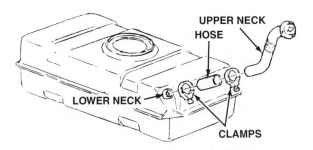

Figure 4-3. Three-piece filler tube assembly.

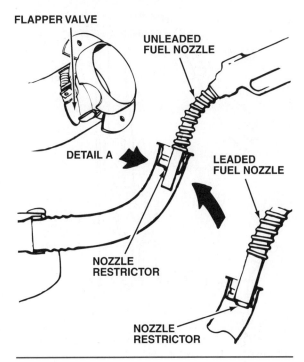

Figure 4-4. Cars that require unleaded fuel have restrictors in the filler tubes so only the smaller unleaded fuel pump nozzles can fit. Restrictors may be spring-loaded flapper valves or simple ring-shaped pieces inside the tube.

have their fuel inlet in other positions. For example, the Type 1 Volkswagen had the fuel inlet under the front hood or in the front body panel. Some 1950s-era GM cars even hid the fuel inlet behind one of the taillights.

Generally, a pair of metal retaining straps holds a fuel tank in place. Underbody brackets or support panels hold the strap ends using bolts. The free ends are drawn underneath the tank to hold it in place, then bolted to other support brackets or to a frame member on the opposite side of the tank.

Filler Tubes

Fuel enters the tank through a large tube extending from the tank to an opening on the outside of the vehicle. There are two types of filler tubes: a rigid, one-piece tube soldered to the tank, figure 4-2, and a three-piece unit, figure 4-3. The three-piece unit has a lower neck soldered to the tank and an upper neck fastened to the inside of the body sheet-metal panel. The two metal necks are connected by a length of hose clamped at both ends.

Federal regulations require that all vehicles equipped with catalytic converters burn unleaded fuel. Lead coats the catalytic material and ruins the converter, so unleaded fuel vehicles have a restrictor in their openings at their filler tubes to prevent using leaded fuel. The smaller-diameter unleaded fuel nozzles at service station pumps fit into filler openings, but the larger nozzles for leaded fuel cannot, figure 4-4. In addition, a spring-loaded flapper valve in the filler tube prevents pouring gasoline into a tank without a small diameter nozzle. A deflector in the filler tube behind the flapper valve prevents fuel splashback during filling. Since 1996, the use of leaded fuel in almost all vehicles sold in the United States is illegal.

Effective September 1993, federal regulations require manufacturers to install a device to prevent fuel from being siphoned through the filler neck. Federal authorities recognized methanol as a poison, and methanol used in gasoline is a definite health hazard. Additionally, gasoline is a suspected carcinogen. To prevent siphoning, manufacturers welded a filler neck check-ball tube in fuel tanks. To drain check-ball–equipped fuel tanks, a technician must disconnect the check-ball tube at the tank and attach a siphon directly to the tank.

Fuel Pickup Tube

The fuel pickup tube is usually a part of the fuel sender assembly or the electric fuel pump assembly, figure 4-5. Since dirt and sediment eventually gather on the bottom of a fuel tank, the fuel pickup tube is fitted with a filter sock or strainer to prevent contamination from entering the fuel lines. The woven plastic strainer also acts as a water separator by preventing water from being drawn up with the fuel.

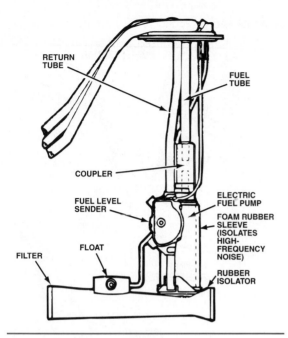

Figure 4-5. RETURN TUBE, FUEL TUBE, COUPLER, FUEL LEVEL SENDER, ELECTRIC FUEL PUMP, FOAM RUBBER SLEEVE (ISOLATES HIGH-FREQUENCY NOISE), FILTER, FLOAT, RUBBER ISOLATOR

Figure 4-5. The fuel pickup tube is part of the fuel sender and pump assembly.

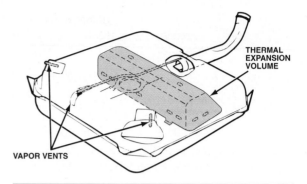

THERMAL EXPANSION VOLUME, VAPOR VENTS

Figure 4-6. This fuel tank has an internal expansion tank to allow for changes in fuel volume due to temperature changes.

Tank Venting Requirements

Fuel tanks must be vented to prevent a **vacuum lock** as fuel is drawn from the tank. As fuel is used and its level drops in the tank, the space above the fuel increases. As the air in the tank expands to fill this greater space, its pressure drops. Without a vent, the air pressure inside the tank would drop below atmospheric pressure, developing a vacuum that prevents the flow of fuel. Under extreme pressure variance, the tank could collapse. Venting the tank allows outside air to enter as the fuel level drops, preventing a vacuum from developing.

Before 1970, tanks were vented directly to the atmosphere with either a length of tubing (vent line) or through a vent in the filler cap. Both systems, however, added to air pollution by allowing fuel vapors into the air. To reduce evaporative hydrocarbon (HC) emissions, evaporative emission control (EVAP) systems have been installed since 1970 on California models and since 1971 on federal vehicles. An EVAP system vents gasoline vapors from the fuel tank directly to a charcoal-filled vapor storage canister, and uses an unvented filler cap. Many filler caps contain valves that open to relieve pressure or vacuum above specified safety levels. Systems that use completely sealed caps have separate pressure and vacuum relief valves for venting.

Because fuel tanks are no longer vented directly to the atmosphere, the tank must allow for

fuel expansion, contraction, and overflow that can result from changes in temperature or overfilling. One way is to use a separate expansion tank, figure 4-6. Another method is to provide a dome in the top of the tank, figure 4-2. As mentioned earlier, some tanks are limited in capacity by the angle of the fuel filler tube, figure 4-1. Many General Motors (GM) vehicles use a design that includes a vertical slosh baffle, which reserves up to 12 percent of the total tank capacity for fuel expansion.

Rollover Leakage Protection

All 1976 and later models have one or more devices to prevent fuel leaks in case of vehicle rollover or a collision in which fuel may spill. Manufacturers have met this requirement by using one of the following:

- Check valve
- Float valve

Variations of the basic one-way **check valve** may be installed in any number of places between the fuel tank and the engine. The valve may be installed in the fuel return line, vapor vent line, fuel tank filler cap, figure 4-7, or carburetor fuel inlet filter, figure 4-8. This type of rollover leakage protection is used primarily by Chrysler and GM.

Ford vehicles use a spring-operated **float valve** in the vapor separator. The float valve closes whenever the chassis is at a 90-degree angle or more.

Vehicles with electric fuel pumps also use these same rollover protection devices, but they have additional features to ensure that the fuel pump shuts off when an accident occurs. Some pumps depend upon an oil pressure or an engine

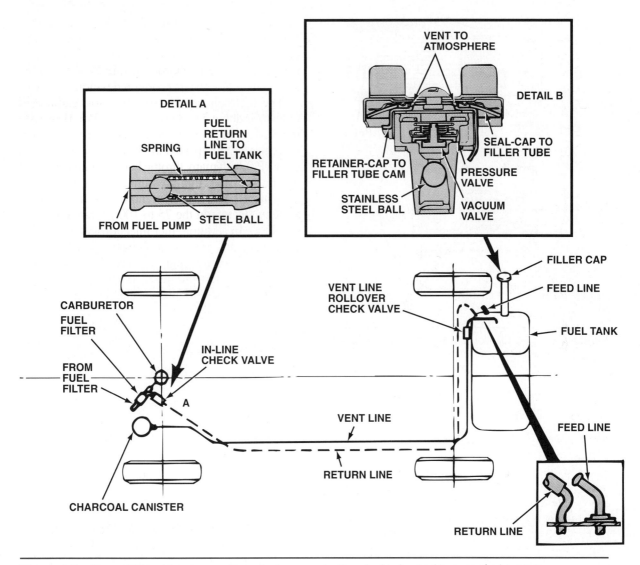

Figure 4-7. Since 1976, at least one rollover leakage protection device is used in every fuel system.

speed signal to continue operating; these pumps turn off whenever the engine dies. On early Bosch L-Jetronic multiport injection systems, a microswitch built into the vane airflow sensor switched on the fuel pump as soon as intake airflow caused the vane to lift from its rest position, figure 4-9.

Ford vehicles with electronic engine controls and fuel injection have another form of rollover leakage protection. An inertia switch, figure 4-10, is installed in the rear of the vehicle between the electric fuel pump and its power supply. As many a motorist has found out the hard way, if the vehicle is involved in any sudden impact, such as a jolt from another vehicle in a parking lot, the inertia switch opens and shuts off power to the fuel

pump. The switch must be reset manually by pushing a button to restore current to the pump.

FUEL LINES

Fuel and vapor lines made of steel, nylon tubing, or fuel-resistant rubber hoses connect the parts of the fuel system. Fuel lines supply fuel to the carburetor, throttle body, or fuel rail. They also return excess fuel and vapors to the tank. Depending on their function, fuel and vapor lines may be either rigid or flexible.

Fuel lines must remain as cool as possible. If any part of the line is located near too much heat, the gasoline passing through it vaporizes and

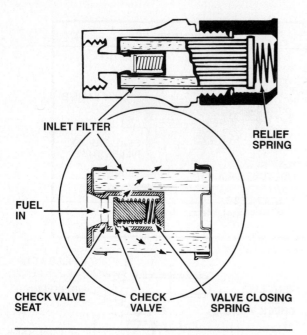

Figure 4-8. The rollover protection check valve is built into the fuel filter used on GM cars.

vapor lock occurs. (See sidebar: "Why Vapor Lock?") When this happens, the fuel pump supplies only vapor that passes into the carburetor or injectors. Without liquid gasoline, the engine stalls and a hot restart problem

The fuel delivery ted engines is abo er, the delivery pr on system is 10 to i (241 kPa) with jection systems. ate with 50 psi (3 systems operate i 7 to 689 kPa). In additi on systems retain residual or rest pres in the lines for a half hour or longer when the engine is turned off to prevent hot engine restart problems. Higher pressure systems such as these require special fuel lines.

NO SIDEBAR ON PAGE 85

Rigid Lines

All fuel lines fastened to the body, frame, or engine are made of seamless steel tubing. Steel springs

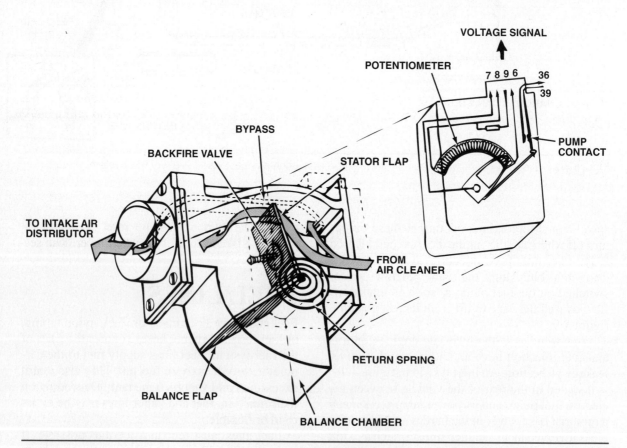

Figure 4-9. On early Bosch systems, airflow closes a microswitch built into the airflow sensor to turn on the fuel pump. In the event of a collision, the switch opens and fuel flow stops.

(a)

(b)

Figure 4-10. (a) Ford uses an inertia switch to turn off the electric fuel pump in an accident. (b) A cut-away of the inertia switch shows a ball held in position by a magnet. Upon impact the ball breaks away from the magnet and opens the electrical contacts.

may be wound around the tubing at certain points to protect against impact damage.

Only steel tubing, or that recommended by the manufacturer, should be used when replacing rigid fuel lines. *Never substitute copper or alu-*

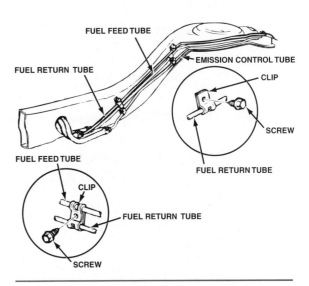

Figure 4-11. Fuel lines are routed along the car frame and secured with clips. Hoses are fastened to steel lines with hose clamps.

minum tubing for steel tubing. These materials do not withstand normal vehicle vibration and could combine with the fuel to cause a chemical reaction.

In some carbureted models, rigid fuel lines are secured along the frame from the tank to a point close to the fuel pump, figure 4-11. The gap between frame and pump is then bridged by a short length of flexible hose that absorbs engine vibrations. Other vehicles run a rigid line directly from tank to pump. To absorb vibrations, the line crosses 30 to 36 inches (750 to 900 millimeters) of open space between the pump and its first point of attachment to the frame, or may contain a "stress loop."

Flexible Lines

Most fuel systems use synthetic rubber hose sections where flexibility is needed. Short hose sections often connect steel fuel lines to other system components. The fuel delivery hose inside diameter (ID) is generally larger (5/16- to 3/8-inches or 8 to 10 millimeters) than the fuel return hose ID (1/4-inches or 6 millimeters).

Replacement fuel hoses should be made of fuel-resistant material. Ordinary rubber hoses such as those used for vacuum lines deteriorate when exposed to gasoline. Fuel-injection systems require special-composition reinforced hoses specifically made for these higher-pressure systems. Similarly, vapor vent lines must be made of materials that resist fuel vapors. Replacement

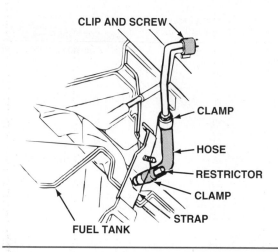

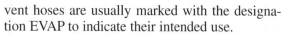

Figure 4-12. Many fuel vent lines have restrictors to control the rate of vapor flow.

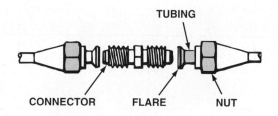

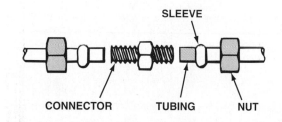

Figure 4-13. Fuel line fittings are either the flare type (top) or the compression type (bottom).

vent hoses are usually marked with the designation EVAP to indicate their intended use.

A metal or plastic restrictor often is used in vent lines to control the vapor flow rate, figure 4-12. These may be installed either in the end of the vent pipe, or in the vapor vent hose itself. When used in the hose instead of the vent pipe, the restrictor must be removed from the old hose and installed in the new one whenever the hose is replaced.

Fuel Line Mounting

Fuel supply lines from the tank to a carburetor, throttle body or fuel rail are routed to follow the frame along the underbody of the vehicle. Vapor and return lines may be routed with the fuel supply lines, usually on the frame rail opposite the supply line. All rigid lines are fastened to the frame rail or underbody with screws and clamps, or clips.

Low-Pressure Fittings and Clamps

Brass fittings used in fuel lines are either the flared type or the compression type, figure 4-13, the flared fitting is more common. The inverted, or SAE 45-degree, flares slip snugly over the connectors to prevent leakage when the nuts are tightened. When replacement tubing is installed, a double flare should be used to ensure a good seal and to prevent the flare from cracking. Compression fittings use a separate sleeve, a ta-

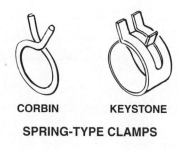

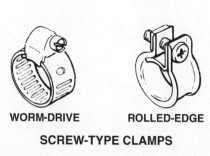

Figure 4-14. Various clamps are used on fuel system hoses.

pered sleeve, or a half-sleeve nut to make a good connection.

Various types of clamps are used to secure fuel hoses on low-pressure fuel systems, figure 4-14. Spring-type clamps are commonly used for original equipment installation, but only screw-type (aircraft) clamps should be reused when hoses are changed. Keystone, Corbin, and other spring-type clamps will not hold securely when reused and should be replaced with new ones if they are removed. Screw-type clamps are made in two styles: worm-drive clamps, and rolled-edge clamps in which the screw and nut stand off from the clamp body, figure 4-14.

Fuel-Injection Lines and Clamps

Hoses used for fuel-injection systems are made of materials with high resistance to oxidation and deterioration. They also are reinforced to with-

stand higher pressures than carburetor system fuel hoses. Replacement hoses for injection systems should always be equivalent to original equipment manufacturer (OEM) hoses.

Do not use spring-type clamps on fuel-injected engines—they cannot withstand the fuel pressures involved. Screw-type clamps are essential on injected engines and should have rolled edges to prevent hose damage. Worm-drive clamps are satisfactory for use on carbureted engines, *but do not work well on fuel-injection systems.* The screw teeth cut and weaken the hose if overtightened.

■ **Safety Cells**

Most automotive fuel tanks are simple steel tanks that hold liquid fuel. Formed to fit the chassis design and containing some baffles to reduce fuel sloshing, they are straightforward devices. A simple steel tank, however, has some serious disadvantages for vehicles used in hazardous operations. Common fuel tanks can be punctured by an impact or leak fuel if overturned. Even with extensive baffling, off-road driving can cause fuel to slosh enough to upset vehicle balance or starve the fuel pickup line. All of these drawbacks to simple fuel tanks led to the development of fuel cells.

A fuel cell is a tank with a rigid shell of steel, aluminum, or some composite material. Inside the shell, a flexible rubber bladder forms a safety liner. The bladder is filled with low-density foam material that absorbs the liquid fuel. The

fuel stays as a liquid within the foam and can be withdrawn easily by the fuel pump and pickup lines. The foam eliminates sloshing by distributing the fuel evenly throughout the tank regardless of the amount of fuel or vehicle motion. The combination of the rubber bladder and the foam prevents—or drastically reduces—leakage in case of impact or tank rupture. Fuel cells also have check valves in the filler and vent lines to prevent leakage in case of rollover.

Fuel cells are mandatory safety equipment in most race cars and are used in many police cars, fire and rescue trucks, ambulances, and some off-road equipment. Special cells with self-sealing, antiballistic ("bulletproof") liners are specified by the U.S. Secret Service as standard equipment in presidential limousines.

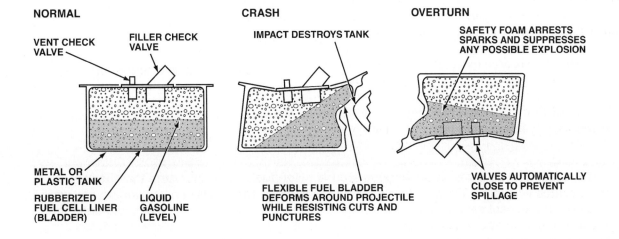

Fuel-Injection Fittings and Nylon Lines

Because of their higher operating pressures, fuel-injection systems often use special kinds of fittings to ensure leakproof connections. Some high-pressure fittings on GM vehicles with port injection systems use O-ring seals instead of the traditional flare connections. When disconnecting such a fitting, inspect the O-ring for damage and replace it if necessary. *Always* tighten O-ring fittings to the specified torque value to prevent damage.

Other manufacturers also use O-ring seals on fuel-line connections. In all cases, the O-rings are made of special materials that withstand contact with gasoline and oxygenated fuel blends. Some manufacturers specify that the O-rings be replaced every time the fuel system connection is opened. When replacing one of these O-rings, a new part specifically designed for fuel system service must be used. The O-rings used in air condition(ing) systems are *not* satisfactory.

Many manufacturers use nylon fuel tubing with several unique push-connect fittings, figure 4-15. Special barbed connectors are required to join sections of nylon tubing together. These fittings may in some cases be released by hand or may require special tools to release the fittings. Some manufacturers may allow repairs to be made to nylon fuel lines in case of damage. There are special barbed fittings available to perform these repairs. Other manufacturers allow no repairs to be made to nylon fuel lines. In case of damage the entire fuel line must be replaced. Be sure to consult the manufacturer's service information before attempting any repairs to nylon fuel lines.

General Motors originally introduced nylon fuel lines with quick-connect fittings at the fuel tank and fuel filter on some 1988 models. Since then, their use has been expanded to more vehicles each model year. Like the GM threaded couplings used with steel lines, nylon line couplings use internal O-ring seals. Unlocking the metal connectors requires a special quick-connector separator tool; plastic connectors can be released without the tool, figures 4-16 and 4-17. Where access to metal connectors is restricted, a special tool is available.

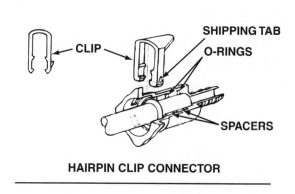

HAIRPIN CLIP CONNECTOR

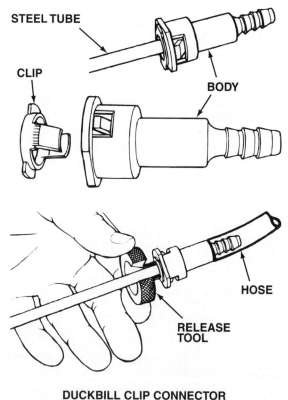

DUCKBILL CLIP CONNECTOR

Figure 4-15. Late-model Ford products use these push-connect fuel line fittings. (Courtesy of Ford Motor Company)

NOTE: Care must be taken when working around nylon fuel lines. Make sure that they are correctly routed away from heat sources such as the exhaust system. When performing any work with an external heat source such as a torch or heat gun, be careful to stay a safe distance from any nylon fuel or vapor lines. When installing or replacing the fuel lines, be careful not to bend the line too sharply or it will kink.

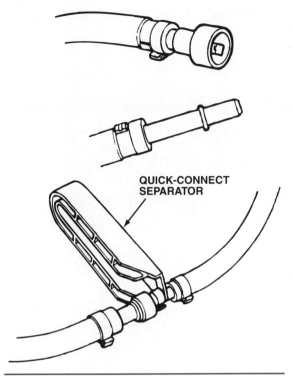

Figure 4-16. The quick-connect separator should be used with all metal GM quick-connect fittings. Plastic fittings can be released by hand without the tool. (Courtesy of General Motors Corporation)

FUEL LINE LAYOUT

With the advent of hotter-running engines for better economy and performance, underhood temperatures have risen to the point that vapor lock and percolation are a real concern to manufacturers. Fuel pressures have tended to become higher to prevent vapor lock, and a major portion of the fuel routed to the carburetor or fuel-injection system returns to the tank by way of a fuel return line. This allows better control, within limits, of heat absorbed by the gasoline as it is routed through the engine compartment. Throttle body and multiport injection systems have typically used a pressure regulator, figure 4-18, to control fuel pressure in the throttle body or fuel rail, and to allow excess fuel not used by the injectors to return to the tank. Heated fuel, once returned to the tank, warms the fuel supply, which tends to assist cold-weather fuel system performance. However, in the summer months, warmer fuel in the tank may create problems, such as an excessive rise in fuel vapor pressures in the tank.

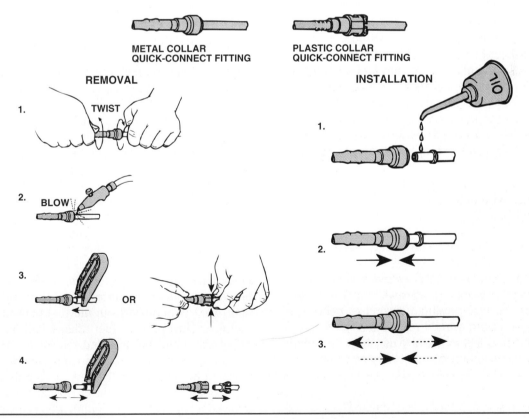

Figure 4-17. Servicing GM quick-connect fittings. (Courtesy of General Motors Corporation)

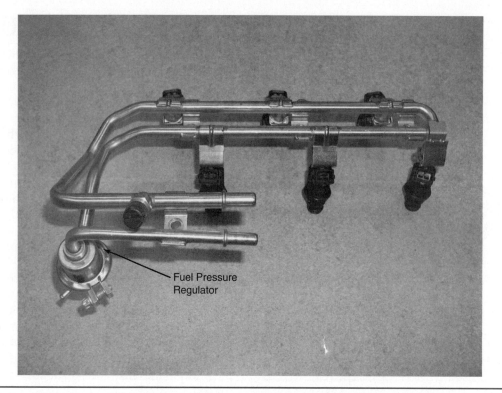

Figure 4-18. The fuel pressure regulator controls the fuel pressure in the fuel rail.

■ Tools for Making Fuel Lines

With few exceptions, replacement fuel lines cannot be bought preformed. Tubing is stocked in large rolls and must be shaped and formed however the technician wants it. Ordinary hand tools cannot be used to properly make a replacement fuel line. Frequently, cutting a tube using a hacksaw leaves a distorted, jagged edge. To ensure a smooth cut, use these special tubing tools:

- Cutter
- Reamer
- Bender
- Flaring device.

The tube cutter uses sharpened metal disks to make a smooth, distortion-free cut. After cutting, a tapered reamer is necessary to remove any burrs that might prevent a good seal. The tube bender shapes the tubing without kinking or bending it. Flaring tools are available to make either single or double flares. It is essential to use them to properly shape the connecting ends of any new fuel line.

Just how much fuel is recirculated? As an example, a passenger vehicle cruising down the road at 60 mph gets 30 mpg. With a typical return-style

fuel system pumping about 30 gallons per hour from the tank, it would therefore burn 2 gallons per hour, and return about 28 gallons to the tank!

With late-model vehicles, there has been some concern about too much heat being sent back to the fuel tank, causing rising in-tank temperatures and increases in fuel vaporization and volatile organic compound (VOC) emissions. To combat this problem, manufacturers have placed the pressure regulator back by the tank instead of under the hood. In this way, returned fuel is not subjected to the heat generated by the engine and the underhood environment. To prevent vapor lock in these systems, pressures have been raised in the fuel rail, and injectors tend to have smaller openings to maintain control of the fuel spray under pressure. These systems are often referred to as **returnless** or demand fuel systems, figure 4-19. The fuel pressure regulator is placed near or in the fuel tank. Fuel is pumped from the fuel pump to the fuel filter. Just past the fuel filter, a fuel line tees off of the main fuel line and returns to the fuel tank. Any excess fuel not needed by the fuel system will be returned to the tank. This return fuel has not been heated from the engine compartment, thus reducing evaporative hydrocarbon emissions.

But fuel pressure is only part of the delivery story. Not only must the fuel be filtered and sup-

■ Fuel System Development

The first automobiles relied on gravity to supply fuel to the engine. These gravity-feed systems mounted the fuel tank higher than the engine, allowing gravity to draw fuel from the tank to the engine. Because these tanks were front-mounted, they had limited capacity and were dangerous.

Moving the fuel tank to the rear of the chassis solved the problems of safety and storage capacity but required the use of a vacuum tank. This was a small fuel tank, still positioned above the engine in the cowl, but connected to the rear tank as well. Suction created by engine vacuum provided fuel for the vacuum tank from the larger rear-mounted tank.

If the vehicle was not driven for a long time, the gasoline in the vacuum tank would eventually evaporate. In this case, it was necessary to prime the engine in order to start it and create vacuum

that would move fuel through the system. With the appearance of the mechanical fuel pump after World War I, the vacuum tank was retired.

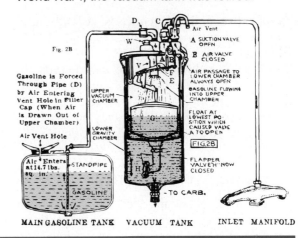

plied under adequate pressure, there must also be a consistent *volume* of fuel to assure smooth engine performance even under the heaviest of loads. The pressure regulator is designed to supply a constant flow or volume of fuel under all conditions. Once fuel volume falls off due to a clogged filter sock in the tank, a loaded filter, or a worn fuel pump, engine performance suffers.

The fuel pressure regulator is usually of the diaphragm type, which compensates for engine load changes to keep a constant pressure drop across the injectors. In this way, high manifold vacuum is less likely to suck fuel from the injectors, and a turbo boost condition is unlikely to prevent fuel from spraying properly and in sufficient quantity into the intake airstream.

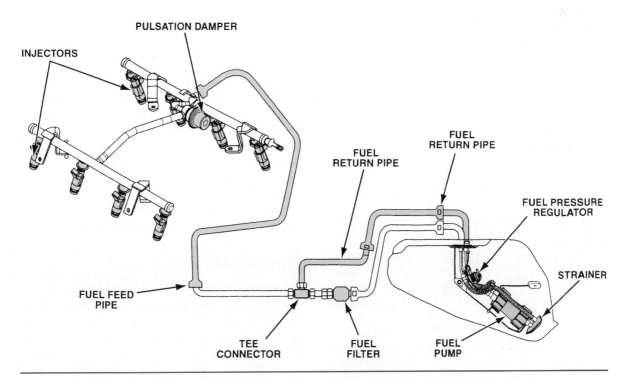

Figure 4-19. Returnless (demand) fuel system components. (Courtesy of General Motors Corporation)

EVAPORATIVE EMISSION CONTROL SYSTEMS

Evaporative emission controls (EVAP) have been an anti-pollution tool since the early 1970s. The purpose of the EVAP system is to trap gasoline vapors—**volatile organic compounds**, or **VOCs**—that would otherwise escape into the atmosphere. These vapors are instead routed into a charcoal canister, from where they go to the intake airflow so they are burned in the engine.

Common Components

The fuel tank filler caps used on vehicles with modern EVAP systems are a special design. Some early GM EVAP systems used an unvented cap with a pressure-vacuum relief valve in the line between the fuel tank and carburetor, figure 4-20, but most EVAP fuel tank filler caps have pressure-vacuum relief built into them, figure 4-21. When pressure or vacuum exceeds a calibrated value, the valve opens. Once the pressure or vacuum has been relieved, the valve closes. If a sealed cap is used on an EVAP system that requires a pressure-vacuum relief design, a vacuum lock may develop in the fuel system, or the fuel tank may be damaged by fuel expansion or contraction.

Various methods protect fuel tanks against fuel expansion and overflow caused by heat. Overfill limiters, or temperature expansion tanks, were used on many early 1970s EVAP systems to prevent filling the tanks to capacity. The limiter attaches to the

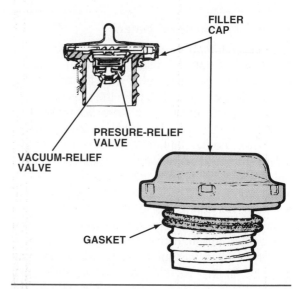

Figure 4-21. Fuel tank caps for modern EVAP systems have vacuum- and pressure-relief valves.

inside of the fuel tank and contains small holes that open it to the fuel area. When the fuel tank appears to be completely full—when it holds no more and the fuel gauge reads full—the expansion tank remains virtually empty. This provides enough space for fuel expansion and vapor collection if the vehicle is parked in the hot sun after filling the tank.

The dome design of the upper fuel tank section used in some late-model cars, or the overfill limiting valve contained within the vapor-liquid separator, eliminates the need for the overfill limiter tank used in earlier systems.

All EVAP systems use some form of **liquid-vapor separator** to prevent liquid fuel from reaching the engine crankcase or vapor storage canister. Some liquid-vapor separators are built into the tank and use a single vapor vent line from the tank to the vapor canister. When the separator is not built in, figure 4-22, it usually is mounted on the outside of the tank or on the frame near it. In this case, vent lines run from the tank to the separator and are arranged to vent the tank regardless of whether the vehicle is level or not. Liquid fuel entering the separator returns to the tank through the shortest line.

Vapor Storage

As explained, an EVAP system traps gasoline vapors from the fuel tank and carburetor and feeds them into the engine intake system or stores them until the engine is started. Almost all late-model EVAP systems store the vapors in a charcoal-granule-filled canister. A few early EVAP systems stored the vapors in the engine crankcase.

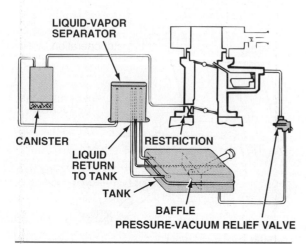

Figure 4-20. Some early GM EVAP systems used an unvented cap with a pressure relief valve in the line between the fuel tank and carburetor.

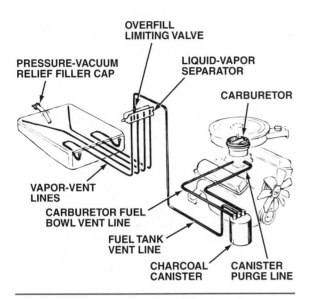

Figure 4-22. This EVAP system has a liquid-vapor separator mounted separately from the tank.

Vapor Canister Storage

Vapor storage canisters have been used on virtually all vehicles since 1972. The canister is usually located under the hood, figure 4-22, and is filled with activated charcoal granules that can hold up to one third of their own weight in fuel vapors, figure 4-23. A vent line connects the canister to the fuel tank. Carburetors with external bowl vents were also vented to the canister. Some Ford and Chrysler vehicles with large or dual fuel tanks may have dual canisters; GM engines may have an auxiliary canister connected to the primary canister purge air inlet to store vapor overflow.

Activated charcoal is an effective vapor trap because of its great surface area. Each gram of activated charcoal has a surface area of 1,100 square meters, or more than a quarter acre. Typical canisters hold either 300 or 625 grams of charcoal *with a surface area equivalent to 80 or 165 football fields.* **Adsorption** attaches the fuel vapor molecules to the carbon surface. This attaching force is not strong, so the system purges the vapor molecules quite simply by sending a fresh airflow through the charcoal.

There are two methods to provide fresh air to the canister for purging. In one design, the bottom of the canister is open to the atmosphere and air enters through a filter, figure 4-24. This design supplies purge air whenever the engine is running. In another design, canisters are closed to the atmosphere and obtain air from the air-injection system. A solenoid controls closed-canister air-flow to purge the vapors during specific engine operating conditions.

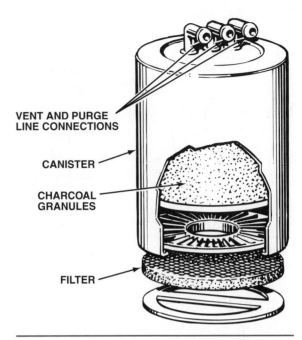

Figure 4-23. A typical vapor storage canister contains 300 or 625 grams of activated charcoal to trap and store fuel vapors.

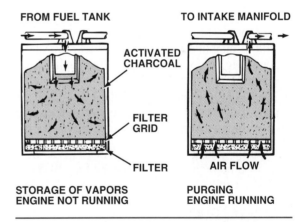

Figure 4-24. Typical vapor canister operation.

A small vapor separator in the supply line between the fuel pump and the carburetor reduces the amount of fuel vapors reaching the carburetor on many vehicles (particularly Ford products), figure 4-25. A vapor return line connects this separator to the fuel tank. Vapors collected in the separator are routed back to the tank to re-condense, or they may travel through the regular vent line to the canister. Continuously venting these vapors back to the fuel tank instead of allowing free travel to the carburetor prevents engine surging from an over-rich fuel-air mixture.

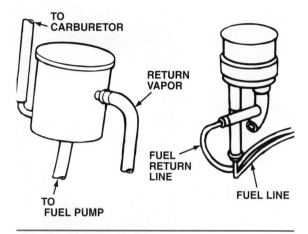

Figure 4-25. Fuel return vapor separators used in some Ford EVAP systems.

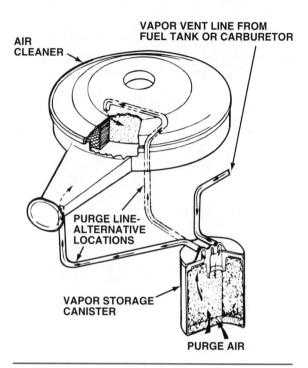

Figure 4-26. Purging the vapor storage canister can be done either through the air cleaner, carburetor, or the throttle body.

Vapor Purging

During engine operation, stored vapors are drawn from the canister into the engine through a hose connected to either the carburetor base, the throttle body, or the air cleaner, figure 4-26. This "purging" process mixes HC vapors from the canister with the existing air-fuel charge. To compensate for the mixture enrichment, carburetors used with an EVAP system are calibrated to take vapor purging into account. If the purge rate is not properly controlled to maintain the correct air-fuel ratio under varying engine operating conditions, engine hesitation and surging can result.

There are several ways to purge the canister. The purging flow rate and method are determined by two things that the process must accomplish:

- Purge the canister
- Minimize the effect on the air-fuel ratio and driveability

Constant Purge

In a constant-purge system, the purge rate remains fixed, regardless of engine air consumption. Intake manifold vacuum draws vapor from the canister by a "tee" in the PCV line at the carburetor or throttle body. Even though manifold vacuum fluctuates, an **orifice** in the purge line provides a constant flow rate.

Variable Purge

In a variable-purge system, the amount of purge air drawn through the canister is proportional to the amount of fresh air drawn into the engine. In other words, the more air the engine takes in, the more purge air is drawn through the canister. A simple

variable-purge system is shown in figure 4-26, which illustrates the system can draw the purge air through the canister by using either:

- A **pressure drop** across the air filter
- The velocity of the air moving through the air cleaner

In both cases, airflow through the air cleaner varies the air flowing through the canister. The simple variable purge often is combined with a constant purge, figure 4-27.

■ Try This Tube Tool

Next time you have a piece of steel or copper tubing break off inside an engine block or other casting, try this simple homemade tool to push it out. Use a wrench socket or bushing, a washer, and a self-tapping screw. Place the socket or bushing—with an inside diameter larger than the broken tubing—over the hole in the block. Then place a large, flat washer on the self-tapping screw and insert it through the bushing or socket into the broken tubing. As you tighten the screw, it will pull the tubing out of the hole.

At idle and low engine speeds, spring force inside the purge valve holds it closed. As the throt-

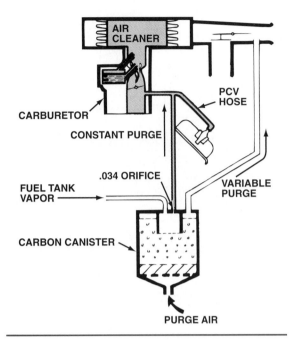

Figure 4-27. In this EVAP system, a variable-purge hose runs from the canister to the air cleaner, and a constant-purge hose runs to the intake manifold.

tle valve moves beyond the vacuum port, vacuum is applied to the purge valve diaphragm, causing the valve to lift off its seat and start the canister purge process.

A purge port may also be used with the constant-purge system. This port is located above the high side of the throttle plate so that there is no purge flow at idle, but the flow increases as the throttle opens.

Computer-controlled Purge

Canister purging on engines with electronic fuel management systems is regulated by the powertrain control module (PCM). Control of this function is particularly important because the additional fuel vapors sent through the purge line can upset the air-fuel ratio provided by a feedback carburetor or fuel injection system. Since air-fuel ratio adjustments are made many times per second, it is critical that vapor purging is controlled just as precisely. This is done by a microprocessor-controlled vacuum solenoid mounted on top of the canister, figure 4-28 and one or more purge valves. Under normal conditions, most engine control systems permit purging only during closed-loop operation at cruising speeds. During other engine operation conditions, such as open-loop mode, idle, deceleration, or wide-open throttle, the PCM prevents canister purging.

Modern fuel systems, which require tighter control of emissions, are likely to use an electron-

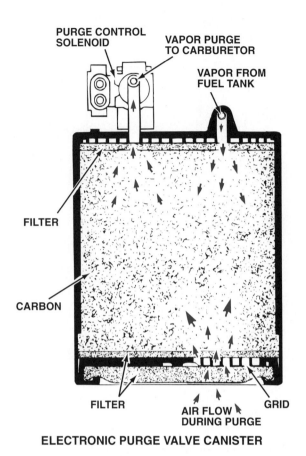

ELECTRONIC PURGE VALVE CANISTER

Figure 4-28. In a computer-controlled purging system, the microprocessor controls purge vacuum with a solenoid.

ically controlled canister purge solenoid that operates on a pulse-width modulated basis. Following a sophisticated strategy stored in the PCM, specific conditions must be met before the PCM enables the pulse-width modulated valve to operate.

All manufacturers use variations of the basic EVAP system described in this section. The system configuration, components, and locations vary according to the specific fuel system and engine, but all function according to the general principles discussed in this chapter.

Non-Enhanced Versus Enhanced Evaporative Systems

Prior to 1996 evaporative systems were referred to as non-enhanced systems. This term refers to evaporative systems that had limited diagnostic capabilities. While they are often PCM controlled, their diagnostic capability is usually limited to their ability to detect if purge has occurred. Many systems have a diagnostic switch that

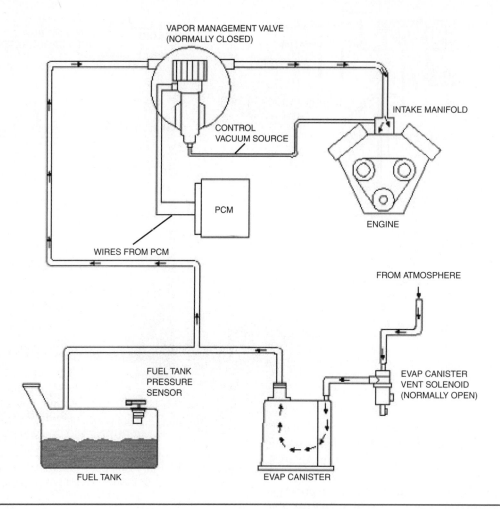

Figure 4-29. An enhanced evaporative system. (Courtesy of Ford Motor Company)

could sense if purge is occurring and set a code if no purge is detected, figure 4-29. On some vehicles the PCM also has the capability of monitoring the integrity of the purge solenoid and circuit. These systems limitations are their ability to check the integrity of the evaporative system on the vehicle. They could not detect leaks or missing or loose gas caps that could lead to excessive evaporative emissions from the vehicle.

Beginning in 1996 with OBD II vehicles the manufacturers were required to install enhanced evaporative systems. These systems have to be able to detect both purge flow and evaporative system leakage. The systems on models produced between 1996 and 2000 have to be able to detect a leak as small as would be created by a hole in the system of .040 inch diameter. Beginning in the model year 2000 the enhanced systems started a phase-in of .020 inch diameter leak detection. Thus, all vehicles built after 1995 have enhanced evaporative systems with the ability to detect

purge flow and system leakage. If either of these two functions fails, the system is required to set a diagnostic trouble code and turn on the MIL light to warn the driver of the failure.

Onboard Diagnostics (OBD)

In the early 1980s, onboard diagnostics (OBD) were introduced by GM to allow technicians to find faults in the engine management system more easily. In 1988, a second generation of onboard diagnostics (OBD II) was mandated for all 1996 and newer vehicles sold in the United States. Some OBD II compliant vehicles started appearing as early as 1994.

There are two major differences between OBD and OBD II. Most importantly, OBD II computer programs run faster for more accurate real-time testing. Also, the software used by the PCM not only detects faults, but also *periodically tests various systems* and alerts the driver before emissions-related components are harmed by system faults. Serious faults cause a blinking Malfunction Indicator Lamp (MIL) or even an engine shutdown; less serious faults may simply store a code but not illuminate the MIL.

The OBD II requirements did not radically affect fuel system design. However, one new component, a fuel evaporation canister purge line pressure sensor, was added for monitoring purge line pressure during tests. The OBD II requirements state that vehicle fuel systems are to be routinely tested *while underway* by the PCM's management system.

All OBD II vehicles—during normal drive cycles and under specific conditions—experience a canister purge system pressure test, as commanded by the PCM. While the vehicle is being driven, the vapor line between the canister and the purge valve is monitored for pressure changes. When the canister purge solenoid is open, the line should be under a vacuum since vapors must be drawn from the canister into the intake system. On the other hand, when the purge solenoid is closed, there should be no vacuum in the line. The pressure sensor detects if a vacuum is present or not, and the information is compared to the command given to the solenoid. If, during the canister purge cycle, no vacuum exists in the canister purge line, a code is set indicating a possible fault, which could be caused by an inoperative or clogged solenoid or a blocked or leaking canister purge fuel line. Likewise, if vacuum exists when no command for purge is given, a stuck open solenoid is evident, and a code is set.

In some states, a periodic inspection and test of the fuel system are mandated along with a dynamometer test. The emissions inspection includes tests on the vehicle before and during the dynamometer test. Before the running test, the fuel tank and cap, fuel lines, canister, and other fuel system components must be inspected and tested to ensure that they are not leaking gasoline vapors into the atmosphere.

First, the fuel tank cap is tested to ensure that it is sealing properly and holds pressure within specs. Next, the cap is installed on the vehicle, and using a special adapter, the EVAP system is pressurized to approximately 0.5 psi and monitored for two minutes. Pressure in the tank and lines should not drop below approximately 0.3 psi.

If the cap or system leaks, hydrocarbon emissions are likely being released, and the vehicle fails the test. If the systems leaks, an ultrasonic leak detector may be used to find the leak.

■ **Why Vapor Lock?**

When gasoline vapors form in the fuel system, vapor lock occurs. This is the partial or complete stoppage of fuel flow to the engine. Partial vapor lock leans the air-fuel mixture and reduces both the top speed and the power of an engine. Complete vapor lock causes the engine to stall and makes restarting impossible until the fuel system has cooled.

Four factors usually cause vapor lock:

- Excessive gasoline temperature
- Lack of sufficient pressure in the fuel system
- Vapor-forming characteristics of a particular gasoline
- The fuel system's inability to minimize vapors

Vapor may form anywhere in the fuel system, but the critical temperature point is at the fuel pump where a vacuum is formed on the inlet side. Engineers have improved fuel pumps and fuel systems to make modern vehicles unlikely to have vapor lock. Oil companies have succeeded in reducing the vaporizing tendencies of gasoline by adjusting its volatility according to seasonal requirements. But vapor lock may still occur in older models during long periods of idle (such as in heavy rush-hour traffic) or when the fuel system is not functioning properly. Periodically inspect the fuel system and correct all air leaks, leaking check valves, or the fuel pressure regulator to prevent vapor lock.

Finally, with the engine warmed up and running at a moderate speed, the canister purge line is tested for adequate flow using a special flow meter inserted into the system. In one example, if the flow from the canister to the intake system when the system is activated is at least one liter per minute, then the vehicle passes the canister purge test.

PUMP OPERATION OVERVIEW

The fuel pump and fuel lines have the job of delivering liquid gasoline from the tank to the carburetor or the fuel-injection system. In most carbureted systems, the fuel pump moves the fuel with a mechanical action that creates a low-pressure or suction area at the pump inlet. This causes the higher atmospheric pressure in the fuel tank to force fuel to the pump. In a mechanical pump, the pump spring applies force on the fuel pump diaphragm within the pump, and delivers fuel under pressure to the carburetor in a reciprocating "push-pull" motion.

On a fuel-injected system, an electrical fuel pump uses a set of rollers, vanes, or a turbine to force fuel upstream to an injector-equipped throttle body or multipoint fuel-injection rail.

Output pressure and volume are two measurements of a fuel pump's performance. When an output pressure or volume is specified for mechanical pumps, it represents the output from the pump at a constant speed. The mechanical fuel pumps, used with carbureted engines, range from 1.75 to 8 psi (12 to 55 kPa) of pressure. Although pump operating pressures used with electronic fuel-injection systems can reach as high as 100 psi (689.5 kPa) depending on system design, most electronic fuel-injection (EFI) systems operate in the 35 to 50 psi (241 to 344 kPa) range. These high pressures are provided by high-volume pumps, controlled by fuel-pressure regulators.

PUMP TYPES

While all pumps deliver fuel through mechanical action, they generally are divided into two groups:

- Mechanical—driven by the vehicle's engine
- Electrical—driven by an electric motor or a vibrating **armature**

To understand how pumps work, let us examine them from most simple to most complex, starting with mechanical pumps.

MECHANICAL FUEL PUMPS

The most common type of fuel pump used by both domestic and foreign automobile manufacturers on carbureted engines was the single-action, diaphragm-type mechanical pump, figure 4-30. The rocker arm is driven by an eccentric lobe on the engine's camshaft. On some overhead-cam engines, the eccentric lobe may be on an accessory shaft. The pump makes one stroke with each revolution of the camshaft. The eccentric lobe—often called simply "the eccentric"—may be part of the camshaft or may be a pressed-steel lobe that is bolted to the front of the camshaft, figure 4-31, along with the camshaft drive gear.

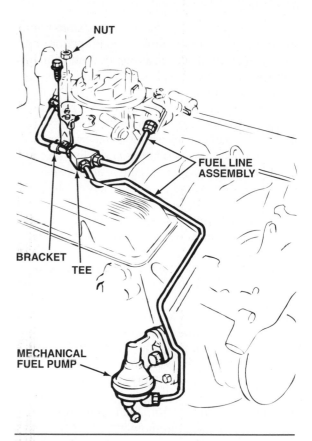

Figure 4-30. Typical mechanical fuel pump and line installation on a carbureted V-8 engine.

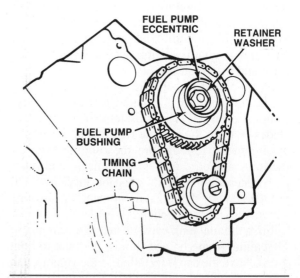

Figure 4-31. This fuel pump eccentric is bolted to the front of the camshaft.

In some applications, the rocker arm is driven directly by the eccentric, figure 4-32. Other engines have a pushrod between the eccentric and the pump rocker arm, figure 4-33. The most common

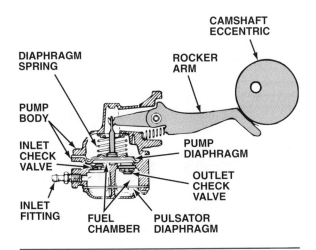

Figure 4-32. Typical diaphragm-type mechanical fuel pump.

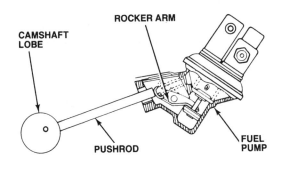

Figure 4-33. Some Chevrolet and Ford engines used a pushrod between the camshaft and pump rocker arm.

example of this arrangement is the small-block Chevrolet V-8 and some Ford 4-cylinder engines.

Mechanical Pump Operation

The fuel pump's intake stroke begins when the rotating camshaft eccentric pushes down on one end of the pump rocker arm. This raises the other end that pulls the diaphragm up, figure 4-32, and compresses the diaphragm spring. Pulling the diaphragm up creates a vacuum, or low-pressure area, in the fuel chamber. The relatively higher atmospheric pressure in the fuel line forces open the inlet check valve and fuel enters the fuel inlet chamber.

As the camshaft eccentric continues to turn, it allows the outside end of the rocker arm to return to its original position by "rocking" back up. Along with the push given by the diaphragm spring, this forces the diaphragm to move back down. This is the start

of the fuel output stroke. As the diaphragm moves down, it causes a pressure buildup in the fuel outlet chamber. Increased pressure closes the inlet check valve and opens the outlet check valve. The fuel flows out of the fuel chamber and into the fuel line on the way to the carburetor. The outlet check valve helps to maintain fuel pressure in the outlet line and prevents fuel from flowing back into the pump. Fuel pump output is judged by the pressure and volume of fuel delivered. Delivery pressure is determined by the strength of the diaphragm spring. Delivery rate or volume is controlled by the float needle and seat in the carburetor fuel bowl and is proportional to engine speed and load. When the carburetor inlet needle valve is open, fuel flows freely from the pump through the lines into the carburetor. When the carburetor fuel bowl is full, the needle valve closes and no fuel flows through the lines.

When the carburetor fuel bowl needle valve is closed, pressure in the fuel line increases, holding the fuel pump diaphragm up, even though the rocker arm continues to move up and down in a "freewheeling" motion. This makes the fuel pump output self-regulating. No fuel is pumped until the fuel level in the carburetor bowl drops enough for the inlet needle valve to open again.

The fuel level in the carburetor bowl varies according to engine operating conditions, so the inlet needle valve's position varies between fully open and fully closed. The needle valve opening and the rate of fuel flowing into the carburetor are proportional to the fuel flow rate out of the carburetor and the volume capacity of the pump.

When an engine with a mechanical fuel pump is shut off, the pump diaphragm spring maintains pressure in the fuel line to the carburetor. If engine compartment heat expands the gasoline in the fuel line too much, the increased fuel line pressure can push the carburetor inlet needle valve open and additional fuel enters the carburetor float bowl. The result is too much fuel in the carburetor bowl, referred to as **percolation.** Excess fuel can also escape the carburetor and go into the vapor storage canister, in most cases saturating it. Also, since fuel evaporates faster when it is hot, it may turn from a liquid into a vapor and cause vapor lock in the pump and/or fuel lines.

Mechanical Pump Applications

Original-equipment mechanical fuel pumps on domestic vehicles are supplied by Carter, AC, or Airtex. Ford used Carter and AC pumps, Chrysler used Carter and Airtex, and GM used AC pumps.

Pumps are so similar in some cases that a manufacturer's production run of the same engine block may use pumps from two different manufacturers. However, replacement pumps must be identical in every respect. Installing a pump that just *looks* like the one removed can result in a broken camshaft or accessory shaft as soon as the engine starts.

■ No One Misses the Good Old Mechanical Fuel Pump

Today's electric impeller-type fuel pump may seem to be a simple and reliable device, but pump manufacturers have worked hard to make it so. Pump designs, capacities, pressures, and performance requirements make the modern pump a rather sophisticated device. This is especially true when you consider that a fuel pump is expected to transport large amounts of gasoline for thousands of miles or kilometers without failure.

Back in the thirties, fuel pump breakdown and replacement were a common occurrence every few thousand miles or several thousand kilometers. The fuel pump of the 1936 Ford V-8 operated from a pushrod. As the pushrod wore, the pump stroke lessened. Most mechanics and a lot of owners kept the fuel pump operating with a wad of chewing gum or tinfoil stuffed into the pushrod cup to compensate for wear. Rather crude, but it worked.

The combination fuel pump/vacuum booster pump was the first big change in pump design. Under wide-open-throttle conditions, when vacuum would drop to nearly zero, no one could keep windshield wipers running at a constant speed. Pump designers provided additional vacuum with a dual pump design. But superhighways, higher horsepower, and emission controls brought new approaches to pump design. Windshield wipers went electric and the vacuum booster fuel pump disappeared. Electric pumps have replaced the traditional mechanical pump design. Automobile manufacturers now build fuel pumps to supply at least 100,000 trouble-free miles (160,000 km).

Mechanical fuel pumps are quite dependable. If they break down, it is usually because of one of these factors:

- A leaking diaphragm
- A worn inlet or outlet check valve
- A worn or broken pushrod
- A worn linkage, which reduces the pump stroke
- Defective seals, causing gasoline contamination of crankcase oil

Occasionally, the camshaft or accessory shaft eccentric may wear enough to reduce the pump stroke, or a bolt-on eccentric may come loose from the camshaft. In these cases, the camshaft or accessory shaft or the bolt-on eccentric must be replaced. It is possible to install an electric fuel pump as a replacement for a defective mechanical pump drive.

ELECTRIC FUEL PUMPS

There are various kinds of electric fuel pumps including:

- Plunger
- Bellows
- Roller cell or vane
- Rotor
- Impeller/turbine

Electric pumps used on carbureted vehicles include the plunger, bellows, and oscillating pump. They were commonly driven by an electromagnet and vibrating armature and were not used as original equipment on fuel-injected vehicles. Earlier fuel-injected vehicles used the roller/vane, rotor style pumps, but modern vehicles used the impeller or turbine pumps, figure 4-34. All electric pumps are driven by a small electric motor, but the turbine pump turns at higher speeds and is quieter than the others.

In the roller cell or vane pump, the **impeller** draws fuel into the pump, then pushes it out through the fuel line to the injection system. Since this type of pump uses no valves to move the fuel, the fuel flows steadily through the pump housing rather than with the **pulsating** motion of other pumps. Since fuel flows through the entire pump, including the electrical portion, the pump stays cool. Usually, only when a vehicle runs out of fuel is there a risk of pump damage. Even though fuel fills the entire pump, no burnable mixture exists inside the pump; there is no air and no danger of commutator brush arcing igniting the fuel.

Most electric fuel pumps are equipped with a fuel outlet check valve that closes to maintain fuel pressure when the pump shuts off. **Residual or rest pressure** prevents vapor lock and hot-start problems on these systems. When servicing these fuel systems, be aware that considerable pressure is held in the fuel line or rail after the vehicle is shut down. Early-model roller pumps used on some import vehicles have a built-in return line to route excess fuel back to the tank. Later models are returnless and simply recirculate excess fuel within the tank.

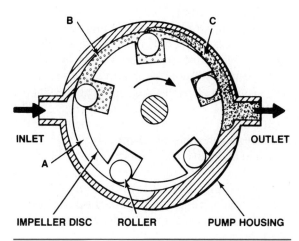

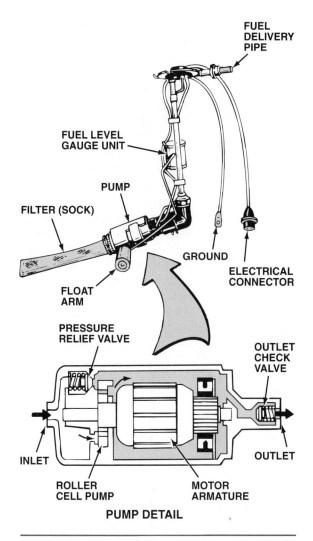

Figure 4-35. The pumping action of an impeller, or rotary vane, pump.

PUMP DETAIL

Figure 4-34. A roller cell–type electric fuel pump.

Figure 4-35 shows the pumping action of a rotary vane pump. The pump consists of a central impeller disk, several rollers or vanes that ride in notches in the impeller, and a pump housing that is offset from the impeller centerline. The impeller is mounted on the end of the motor armature and spins whenever the motor is running. The rollers are free to slide in and out within the notches in the impeller to maintain sealing contact. Unpressurized fuel enters the pump and fills the spaces between the rollers, figure 4-35. As the impeller rotates, a portion of the fuel is trapped between the impeller, the housing, and two rollers, figure 4-35. Further rotation toward the offset side of the housing pressurizes the fuel and forces it out of the pump, figure 4-35.

Electric Pump Location

The electric fuel pump is a pusher unit. Unlike the mechanical fuel pump on carbureted engines that

pulls fuel from the tank, the electric pump pushes the fuel through the supply line. Because it does not rely on the engine camshaft for power, an electric pump can be mounted in the fuel line anywhere on the vehicle—often inside the fuel tank.

Pusher pumps are most efficient when they are mounted in or near as possible to the fuel tank and at or below its level, so they can use gravity to transfer fuel from the tank to the pump inlet. This pump mounting also eliminates the problem of vapor lock under all but the most severe conditions. With the pump mounted in the tank, the entire fuel supply line to the engine can be pressurized. Because the fuel when pressurized has a higher boiling point, it is unlikely that vapor will form to interfere with fuel flow. Having the pump close to or inside the tank also allows the pump to remain cooler because it is away from engine heat, so it is less likely to overheat during hot weather.

In-tank electric fuel pumps generally are combined with the fuel gauge sending unit, figure 4-36, to form a single assembly. Some have two electric fuel pumps:

- A low-pressure, in-tank impeller pump
- A high-pressure, chassis-mounted impeller pump

The low-pressure in-tank pump provides fuel to the high-pressure pump to prevent vapor lock on the suction side of the fuel system. In the two-pump system, the in-tank pump sometimes is called a booster pump. The high-pressure pump provides the injection system with more than adequate fuel pressure, although, with any pump, the higher the pressure, generally, the noisier the pump tends to be. For this reason, fuel pump mounting is critical to prevent the vehicle chassis from picking up and amplifying either mechanical or hydraulic fuel line noise.

Generally, fuel pumps located outside the fuel tank are placed before the fuel filter. While this may seem contrary to the pump's welfare, engineers figure it is better that a pump not be starved for fuel in the event of a filter being clogged. The sock filter in the fuel tank catches the coarse debris before it reaches the pump.

Water and rust are the worst enemies of the fuel pump. When the system is at rest, water in the fuel will settle in the low spots of the fuel system. In the event of a fuel pump failure, check for rust in the system. Rust is a sure sign that water is, or has been, present.

Modular Fuel Sender Assembly

General Motors introduced the modular fuel sender assembly, figure 4-36, on some 1992 models in an effort to standardize fuel sender design. The modular fuel sender consists of a replaceable fuel-level sensor, a roller vane pump, and a fuel strainer. The

reservoir housing is attached to the cover containing fuel pipes and the electrical connector. Fuel is transferred from the pump to the fuel pipe through a convoluted (flexible) fuel pipe. The convoluted fuel pipe eliminates the need for rubber hoses, nylon pipes, and clamps. The reservoir dampens fuel slosh to maintain a constant fuel level available to the roller vane pump; it also reduces noise.

Some roller vane pumps have a two-stage pumping action with a low-pressure turbine section and a high-pressure roller vane section. The pump motor and end cap complete the pump assembly. The turbine impeller has a staggered blade design to minimize pump harmonic noise and to separate vapor from the liquid fuel. The roller vane section creates the high pressure required for fuel injection. The end cap assembly contains a pressure relief valve and a radio frequency interference (RFI) suppression module. The check valve generally used with roller vane pump designs is located in the upper fuel pipe connector assembly, figure 4-37.

After it passes through the strainer, fuel is drawn into the lower housing inlet port by the impeller, which sends it to the first stage of the roller

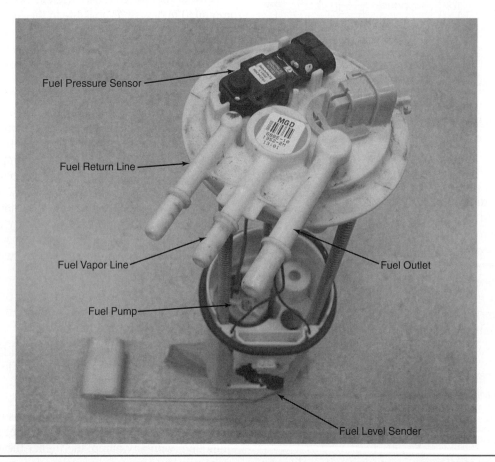

Figure 4-36. A modular fuel sending unit contains the pump, fuel level sensor, fuel pressure sensor, and fuel strainer in one package.

vane pump, figure 4-37. Here, it is pressurized and delivered to the convoluted fuel tube for transfer through a check valve into the fuel feed pipe. Some of the flow, however, is returned to the jet pump for recirculation. Excess fuel is returned to the reservoir through one of the three hollow support pipes. The hot fuel quickly mixes with the cooler fuel in the reservoir and this minimizes the possibility of vapor lock.

Electric Pump Control Circuits

Most original-equipment electric fuel pumps of the 1970s were controlled by an oil pressure switch. This switch opens the electric circuit to the pump motor when the engine is off and controls the operation of the pump when the engine is started and while it is running.

The oil pressure switch has two sets of contacts. One is normally closed, allowing battery voltage to flow from the starter solenoid or relay to the fuel pump. The second set of contacts is normally open until oil pressure builds up enough to close them. Turning the ignition key to Start energizes the pump through the normally closed contacts, figure 4-38. Once the engine is running, the pump receives power through a sec-

ond set of contacts that are closed by oil pressure, figure 4-39.

More recent fuel pump circuits feature an electrical fuel pump, controlled by the fuel pump relay. Fuel pump relays are activated initially by turning the ignition key to ON; this allows the

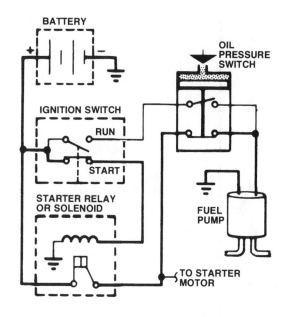

Figure 4-38. During cranking, the fuel pump receives current through the starter relay (solenoid) and the normally closed contacts of the oil pressure switch.

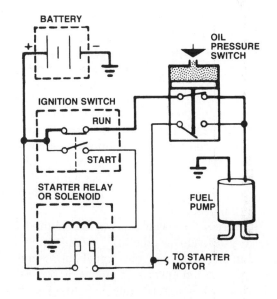

Figure 4-39. When the engine starts, the oil pressure opens one set of switch contacts and closes another. Current then flows through the ignition switch, through the oil pressure switch, and through the fuel pump—as long as the oil pressure remains above a minimum level.

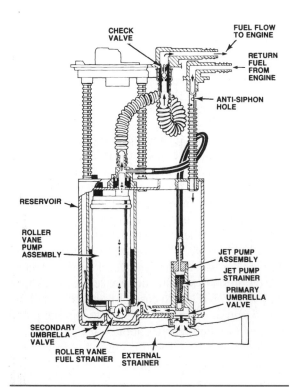

Figure 4-37. Fuel flow through the modular fuel sender assembly. (Courtesy of General Motors Corporation)

FUEL PUMP RELAY CIRCUIT

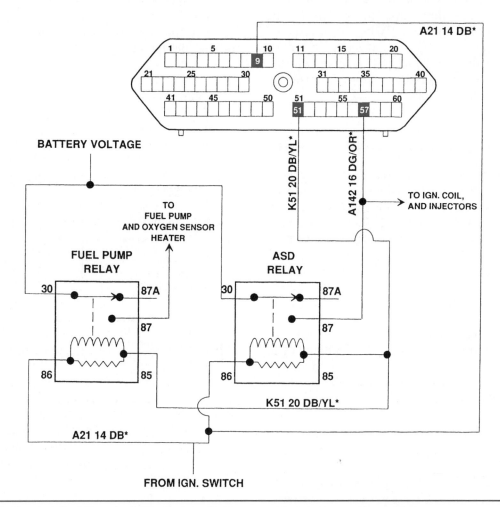

Figure 4-40. A typical Chrysler fuel pump circuit. (Courtesy of DaimlerChrysler Corporation)

pump to pressurize the fuel system. As a safety precaution, the relay de-energizes after a few seconds until the key is moved to the Crank position. On some systems, once an ignition coil signal, or "tach" signal, is received by the engine control computer indicating the engine is rotating, the relay remains energized even with the key released to the Run position.

Here are some specific examples, figure 4-40: On Chrysler vehicles, the logic module must receive an engine speed (rpm) signal during cranking before it can energize a circuit driver inside the power module to activate an automatic shutdown (ASD) relay to power the fuel pump, ignition coil, and injectors. As a safety precaution, if the rpm signal to the logic module is interrupted, the logic module signals the power module to de-

activate the ASD, turning off the pump, coil, and injectors. In some vehicles, the oil pressure switch circuit may be used as a safety circuit to activate the pump in the ignition switch run position.

GM systems energize the pump with the ignition switch to initially pressurize the fuel lines, but then deactivate the pump if an rpm signal is not received within one to two seconds. The pump is reactivated as soon as engine cranking is detected. The oil pressure sending unit serves as a backup to the fuel pump relay. In case of pump relay failure, the oil pressure switch will operate the fuel pump once oil pressure reaches about 4 psi (28 kPa).

Most Fords with fuel injection have an inertia switch in the trunk, figure 4-41, between the fuel pump relay and fuel pump. When the ignition switch is turned to the ON position, the electronic

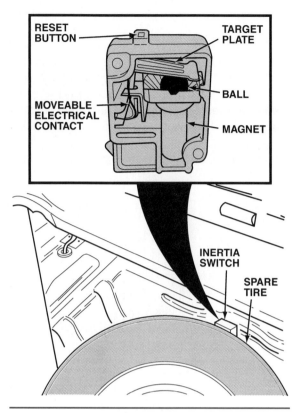

Figure 4-41. Depressing a button can reset the inertia switch found on many Ford vehicles. (Courtesy of Ford Motor Company)

engine control (EEC) power relay energizes, providing current to the fuel pump relay and a timing circuit in the EEC module. If the ignition key is not turned to the Start position within about one second, the timing circuit opens the ground circuit to de-energize the fuel pump relay and shut down the pump. This circuit is designed to pre-pressurize the system. Once the key is turned to the Start position, power to the pump is sent through the relay and inertia switch.

The inertia switch opens under a specified impact, such as a collision. When the switch opens, current to the pump shuts off because the fuel pump relay will not energize. The switch must be reset manually by opening the trunk and depressing the reset button before current flow to the pump can be restored.

Vehicles that use Bosch port fuel-injection systems with airflow sensors use fuel pump relays activated by a microswitch in the system's vane airflow sensor. That design was followed by the more standard practice of relying on the engine speed signal during engine startup, and while the engine is running to keep the relay energized. In the event of a crash, the loss of the rpm reference

signal would open the relay and the pump would shut off immediately. Thus, a fire would not be fed by continued fuel delivery.

Modern vehicles use fuel-pump circuits that are controlled by the engine computer. Although the fuel pump relay's power circuit has ignition power when the ignition is on, the relay ground path is computer controlled depending on a momentary "key on," or a engine speed signal for the computer to keep the relay activated. On turbocharged vehicles, there may be two separate circuits to the pump—one for low output, another for full output under boost conditions. This tends to prolong pump life and limit noise while under cruise conditions.

With modern vehicles using returnless fuel systems, higher fuel pressures are regulated to avoid vapor lock and hot-start problems. With higher pressures come louder noises generated as pressure pulses travel from the fuel pump through the fuel lines and rail. Some manufacturers use an accumulator in the system to reduce pressure pulses and noise. Another way to help reduce noise, current draw, and pump wear is to reduce the speed of the pump when less than maximum output is required. Pump speed and pressure can be regulated by controlling the voltage supplied to the pump with a resistor switched into the circuit, or by letting the engine-control computer pulse width modulate (PWM) the voltage supply to the pump, through a separate fuel pump driver electronic module. With slower pump speed and pressure, less noise is produced.

FUEL FILTERS

Despite the care generally taken in refining, storing, and delivering gasoline, some impurities get into the automotive fuel system. Fuel filters remove dirt, rust, water, and other contamination from the gasoline before it can reach the carburetor or fuel injectors.

The useful life of all filters is limited, although Ford specifies that its filters used with some fuel-injection systems should last the life of the vehicle. If fuel filters are not cleaned or replaced according to the manufacturer's recommendations, they can become clogged and restrict fuel flow.

In addition to using several different types of fuel filters, a single fuel system may contain two or more filters. Automobile manufacturers locate these filters in several places.

■ Buzzzzz, Click, Smack

A rotary electric fuel pump is virtually standard equipment on late-model fuel-injected vehicles. The roller vane or roller cell pump delivers the steady volume and pressure needed for reliable injection operation.

The other general type of electric fuel pump—the vibrating-armature pump—has always been a popular aftermarket item for high-performance engines. Many also were used on older imported cars. Vibrating-armature pumps make a distinctive buzzing and clicking sound as they pressurize the fuel lines when the ignition key is turned on.

The vibrating-armature pump operates a plunger, a bellows, or a diaphragm in pulses. The action is similar to that of a mechanical pump. An armature within an electromagnetic coil operates the bellows, plunger, or diaphragm. Current is applied to the coil to move the armature downward and develop pump suction. This opens an inlet check valve, similar to the action in a mechanical pump.

When the armature nears the bottom of its stroke, it opens an electrical contact that de-energizes the coil. A spring then forces the plunger, diaphragm, or bellows upward to force fuel through the outlet check valve.

Fuel output is regulated as it is in a mechanical pump. As pressure rises in the outlet line, it overcomes spring force and keeps the plunger, diaphragm, or bellows from moving on an output stroke.

The electrical contacts on some older pumps of this kind would pit and stick together from the continuous arcing as they operated. Many owners of old British sports cars wondered why the small ball-peen hammer was included in their car's tool kit. They wondered, that is, until the first time they had to beat on an electric fuel pump to free the stuck contacts so they could drive home.

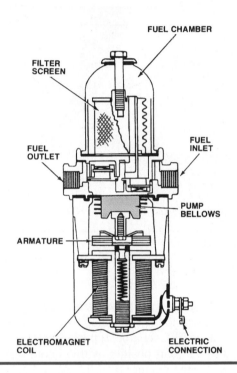

Fuel Tank Filters and Strainers

A sleeve-type filter of woven saran is usually fitted to the end of the fuel pickup tube inside the fuel tank, figure 4-42. This filter "sock" prevents sediment, which has settled at the bottom of the tank, from entering the fuel line. It also protects against water contamination by plugging itself up. If enough water enters the fuel tank, it accumulates on the outside of the filter and forms a jelly-like mass. If this happens, the filter must be replaced. Otherwise, no maintenance is required for this filter.

Inline Filters

The inline filter, figure 4-43, is located in the line between the fuel pump and the carburetor, throttle body, or fuel rail. This filter protects the system from contamination, but does not protect the fuel pump. The inline filter usually is a metal container with a pleated paper element sealed inside. Carbureted systems may use filters with a plastic housing.

Some fuel-injection systems use inline filter canisters. These are larger units approximately the size of a soda can, figure 4-44. They may be mounted on a bracket on the fender panel, a shock tower, or another convenient place in the engine compartment. They may also be installed under the vehicle near the fuel tank.

Inline filters must be installed so that gasoline flows through them in the proper direction, as indicated by an arrow on the filter. If an inline filter is installed backwards, it restricts fuel delivery.

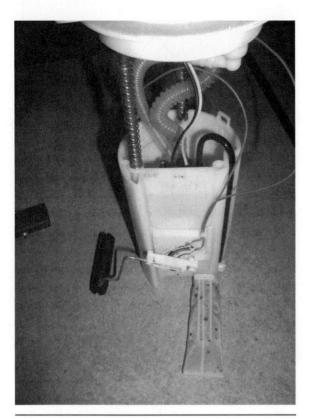

Figure 4-42. Almost all fuel systems have a filter or strainer attached to the pickup tube inside the tank. This one is part of a modular fuel assembly.

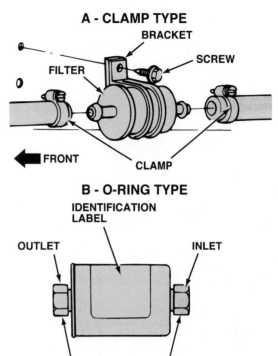

Figure 4-43. Inline fuel filters may be attached to the fuel line with screw clamps or threaded connections. They must be installed in the proper direction.

Types of Fuel Filters

Older filter elements were made of sintered metal. They could be washed and reused. Modern filters are generally made of plastic or metal, with pleated paper filter elements inside, and are disposable. Filters come in a wide variety of shapes, sizes, and capacities and may have a number of inlet and outlet fittings. When replacing a filter, always make certain that the correct replacement is used and properly installed.

Applications

Ford and GM equipped most of their carbureted engines with carburetor fuel inlet filters. Some Motorcraft, Holley, and Rochester carburetors used a throwaway pleated paper element, figure 4-45.

Most fuel-injected vehicles have an inline fuel filter mounted somewhere between the fuel tank and the engine. These filters may be near the gas

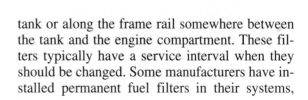

Figure 4-44. Fuel-injection systems use large-capacity inline fuel filters.

tank or along the frame rail somewhere between the tank and the engine compartment. These filters typically have a service interval when they should be changed. Some manufacturers have installed permanent fuel filters in their systems,

Figure 4-45. Later-model Rochester carburetors use a pleated paper inlet filter with a check valve.

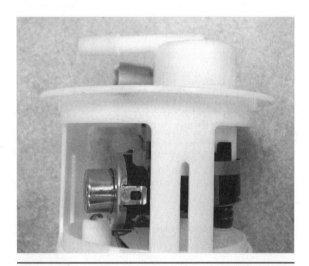

Figure 4-46. The fuel filter is located in the top of the fuel modular unit. It cannot be serviced separately. If the filter is restricted, the modular sending unit must be replaced.

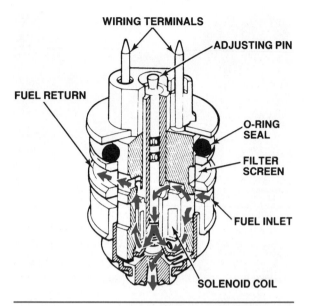

Figure 4-47. Injectors used in throttle-body units have one or more filter screens. (Courtesy of Daimler-Chrysler Corporation)

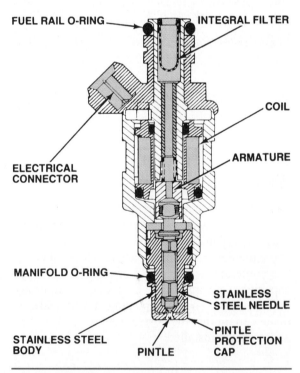

Figure 4-48. Port fuel injectors generally use an integral filter screen. (Courtesy of Ford Motor Company)

such as the one shown in figure 4-46. The filter is located in the fuel assembly module and cannot be serviced separately.

Fuel Injection and Filters

With fuel injection, proper filtering of gasoline is essential, because particles larger than ten **microns** (0.000039 in.) can interfere with the close tolerances in the fuel injectors. Thus, fuel-injection systems tend to use various types of larger-capacity fuel filters:

- The fuel tank filter removes particles larger than 50 microns (0.00197 in.) in size.
- A large-capacity inline filter, figure 4-44, removes particles greater than 10 to 20 microns (0.00039 to 0.00079 in.) in size.

All injectors, throttle body or port, are fitted with one or more filter screens or strainers to remove any particles (generally 10 microns or 0.00039 in.) that might have passed through the other filters. These screens, which surround the fuel inlet, are on the side of throttle-body injectors, figure 4-47, and internal to port injectors, figure 4-48.

SUMMARY

Automotive fuel tanks may be mounted either horizontally or vertically, depending on vehicle configuration. Filler tubes may serve as fill limiters as well as allowing the tank to be filled. Vehicles using unleaded gasoline require a smaller nozzle to supply fuel to the tank in order to prevent use of leaded fuel.

Tanks must be vented, but fuel vapors must not be allowed to escape into the atmosphere. Fuel system vapors are stored in a canister for burning once the EVAP system enables the canister to be purged. Provisions for fuel system shutdown in the event of a crash or rollover are provided as part of the fuel system. Computer-controlled engine management systems use vacuum or electrical canister purge valves to control the purge function.

Older fuel systems often used return lines to route fuel back to the tank and carry away heat, but more modern designs use returnless systems and higher rail pressures to prevent vapor lock. OBD II–equipped vehicles perform computer-controlled canister purge systems tests during the normal drive cycle. Fuel systems on these vehicles are tested in the service bay for their ability to hold pressure and prevent fuel vapor leakage into the atmosphere.

Fuel pumps move fuel from the tank to the carburetor or fuel-injection system. Mechanical pumps use a spring-loaded diaphragm and check valves; they are driven by the engine camshaft or eccentric on the front of the camshaft. All fuel-injection vehicles today use an electric fuel pump to pressurize the fuel, the most common type being the impellor type. Electric fuel pumps are generally relay or electronically controlled. The fuel pump relay is controlled by an oil pressure switch, an airflow sensor microswitch, or more commonly, by the engine speed signal.

Filters come in a variety of types, shapes, and sizes. They are generally of the throwaway type.

Review Questions

Choose the letter that represents the best possible answer to the following questions:

1. Technician A says that a fuel tank filler tube may be one piece welded to the tank. Technician B says a fuel tank filler tube may be two pieces with a hose between them. Who is correct?
   a. A only
   b. B only
   c. Both A and B
   d. Neither A nor B

2. In order to prevent any unwanted substance from being picked up by the fuel pickup tube, it is equipped with:
   a. A water separator
   b. A filter sock
   c. Both A and B
   d. Neither A nor B

3. If the fuel tank is not vented properly, fuel delivery may be prevented due to:
   a. Vapor lock
   b. Vacuum lock
   c. Pressure lock
   d. None of the above

4. High-pressure fuel delivery lines should be equipped only with what type of clamps?
   a. Worm
   b. Keystone
   c. Rolled edge
   d. Corbin

5. EVAP systems were mandated on all cars beginning in the:
   a. 1960s
   b. 1970s
   c. 1980s
   d. 1990s

6. The vapor canister is used to:
   a. Collect fuel vapors from the tank
   b. Collect fuel vapors from the carburetor float bowl
   c. Allow fuel vapors to be burned in the engine
   d. All of the above

7. Vapor lock in the fuel lines can be minimized by:
   a. Minimizing the amount of suction from the tank and the pump
   b. Using rest or static pressure
   c. Keeping the fuel in the fuel lines cool
   d. All of the above

8. Electric fuel pumps for fuel-injected engines:
   a. Are generally located near the engine
   b. Are usually of the roller, vane, or impeller type
   c. Are more tolerant of water and rust than the diaphragm type
   d. Are prone to failure from voltage spikes

9. Electric fuel pumps on late-model vehicles use safety circuits so that:
   a. In the event of a crash, fuel does not feed a fire
   b. Too much fuel does not flood the engine
   c. The pump life is extended
   d. All of the above

10. With electronically managed fuel systems, canister purging may be:
    a. Controlled by the PCM
    b. Activated by a vacuum solenoid valve
    c. Activated by a pulse-width modulated electrical solenoid
    d. All of the above

11. Late-model Ford vehicles equipped with electronic engine controls and fuel injection disable the fuel pump circuit in the event of a crash by using:
    a. An airflow-controlled microswitch
    b. An inertia switch
    c. A spring-loaded ball check valve
    d. None of the above

12. Technician A says that carbureted vehicles require a fuel delivery pressure of about 10 to 15 psi. Technician B says some fuel systems operate at pressure of 75 to 100 psi. Who is correct?
    a. A only
    b. B only
    c. Both A and B
    d. Neither A nor B

13. Vapor lock may be caused by:
    a. A loss of fuel pressure in the delivery system
    b. A leaking check valve or pressure regulator
    c. Excessive underhood temperatures
    d. All of the above

14. OBD II–equipped vehicles periodically run a test of the EVAP system while underway to make sure that:
    a. The purge valve is working
    b. The bowl vent valve is working
    c. The rollover valve is working
    d. None of the above

15. An in-shop inspection of the fuel system of OBD II–equipped vehicles may include:
    a. Testing of the fuel system to make sure that it holds pressure
    b. Testing of the fuel cap
    c. Flow testing of the canister purge system
    d. All of the above

16. The fuel rail assembly includes which components?
    a. Fuel inlet
    b. Pressure regulator
    c. Fittings for injectors
    d. All of the above

17. Returnless fuel systems were developed mainly to:
    a. Save manufacturing cost of fuel lines
    b. Help prevent excessive fuel tank vapors
    c. Help prevent vacuum lock
    d. All of the above

18. The fuel pump in an electronic engine control management system will run for one or two seconds after the ignition is turned on and then shut down unless the computer:
    a. Receives a key-on signal
    b. Receives an engine speed signal
    c. Receives a fuel pump relay signal
    d. Receives a coolant temperature signal

19. Inline fuel filters used with electric fuel pumps on fuel-injected engines are located:
    a. In the tank
    b. Between the tank and the fuel pump
    c. Between the fuel pump and the throttle body or fuel rail
    d. At the highest point in the fuel system

20. Fuel filtration may occur at the:
    a. Tank
    b. Inline filter
    c. Injectors
    d. All of the above

5

Engine Control Systems

OBJECTIVES

After completion and review of this chapter, you will be able to:

- Define *voltage, current,* and *resistance.*
- Have knowledge of the four basic computer functions.
- Explain the different types of memories that are used in computers.
- List the four standard communication protocols used on OBD II vehicles.
- Have knowledge of the differences between Class A, B, and C networks.
- Explain the different data line configurations.

KEY TERMS

actuator
alternating current (AC)
ampere
analog
analog-to-digital
binary
CAN
CARB
central processing
  unit (CPU)
circuit
conductors
conventional theory
  of current flow
digital
digital-to-analog
direct current (DC)
DLC
electromotive
  force (EMF)
electron theory of
  current flow
ground path
infinite resistance

input conditioning
keep-alive
  memory (KAM)
multiplexing
ohm
open circuit
parallel circuit
potential
programmable read-
  only memory
  (PROM)
random-access
  memory (RAM)
read-only
  memory (ROM)
rectified
resistance
series circuit
series-parallel circuit
UART
voltage
voltage drop
voltage spike
volts

INTRODUCTION

The arrival of electronic fuel metering resulted in a gradual modification of the traditional carburetor and how it works. By the 1980s, electronic feedback carburetors were rapidly replaced by fuel-injection systems, which were also electronically controlled. Today, all new cars and trucks use computer-controlled electronic fuel injection.

In this chapter, we cover the basic operating principles of an electronic fuel-injection metering system, how the components work, and how they interact. In later chapters we will continue our study of electronic engine management, dealing with electronic engine controls, variations on fuel-injection systems, supercharging, and turbocharging.

ELECTRONIC CONTROL SYSTEMS

To meet stringent emission control requirements in the early 1970s, automotive engineers began to apply electronic controls to basic automotive systems. The use of electronics was first applied to ignition timing and later to fuel management. Electronic control systems introduced a degree of precision that earlier electromechanical and vacuum-operated systems could not achieve. Fuel delivery and ignition timing had to be more precisely managed to meet ever-changing engine load and speed requirements. With electronic controls came a significant decrease in emission levels, major improvements in driveability, and increased reliability of the systems.

In an attempt to meet ever-tightening tailpipe emission requirements imposed by federal and state governments, high-energy electronic ignition systems, such as GM's High Energy Ignition (HEI), were the first to appear. Electronically controlled fuel metering systems were introduced as well, notably on the Volkswagen in the late 1960s. Actually, both so-called electronically pulsed (controlled "on-time") and "continuous" (always spraying fuel) injection systems were introduced. Most vehicles were, and are, equipped with electronically pulsed systems.

It did not take long before electronic ignition and fuel systems were integrated to form the early engine management systems such as the Bosch Motronic System. By the early 1980s, even more automotive systems were under the control of an onboard computer, recognizing that total systems management helps reduce cost and emissions, while improving driveability.

ELECTRICAL REVIEW

A good understanding of basic electricity and how an automotive onboard computer works leads to the understanding of how electronic control systems function. The following section is intended as a *review* of basic electrical principles and terms.

Current

We will begin with a review of basic electrical theory because electronic fuel metering is based on the simple principles of electricity. This may have already been covered in an automotive electrical and electronics class. By reviewing and then moving to a study of electronic engine controls, it becomes clear that the most complex computer systems are based on fundamental laws of science and engineering.

Electricity is nothing more than the flow and control of electrons. Current is the flow of electrons through a circuit, and can be described by the **conventional theory of current flow** or the **electron theory of current flow,** figure 5-1. Either theory may be used with equal accuracy, as long as it is used consistently. When scientists began to make discoveries about electricity, they thought electrons traveled from positive to negative—the conventional theory of current flow. In the past, automobile electrical systems were always described using the conventional theory, as if current traveled from the positive battery terminal to the negative terminal. In the automotive trade, it is still the more common method of describing current flow.

The electron theory of current flow states that current travels from negative to positive. This theory is generally used in electronic communications, computers, and all other areas of the electronics industry. In recent years, however, the electron flow theory has also been used to describe some automotive electronic control systems. Our review of current, voltage, and resistance in this chapter is based on the conventional current theory, unless stated otherwise.

Direct Current Versus Alternating Current
Most, if not all, circuits in the vehicle carry current in only one direction. These circuits carry **direct current,** or **DC.** This means that electrons travel directly from positive to negative (conventional theory). If current could reverse itself for a time and travel in the opposite direction, from negative back to positive, then the circuit would be carrying **alternating current,** or **AC.** Batteries supply DC current, whereas AC generators ("alternators") produce AC current. The AC current is changed (**rectified**) to DC before leaving the generator.

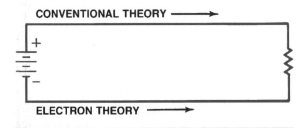

Figure 5-1. Two theories of current flow.

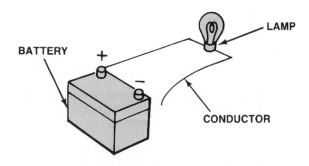

Figure 5-2. An incomplete (open) circuit.

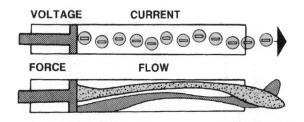

Figure 5-3. Voltage pushes current flow similar to pressure pushing water flow.

Circuits

An electrical current needs a path along which to travel. This path is called a **circuit.** If current is to flow, the path must be complete, forming a complete circuit. Any break in the circuit creates an **open circuit.** Current cannot travel because electrons cannot get through. A technician does not always want electrons to flow. At these times, an intentional break in the circuit is necessary to create an open circuit.

A circuit must have two components, at minimum, for managed current to flow. These components are a battery or other power source, and a load device (lamp, motor, etc.) to create work. **Conductors,** usually wires, connect the two components, battery and load. The load device is often bolted to the vehicle's metal chassis, which serves as a conductor for electrons to travel back to the battery. The vehicle chassis and frame serve as a **ground path.**

In this example, current travels from the positive side (B+ terminal) of the battery, through the wires to the lamp, and back to the negative terminal (B−) of the battery through the chassis and frame (and negative battery cable). If one wire is taken off the battery terminal, figure 5-2 the circuit is open. Current cannot travel and the lamp does not light. Breaking the circuit by removing one wire from its terminal is the same as using a switch to break the circuit.

Amperage

The amount of electrons, or current, is measured in amperes. Current through a conductor is comparable to water flowing through a pipe, figure 5-3. Whereas water flow is measured by counting how many gallons or liters flow past a point within a certain time, electrical current is measured by counting electrons. It takes 6.28 billion electrons passing one point in one second to equal one **ampere,** or amp, of current. Amperage is sometimes referred to as intensity, or "I." Current must

travel through a complete circuit and requires a return path to the source voltage.

Voltage

Current cannot flow unless some force pushes the electrons in one direction through the conductor (metal wire, frame, etc.). This pushing force is called **electromotive force,** also called **EMF** or **E.** EMF is measured in units of **volts** and is sometimes referred to as **voltage,** or **potential.** Voltage can be compared to the pressure that moves water through a pipe, figure 5-3. Voltage also measures a potential difference in force existing between two points. One point may be negatively charged and the other may be positively charged, figure 5-4, such as the B+ and B− terminals of a battery.

The potential difference between two positive (or two negative) voltage points within a circuit may also be measured. Such a difference in potential measured across a resistance is known as **voltage drop.** The relative strength, or loss, of EMF depends on the strength of the charges at each point measured, such as on the two sides (before and after) of a load.

Potential voltage exists even if no current travels. A good example of this is when disconnecting one wire from a battery terminal. Current is dangerous because voltage is present, but not seen

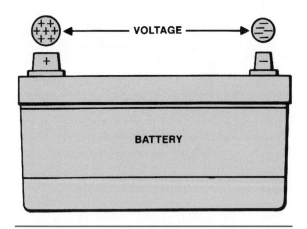

Figure 5-4. Voltage is a potential difference in electromotive force.

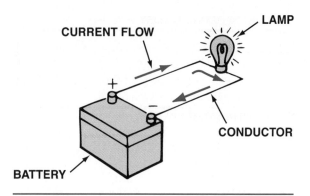

Figure 5-5. A simple circuit.

doing any work, figure 5-4. Laying a wrench across the battery terminals is one way to complete the circuit with devastating results!

Resistance

Voltage is required to force current through a circuit consisting of conductive materials. Yet all conductive materials oppose current to some extent. This opposition is called **resistance.** There are no perfect conductors on the vehicle. This means all conductors connecting the power source to the load devices offer resistance and a resulting loss of EMF (voltage). There must be resistance if current is to be regulated or limited to manageable limits. The resistance of an electrical load device is much more important than that of a conductor. The lamp in figure 5-5 is a load device with much more resistance than the wire conductors.

Both wanted or unwanted resistance may be present in places other than the wires and devices of a circuit. A break in the circuit, such as in figure 5-2, creates **infinite resistance** (too high to measure). The following are causes of unwanted resistance that impedes current: loose or corroded connections, chafed, worn or corroded wiring. The higher the current, the greater the voltage drop.

Resistance is measured in a unit called an **ohm.** Load devices that create work must have resistance in order to perform their work. The loss of force, or voltage, across a load is exchanged for work done by the amps. Amps do the work, not volts. The loss of pressure is the result of resistance. If unwanted resistance is present in a circuit, it causes a drop of voltage where it is not wanted.

Power

Power refers to the amount of actual work done and is measured in watts. Wattage is calculated by multiplying volts times amps. All other factors being the same, a 120-volt lamp drawing 0.5 amp consumes 60 watts of power, just as a 12-volt lamp drawing 5 amps draws 60 watts of power. Both lamps do the same amount of work. Wattage is a useful way to measure and compare work done despite variations in voltage, amperage, and resistance. Most generating devices and lamps come with specifications in watts. Today, even engine power output is specified in watts, along with the traditional measurement in horsepower.

Circuit Components

Before discussing types of electrical circuits, we should mention the components used within typical automotive circuits. Categories of circuit components include power sources, circuit protection devices, control devices, loads, and grounds.

Power Sources

Onboard power sources delivering high current are the battery and the ac generator. Small current generators are also common, such as permanent magnetic generators and oxygen sensors, which are used as input devices.

Circuit Protection Devices

Circuit protection devices prevent excessive current from heating and destroying wiring and components. They are usually found in series after the power source to protect the entire circuit. Protection devices include fuses, circuit breakers, and fusible links.

Fuses may be the AGE all-glass, tubular type used on older American vehicles, the ceramic type used on older European vehicles, or the other rectangular designs seen on Asian vehicles. Today, the "blade" type fuse is common with mini, standard, and maxifuses found in various onboard locations.

Circuit breakers are used in critical circuits and most breakers are self-resetting, such as those built into the headlight switch or used for the wiper motor.

Fusible links are an inexpensive substitute for circuit protection. A fusible link is a piece of wire two sizes smaller than the normal wiring in a circuit, and thus will "burn out" if current draw is excessive. Some fusible links are buried within wiring harnesses and may be difficult to find without consulting a parts locator. When burned through, the fusible link must be cut out of the circuit and a replacement soldered into place. Other fusible links are the "plug in" variety and easily found in the engine compartment.

Control Devices

Control devices turn the flow of electrons on or off in a circuit. They may be placed in a series following the protection device in the circuit. Control devices may also be last in the series as electrons return through "ground" to the battery. So-called "ground-side" switching protects expensive computers from damage in the event of a "short" or "ground."

Types of control devices include the mechanical switch, the relay (an electrically controlled switch), the transistor (an electronically controlled switch), and the diode (a one-way current check valve). Diodes are also used as protection devices in solid-state, transistorized circuits.

Load Devices

Load devices do the work, but by offering resistance they drop voltage in order to put the electrons to work. As a technician troubleshooting with an oscilloscope, it is important to understand how the load devices react during dropping voltage.

Load devices may draw very high current, as much as 80 amps or more, or as low as the milliampere range. Low current devices that draw less than one amp are usually referred to as *electronic devices* rather than electrical devices. High current load devices include the starter motor, blower motor, headlamps, electric rear defroster, air conditioning compressor clutch coil, and other motorized accessories. More normal current loads include the ignition coil, and other electrical devices, interior and marker lighting,

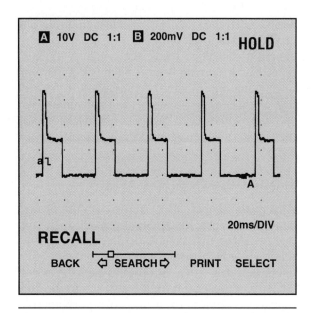

Figure 5-6. A voltage spike is a sudden, sharp, significant rise in voltage.

radio, and some actuators. Relatively low current is drawn by instrument lamps, solid-state control circuits/control modules, and similar instruments.

Besides being categorized by how much current a load draws, load devices are also categorized by their design or operating principle. From a design perspective, there are basically three load types based on how the loads *react* to current: resistive loads, inductive loads, and capacitive loads. To keep it simple, a load's *reactance* describes how current is handled when it meets a load device and what the resulting voltage fluctuations might look like on an oscilloscope.

Resistive loads offer "nonreactive" resistance to the circuit. Even corroded wire, as undesirable as it is, is nonreactive to the current. On a scope, a simple drop of voltage would be observed across a resistive load.

Inductive loads work on the principle of magnetic induction. These loads include motors, coils, relay windings, and electrical solenoids including fuel injectors. Inductive loads do not tolerate changes of current like resistive loads do, and offer "inductive reactance" to current change when the circuit is turned on and off. By design, inductive loads build up magnetism and when current shuts off, the collapsing magnetic field creates a **voltage spike,** a sudden and often extreme rise of voltage in the circuit, figure 5-6. This may be desirable or undesirable, depending on the purpose of the circuit.

Capacitive loads are those that offer capacitive reactance to the flow of current. Capacitors used for radio interference suppression or in ignition circuits are prime examples of this type of load. To oversimplify, capacitive loads allow voltage fluctuations, like ac current, to pass, but block dc current. Older point-type ignition systems relied on a capacitor, or condenser, to suppress and limit arcing and subsequent burning of the ignition contact points as the ignition coil's magnetic field collapsed.

Conductors and Circuits

Any material that easily transports current is a conductor. Where electricity travels through conductive wires, they do no work because they have relatively low resistance. Loads, on the other hand, perform work. In order to perform work, something must offer high resistance to the current. Load devices exchange voltage (potential) for work done. For example, the filament of a bulb or the windings of a motor are both conductors with relatively high resistance.

Resistance of a wire conductor or of a load device may change with varying conditions. The actual resistance of a conductor depends upon five factors:

- **The conductive material**—Any material with few free electrons (rubber, plastic, wood, etc.) is a poor conductor since its resistance to current is high. All conductors offer some resistance, but the resistance of a good conductor is so small that a fraction of a volt allows current to flow.
- **The length of the conductor**—Electrons in motion meet resistance in the conductor. The longer a piece of wire, the farther the electrons must travel, and the greater the collective amount of resistance.
- **The cross-sectional area of the conductor**—The thinner a piece of wire, the higher its resistance.
- **The temperature of the conductor**—In most cases, the higher the temperature of the conducting material, the greater its resistance. That is why most electrical tests must be performed at normal operating temperature for accurate readings.
- **The condition of the conductor**—If a wire is partially cut, it acts almost as if the entire wire is a smaller diameter, offering high resistance at the damaged point. Loose or corroded connections have the same effect. High resistance at connections is a major cause of electrical problems, including burnouts and fires.

Every electrical load in a circuit offers some resistance. Voltage is reduced as it moves the current through each load. Voltage is electrical energy and as it moves current through a load, some of the electrical energy is changed into another form of energy, such as light, heat, or motion. As mentioned earlier, the amount of voltage lost in moving current through each load is called its voltage drop. If you measure the voltage drop at every load in a circuit and add the measurements, they equal the original voltage available. Potential is lost, but the current does not disappear. The resistance converts the voltage into a different form of energy.

The resistance of any electrical part, load or conductor, is measured in three ways:

- Direct measurement with an ohmmeter, which measures the ohms of resistance offered by the load
- Indirect measurement with a voltmeter, which measures the voltage drop through the load
- Indirect measurement with an ammeter, which measures the current through the load.

Basic Circuits

There are three basic circuits:

- Series
- Parallel
- Series-parallel

In a **series circuit,** the current has only one path to follow. Using conventional current flow theory, the current in figure 5-7 must flow from the battery, through the resistor, and back to the battery. The circuit must be continuous. If one wire is disconnected from the battery, the circuit is broken and there is no current. If electrical loads are wired in series, they must all be switched on and working or the circuit is broken.

In a **parallel circuit,** current can follow more than one path to complete the circuit. The points where current paths split and rejoin are called junction points. The separate paths that split and meet at junction points are called branch circuits or shunt circuits. Figure 5-8 shows a parallel circuit.

As the name suggests, a **series-parallel circuit** combines the two types of circuits already discussed. Some of the loads are wired in series, but some loads are wired parallel, figure 5-9. The fuel pump circuit shown in figure 5-10 is a parallel circuit with an inertia cut off switch in wired in series. Most of the circuits in an automobile electrical system are series-parallel.

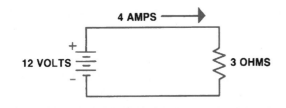

Figure 5-7. A simple series circuit.

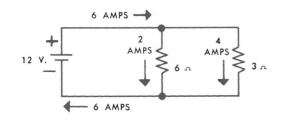

Figure 5-8. A simple parallel circuit.

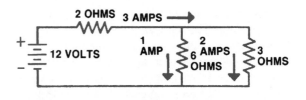

Figure 5-9. A series-parallel circuit.

COMPUTER CONTROL

A computer makes decisions based on the information it receives. It cannot think on its own, but instead uses a detailed set of instructions called a program.

Computers use voltage to send and receive information. As previously discussed, voltage is electrical pressure and does not travel through circuits. It causes current, which does the real work in an electrical circuit. However, voltage is used as a signal of how much work the circuit is performing. A computer converts input information or data into

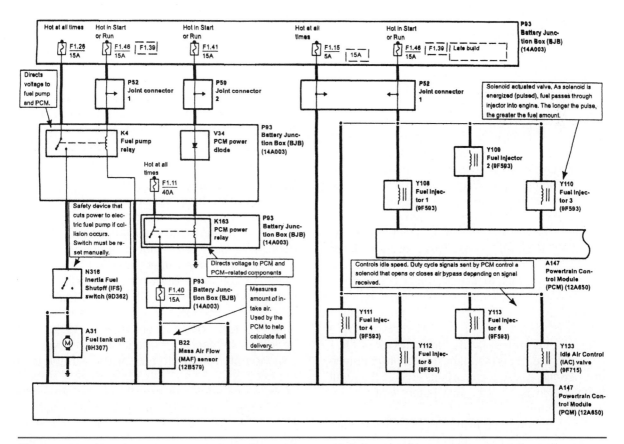

Figure 5-10. This fuel pump circuit illustrates a series-parallel circuit. Note the inertia switch in series with the fuel pump. (Courtesy of Ford Motor Company)

voltage signal combinations that represent number combinations. The number combinations represent a wide variety of information—temperature, speed, pressure, or even words and letters. A computer processes the input voltage signals it receives by computing what they represent and delivering the data in computed or processed form.

The Four Basic Computer Functions

Regardles of the size or use to which it is put, the operation of every computer can be divided into four basic functions, figure 5-11:

- Input
- Processing
- Storage
- Output

These basic functions are not unique to computers; they are also found in many noncomputer systems. However, we need to know how the computer handles these functions.

First, the computer receives a voltage signal from an input device. The device may be as simple as a button or a switch on an instrument panel, or a sensor on an automotive engine. Typical types of automotive sensors are shown in figure 5-12. The keyboard on your personal computer or the programming buttons of a video cassette recorder are other examples of an input device.

Modern automobiles use various electromechanical, electrical, and magnetic sensors to measure factors such as vehicle speed, engine rpm, air pressure, oxygen content of exhaust gas, airflow, and temperature. Each sensor transmits its information in the form of voltage signals. The computer receives these voltage signals, but before it can use them, the signals must undergo a process called **input conditioning.** This process includes amplifying voltage signals that are too small for the computer circuitry to handle. Input conditioners generally are located inside the computer, but a few sensors have their own input conditioning circuitry.

Second, input voltage signals received by a computer are *processed* through a series of electronic logic circuits maintained in its programmed instructions. These logic circuits change the input voltage signals, or data, into output voltage signals, or commands.

Third, the program instructions for a computer are *stored* in electronic memory. Some programs

may require cetain input data be stored for later reference or future processing. In others, output commands may be delayed or stored before they are transmitted to devices elsewhere in the system. Computers use a number of different memory devices, which we look at later in this chapter.

Fourth, after the computer has processed the input signals, it sends *output* voltage signals or commands to other devices in the system, such as a system actuator figure 5-12. The computer grounds the voltage signal to complete the circuit. An **actuator** is an electrical or mechanical device that does the desired operation, such as adjusting engine idle speed, altering suspension height, or regulating fuel metering.

Computers also communicate with, and control, each other through their output and input functions. This means that the output signal from one computer system is an input signal for another computer system. General Motors introduced a body computer module (BCM) on some 1986 models. This acts as a master control unit by managing a network containing all sensors, switches, and other vehicle computers, figure 5-13.

As an example, suppose the BCM sends an output signal to disengage the air conditioning compressor clutch. That same output signal becomes an input signal to the electronic control module (ECM) that controls engine operation. Based on the signal from the BCM, the ECM signals an actuator to reduce engine speed to account for the decreased load of the compressor. This affects the fuel metering system.

■ **Data Priority**

Some kinds of data take priority over other kinds of data. Engine temperature input data takes priority over climate control, for example. Because it's more important to cool the engine than to cool the passengers, engine temperatures are sampled more often than selections on the climate control panel. Should the computer detect engine overheating, fixing the problem takes priority, and a cooling "fan-on" command is executed before other more-routine tasks are handled. In fact, air conditioning may be switched off in order to further unload and cool the engine.

The four basic functions described earlier are common to all computers, regardless of size or purpose. They also form an organizational pattern

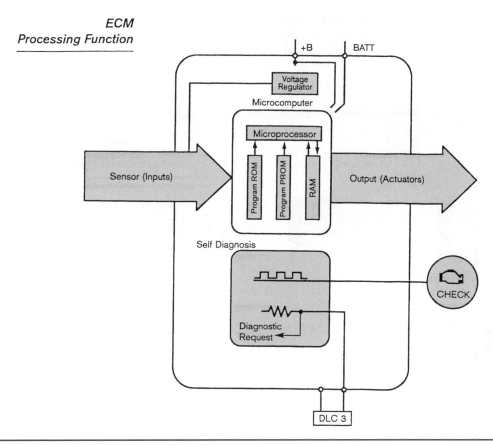

Figure 5-11. The internal components of an engine control module that perform the input conditioning, processing, storage, and output functions. (Provided courtesy of Toyota Motor Sales U.S.A., Inc.)

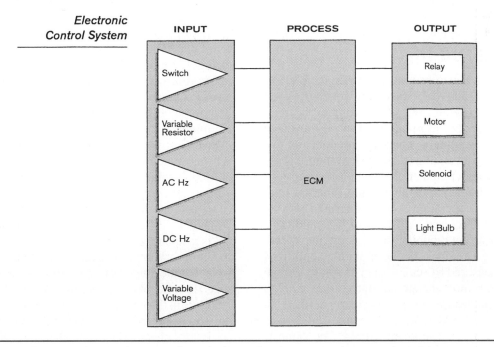

Figure 5-12. There are four basic types of sensor input signals to an engine control module. (Provided courtesy of Toyota Motor Sales U.S.A., Inc.)

109

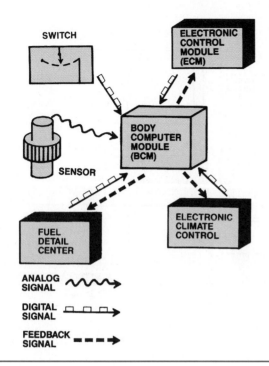

Figure 5-13. Cadillac's BCM accepts inputs from a variety of sources and manages the other onboard computer systems. (Courtesy of General Motors Corporation)

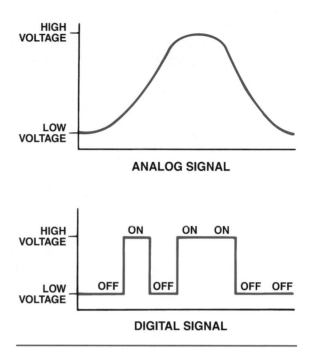

Figure 5-14. An analog signal is continuously variable. (Courtesy of Ford Motor Company)

to troubleshoot a malfunctioning system. While most input and output devices can be adjusted or repaired, the processing and storage functions can only be replaced.

Analog and Digital Systems

A computer has to be told how to do its job. The instructions and data necessary to do this are called the program. Because a computer cannot read words, the information must be translated into voltage signals. This can be done in two ways, using an analog or a digital system.

An **analog** computer uses a continuously variable voltage signal or processing function relative to the operation being measured or the adjustment required. Most operating conditions affecting an automobile, such as engine speed, are analog variables. These operating conditions can be measured by sensors. For example, engine speed does not change abruptly from idle to wide-open throttle. It varies in clearly defined, finite steps—1,500 rpm, 1,501 rpm, 1,502 rpm, and so on—that can be measured. The same is true for temperature, fuel metering, airflow, vehicle speed, and other factors.

If a computer is to measure engine speed changes from 0 rpm through 6,500 rpm, it can be programmed to respond to an analog voltage that varies from 0 volts at 0 rpm to 6.5 volts at 6,500 rpm. Any analog signal between 0 and 6.5 volts represents a proportional engine speed between 0 and 6,500 rpm.

Analog computers have several shortcomings, however. They are affected by temperature changes, supply voltage fluctuations, and signal interference. They are also slower in operation, more expensive to manufacture, and more limited in what they can do compared to digital computers.

In a **digital** computer, the voltage signal or processing function is a simple high/low, yes/no, on/off. The digital signal voltage is limited to two voltage levels. One is a positive voltage, the other is no voltage. Since there is no stepped range of voltage or current in between, a digital binary signal is a square wave, figure 5-14.

Using our engine speed example above, suppose that the computer needs to know that engine speed is either above or below a specific level, say 1,800 rpm. The computer does not need to know the exact engine speed, but only whether it is above or below 1,800 rpm. Therefore the digital signal can be no voltage below 1,800 rpm and any arbitrary voltage when engine speed is above

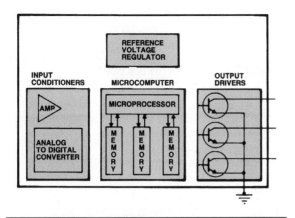

Figure 5-15. Computers need to change inputs into digital signals or binary fits. This occurs in the analog-to-digital A/D converter or what may be called the input interface. (Courtesy of Ford Motor Company)

1,800 rpm. As you can see, a digital signal acts like a simple switch to open and close a circuit.

An engineer can reverse the switch functions to provide a high input signal below 1,800 rpm and a low (zero voltage) signal above 1,800 rpm. The result at the computer is the same. The computer receives a simple digital input signal that represents a *change in operating conditions.*

The signal is called "digital" because the on and off signals are processed by the computer as the digits or numbers 0 and 1. The number system containing only these two digits is called the **binary** system. Any number or letter from any number system or language alphabet may be translated into a combination of binary 0s and 1s for the digital computer.

A digital computer changes the analog input signals (voltage) to digital bits (binary digits) of information through an **analog-to-digital** (A/D) converter circuit, figure 5-15. The binary digital number, figure 5-16, is used by the computer in its calculations or logic networks. Output signals are usually digital signals that turn system actuators on and off. Some digital signals are changed to an analog output signal through a **digital-to-analog** (D/A) converter. This is the opposite of the A/D converter circuit that changes analog input signals. More often, however, a digital output signal is made to approximate an analog signal through a variable duty cycle, which you learn more about later.

The digital computer processes thousands of digital signals per second because its circuits are able to switch voltage signals on and off in billionths of a second.

■ Computer Communication

As an example of how two or more computers might "talk to one another" and exchange inputs and outputs, suppose the vehicle driver selects air conditioning "on," at the climate control panel.

First, the climate control panel inputs "AC on" to the BCM. Before the BCM turns on the AC however, it must determine if engine rpm can support the compressor load without engine stalling, so it checks engine rpm input data from the distributor. If the BCM determines engine rpm to be too low, it delays "AC on," and requests an increase of engine rpm (to prevent stalling) from the powertrain control module (PCM). The PCM commands an idle-up condition from the idle speed control motor (ISC) or idle air control (IAC) device.

The rpm is monitored by the BCM, and once it is sufficient, the BCM commands the AC compressor clutch relay to turn on. With the relay on, the compressor clutch coil gets power, and the AC compressor is activated. All of this happens in less than a second or so, but it illustrates the processing of data that takes place for a "simple" operation like AC on.

Analog-to-Digital Conversion

As mentioned earlier, most operating conditions that affect an automobile are analog variables. When the computer needs to know whether an operating condition is above or below a specified point, a digital sensor can be used to act as a simple on/off switch. Below the specified point, the switch is open. The computer receives no voltage signal until the condition reaches the specified point, at which time the switch closes. This is an example of a simple digital off/on circuit: off = 0, on = 1.

Let's use the engine coolant temperature (ECT) as an example: Consider that the computer needs to know the exact temperature within one degree. Now suppose the sensor measures temperature from 0°F to 300°F and sends an analog signal that varies from 0 to 6. Each 1-volt change in the sensor signal is the equivalent of a 50-degree change in temperature. If 0 volts equals a temperature of 0°F and 6 volts equals 300°F:

- 1.00-volt change of output from the sensor equals a 50°F change of coolant temperature
- 0.50-volt change equals a 25°F change
- 0.10-volt change equals a 5°F change
- 0.02-volt change equals a 1°F change

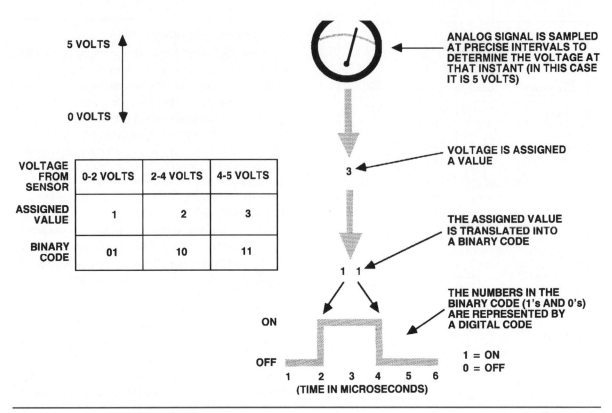

Figure 5-16. The process of analog-to-digital conversion performed by the converter section of the computer. (Courtesy of Ford Motor Company)

Conversely, in order for the computer to determine a temperature change of 1°F, it must react to sensor voltage changes as small as 0.020 volt or 20 millivolts. For example, if coolant temperature is 125°F, ECT output to the computer would be 2.5 volts. If the temperature rises to 126°F, sensor voltage increases to 2.52 volts. Again, this is a simplified example.

Actually, temperature does not pass in steps from one degree to another; it passes smoothly through a range of temperatures. But the automobile computer converts a sweeping analog voltage input into small digital voltage steps. To do this, the computer sends the signal through analog-to-digital (A/D) conversion circuits, where the sweeping analog sensor voltage is converted to a series of voltage changes for each degree of coolant temperature change. This digitizes conditions of an analog signal into a binary signal for the processor to handle.

PARTS OF A COMPUTER

We have discussed the functions, logic, and software used by a computer. The software consists of the programs and logic functions stored in the computer circuitry. The hardware is the mechanical and electronic parts of a computer. Figure 5-17 shows the basic structure of a computer.

Central Processing Unit

The microprocessor is called the **central processing unit (CPU)** of a computer, but is referred to as the electronic control module (ECM) in automotive applications. Because it does the essential mathematical operations and logic decisions, the CPU can be considered the heart of a computer.

Computer Memory

Other circuits perform the storage and memory functions. These are not processing circuits, but simply store the computer operating program, the system sensor input data, and the system actuator output data for use by the ECM.

Figure 5-17. This photo shows the inside of a typical engine control computer.

■ Sensor Outputs

In reality, sensor outputs are not linear, nor do outputs ever drop to zero volts or reach 5 volts unless there is a wiring or sensor problem. A 0-volt or 5-volt signal would cause the malfunction indicator lamp (MIL) in the dash to illuminate, and the computer would store a diagnostic trouble code (DTC) indicating a high- or low-voltage ECT sensor circuit fault. The technician would retrieve such a code from the ECM and troublshoot the circuit.

ROM, PROMs, EPROM, and EEPROMs

Computer memory is composed of two different types: permanent and temporary memory. Permanent memory is called **read-only memory (ROM)** because the computer only reads the contents. Any ROM data that is programmed cannot normally be changed. This data is retained even when power to the computer is shut off. Part of the ROM is built into the computer, and some is programmed into an IC (integrated circuit) chip commonly referred to as the **PROM,** for **programmable read-only memory.** The PROM contains the read-only memory for operating a vehicle with a prescribed engine/transmission/axle ratio assembly. Even accessories installed on the vehicle at the factory affect the information recorded in the PROM.

The PROM normally stays with the vehicle for life, yet a PROM change may be required to im-

prove the vehicle emissions or driveability. Should the prerecorded instructions need updating, the entire computer often must be exchanged. This is an expensive proposition. Swapping either a PROM or an entire computer from one vehicle to another can adversely affect vehicle performance and driver satisfaction. Unfortunately, PROM updates are quite common and GM vehicles for years used a removable PROM in their engine control module to make PROM updates easier and less expensive.

However, PROMs are being replaced gradually by another type of chip with read-only memory called erasable, programmable, read-only memory (EPROM). Data on an EPROM is permanent storage that cannot be erased in the field. But, its most important feature is the ability of the EPROM to be erased and reprogrammed by the manufacturer. Because EPROM memory is erasable, it can be changed by exposure to ultraviolet light.

If the memory can be electrically reprogrammed one byte at a time, it is called an electrically erasable PROM or EEPROM. After the chip's memory is cleared, it can be reprogrammed with new data in less than four seconds via modem linked to the manufacturer. An EEPROM allows a dealership to update or change instructions in the computer memory without having to replace the entire module. Since 1996, the EEPROM is the industry standard.

Another form of memory is temporary memory, called **random-access memory (RAM).** The microprocessor can write or store new data into it as directed by the computer program, as well as read the data already in it. "Scratchpad" is another name for RAM because the computer uses it as a notepad. Automotive computers use two types of RAM memory: volatile and nonvolatile.

Volatile RAM memory is lost whenever the ignition is turned off. However, a type of volatile RAM called **keep-alive memory (KAM)** is wired directly into battery power and prevents the data from being erased when the ignition is turned off. Both RAM and KAM have the disadvantage of losing their memory when disconnected from their power source. One example of RAM and KAM is the loss of station settings in a programmable radio when the battery is disconnected. Because all the settings are stored in RAM, they must be reset when the battery is reconnected. System diagnostic trouble codes (DTCs) have traditionally been stored in RAM and are erased by disconnecting the battery.

Nonvolatile RAM memory retains its information even when the battery is disconnected. One use for this type of RAM is the storage of warranty related data. Another is the storage of odometer readings in the electronic speedometer. The memory chip retains the mileage accumulated by the vehicle. When speedometer replacement is necessary, the odometer chip is removed and installed in the new speedometer unit. The KAM is used for such diverse things as radio settings, mirror positions, and engine or transmission adaptive strategies, which are covered in the next chapter.

Input and Output Circuits

We have already discussed the circuits that condition input signals from sensors into binary language for the ECM. Also discussed is how an output is converted from binary digital code into an analog signal in preparation for output to an actuator. But the computer is not always connected directly to an input or output device. The signals are often received by and sent to other computers. Signals from one computer to another are often transmitted over a buss (common wires) with parallel connections. Parallel data transmission over a buss allows much more data to be handled in a shorter time. Also, multiple input signals can be received at once, even while the computer is sending multiple output signals.

Clock Rates and Timing

Remember, the microprocessor receives sensor input voltage signals from the A/D converter. It processes incoming data after communicating with various memory circuits and instructions in ROM. After processing the data, the microprocessor signals the appropriate actuators by sending output voltage signals to the DA converter. All this communication is accomplished through transmission of long strings of 0s and 1s in binary code. Because the microprocessor must have some way of knowing when one signal ends and another begins, it contains a crystal oscillator called a clock generator.

Clock Pulses

The crystal oscillator is a component of the computer used to generate a steady stream of constant voltage pulses. Both the microprocessor and the memory circuits monitor the clock pulses while they are communicating, figure 5-18. They are able to distinguish between the binary number 01 and the binary number 0011 because they know how long each voltage pulse should be, figure 5-19. To complete the process, the input and output circuits also monitor the clock pulses, figure 5-20.

■ Fiber Optics

A number of automobiles are handling the onboard transfer of data without the use of expensive wiring. Instead, data is being sent and received from one computer to another, or between various onboard components, using fiber optics. Digitized data is transferred throughout the vehicle faster, with greater reliability and savings, through fiber-optic strands. Digital data is encoded at the transmitter, such as the BCM, to notify the light switch pod to activate a left turn signal. The light switching pod, in turn, commands the left turn lamp assembly to start blinking. The command is decoded at the lamp housing, and the left turn signal blinks.

Baud Rate

Serial data is transmitted at precise intervals. The speed at which a computer transmits data is called the baud rate, or bits per second. Not all computers operate at the same speed; some are faster than others. The speed at which a computer operates is specified by the cycle time, or clock speed, required to perform certain mea-

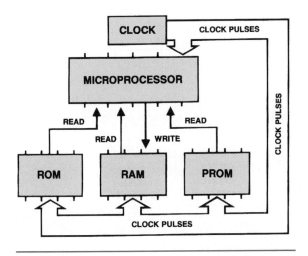

Figure 5-18. The clock pulses are monitored by the microprocessor and memories. (Courtesy of General Motors Corporation)

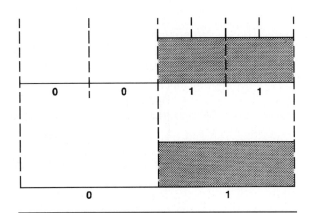

Figure 5-19. Clock pulses are used by the computer to distinguish between different signals. (Courtesy of General Motors Corporation)

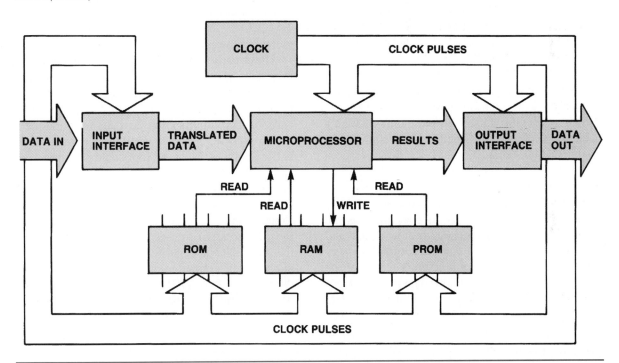

Figure 5-20. The clock pulses also are monitored by input and output circuits. (Courtesy of General Motors Corporation)

surements. Just as miles per hour (mph) helps in estimating the length of time required to travel a certain distance, the baud rate is useful in estimating how long a given computer needs to transmit a specified amount of data to another computer. Automotive computer baud rates have increased impressively over the last few years. The speed of data transmission is an important factor both in system operation and in system troubleshooting. The higher the baud

rate, the more precision the computer has when controlling system operation and communicating with actuators or other computers. With higher computer power, many more onboard systems can be monitored and more executive functions can be included to reach and execute fail-safe decisions. These include commanding limp-home mode, or even engine shutdown, to prevent damage to the engine or the catalytic converter.

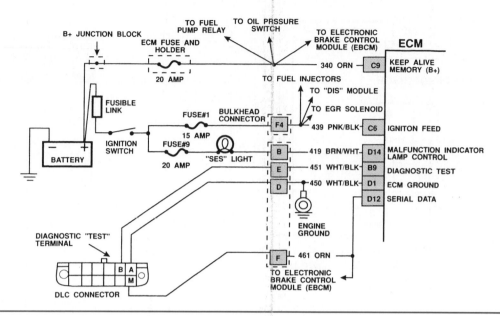

Figure 5-21. The serial data line connects to the DLC so the scan tool can access it. (Courtesy of General Motors Corporation)

DATA LINES

Computers used on today's vehicles need to be able to communicate with other modules used on the vehicle or need to be able to communicate with a handheld diagnostic tool such as a scan tool. To accomplish this, computers use a variety of serial data lines and communication protocols. Some early onboard engine controllers used a data line to connect to the diagnostic connector provided by the manufacturer to give service technicians the ability of reading diagnostic trouble codes and viewing data parameters that the computer is seeing. One of the first manufacturers to allow the technician to have access to this data line was General Motors. As seen in figure 5-21, the serial data line was connected to pin M of the data link connector (**DLC**). In addition, pin B of the DLC was the diagnostic request wire. In order to enter the diagnostic mode, pin B is grounded to pin A either manually by jumping the A to B connector or using the scan tool. When pin B is grounded, the computer would send codes out on the serial data line to the scan tool. In addition, the scan tool could "listen in" to the computer's data circuit and translate the serial data into a readable value for the technician. For instance, if the technician needs to know what the engine coolant sensor value is, the scan tool would display what the computer was interpreting the value of the engine coolant temperature in degrees F or C. This allowed the technician to quickly verify input data and its accuracy quickly without having to perform individual voltage or resistance measurements on these circuits.

These early data circuits were very limited in their capability. They were relatively slow. The early General Motors system communicated at a baud rate of 160 bps. In addition, most of these early data communications circuits were unidirectional, meaning that they could only communicate to the scan tool. The scan tool could not communicate with the computer. Later systems became bidirectional, allowing two-way communications from the scan tool to the computer or from the computer to the scan tool. As the amount of information in the data stream increased, the speed (baud rate) of the serial data also had to increase. As new generation vehicle computers were introduced, the baud rate increased to handle the increased amount of data transmission and provide faster updates of data parameters to the scan tool.

These early computers with data stream and scan and tool access allowed technicians to more easily diagnose engine and emissions related malfunctions. However, many manufacturers did not provide data stream capability on their vehicles. The California Air Resources Board (**CARB**) recognized the advantage of allowing technicians access to the data stream and an easy method of displaying diagnostic trouble codes. Accordingly, CARB introduced onboard diagnostics one (OBD I) regulations that required manufacturers to provide this access to technicians. Although this was an improvement for the repair industry, it also created some difficulties due to the lack of standardization. Communication protocols were not standardized; diagnostic connections were different configurations and in different locations on the vehicle. This

OBD II Communication Protocols

SAE/ISO Designation	Name	Network Communications & Diagnostics	Speed
SAE J1850 (1)	Class 2/PCI	Both	10.4 Kbps VPW
SAE J1850 (2)	SCP	Both	41.6 Kbps PWM
ISO 9141-2 (K Line)	K Line	Diagnostics	10.4 Kbps
ISO 14230-4 KWP 2000	Keyword Protocol 2000	Diagnostics	10.4 Kbps
SAE J2284 ISO 15765-4	Controller Area Network (CAN)	Both	125 Kbps 1,000 Kbps
J2534	Pass-Thru Programming	Programming	

Figure 5-22. This table shows the different communication protocols that may be used on OBD II vehicles. Some of these protocols support diagnostics only, while others support both diagnostics and networking.

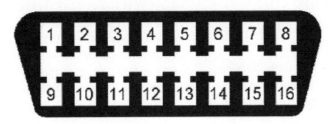

Pin 1   Discretionary   Pin 9 Discretionary
Pin 2   Bus + J1850   Pin 10 Bus – J1850
Pin 3   Discretionary   Pin 11 Discretionary
Pin 4   Chassis Ground   Pin 12 Discretionary
Pin 5 Signal Ground   Pin 13 Discretionary
Pin 6 CAN High   Pin 14 CAN Low
Pin 7 K-Line ISO 9141/KWP   Pin 15 L-Line ISO9141/KWP
Pin 8 Discretionary   Pin 16 Unswitched Battery +

Figure 5-23. Shown are the pin assignments for the OBD II connection. Any pins that are marked "discretionary" means that the manufacturer may use these pins however they want. (Courtesy of the U.S. EPA)

led to scan tools needing a wide variety of scan tool software and connectors to be able to connect to the variety of models in the marketplace. Many scan tools only worked on certain model vehicles. These differences once again led CARB and the EPA to introduce legislation that would help standardize the systems to provide easier access. These regulations are what are known as onboard diagnostics two (OBD II), which were implemented for 1996 and newer models. For a review of these regulations, refer to Chapter 1 of this manual.

OBD II standards allow five data line communications standards that are either Society of American Engineers (SAE) or International Stan-

dards Organization (ISO) approved; refer to the chart in figure 5-22. In addition, the DLC is standardized as to the relative location in the vehicle and the configuration of the pins, figure 5-23. Thus, any scan tool that is designed to work on the generic OBD II system should be able to plug into the connector and communicate via the standardized communication line to access diagnostic trouble codes and emission-related information.

Multiplexing

Multiplexing is the term used to describe computer networking. As the number of computers has grown on the vehicle, the need to allow these computers to communicate and share information with each other has also grown. By connecting the computers together on a communications network, they can easily share information back and forth.

This multiplexing has a number of advantages. These advantages are:

- The elimination of redundant sensors and dedicated wiring for these multiple sensors
- The reduction of the number of wires, connectors, and circuits
- Allows more features and option content to be added to new vehicles
- Weight reduction, thus increased fuel mileage
- Allows features to be changed with software upgrades instead of component replacement.

SAE Classification of Networking Protocols

The SAE has established three classes for network protocols. These are Class A, B, and C.

- Class A are low-speed (less than 10 Kbps) and used mostly for convenience features
- Class B are medium-speed (10 Kbps to 125 Kbps) and used for powertrain and emissions
- Class C are high-speed (125 Kbps to 1 Mbps) and used for real-time control

Class A multiplexing are low-speed communication protocols such as universal asynchronous receiver transmitter (**UART**). These communication protocols are manufacturer-specific; therefore, there is no standardization between manufacturers. Class B multiplexing meets the SAE J1850 or ISO 9141-2 standards. This is a medium-speed communications data line 10,400 Kbps or greater. Currently Class B multiplex data lines are required for all OBD II vehicles. Class C multiplex data lines are high-speed lines typically using the **CAN** protocol. The CAN protocol was developed by Robert Bosch GmbH in the 1980s. The 1991 Mercedes S-class was the first automobile to utilize the CAN protocol. Domestic manufacturers are now starting to change to the CAN protocol to standardize with the European manufacturers and because by 2008 CARB will require the use of CAN on vehicles.

Network Configurations

Multiplex networks can be configured differently. One method of configuring networks is the loop configuration, figure 5-24. The loop configura-

tion has two data lines attached to each controller. The controllers can communicate with other controllers in either direction. The advantage of this type of configuration is that an open in the data line will not affect the operation of the network. However, two opens in the data line will cause certain controllers to lose communication, depending on where the opens are. The disadvantage of this type of configuration is that if the entire network stops communicating due to a ground in the data line, or a short to ground within a controller, each controller will have to be isolated individually by disconnecting to diagnose the fault.

Another more common method of configuring networks is the star configuration, figure 5-25. In this configuration, each controller is connected to a single data line wire that then is routed back to a common junction. This junction may be in the form of a bus bar or splice clip. If a splice clip is used, it is easier to diagnose a grounded data circuit due to the fact that each controller and circuit can be isolated from the network at a central point in the circuit. The disadvantage of this type of circuit is that any open in a data line will cause that controller to lose communication with the rest of the network.

In some applications the manufacturer may use a combination of the two configurations. That is, a portion of the network may be configured in the loop configuration and another portion of the network may be configured in the star configuration, figure 5-26.

Control Module Names and Locations

The onboard automotive computer has had many names. Early on, it was referred to as "the brain" or the "black box" because no one in the field understood its functioning. In recent years, it has been called an electronic control unit (ECU), or electronic control module (ECM), a controller, or an assembly, depending upon the manufacturer and the computer application. The Society of Automotive Engineers (SAE), an organization that publishes recommended standards for the industry, recommends using the term ECM. If the computer controls both the engine and transmission functions, the proper term is PCM, for powertrain control module.

The internal computer components and circuit boards are installed in a metal case to shield them from electromagnetic interference (EMI). Without

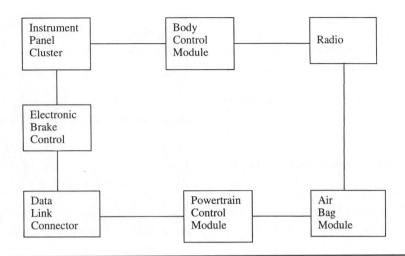

Figure 5-24. A network configured in the loop configuration.

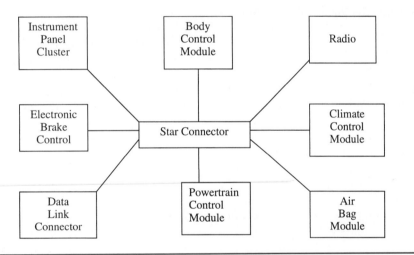

Figure 5-25. A network configured in the star configuration.

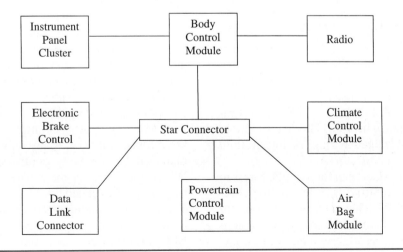

Figure 5-26. A network configured in the star and loop configuration.

such shielding, computers are affected by things like cell phones, CB radios, and microwaves. Wiring harnesses link the onboard computer to sensors and actuators and terminate with multi-pin weather-proof connectors. These connectors should be treated carefully to avoid bent pins or torn weather seals.

Many problems start after a technician has mishandled or abused the often hard-to-access connectors of the wiring harnesses. Once connector pins get corroded, it is difficult to locate the weather-related malfunctions. Sometimes computer data is transmitted on shielded wiring. These wires are surrounded by a grounded metal sleeve. Wiring of this type is called coaxial cable, and requires special repair techniques to avoid future problems. Never attempt to splice into a computer wiring harness. When mounting accessories or hardware under the hood or instrument panel, do not drill or screw into the panel without knowing what is behind it.

Types of Computers

Onboard computers range from single-function modules that control a single operation, to multifunction units that manage many of the separate, but linked, electronic systems in the vehicle. They vary in size from a miniature assembly to a notebook-sized box. Early Chrysler computers were attached to the air clearner housing in the engine compartment. The computers used on Chrysler modular control systems in the mid-1980s were two-piece units with the power module installed between the battery and the fender in the engine compartment, and the logic module behind a kick panel in the passenger compartment. The Chrysler SMEC (single module engine controller) replaced the two-piece. At one time the majority of powertrain control computers were installed inside the vehicle. They were usually mounted under the dash where they were protected from moisture, dirt, and extreme heat. However, mounting them under the dash means more wiring has to be used to connect to the under-hood sensors. In addition, many times inline connectors had to be used. This leads to more possibilities of damaged wiring, or intermittent connections at the connectors.

The current trend has been for manufacturers to locate the engine controllers in the engine compartment, figure 5-27. This reduces the amount of wiring needed and eliminates many of the connectors used, thus increasing reliability of the ex-

Figure 5-27. At one time, most engine computers were mounted inside the vehicle. Now many manufacturers are mounting the PCM in the engine compartment to save on wiring and connectors.

ternal circuits to and from the computer. However, when the PCM is located under the hood, it is subject to more vibration, heat, and moisture than when located inside the vehicle. The computers have to be sealed well and located in a position that ensures they receive an adequate airflow to cool them. In addition, they have to be located away from road spray, which can cause corrosion.

Computers and Programs

It has been said that modern vehicle computers have more processing power than the computers that controlled the first lunar module. With today's sophisticated onboard systems, computers control almost all functions to some extent. Some of the automotive systems managed by computers include climate, seating and mirror position, acceleration, braking, ride, suspension, aerodynamics, engine and powertrain, exhaust, steering, lighting, passive restraint, entertainment, security, theft recovery, and navigation. The list goes on. Quite possibly, collision avoidance, lane control, smart highway systems and other features will become standard features mandated by the federal government.

Where once many computers and modules were found onboard, many subsystems are now managed by the body control module (BCM). Where the BCM is the master, other computer modules are managed by, and report to, the BCM. These "slave" modules may include the PCM, the TCM (transmission control module), and the EBCM (electronic brake control module).

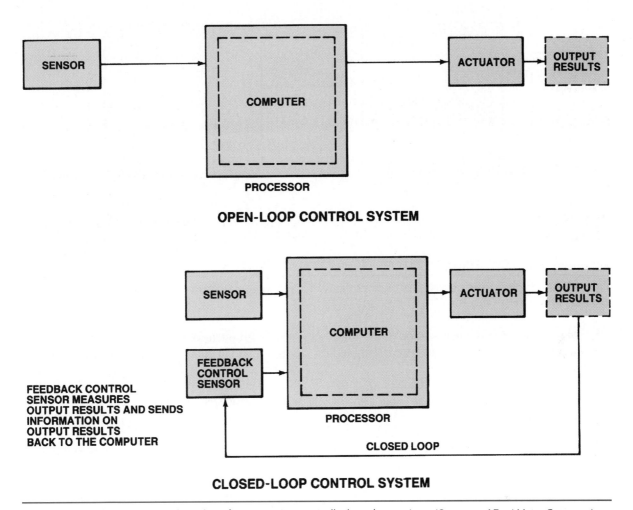

OPEN-LOOP CONTROL SYSTEM

FEEDBACK CONTROL
SENSOR MEASURES
OUTPUT RESULTS AND SENDS
INFORMATION ON
OUTPUT RESULTS
BACK TO THE COMPUTER

CLOSED-LOOP CONTROL SYSTEM

Figure 5-28. The two control modes of a computer-controlled engine system. (Courtesy of Ford Motor Company)

FUEL CONTROL SYSTEM OPERATING MODES

A computer-controlled fuel metering system can be selective; depending upon the computer program, it may have different operating modes. The onboard computer does not need to respond to data from all of its sensors, nor does it have to respond to the data in the same way each time. Under specified conditions, it may ignore sensor input. Or, it may respond in different ways to the same input signal, based on inputs from other sensors. Most current control systems, figure 5-28, have two operating modes: open-loop and closed-loop. The most common application of these modes is in fuel-metering feedback control, although there are other open- and closed-loop functions. Air conditioning automatic tempera-

ture control is an example. Control logic programmed into the computer determines the choice of operating mode according to engine operating conditions.

Open-Loop Control

Open-loop control means that the onboard computer works according to established conditions in its program. It gives the orders and the output actuators carry them out. The computer ignores sensor feedback signals indicating an error in or a deviation from the results of an actuator operation as long as the preestablished conditions exist.

For example, the computer is programmed to provide a specific amount of fuel according to the coolant temperature and a fixed spark timing when the engine starts. Because these factors are predetermined in the program, regardless of other factors, the computer ignores signals, or feedback,

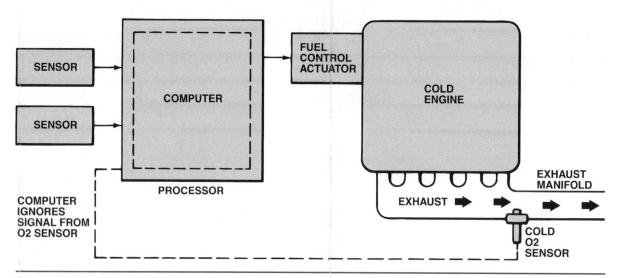

Figure 5-29. In open-loop, the computer ignores the O_2 sensor signal and operates on a predetermined program. (Courtesy of Ford Motor Company)

from the exhaust gas oxygen sensor (O_2S) until coolant temperature reaches the predetermined level, figure 5-29.

Open-loop control is not restricted to computer systems; it is present in many other control systems. For example, suppose you are reading in a comfortable living room chair when you begin to feel a chill in the room. You get up and turn on the switch controlling the furnace. The furnace warms the room to a point where you now feel too warm. You must get up again and shut the furnace switch off. The room cools back down until it gets chilly again and you must repeat the procedure. This is an example of an open-loop control system.

Closed-Loop Control

Closed-loop control means that once certain conditions (such as coolant temperature) have been met, the computer now reads and responds to signals from *all* of its sensors. When the engine starts (open loop), the computer ignores input from the O_2S. As soon as the preconditions specified in the computer program are met, the computer accepts the sensor input and adjusts the fuel accordingly. The computer is responding to a "feedback" signal; that is, the sensor is telling the computer of an error factor in its operation that must be corrected, figure 5-30. Based on the feedback signals, the computer constantly corrects its output signals in an effort to eliminate the error factor.

As with open-loop control, closed-loop operation is found in other forms of control systems. For example, suppose you had a temperature control thermostat instead of the on/off switch in the furnace control system discussed above. With the thermostat set at a certain temperature, the thermostat controller turns on the heat when room temperature drops below the setting, and shuts it off when the temperature rises above the setting. This is called a limit-cycle control system because it tries to maintain an average temperature. It does this by operating only when the temperature exceeds or falls under the preselected limits. The limits used depend on system design; for the simple temperature control system discussed, the limits might be $+/-5$ percent. Thus, if the thermostat is set for a temperature of 70 degrees, the controller turns the furnace on at approximately 66 degrees and off at about 74 degrees. In this way, the system maintains an average room temperature of approximately 70 degrees.

Another type of closed-loop control is called proportional control. In this type of control system, the computer subtracts the feedback signal from the previous output signal to determine the error signal. The error signal is then used by the computer to change the next output signal. If the computer determines that its previous output was 10 percent greater than it should be, it reduces the next output signal by 10 percent. Proportional closed-loop control is used with fuel metering, spark timing, and many other outputs in an engine control system.

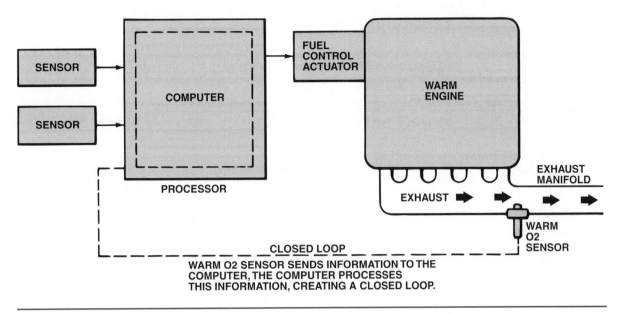

Figure 5-30. In closed-loop, the computer accepts the O_2 sensor signal and readjusts the air-fuel mixture accordingly. (Courtesy of Ford Motor Company)

SUMMARY

All computers must perform four basic functions: input, processing, storage, and output. Engine and powertrain control computers use various sensors to receive input data. This data is compared to lookup tables in the computer memory. Some data may be stored in memory for future use. The computer output takes the form of voltage signals sent to actuators.

Computers use data lines to communicate to the scan tool via the data link connector. The speed of this communication is referred to as the baud rate. There are four different communication protocols allowed on OBD II equipped vehicles. Many vehicles use multiplexing to allow multiple computers to communicate with each other. This reduces redundancy of sensors and wiring.

Review Questions

Choose the letter that represents the best possible answer to the following questions:

1. The electromotive force that moves electrons is called:
   a. Resistance
   b. Voltage
   c. Amperes
   d. Ohms

2. The rate of current flow in a circuit is measured in:
   a. Ohms
   b. Volts
   c. Amperes
   d. Watts

3. Resistance is measured in:
   a. Ohms
   b. Amperes
   c. Watts
   d. Volts

4. Technician A says that an engine control module (ECM) receives input information from its actuators, processes data, stores data, and sends output information to its outputs. Technician B says that today's automotive computers rely on feedback data to make decisions. Who is correct?
   a. Technician A
   b. Technician B
   c. Both a and b
   d. Neither a nor b

5. The operational program for a specific engine and vehicle is stored in the computers:
   a. Logic module
   b. Programmable read-only memory
   c. Random-access memory
   d. Keep-alive memory

6. Today's onboard computers calculate data in a language consisting of:
   a. Digital "ons and offs"
   b. Analog waveforms
   c. Pluse-width modulated data
   d. None of the above

7. The computer can read but not change the data stored in:
   a. ROM
   b. RAM
   c. KAM
   d. PRAM

8. Technician A says analog input data must be digitalized by an A/D converter. Technician B says output data must be changed to analog signals by a DA converter. Who is correct?
   a. Technician A
   b. Technician B
   c. Both a and b
   d. Neither a nor b

9. Clock pulses are used in computers to:
   a. Keep track of time
   b. Generate the on and off bits
   c. Synchronize the on and off bits
   d. None of the above

10. Which of the following is the "scratch pad" of the computer where information can be written into as well as erased?
    a. ROM
    b. PROM
    c. RAM
    d. PRAM

11. Baud rate refers to the:
    a. Speed that the computer operates
    b. Speed that decisions are made by the computer
    c. The speed that serial data is transmitted
    d. None of the above

12. Multiplexing allows:
    a. The elimination of a lot of wiring
    b. Computers to communicate with each other
    c. The elimination of redundant sensors
    d. All of the above

13. The communication protocol for OBD II
    equipped vehicles is:
    a. J1850
    b. CAN
    c. Keyword 2000
    d. Any of the above

14. Data lines may be configured in which
    configuration?
    a. Star
    b. Loop
    c. Combination of star and loop
    d. Any of the above

15. In the loop configuration if an open occurs
    in the data line, the entire data circuit will
    not operate.
    a. True
    b. False

6

Engine Management Input Devices

OBJECTIVES

After completion and review of this chapter, you will be able to:

- List the critical sensor inputs that an engine management system needs in order to calculate and deliver the correct amount of fuel for good driveability and emissions.
- Describe the three ways that input sensors work electrically.
- Define the term *analog*.
- Identify the difference between a pull-up and pull-down circuit.
- Describe the operation of three types of oxygen sensors.

KEY TERMS

air-fuel ratio sensor   MAP
ECT O_2S
HO_2S

INTRODUCTION

All engine management computers use input devices to send information to the computer so the computer can determine engine operating conditions. These input devices are similar on all fuel-injection systems. This sensor information is used by the computer for startup, open-loop operation, closed-loop operation, acceleration enrichment, deceleration enleanment, and other operating modes. Some engine management systems may use a large number of these input sensors, while other systems may use a minimal number of input devices. Some of these input devices are critical for correct fuel calculation and delivery. Many other input sensors may simply be used by the computer to operate other non-fuel delivery functions. Figure 6-1 shows what could be considered some typical inputs used by an engine management system. This does not mean that all engine management systems use all these inputs or that they may not have other inputs that are not shown in this figure.

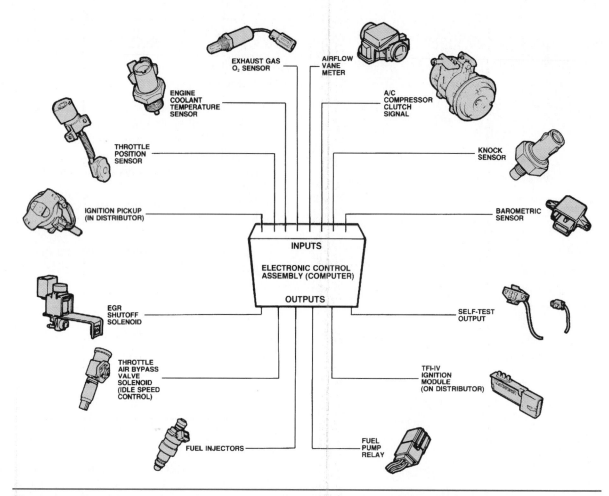

EXHAUST GAS
O₂ SENSOR

AIRFLOW
VANE
METER

ENGINE
COOLANT
TEMPERATURE
SENSOR

A/C
COMPRESSOR
CLUTCH
SIGNAL

THROTTLE
POSITION
SENSOR

KNOCK
SENSOR

IGNITION PICKUP
(IN DISTRIBUTOR)

BAROMETRIC
SENSOR

INPUTS

ELECTRONIC CONTROL
ASSEMBLY (COMPUTER)

OUTPUTS

EGR
SHUTOFF
SOLENOID

SELF-TEST
OUTPUT

THROTTLE
AIR BYPASS
VALVE
SOLENOID
(IDLE SPEED
CONTROL)

TFI-IV
IGNITION
MODULE
(ON DISTRIBUTOR)

FUEL INJECTORS

FUEL
PUMP
RELAY

Figure 6-1. Typical components of a late-model Ford electronic engine control (EEC-IV) system. (Courtesy of Ford Motor Company)

ELECTRICAL OPERATION OF INPUT DEVICES

Regardless of the number or type of input devices that are used on an engine management system, input devices usually operate electrically by one of three methods. These methods are:

- Switched input
- Analog
- Digital

Before we learn how these individual sensors operate, we need to have a thorough understanding of these electrical concepts of sensor operation.

Switched Inputs

A switched input is an input that has either one of two states, on or off. These types of inputs are often used for inputs such as AC request, transmission gear position, or ignition input. Figure 6-2 illustrates a switched input. This type of switched input is referred to as a pull-up or power-side switch. These terms refer to the fact that the voltage in the circuit is provided from an external source. When the switch is open, no voltage would be present at the pin of the PCM. However, when the switch closes, this voltage is applied to the PCM terminal. Internally the PCM has a dropping resistor connected to ground. Internally the PCM senses the voltage before it is dropped across the internal dropping resistor. An example of how this input may be used by the PCM may be the AC request signal. Whenever the PCM senses no voltage at this input, it would represent no request to enable the AC compressor. Whenever the switch is closed, the PCM would sense system voltage applied to this terminal and would indicate to the PCM that the driver has requested operation of the air conditioning compressor.

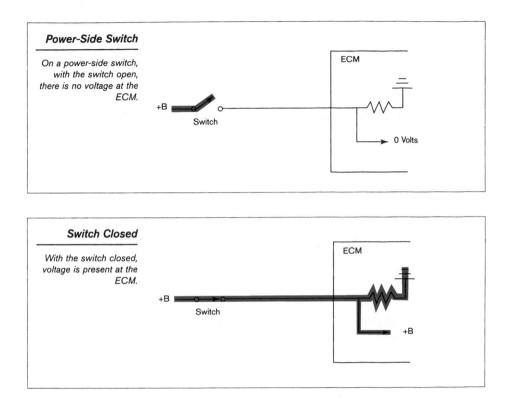

Figure 6-2. A switched input configured as a power-side switch or a pull-up circuit. (Provided courtesy of Toyota Motor Sales, U.S.A., Inc.)

NOTE: Manufacturers have used various names for their controllers over the years such as electronic control assembly (ECA), electronic control module (ECM), and logic module. Throughout this text the term powertrain control module (PCM) will be used. This is the correct OBD II terminology for a controller that controls both engine and transmission operation.

Another type of switched input circuit is shown in Figure 6-3. This type of switched input is referred to as a pull-down or ground-side switched input. The difference between this type of switched input circuit and the switched input circuit discussed previously is that the voltage is sourced internally within the PCM and grounded by the switch when the switch is closed. The source voltage is provided by the PCM through a dropping resistor. Whenever the switch is open, the PCM would sense this voltage within the PCM. When the switch is closed, this voltage would drop to near 0 volts within the PCM, indicating to the PCM that the switched input has been actuated. An example of this type of switched input could be the power-steering pressure switch. Whenever the PCM senses voltage in the circuit to the power-steering pressure switch, it would indicate that little steering load is present. However, when the switch closes, the circuit voltage drops to near 0 volts, which would indicate to the PCM that high power-steering load is present.

As can be seen in the previous examples, switched inputs are very simple input circuits to the PCM. In order to diagnose switched inputs, the technician simply needs to know what type of switched input circuit is being used. If the voltage is externally sourced, the switched input is a pull-up or power-side switched circuit. If the voltage is internally sourced from within the PCM, the switched input is a pull-down or gound-side switched circuit.

Analog Inputs

An analog input is an input whose voltage is constantly variable. This means that the voltage output from the sensor can be any voltage level within its operating range at any given instant. This does not mean that the voltage output from the sensor is always varying. For instance, a throttle position sensor can provide a signal between 0 and 5 volts

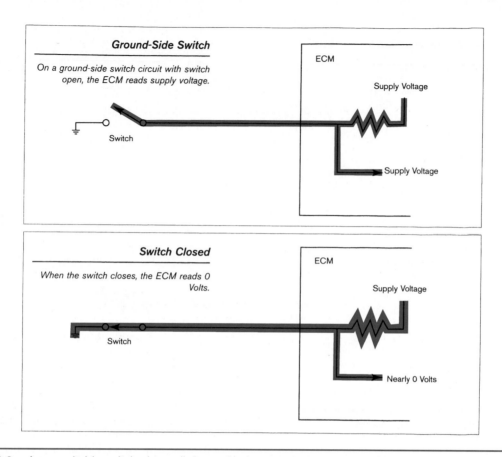

Figure 6-3. A ground-side switched or pull-down circuit. (Provided courtesy of Toyota Motor Sales U.S.A., Inc.)

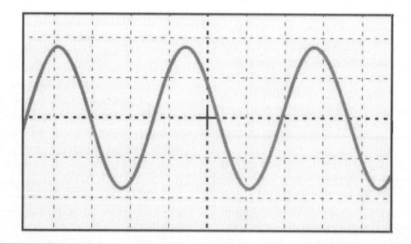

Figure 6-4. Analog signal.

depending on the throttle position. If the throttle is not moving, the voltage output from the throttle position sensor will be constant. However, the output from the sensor can vary between 0 and 5 volts; therefore, it is classified as an analog sensor. Analog sensors may operate with either AC or DC voltage. As shown in figure 6-4, many vehi-

cle speed sensors produce an AC voltage output. This AC voltage is an analog voltage because it is constantly variable. How much voltage that is produced by the sensor varies with vehicle speed, strength of the magnet in the sensor, and temperature of the sensor. The computer looks at the frequency of this analog signal instead of the

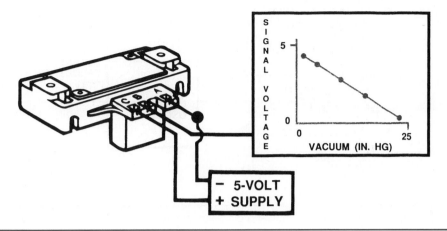

Figure 6-5. Analog sensor. (Courtesy of General Motors Corporation)

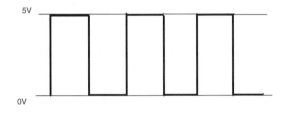

Figure 6-6. Digital signal.

amount of voltage produced by the sensor. An analog sensor that produces DC voltage is shown in figure 6-5. This sensor senses engine vacuum. It produces a 0 to 5 volt signal based on the engine vacuum. This is an analog signal due to the fact that it is constantly variable and can produce any voltage between 0 and 5 volts. Analog sensors may be of the resistive type, voltage-generating type, or potentiometer type.

Digital Inputs

The third type of sensor input signal is the digital generating sensor input. These types of sensors generate a digital signal, figure 6-6. This type of signal can be read directly by the computer without having to be converted by the A/D converter. These digital signals may be produced at different voltage levels, such as 0 to 5 volts, 0 to 7 volts, and 0 to 12 volts. The computer reads the frequency of these digital signals to determine the sensor input information. Figure 6-7 shows a digital type of mass air flow sensor. This type of air flow sensor uses a resistive element that is heated. As the incoming air

passes the heated element, the element tries to cool down. The current in the wire is increased to maintain the temperature of the resistive element. The amount of current needed to maintain this temperature is then converted to a digital signal and sent to the PCM. The frequency of this digital signal is then translated by the PCM into a measurement of the amount of air entering the engine in grams per second. Digital input signals can be produced by a variety of different input sensors that are used by engine management systems.

Virtually all computer inputs work electrically on one of the three methods described previously. The difference between the same sensors used by different manufacturers is the type of signal that it produces. For instance, some mass air flow sensors produce an analog signal to the engine controller, while other mass air flow sensors produce a digital signal to the engine controller. Obviously, before testing any of these sensors, the technician must know what type of signal is produced by the input sensor being diagnosed.

The accuracy of many of these resistive types of input sensors is determined by the accuracy of the voltage applied to them. Since resistive-type sensors do not generate a voltage, they must be provided with a stable accurate voltage provided to them. This voltage is called the reference voltage. To ensure an accurate voltage, a 5-volt reference voltage is supplied by the computer to the various sensors, figure 6-8. This voltage remains very stable despite the changes in battery voltage or charging system voltage. If the 5-volt reference signal changes due to an internal or external circuit problem, the accuracy of the signal from the sensor is affected.

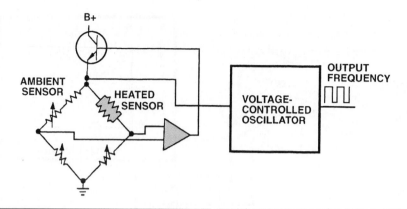

Figure 6-7. A mass air flow sensor that produces a digital output. (Courtesy of General Motors Corporation)

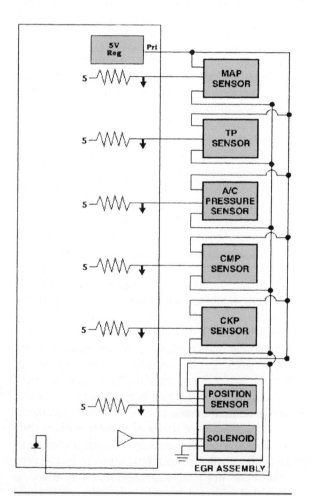

Figure 6-8. 5-volt power supply. (Courtesy of Daimler-Chrysler Corporation)

Characteristics and Features

Automotive sensors must operate in a severe environment. For this reason, they must be designed for long-term, dependable operation while providing reliable signals. The simpler the sensor design, the more reliable it is. Because many sensors have dif-

ferent uses in different systems, they may have certain characteristics that apply to specific usage. But all sensors must have four major characteristics, or operating features, if they are to function properly. These characteristics affect the selection of a particular sensor for a given function, and establish the specifications for troubleshooting and service. The four important characteristics are:

- **Repeatability:** An analog sensor produces a varying signal proportionate to the condition it measures. It must do so for each increment of change throughout its operating range. The signals produced at each increment must be repeatable in both directions, up and down.
- **Accuracy:** The sensor must work properly within the tolerances or limits designed into it. A digital sensor switch may close at 195 degrees ± 1 degree, or it may close at 195 degrees ±10 degrees. The tolerances depend on how the sensor is used, but once established, the sensor must work consistently within them. These design tolerances are used to establish the test specifications used to troubleshoot the sensor operation.
- **Operating Range:** An operating or dynamic range within which the sensor must function is established. The operating range of a digital sensor consists of only one or two switching points. An analog sensor, however, has a wider operating range and must produce accurate, proportional signals throughout that range. When the signal is not within its range limits, it is not proportionate to the value being measured and is ignored by the computer.
- **Linearity:** This refers to the sensor accuracy throughout its dynamic range. Within this range, an analog sensor must be consistently

proportionate as possible to the measured value. An ideal linear sensor signal would appear as a diagonal straight line positioned at a 45-degree angle on a graph. But no sensor has perfect linearity. In actual practice, the sensor signal deviates slightly above and below the 45-degree line on the graph, with the greatest accuracy appearing near the center of its dynamic range. The computer program contains memory data to accommodate such minor deviations in a sensor range. Signal processing circuits make the necessary compensations to provide the computer with an accurate representation of the factor measured by the sensor.

Proper design and installation of sensors includes the use of electromagnetic interference (EMI) suppression and shielding. Sensors produce a low-voltage, low-current signal that makes them especially prone to EMI. The use of low-resistance connections and proper wiring location are important factors in the transmission of sensor signals.

CRITICAL SENSOR INPUTS

Every engine management system has a variety of sensor inputs. The number and uses of these sensor inputs varies with the complexity of the engine management system. However, all engine management systems must have certain critical sensors that provide the necessary information to the engine management system for the calculation of the correct fuel delivery and control of timing. These sensors are common to most systems. The following material will describe these sensors, how they work, and how the engine controller uses the information from the sensor. While there always will be some variations between systems, for the most part these sensors are used by all manufacturers.

ENGINE COOLANT TEMPERATURE SENSOR (ECT)

The engine coolant temperature sensor is a two-wire sensor that threads into the coolant passage so it can accurately measure the coolant temperature, figure 6-9. This sensor is a thermistor, which is a solid-state variable resistor whose resistance

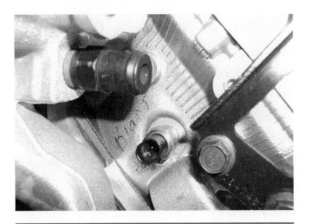

Figure 6-9. The engine coolant sensor (ECT) is typically a two-wire sensor that screws into a coolant passage.

changes with temperature. Thermistors are classified into two groups: positive temperature coefficient (PTC) resistors and negative temperature coefficient (NTC) resistors. These names simply mean that:

- The resistance of a PTC thermistor increases as the temperature increases.
- The resistance of an NTC thermistor decreases as the temperature increases.

Both kinds are used in automotive systems; however, the NTC is the more common. A typical **ECT** sensor circuit consists of the PCM, a 5-volt reference, wiring, and the ECT sensor, figure 6-10. A 5-volt reference is applied to a fixed resistor within the PCM. It then is connected in series with the thermistor (ECT). The PCM measures the voltage drop after the internal resistor. The 5 volts must divide proportionately across the two resistances. Therefore, any change in the resistance of the ECT due to temperature causes a corresponding change to the voltage drop across the internal resistance. As the resistance of the ECT goes up, voltage drop across the internal resistance goes down. This allows more of the voltage to be applied to and drop across the ECT sensor. Therefore, as the temperature changes at the engine, the voltage that the PCM sees in the ECT circuit changes also. The range of the sensor is from 0 to 5 volts; however, the only time the voltage could read 5 volts is if the circuit is open. The only time that the circuit could read 0 volts is if the circuit was grounded.

The information from the engine coolant temperature is a critical input to the PCM. In order for the PCM to calculate the correct amount of fuel for start-up, it must know the temperature of the

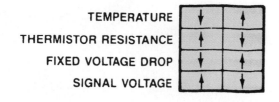

Figure 6-11. Most fuel-injection systems use a throttle position sensor (TPS) that mounts directly to the throttle body.

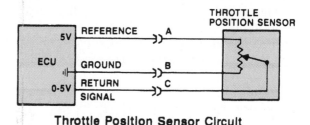

Throttle Position Sensor Circuit

Figure 6-12. A typical throttle position sensor (TPS) circuit. (Courtesy of DaimlerChrysler Corporation)

THROTTLE POSITION SENSOR

The throttle position sensor is connected to the throttle body, figure 6-11, to sense the movement of the throttle by the driver. It usually is a potentiometer that operates between 0 and 5 volts. A typical throttle position sensor (TPS) circuit consists of the PCM, a 5-volt reference, circuit wiring, and the throttle position sensor attached to the throttle body, figure 6-12. A 5-volt reference and ground is applied to the TPS. The third circuit is the signal circuit, which feed backs to the PCM. Since the TPS is a potentiometer or variable resistor, as the contacts change position across the internal resistance, the voltage on the signal wire changes. This voltage can range between 0 and 5 volts; however, by design the voltage is never allowed to go under about 0.3 volts or above 4.8 volts. This is to allow detection of either an open circuit, which would cause the signal voltage to drop to 0 volts, or a shorted circuit, which could allow the voltage to rise to 5 volts. Either of these two conditions would allow the computer to detect a problem with the circuit or sensor and use default values for operation.

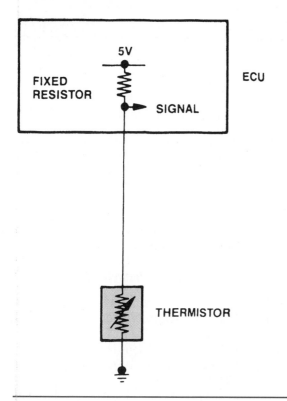

Figure 6-10. A typical ECT sensor circuit. (Courtesy of DaimlerChrysler Corporation)

engine. A cold engine needs a very rich mixture to start properly. If this coolant temperature information is not accurate or missing, the engine may not start when cold. The PCM uses the engine coolant information for purposes other than just start-up enrichment. These may include electric fan operation, closed-loop operation, control of the temperature warning lamp, and control of the temperature gauge on the dash. Start-up enrichment is the most critical function of the ECT on a fuel injection vehicle. Due to the critical function of the ECT, most engine management systems have built-in default values that the computer will use if it detects a loss of the ECT signal. These may include substituting a fixed value for the ECT, turning the engine coolant fan on to prevent overheating, and disabling the AC if it is turned on.

The TPS is another of the critical sensor inputs that the computer needs for the proper operation of the engine management system. It could be compared to the accelerator pump of a carbureted vehicle. Anytime that the driver needs to accelerate the vehicle, the engine needs extra fuel enrichment so that the engine does not bog down or even stall. The engine also needs extra fuel to compensate for increased load on the engine such as when driving up an incline. Anytime the engine controller senses an increase in the opening of the throttle, it increases the injector on time momentarily to allow extra fuel to be delivered to the engine. While this is not the only purpose of the TPS, it is one of the most critical roles. Other uses for the TPS reading by the engine controller might include transmission shifting, disengagement of the transmission converter clutch, and fuel enleanment when decelerating the vehicle.

MANIFOLD ABSOLUTE PRESSURE SENSOR

A fuel injection vehicle needs to sense the load on the engine in order to correctly calculate the proper fuel delivery. Two methods are used to sense engine load. The first method is by sensing changes in manifold pressure (vacuum) by using a manifold absolute pressure sensor. The second method is by using a mass air flow sensor to directly mea-

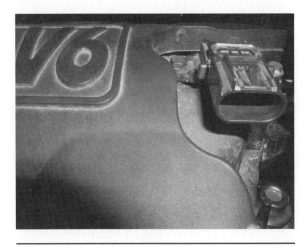

Figure 6-13. A manifold absolute pressure (MAP) sensor may mount directly to the manifold or may be connected to the manifold with a vacuum hose.

sure the amount of air entering the engine. The mass air flow sensor will be covered separately.

The manifold absolute pressure (**MAP**) sensor connects directly to the intake manifold, figure 6-13, or indirectly to the intake manifold using a vacuum hose. The MAP sensor is a pressure sensor that utilizes a small silicon chip that forms a small diaphragm that flexes with changes in pressure. One side of the diaphragm has a reference pressure sealed in a chamber, acting against the diaphragm. The other side of the diaphragm connects to the intake manifold, figure 6-14. As the manifold pressure changes, it causes the diaphragm to flex. There are a series of resistors around the edge of the diaphragm, which will

Manifold Absolute Pressure (MAP) Sensor

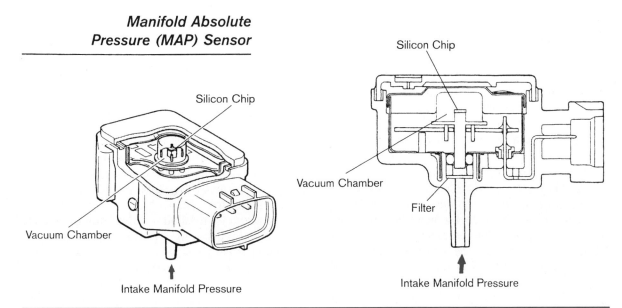

Figure 6-14. The internal construction of a typical manifold absolute pressure (MAP) sensor. (Provided courtesy of Toyota Motor Sales U.S.A, Inc.)

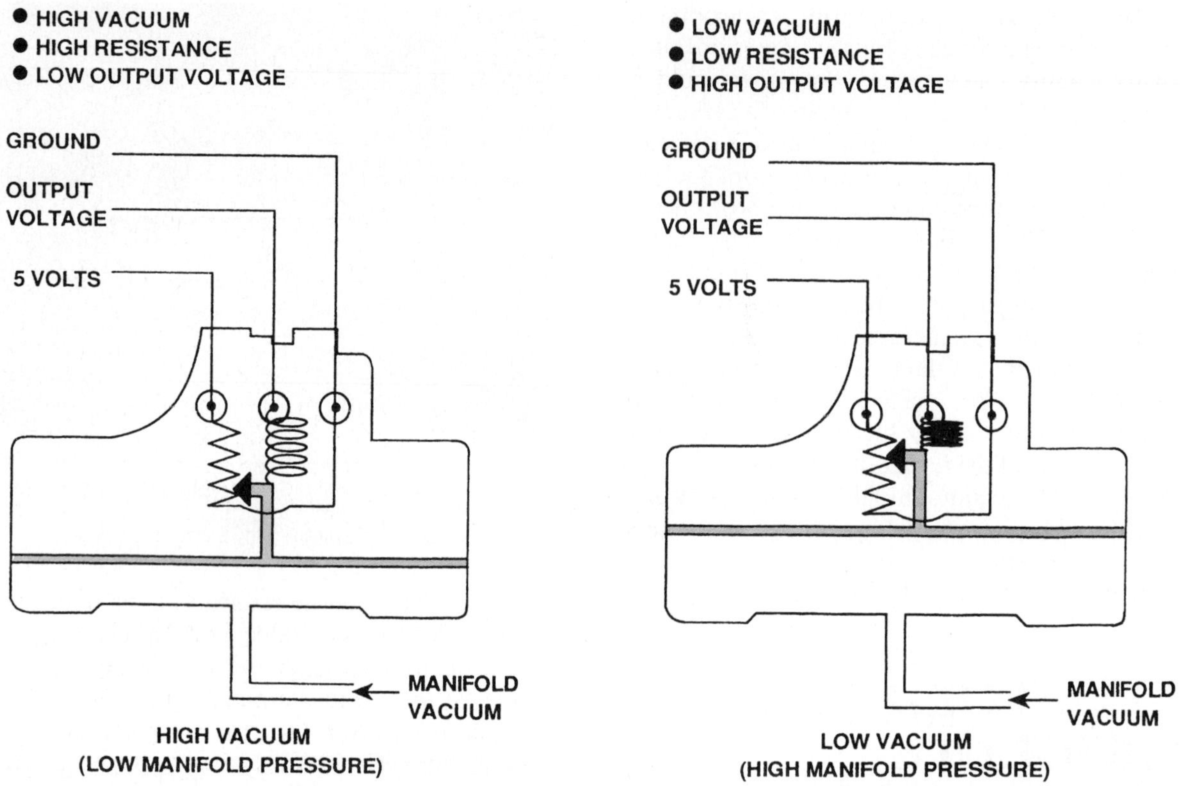

- HIGH VACUUM
- HIGH RESISTANCE
- LOW OUTPUT VOLTAGE

- LOW VACUUM
- LOW RESISTANCE
- HIGH OUTPUT VOLTAGE

GROUND

OUTPUT
VOLTAGE

5 VOLTS

GROUND

OUTPUT
VOLTAGE

5 VOLTS

MANIFOLD
VACUUM

MANIFOLD
VACUUM

HIGH VACUUM
(LOW MANIFOLD PRESSURE)

LOW VACUUM
(HIGH MANIFOLD PRESSURE)

Figure 6-15. A MAP sensor generates a varying voltage based on vacuum. (Courtesy of DaimlerChryler Corporation)

change resistance as the diaphragm flexes. This change in resistance produces a varying voltage that is proportional to the change in pressure that the MAP sensor is sensing.

The MAP sensor circuit consists of the PCM, a 5-volt reference, sensor ground, and the MAP sensor signal, figure 6-15. The 5-volt reference is supplied to the MAP sensor to ensure a steady, accurate voltage supply. The voltage signal from the MAP sensor can vary between 0 and 5 volts. The MAP sensor signal voltage is high when the intake manifold pressure is high, such as no vacuum or very low vacuum. The MAP sensor signal voltage is low when the intake manifold pressure is low, such as high vacuum that occurs at idle or under closed-throttle deceleration.

The MAP sensor is a critical input to the PCM for the correct fuel calculation and delivery. When used on an engine without a Mass Air Flow Sensor, it provides the necessary signal to the PCM for it to calculate the air flow entering the engine. Without an accurate signal from the MAP sensor, the fuel calculations will be incorrect leading to excessively rich or lean mixtures. In many cases, the engine may not even run. The

MAP sensor may also be used for other purposes such as detecting if exhaust gas recirculation is occurring. However, on any fuel-injection vehicle that does not use a mass air flow sensor, it is a critical input to the PCM for the calculation of fuel delivery.

Mass Air Flow Sensor

A more accurate way for the PCM to determine the engine load is through the use of a mass air flow (MAF) sensor, figure 6-16. All of the air entering the engine must flow through the mass air flow sensor. The sensor can measure the amount of air entering the engine and output a signal to the PCM allowing the PCM to determine exactly how much air is entering the engine. This is a more accurate method of determining engine load than by using the MAP sensor. A variety of different mass air flow sensors are used by different manufacturers such as the Karman Vortex air flow meter, the hot wire air flow sensor, and the heated film air flow sensor. The signal output from these air flow sensors may be an analog

Figure 6-16. The mass air flow (MAF) sensor is usually mounted between the air filter and the throttle body and connected by ducting.

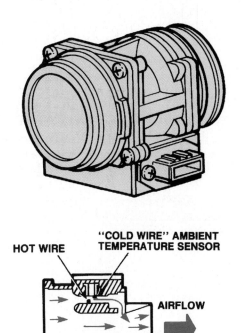

Figure 6-17. Operation of the hot wire MAF sensor. (Courtesy of Ford Motor Company)

voltage or a digital frequency signal, depending on the design.

Figure 6-17 illustrates a hot wire air flow sensor used by Ford. The air flow sensor is located between the air filter element and the throttle body. Usually it is connected with ducting, although in some applications it may be incorporated into the throttle-body assembly. It is important that no air is allowed to enter the engine that does not flow through the air flow sensor. For instance, if there is a hole in the ducting between the air flow sensor and the throttle body then unmetered air will enter the engine and cause lean operation.

As mentioned previously, the input signal to the PCM may be either an analog signal voltage that varies between 0 volts to 5 volts or, in some cases, 0 volts to 12 volts. Other MAF sensors produce a digital signal that varies in frequency as the airflow through the sensor changes, figure 6-7. The output frequency of this type of sensor may range from as low as 10 Hz to above 10,000 Hz. Any change in the amount of airflow through the sensor causes a corresponding change in the frequency output from the MAF.

As mentioned previously, when describing the importance of the MAP sensor input for fuel calculations, the MAF input is also a very critical input to the PCM for the correct calculation of fuel delivery. A malfunctioning MAF sensor may cause an extremely rich or lean condition. Without the MAF sensor input the engine will not run; therefore, default values for mass air flow sensors are generally built into the PCM's memory. If the PCM detects a defect in the MAF sensor or circuit, it will substitute airflow values based on

other operating parameters such as throttle position and engine rpm.

IGNITION REFERENCE SIGNAL

In order for the PCM to initialize and control the injectors, the PCM needs to receive a signal from the engine. This input may have a variety of different names dependent on the manufacturer. It may be called the reference signal (OBD II terminology), engine speed signal, crankshaft reference signal, profile ignition pick-up, or some other name. This signal may originate from a dedicated sensor such as a crankshaft position sensor and be sent directly to the PCM, or it may be generated by the ignition system and be sent to the PCM. Regardless of where this signal originates from, it is a critical input to the PCM for any fuel-injection system. Without this signal the PCM cannot pulse the injectors or on some systems will not energize the fuel pump. Basically, if this reference

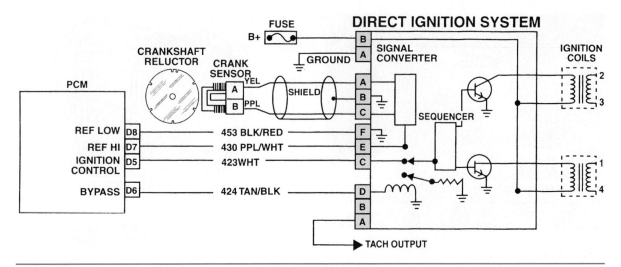

Figure 6-18. This is an example of a reference signal being generated as an AC signal by the crankshaft sensor. It is then converted to a digital signal by the ignition module. (Courtesy of General Motors Corporation)

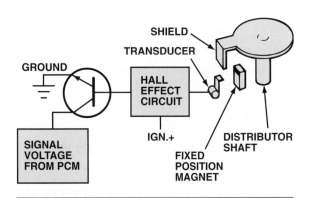

Figure 6-19. When the Hall Effect switch is on, a permanent magnet mounted beside the semiconductor wafer induces a voltage across the wafer. (Courtesy of General Motors Corporation)

signal is missing on a fuel injection vehicle, the engine will not start or run.

Different types of sensors are used to generate these signals. Many manufacturers use magnetic-type sensors that generate an AC signal. This AC signal may be sent directly to the PCM or may be sent to the ignition control module. Usually if this signal is first sent to the ignition module as an AC signal, it is then modified by the ignition module and sent to the PCM as a digital input, figure 6-18. Other systems use Hall Effect sensors, which generate a digital signal that is then sent to the PCM, figure 6-19. Using a Hall Effect type of sensor eliminates the need to convert the signal from an AC signal to a digital signal.

OXYGEN SENSORS

The oxygen sensor is used on fuel-injected vehicles to provide feedback control, figure 6-20. The PCM uses the previously mentioned sensors to determine the correct amount of fuel delivery necessary to make the engine run properly. However, in order to maintain the lowest emissions levels from the vehicle the air-fuel ratio needs to be maintained as close to 14.7:1 as possible. The oxygen sensor is used to sense the amount of oxygen left in the exhaust stream and provide an input signal to the PCM. The PCM uses this signal from the oxygen sensor to adjust the calculated fuel mixture either slightly richer or slightly leaner as needed. This essentially provides feedback control to the PCM so the PCM can "fine tune" the fuel mixture to maintain the desired air-fuel ratio. Whenever the PCM is using the oxygen sensor to correct the fuel mixture, it is referred to as closed loop. Without oxygen sensor feedback the PCM cannot go into closed-loop mode and will remain in the open-loop mode.

The oxygen sensor may be installed in the exhaust manifold, or it may be located downstream from the manifold in the exhaust pipe. This places it directly in the path of the exhaust gas stream where it monitors oxygen level in both the exhaust stream and the ambient air. In a zirconia oxygen sensor the tip contains a thimble made of zirconium dioxide (ZrO_2), an electrical conduction material capable of generating a small voltage in the presence of oxygen.

Figure 6-20. The oxygen sensor is usually mounted as close to the engine as possible. Often it is mounted in the exhaust manifold.

Exhaust from the engine passes through the end of the sensor where the gases contact the outer side of the thimble. Atmospheric pressure enters through the other end of the sensor or through the wire of the sensor and contacts the inner side of the thimble, figure 6-21. The inner and outer surfaces of the thimble are plated with platinum. The inner surface becomes a negative electrode; the outer surface is a positive electrode. The atmosphere contains a relatively constant 21 percent of oxygen. Rich exhaust gases contain little oxygen. Exhaust from a lean mixture contains more oxygen.

Negatively charged oxygen ions are drawn to the thimble where they collect on both the inner and outer surfaces, figure 6-22. Because the oxygen present in the atmosphere exceeds that in the exhaust gases, the air side of the thimble draws more negative oxygen ions than the exhaust side. The difference between the two sides creates an electrical potential, or voltage. When the concentration of oxygen on the exhaust side of the thimble is low, a high voltage (0.60 to 1.0 volts) is generated between the electrodes. As the oxygen concentration on the exhaust side increases, the voltage generated drops low (0.00 to 0.3 volts).

This voltage signal is sent to the computer where it passes through the input conditioner for amplification. The computer interprets a high-voltage signal (low oxygen content) as a rich air-fuel ratio, and a low-voltage signal (high oxygen content) as a lean air-fuel ratio. Based on the O_2S signal (above or below 0.45 volts) the computer compensates by making the mixture either leaner or richer as required to continually vary close to a 14.7:1 air-fuel ratio to satisfy the needs of the three-way catalytic converter. The O_2S is therefore the key sensor of an electronically controlled fuel-metering system for emission control.

An O_2S does not send a voltage signal until its tip reaches a temperature of about 572°F (300°C). Also O_2 sensors provide their fastest response to mixture changes at about 1,472°F (800°C). The computer cannot rely on the O_2 sensor's signal when the engine starts and the O_2S is cold, or when the engine is left idling for an extended period of time, allowing the O_2S to cool off. So the computer runs the engine in the open-loop mode, drawing on prerecorded data in the PROM for

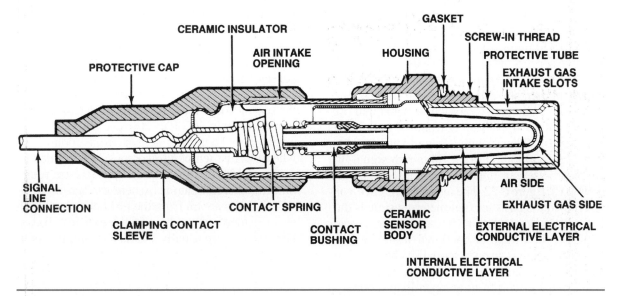

Figure 6-21. The components of a typical O_2 sensor. (Courtesy of Ford Motor Company)

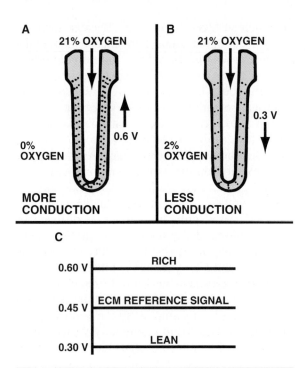

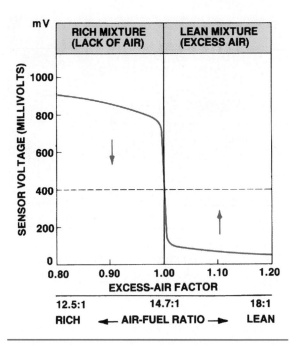

Figure 6-23. The O$_2$S provides its fastest response at the stoichiometric air-fuel ratio of 14.7 to 1.

Figure 6-22. A difference in oxygen content between the atmosphere and exhaust will generate varying voltage from an oxygen sensor.

fuel control on a cold engine, or when O$_2$S output is not within certain limits. Figure 6-23 shows the operating range of an O$_2$S at 1,472°F (800°C). Sensor voltage changes fastest (switches) at an air-fuel ratio of 14.7:1 at this temperature.

An O$_2$S measures oxygen, but it does not measure air-fuel ratio. If the engine misfires, exhaust oxygen content is high because less oxygen is consumed in the combustion. There is a large amount of oxygen in the unburned exhaust mixture, and the sensor delivers a false "lean mixture" signal. Keep this in mind when trouble shooting. Unlike resistive sensors, an O$_2$S (as well as any generator sensor) does not require a reference voltage. The GM computer, however, uses an internal reference voltage as a comparison for the sensor signal. Because the O$_2$S signal ranges from 0.1 to 0.3 volts (100 mV to 300 mV) with a lean mixture to 0.6 or 0.9 volts (600 to 900 mV) with a rich signal, the computer uses an internal reference of .45 volt (450 mV), figure 6-22. The internal reference voltage is also the basis for fuel-metering signals during open-loop operation. If the computer reads 450 mV, then there is no signal from the O$_2$S; therefore, there is an open loop. If it changes from 450 mV, the computer reads this change and changes the circuit from an open-loop mode to a closed-loop mode.

To make the system more responsive, manufacturers went first to the concept of installing a separate O$_2$S in each manifold of a V-type engine. This arrangement was not always used, due to cost, until the introduction of OBD II second-generation onboard diagnostics. Yet the introduction of the heated exhaust gas sensor (HO$_2$S), figure 6-24, ensured quick warm-up of the HO$_2$S and rapid change to closed-loop operation, even if the engine was left idling after start-up.

An **HO$_2$S** is constructed and operates the same as an O$_2$S, but contains a built-in heater powered by the vehicle battery whenever the ignition is in the run position. A third wire to the sensor delivers battery current (3 amperes or less) to the sensor electrode. This helps warm the sensor to operating temperature more quickly and permits the sensor to operate at lower exhaust temperature (approximately 392°F or 200°C). The heating element also keeps the sensor from cooling off when exhaust temperature drops, such as during prolonged idling. In the past, HO$_2$Ss were found on turbocharged engines where the sensor is installed downstream from the turbocharger, which absorbs much of the heat in the exhaust. Today, HO$_2$ sensors are usually standard equipment both before and after the catalytic converter.

When three wires are present, one is signal, one is sensor ground, and one is for the heater. When four wires are present, one is signal, one is

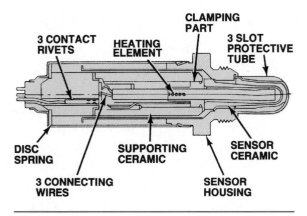

Figure 6-24. The components of a typical HO₂S. (Courtesy of Ford Motor Company)

sensor ground, one is heater voltage, and one is heater ground.

Titania Exhaust Gas Oxygen Sensors

While zirconia O_2 sensors have proven themselves, O_2 sensors containing titanium dioxide, or titania-sensing elements, are also used. Both types of sensors provide a feedback signal indicating the relative rich or lean condition of the exhaust. Unlike the zirconia sensor, the titania sensor does not generate the voltage. Instead it detects exhaust oxygen content by acting as a variable resistor.

The computer provides a constant reference voltage to the sensor. As the oxygen content of the exhaust changes, the resistance of the titania-sensing element varies. Because the resistance of the titania is also affected by temperature, a heating element is incorporated in the sensor to maintain a constant temperature of approximately 1475°F (850°C).

Figure 6-25 shows the heated titania O_2S as used on 4.0 L Jeep engines. The computer provides the sensor with a 5-volt reference voltage. The signal voltage is measured between the sensor and a fixed resistor in a series with the sensor. As sensor resistance changes, the signal voltage (voltage drop) also changes. With a rich mixture (low oxygen content), the sensor signal voltage is below 2.5 volts. With a lean mixture (high oxygen content), sensor signal voltage exceeds 2.5 volts. Although signal voltage direction, relative to oxygen content of the exhaust gas, is the opposite of a zirconia O_2S, the computer program allows it to read the signal correctly and makes the necessary adjustments in fuel metering.

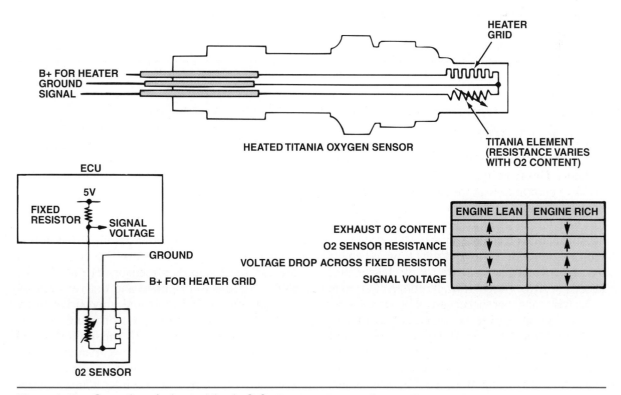

	ENGINE LEAN	ENGINE RICH
EXHAUST O2 CONTENT	↑	↓
O2 SENSOR RESISTANCE	↓	↑
VOLTAGE DROP ACROSS FIXED RESISTOR	↓	↑
SIGNAL VOLTAGE	↑	↓

Figure 6-25. Operation of a heated titania O_2S. (Courtesy of DaimlerChrysler Corporation)

A/F Sensor Detecting Circuit

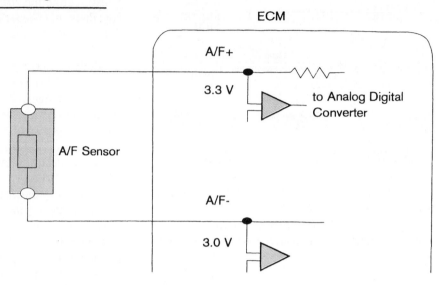

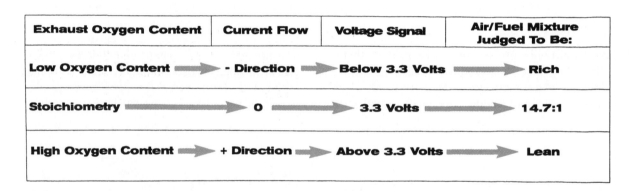

Exhaust Oxygen Content	Current Flow	Voltage Signal	Air/Fuel Mixture Judged To Be:
Low Oxygen Content ➡	– Direction ➡	Below 3.3 Volts ➡	Rich
Stoichiometry ➡	0 ➡	3.3 Volts ➡	14.7:1
High Oxygen Content ➡	+ Direction ➡	Above 3.3 Volts ➡	Lean

Figure 6-26. The air fuel (A/F) sensor is used by some manufacturers. (Provided courtesy of Toyota Motor Sales U.S.A., Inc.)

Air-Fuel Ratio Sensor

The **air-fuel ratio sensor** is used by some manufacturers. This type of sensor produces a positive or negative current flow as the oxygen content of the exhaust stream changes, figure 6-26. At 14.7:1 air-fuel ratio the sensor produces no current. If the exhaust stream oxygen content decreases (rich), the sensor produces a negative current flow. When the oxygen content increases (lean), the sensor produces a positive current flow. It is often referred to as the wide range sensor because of its ability to detect air-fuel ratios over wider ranges than the zirconia or titania sensors. This sensor operates at approximately 1200°F (650°C). The detection circuit in the PCM measures the direction and how much current is being produced. This means that

the PCM also knows what the air-fuel ratio is and can adjust the fuel mixture much quicker than systems that use other types of O_2 sensors.

Other Sensor Inputs

While there are many other inputs used on fuel injection systems, the inputs listed previously are the most important sensors needed by the PCM in order to calculate and deliver the correct amount of fuel to ensure good driveability and low emissions. Other sensor inputs such as intake air temperature (IAT) and vehicle speed sensor (VSS) are used by most manufacturers. How important of a role these sensors play on fuel delivery varies between systems. For instance, some systems rely more heavily on the IAT. If the IAT were to become inoperative, poor driveability and higher emissions may result. On other systems, if the IAT were to become inoperative, there may be no noticeable symptoms. If the vehicle is equipped with an electronically shifted transmission, then the VSS plays a more important role as an input to the PCM. Other systems may use the VSS more to aid in diagnostics and setting code parameter conditions.

What is important to remember is the role that these sensors play in providing critical input information to the PCM and their impact on driveability and emissions. While there are unique differences between manufacturers' systems, there are similarities in the role these sensors play on all systems. Understanding how these sensors work electrically is important knowledge that you must have in order to diagnose engine management systems.

SUMMARY

All inputs to the computer work in one of three ways electrically. These three ways are switched, analog, and digital. While there are many inputs to an engine management system, certain of these inputs are more critical than others. These sensors provide the necessary information that the PCM needs in order to calculate the correct amount of fuel delivery for the best driveability and lowest emissions. The oxygen sensor serves as the feedback sensor for an engine management system. The PCM calculates how much fuel is needed but uses the oxygen sensor as a feedback device to determine if the fuel delivery is correct or if it needs to be modified. When the engine management system is using the oxygen sensor to modify the fuel delivery, the system is said to be operating in closed-loop mode. In the open-loop mode the PCM utilizes all the other important input sensors; however, it ignores the oxygen sensor.

Review Questions

Choose the letter that represents the best possible answer to the following questions:

1. An exhaust gas oxygen sensor (O_2) is an example of a:
   a. Resistor
   b. Potentiometer
   c. Generator
   d. Solenoid

2. The reference value sent to a sensor by the computer must be:
   a. Above battery voltage
   b. Exactly the same as battery voltage
   c. Less than minimum battery voltage
   d. Battery voltage plus or minus 5 volts

3. The simplest digital sensor is a:
   a. Solenoid
   b. Switch
   c. Timer
   d. Relay

4. A variable resistance sensor is called a:
   a. Potentiometer
   b. Thermistor
   c. Transformer
   d. Generator

5. Technician A says throttle position sensors are potentiometers. Technician B says throttle position sensors are analog. Who is correct?
   a. Technician A
   b. Technician B
   c. Both a and b
   d. Neither a nor b

6. A digital signal is:
   a. A varying voltage signal
   b. An AC signal
   c. A repeating on/off DC signal
   d. None of the above

7. A switched input produces a(n) _____ input signal.
   a. Analog
   b. Varying
   c. Digital
   d. None of the above

8. In a pull-down or ground-side switched input the voltage is sourced from:
   a. An external circuit
   b. Within the PCM
   c. Either of the above
   d. Neither of the above

9. Which of the following inputs **does not** produce an analog input signal?
   a. Throttle position sensor
   b. Engine coolant temperature sensor
   c. Manifold air pressure sensor
   d. Power-steering pressure switch

10. Mass air flow sensors generate what kind of input signals?
    a. Digital
    b. Analog
    c. Both a and b
    d. Neither a nor b

11. Which of the following sensors would not be as important to the engine controller for the fuel delivery calculation?
    a. MAP
    b. MAF
    c. IAT
    d. ECT

12. Most engine coolant temperature sensors are:
    a. Positive temperature coefficient thermistors
    b. Digital
    c. Negative temperature coefficient thermistors
    d. None of the above

13. Technician A says that the resistance of an NTC-type thermistor decreases as the temperature increases. Technician B says that the resistance of an NTC-type thermistor increases as the temperature increases. Who is correct?
    a. Technician A
    b. Technician B
    c. Both a and b
    d. Neither a nor b

14. The computer uses the throttle position sensor's input mainly for:
    a. Cruise control operation
    b. Acceleration enrichment
    c. Steady-state cruise fuel delivery
    d. Start-up enrichment

15. Which of the following sensors does the engine management computer use to determine engine load?
    a. Mass air flow sensor
    b. Manifold air pressure sensor
    c. Either a or b
    d. Neither a nor b

16. Technician A says that high engine vacuum causes a high voltage output from the MAP sensor. Technician B says high engine vacuum causes a low voltage output from the MAP sensor. Who is correct?
    a. Technician A
    b. Technician B
    c. Both a and b
    d. Neither a nor b

17. On many fuel-injection systems if the reference signal to the computer is lost:
    a. The injectors will not be pulsed
    b. The fuel pump will not run
    c. Both a and b are correct
    d. Neither a nor b is correct

18. When the engine control computer is using the oxygen sensor signal to adjust the fuel mixture, it is known as:
    a. Open-loop mode
    b. Closed-loop mode
    c. Stoichiometric
    d. Run mode

19. All oxygen sensors produce a varying voltage based on the oxygen content in the exhaust stream.
    a. True
    b. False

20. Technician A says that a zirconia-type oxygen sensor produces a high voltage (above 600 mV) when the exhaust is rich. Technician B says that the zirconia oxygen sensor produces a low voltage (below 300 mV) when the exhaust is lean. Who is correct?
    a. Technician A
    b. Technician B
    c. Both a and b
    d. Neither a nor b

7

Engine Management Output Devices

OBJECTIVES

After completion and review of this chapter, you will be able to:

- List the most common outputs on a typical fuel engine vehicle.
- Describe the ways outputs work electrically.
- Define the term *pulse width modulation*.
- Describe the differences between high-side and low-side drivers.

KEY TERMS

high-side drivers   serial data
low-side drivers

INTRODUCTION

In the previous chapter you looked at the inputs to engine management computers. You learned how inputs work electrically and which inputs are critical inputs for the correct fuel delivery calculation by the PCM. This chapter will focus on the output side of the engine management system. Typically, most engine management systems have more input devices than output devices. The computer needs a lot of information about engine operating conditions in order to deliver the correct amount of fuel to the engine for the best driveability and the lowest emissions. Figure 7-1 shows some typical outputs of an engine management system. A given system may use all of these outputs or only some of these outputs depending on the complexity of the system. As with input devices however, all systems use some of these outputs. For example, every fuel injection system has to have an injector or injectors to deliver fuel to the engine. All engine management computers control ignition timing to help achieve the lowest emissions levels. Even the most basic early engine management systems controlled fuel delivery and ignition timing. As the complexity of the systems has increased, the number of output devices controlled by the engine management system has also increased. Many of these outputs may have no impact on fuel delivery or emissions. However, since many of these outputs could affect engine operation such as increased engine load, it has become more common to have the engine management system control these outputs.

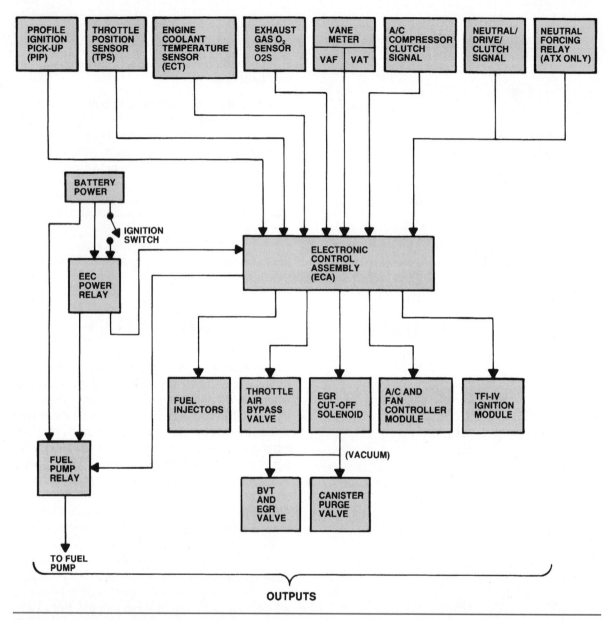

Figure 7-1. Typical engine control system inputs and outputs. (Courtesy of Ford Motor Company)

Another reason for the increasing number of outputs that are controlled by the computer is multiplexing. In Chapter 5 you learned how computers communicate with each other using data lines. One of the advantages of multiplexing is the elimination of redundant sensors and wiring. A good example would be the engine temperature warning lamp. The engine temperature lamp was commonly controlled by a separate temperature switch or sensor installed on the engine. It was hardwired to the instrument cluster. At the same time the engine management computer needed to know the coolant temperature, so an engine coolant tempera-

ture sensor served as an input to the computer. By redesigning the circuit of the engine temperature warning lamp and connecting it to the output side of the PCM, the coolant temperature switch or sensor and wiring could be eliminated. Since the PCM already knows the engine temperature, it can control the operation of the temperature warning lamp.

Most all outputs work electrically in one of three ways. These are:

- Switched
- Pulse width modulated
- Digital

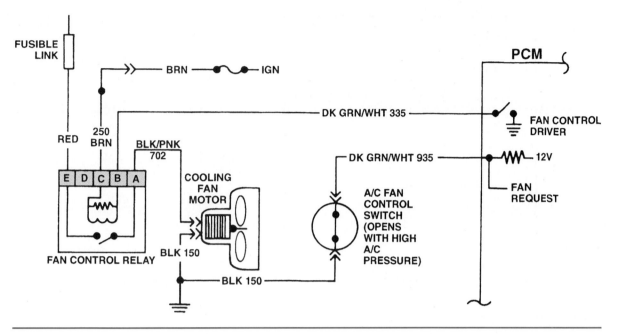

Figure 7-2. Electric cooling fan control circuit. (Courtesy of General Motors Corporation)

A switched output is an output that is either on or off. An example would be a relay. In many circuits the PCM uses a relay to switch a device on or off. This is because the relay is a low-current device that can switch a higher-current device. Most computer circuits cannot handle a lot of current. By using a relay circuit as shown in figure 7-2, the PCM provides the output control to the relay that in turn provides the output control to the device. The relay coil that the PCM controls typically draws less than 0.5 amps. By contrast the device that the relay controls may draw 30 amps or more.

These switches that are used are actually transistors. They are often called output drivers. Whenever you see the term *driver,* you can safely assume that a transistor is being used to complete the circuit or "switch it on or off." There are two kinds of drivers used to control outputs; these are low-side and high-side drivers.

LOW-SIDE DRIVERS

Low-side drivers are transistors that complete the ground path in the circuit. Figure 7-3 illustrates a fuel pump relay circuit that is ground-side switched by the engine controller. Ignition voltage is supplied to the relay as well as battery voltage. The engine controller's output is connected to the ground side of the relay coil. The controller energizes the fuel pump relay by turning the transistor on and com-

pleting the ground path for the relay coil. A relatively low current flows through the relay coil and transistor that is inside the controller. This causes the relay to switch and provides the fuel pump with battery voltage. The majority of switched outputs have typically been low-side drivers. Typically, high-side drivers have only been used in limited applications.

HIGH-SIDE DRIVERS

High-side drivers are beginning to be used in more applications. High-side drivers are being used to control high-current devices. As can be seen in figure 7-4, the outputs shown on the right are all controlled on the power side of the circuit using high-side drivers. In these applications when the transistor is switched on, voltage is applied to the device. A ground has been provided to the device, so when the high-side driver switches, the device will be energized. In some applications high-side drivers are used instead of low-side drivers to provide better circuit protection. General Motors vehicles have used a high-side driver to control the fuel pump relay instead of a low-side driver. In the event of an accident, should the circuit to the fuel pump relay become grounded, a high-side driver would cause a short circuit, which would cause the fuel pump relay to de-energize.

Using a schematic should help you determine what type of driver is being used to control an

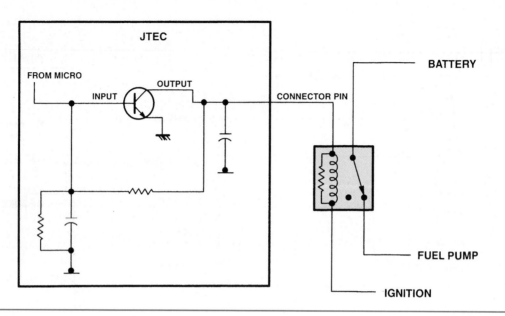

Figure 7-3. Fuel pump relay circuit. (Courtesy of DaimlerChrysler Corporation)

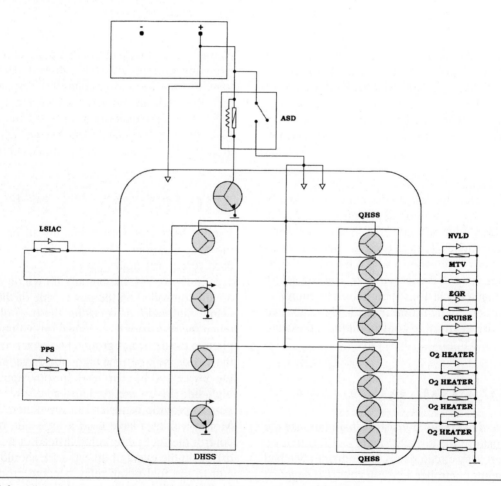

Figure 7-4. High-side controlled outputs. (Courtesy of DaimlerChrysler Corporation)

output device. If the schematic is drawn properly, you should be able to determine if the driver is supplying a ground to the output device or if the driver is switching voltage to the device. If the schematic does not show the driver, then you will need to study the circuit to determine if the output device is provided a ground from the external circuit or if the output device is provided voltage from the external circuit. From this you should be able to determine if the driver is a low-side or high-side driver. You will need to determine this before you can diagnose faults in the output device or circuit.

PULSE WIDTH MODULATION

Pulse width modulation (PWM) is a method of controlling an output using a digital signal. Instead of just turning devices on or off, the engine controller needs to control output devices more precisely. An example would be a vacuum solenoid. If the vacuum solenoid were simply controlled by a switched driver, it would have one of two states, either on or off. That would mean that either full vacuum would flow through the solenoid or no vacuum would flow through the solenoid. However, if we want to control how much vacuum will flow through the solenoid, then we can use pulse width modulation. A PWM signal is a digital signal usually 0 volts and 12 volts that is cycling at a fixed frequency. By varying the length of time that the signal is on, as opposed to how long of time the signal is off, provides a signal that can vary the on and off time of an output. The number of on and off times is referred to as duty cycle. Figure 7-5 shows an oscilloscope pattern of a PWM signal that has a pulse width of 10 percent on and 90 percent off. Depending on the frequency of the signal, which is usually fixed, this signal would turn the device on and off a fixed number of times per second. However, each time the voltage is high (12 volts) would represent only 10 percent. The other 90 percent of the time the signal would be low (0 volts). Figure 7-6 shows the same PWM signal; however, in this case, the time that the voltage is high (12 volts) represents 90 percent and the time that the signal is low (0 volts) is only 10 percent. In other words, if this signal were applied to the vacuum solenoid, the solenoid would be on 90 percent of the

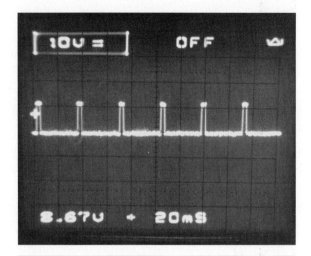

Figure 7-5. This oscilloscope pattern shows a 0 volt to 12 volt pulse width modulated signal that has a 10 percent on time and a 90 percent off time.

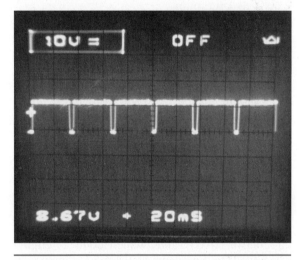

Figure 7-6. This oscilloscope pattern shows a 0 volt to 12 volt pulse width modulated signal that has a 90 percent on time and a 10 percent off time.

time. This would allow more vacuum to flow through the solenoid. The computer has the ability to vary this on and off time or pulse width modulation at any rate between 0 percent and 100 percent.

By using this method, a computer can very precisely control an output device. A good example is a fuel injector. The amount of fuel delivered to the engine from an injector is determined by the amount of injector on time of the injector. The computer varies the pulse width of the injector to vary the amount of fuel delivered to the engine. PWM may be used to control vacuum through a solenoid, the amount of purge

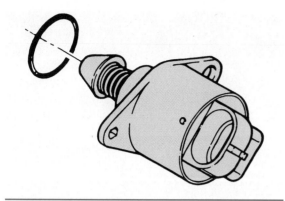

Figure 7-7. The stepper motor is a DC motor that moves in specific increments. (Courtesy of Ford Motor Company)

of the evaporative purge solenoid, the speed of a motor, control of a linear motor, or even the intensity of a lightbulb.

Digital Outputs

Another type of output signal is a digital output. This should not be confused with pulse width modulation. A digital output is used to control stepper motors. A stepper motor, figure 7-7, is also a digital actuator. Stepper motors are direct current motors that move in fixed steps or increments from de-energized (no voltage) to fully energized (full voltage). A stepper motor often has as many as 120 steps of motion.

A common use for stepper motors is as an idle air control (IAC) valve, which controls engine idle speeds and prevents stalls due to changes in engine load. When used as an IAC, the stepper motor is usually a reversible DC motor that moves in increments, or steps. The motor moves a shaft back and forth to operate a conical valve. When the conical valve is moved back, more air bypasses the throttle plates and enters the engine, increasing idle speed. As the conical valve moves inward, the idle speed decreases.

When using a stepper motor that is controlled by the PCM, it is very easy for the PCM to keep track of the position of the stepper motor. By counting the number of steps that have been sent to the stepper motor, the PCM can determine the relative position of the stepper motor. While the PCM does not actually receive a feedback signal from the stepper motor, it does know how many steps forward or backward the motor should have moved.

CRITICAL OUTPUTS

As with inputs certain critical outputs are common to most all engine management systems. Figure 7-1 illustrates some typical outputs that may be found on an engine management system. Remember that the goal of an engine management system is to provide good driveability and maintain low emissions. Obviously, if this is the main intent of an engine management system, then fuel control and timing control are the most important functions of these systems.

FUEL INJECTORS

The fuel injector is a solenoid that is controlled by the PCM. Throttle body systems use one or two injectors that inject fuel into the engine from above the throttle plates. Port fuel-injection systems use individual injectors that inject fuel in the intake ports. Pressurized fuel is provided to the injector(s) using an electric fuel pump. The injector driver, usually a low-side driver, energizes the fuel injector, which allows fuel to flow through the injector. The amount of fuel that flows through the injector is primarily controlled by the injector on time or pulse width. The PCM calculates this pulse width based on inputs from the various input sensors such as:

- Engine coolant temperature (ECT)
- Engine rpm
- Manifold air pressure (MAP) or mass air flow (MAF)
- Throttle position (TPS)
- Oxygen sensor (O_2S)
- Intake air temperature (IAT)

In addition to controlling pulse width of the injector, the PCM also varies the frequency of the injector or how often the injector is actuated, figure 7-8. The injection frequency varies according to engine speed. As engine rpm increases, the volume of air entering it also increases. Even if the same air-fuel ratio is required, fuel injection frequency must increase with engine speed to be maintained. In figure 7-8 the middle waveform is of a lower engine speed and injection frequency than the bottom waveform.

IGNITION CONTROL

Ignition control (IC) is the OBD II terminology for the output signal from the PCM to the ignition system that controls engine timing. Previously

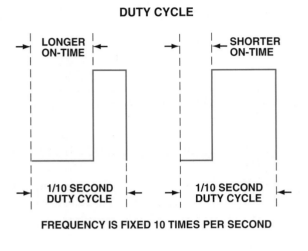

DUTY CYCLE

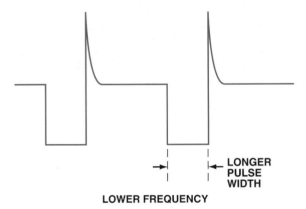

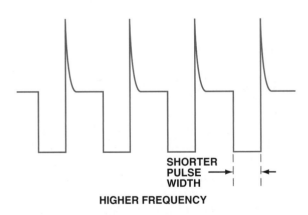

Figure 7-8. The top of this illustration shows oscilloscope waveforms comparing a longer on time to a shorter on time, each with the same duty cycle. The bottom fuel injector waveform has a shorter pulse width than the middle waveform. The bottom waveform has a higher frequency of injection.

each manufacturer used a different term to describe this signal. For instance, Ford referred to this signal as spark output (Spout) and General Motors referred to this signal as electronic spark timing (EST). This signal is now referred to as the ignition control (IC) signal. The ignition control signal is usually a digital output that is sent to the ignition system as a timing signal. If the ignition system is equipped with an ignition module, then this signal is used by the ignition module to vary the timing as engine speed and load changes. If the PCM directly controls the coils, such as most coil-on-plug ignition systems, then this IC signal directly controls the coil primary. As can be seen in figure 7-9, there is a separate IC signal for each ignition coil. The IC signal controls the time that the coil fires; that is to say, it either advances or retards the timing. In addition, on many systems this signal controls the duration of the primary current flow in the coil, which is referred to as the dwell.

ELECTRONIC THROTTLE CONTROL

One of the latest outputs of a critical nature is the electronic throttle control (ETC) system used by some manufacturers. This system eliminates the mechanical linkage between the accelerator pedal and the throttle plates. The throttle plates are controlled by an electric actuator motor and the PCM, figure 7-10. Incorporated into the electronic throttle body are also dual throttle position sensors that provide throttle position input to the PCM. The output is the electric actuator motor that is controlled by a duty-cycle signal from the PCM. The throttle plate is held closed by spring tension to a preset "limp home" position. This allows the throttle plates to always go closed if there should be a loss of power to the throttle actuator. Dual accelerator pedal position sensors are used to signal the PCM of the driver's requested accelerator pedal position. In response to these signals the PCM duty-cycles the actuator motor to the desired throttle position. Dual throttle position sensors incorporated into the throttle actuator provide throttle position feedback to the PCM, figure 7-11.

CAUTION: Do not manually open the throttle plates on these throttle actuators. Manually trying to open the throttle plates may damage the actuator. On most applications a scan tool must be used to command the actuator to move.

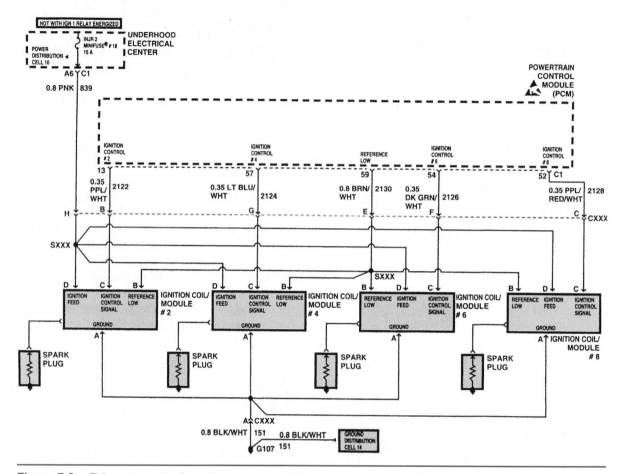

Figure 7-9. This schematic of a coil-over-plug ignition system shows the IC signal from the PCM to the coil. (Courtesy of General Motors Corporation)

Figure 7-10. An electronic throttle control used on a 5.7L Chrysler engine. (Courtesy of DaimlerChrysler Corporation)

IDLE SPEED CONTROL

Most fuel-injection systems use some form of idle air control. Idle air control is usually an idle air bypass circuit that directs air around the throttle plates when they are closed, figure 7-12. The PCM controls the idle speed within a range based on other inputs and the load on the engine. Usually a stepper motor is used for the idle air control valve. The PCM can be programmed to move the idle air control valve a certain amount of steps such as for start-up. This provides the fast idle needed for the initial start-up of the engine. As long as the driver is not controlling the throttle plate position by having their foot on the accelerator, the PCM will control the amount of air entering the engine with the idle air control valve. This provides the correct idle speed based on load and engine temperature.

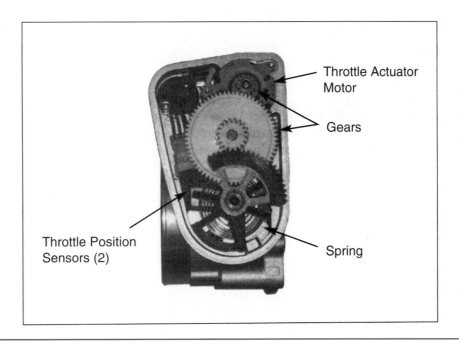

Figure 7-11. 5.7L ETC throttle actuator motor, throttle position sensors, spring, and gears. (Courtesy of Daimler-Chrysler Corporation)

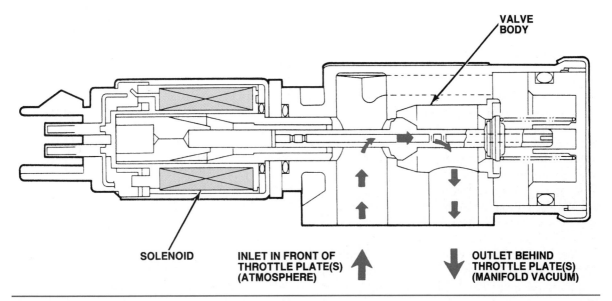

Figure 7-12. A solenoid-operated air bypass valve is used with most domestic multipoint systems to provide idle airflow around the throttle plate. (Courtesy of Ford Motor Company)

FUEL PUMP

On most engine management systems the fuel pump is a controlled output from the PCM, figure 7-13. Upon initial key on input, the PCM will energize the fuel pump for about two seconds. This provides a prime pulse to pressurize the fuel rail so that there is an adequate supply of fuel at the injectors. After the initial key on prime pulse, the PCM will shut the fuel pump off and wait to receive an engine cranking signal from the engine. Once this rpm signal reaches a certain value, the PCM will re-energize the fuel pump circuit allowing the fuel pump to run continuously. As long as an rpm signal is present at the PCM input, the fuel pump will remain energized. If this

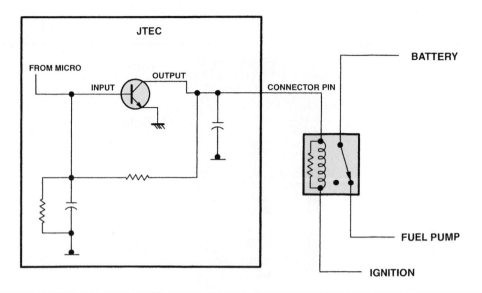

Figure 7-13.   Fuel pump relay circuit. (Courtesy of DaimlerChrysler Corporation)

rpm signal is lost when the engine is running, the fuel pump will shut down on most vehicles.

SERIAL DATA

While not always thought of as an output, **serial data** is an important output from the PCM. Earlier systems usually only had one data line, which was a direct output from the PCM to the data link connector (DLC). This data line allows a scan tool to interface to the PCM and receive data about operating conditions, read diagnostic trouble codes, and in some cases provide bidirectional communication between the scan tool and the PCM. As previously discussed in Chapter 5, most vehicles today use multiplexing to allow this data line to connect to other modules on the vehicles. In these applications all the modules may be able to send data as an output on the data line. Regardless of the number of modules sharing the data line, the PCM still sends data as an output on this data line and can be accessed at the DLC using a compatible scan tool.

MALFUNCTION INDICATOR LAMP

Malfunction indicator lamp (MIL) is the correct OBD II terminology for the warning lamp that the PCM controls, used to indicate a malfunction that would cause emissions to exceed certain levels. In

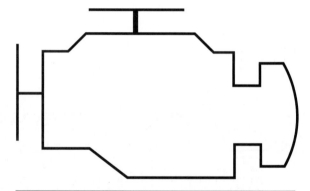

Figure 7-14.   This engine symbol is the malfunction indicator lamp on some OBD II equipped vehicles.

the past manufacturers used different terms to describe this lamp such as power loss, check engine, or emissions. Also on non-OBD II vehicles this lamp may have illuminated for failures in other systems that were not emissions related. OBD II vehicles may only illuminate this light when a failure occurs in the OBD II system that would cause emissions from the vehicle to become excessive. Although the lamp is called a malfunction indicator lamp, the indicator on the dash may say "check engine" or "service engine soon." Other vehicles may show an amber symbol of an engine, figure 7-14. As mentioned previously, this lamp is controlled by the PCM, figure 7-15, and is commanded on when a failure in the OBD II systems exists that would cause excessive emissions from the vehicle. Whenever the lamp illuminates, a diagnostic trouble code

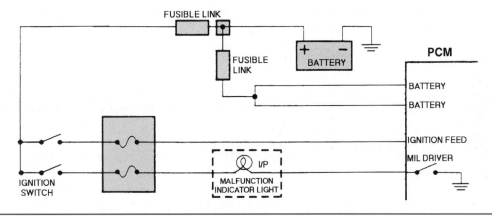

Figure 7-15. MIL circuit. (Courtesy of General Motors Corporation)

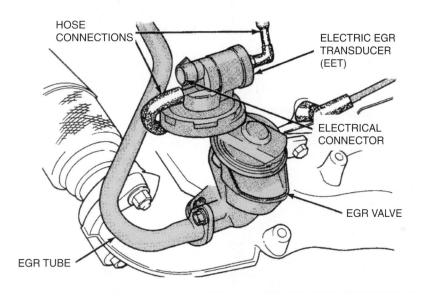

Figure 7-16. PCM controller EGR transducer (solenoid). (Courtesy of DaimlerChrysler Corporation)

will be stored in the PCM to indicate what system or circuit the failure is in.

EXHAUST GAS RECIRCULATION

Since the introduction of exhaust gas recirculation (EGR) valves in the early 1970s, various types of valves and controls have been used. These will be covered in more detail in Chapter 13. Most current EGR valves are directly controlled by the PCM. Accurate control of the EGR valve using engine vacuum was difficult at best. As manufacturers added engine management computers, the EGR system typically became a controlled output of the PCM. How this control is accomplished varies between different manufacturers. Early systems typically used a vacuum-operated EGR valve and the PCM controlled the vacuum to the valve through the use of a solenoid, figure 7-16. The EGR solenoid is pulse width modulated (PWM) by the PCM to control the amount of vacuum that reaches the EGR diaphragm. This allows the valve to meter the amount of exhaust gas recirculation occurring to maintain good driveability while lowering NO_x emissions. While this method provided more accurate control of the EGR system, vacuum-actuated EGR valves cannot be controlled as accurately as electronic valves. Many manufacturers have moved away from vacuum-actuated EGR valves and instead are using electric-actuated EGR valves, figure 7-17. These valves may be a linear-type electric actuator or a solenoid-type design. The advantage to using a linear actuator is that it can respond to changes up to 10 times faster than the vacuum-actuated valve. They are also more precise and can more accurately meter the exhaust gases into the engine. The PCM controls the linear EGR valve by using pulse width modulation to the coil. Incorporated into the valve is also a pintle position feedback sensor that is an input to the PCM, figure 7-18. This provides a pintle position signal to the PCM so that the PCM can position the pintle accurately and also provide diagnostic capabilities for the EGR system.

OTHER OUTPUTS

Most engine management systems use the outputs that were previously discussed. Many systems have additional outputs that vary from vehicle to vehicle. For instance, most AC com-

pressors are PCM controlled. On many vehicles the PCM controls operation of the cruise control. Many electrically shifted transmissions use the PCM to control the shift solenoids or pressure solenoids in the transmission. Other manufacturers use a separate transmission control module to operate the transmission. The trend has been to allow the PCM to control more devices on the vehicle and engine that traditionally have not been PCM controlled. In some instances, the starter is now PCM controlled. Undoubtedly this trend will continue in the future.

EVAPORATIVE AND PURGE SOLENOIDS

The evaporative system is used to prevent evaporative emissions from escaping the vehicle's fuel system into the atmosphere. This system will be covered in detail in Chapter 13. However, the evaporative controls are now usually PCM output controlled devices on both OBD II equipped vehicles and many newer non-OBD II equipped vehicles.

Referring to figure 7-19 of an enhanced evaporative system, there are two controlled outputs from the PCM. There is a normally closed purge solenoid that is pulse width modulated by the PCM when purging of the vapor canister is needed. This purge will occur after the engine is warm and certain other parameters are met such as throttle angle, vehicle speed, or engine rpm.

In addition to the purge solenoid, this system is equipped with a vent solenoid. The vent solenoid is a normally open solenoid that allows fresh air to be drawn into the vapor canister when purge is oc-

Figure 7-17. A linear EGR valve is controlled by a pulse width modulation signal from the PCM.

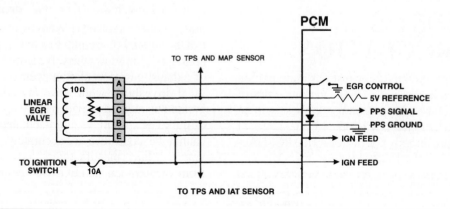

Figure 7-18. Linear EGR circuit. (Courtesy of General Motors Corporation)

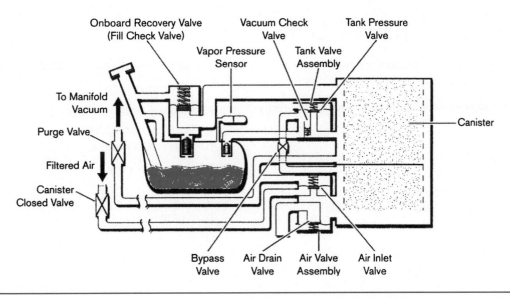

Figure 7-19. An enhanced evaporative system. (Provided courtesy of Toyota Motor Sales U.S.A., Inc.)

curring. The PCM will close the vent solenoid under certain operating conditions to test the integrity and functionality of the evaporative system.

SUMMARY

Electrically PCM outputs work one of three ways—switched, pulse width modulated, and digital. Transistor drivers are used to switch output devices. These drives can be low-side or high-side drivers. Low-side drivers switch the ground to the device and high-side drivers switch voltage to the device. Pulse width mod-

ulation can provide a variable output to a device by controlling the on and off time of a digital signal. Pulse width modulation is used to control injectors, solenoids, and motors among other things.

All engine management systems share critical outputs such as fuel injectors, ignition control, electronic throttle control, idle speed, serial data, and the malfunction indicator lamp. Many emission control devices such as exhaust gas recirculation and evaporative purge and vent solenoids are controlled by the PCM instead of using vacuum as done previously. This improves the accuracy and response rate of these devices or systems.

Review Questions

Choose the letter that represents the best possible answer to the following questions:

1. Which way electrically do PCM outputs work?
   a. Pulse width modulation
   b. Switched
   c. Digital
   d. Any of the above

2. Transistors used as high-side drivers switch the _____ side of the circuit.
   a. Ground
   b. B+
   c. Both ground and B+
   d. None of the above

3. A digital output would be used to control which of these devices?
   a. IAC
   b. Purge solenoid
   c. Relay
   d. Fuel injector

4. What type of signal could be used to control how much vacuum flows through a solenoid?
   a. Switched
   b. Analog
   c. Pulse width modulation
   d. Digital

5. The amount of fuel that flows through a fuel injector is controlled by:
   a. The amount of voltage to the injector
   b. The length of time the injector is turned on
   c. The frequency that the injector is turned on
   d. Both b and c

6. Electronic throttle controls use an actuator to:
   a. Measure throttle position
   b. Open and close the throttle plates
   c. Move the accelerator pedal
   d. None of the above

7. Technician A says that you can open and close the throttle plates manually on a vehicle that is equipped with a throttle actuator without causing damage. Technician B says that a scan tool should be used to move the throttle plates. Who is correct?
   a. Technician A
   b. Technician B
   c. Both a and b
   d. Neither a nor b

8. Transistors used as low-side drivers switch the _____ side of the circuit.
   a. B+
   b. Ground
   c. Both ground and B+
   d. None of the above

9. The idle air control valve is used to control the position of the throttle plates.
   a. True
   b. False

10. On most fuel-injected vehicles the fuel pump should run:
    a. When the ignition is on
    b. For 2 seconds after the ignition is turned on
    c. When an ignition reference signal is received by the PCM
    d. Both b and c

11. Technician A says that serial data is an output from the PCM. Technician B says that serial data may also be an input to the PCM. Who is correct?
    a. Technician A
    b. Technician B
    c. Both a and b
    d. Neither a nor b

12. On an OBD II equipped vehicle the malfunction indicator lamp may also be called the:
    a. Power loss lamp
    b. Emissions lamp
    c. Check engine lamp
    d. Any of the above

13. Many newer evaporative systems use solenoids that are controlled by:
    a. Vacuum
    b. The PCM
    c. The evaporative canister
    d. None of the above

14. Technician A says that many EGR valves are PCM controlled using vacuum solenoids. Technician B says that many EGR valves are electrically controlled by the PCM. Who is correct?
    a. Technician A
    b. Technician B
    c. Both a and b
    d. Neither a nor b

8

Electronic Engine Control Systems

OBJECTIVES

Upon completion and review of this chapter, you will be able to:

- Describe open-loop and closed-loop fuel control.
- List six engine operating modes of a fuel control system.
- Define the term *stoichiometric.*
- Describe the purpose of short-term and long-term fuel control.
- Explain the difference between a speed density and mass air flow system.
- Have knowledge of the difference between OBD I and OBD II systems.
- Define *enabling criteria.*
- Explain the OBD II code structure.

KEY TERMS

adaptive memory
block learn
bus link
data link connector
  (DLC)
Diagnostic Executive
diagnostic trouble
  codes (DTCs)
enabling criteria
engine mapping
flag
freeze-frame record

I/M readiness test
intake air temperature
  (IAT)
integrator
malfunction indicator
  lamp (MIL)
powertrain control
  module (PCM)
trip
vane-type air flow
  sensor (VAF)

INTRODUCTION

The computers used to control various electrical systems, including engine operation, may be called modules, assemblies, or electronic control units, as explained in Chapter 13. Some are single-function devices that control a single system. Others are multiple-function devices that regulate more than one system. A few, such as the BCM (Body Control Module), even act as master units, supervising a network of computer-controlled systems. These computers use input signals from various sensors to control a given system through a series of actuators, figure 8-1, and the output signal from one computer may also act as an input signal to another computer.

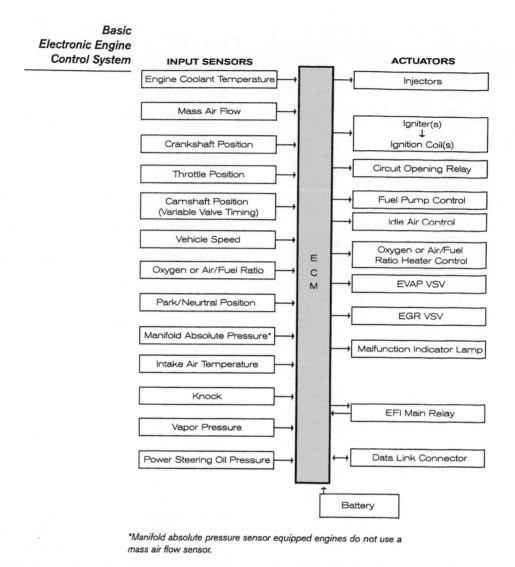

*Basic
Electronic Engine
Control System*

Figure 8-1. This chart shows a basic electronic engine control system. Sensors provide the needed data. The ECM will send the appropriate signal to the actuators. (Provided courtesy of Toyota Motor Sales U.S.A., Inc.)

COMPUTER FUNCTIONS— A REVIEW

As explained in previous chapters, every computer performs four basic functions. Electronic engine control systems provide some of the best examples of these functions:

- **Input:** Variable voltage signals provided by sensors are the computer input data.
- **Processing:** Processing begins with receiving input signals and continues as the computer evaluates and compares multiple signals and makes decisions for output commands.

- **Storage:** A computer holds its program, or operating instructions, and basic data about vehicle design (weight, engine, transmission, and accessory combinations) in read-only memory (ROM). Input and output signals are stored in random access memory (RAM).
- **Output:** After receiving and processing input data, the computer sends output voltage signals to various actuators in the engine control system.

In an engine control system, the actuators are electromechanical or electronic devices controlling fuel metering, ignition timing, emission control operation, and other engine operations.

Open- and Closed-Loop Operation

Every engine control system has two basic operating modes: open loop and closed loop. In open-loop operation, the computer does not respond to a feedback signal from the oxygen sensor (O_2S). The computer simply bases its fuel decisions on engine speed, temperature, and load sensors, and ignores O_2S input signals.

In closed-loop operation, a computer receives and responds to a signal from the O_2S that measures output results. The principal feedback signal in an engine control system is the O_2S. The oxygen sensor measures oxygen in the exhaust and sends a corresponding voltage signal to the computer. The computer interprets this signal as a measurement of the air-fuel ratio. The oxygen sensor signal closes the loop. The computer measures the feedback results and adjusts the output signals if the results are out of the desired limits.

Although the oxygen sensor is the principal closed-loop feedback sensor, engine control computers receive or infer feedback signals from other sensors. A detonation, or knock sensor, for example, sends a signal when the combined results of engine temperature, ignition timing, and air-fuel ratio cause engine pinging. Although a detonation signal is not the result of a single output, the sensor measures the results of outputs contributing to detonation, and allows the computer to correct its signals accordingly. Exhaust gas recirculation (EGR) sensors provide similar feedback signals that indicate the EGR valve position.

Idle speed control is another example of a feedback signal. If idle speed changes from a programmed value, an ignition (tachometer) signal informs the computer. The computer then directs an actuator (an idle air control motor or stepper motor) to adjust airflow or throttle position and return the engine to the desired idle speed.

Essentially, a computer operating in closed loop is constantly "retuning" the engine to keep it within programmed limits. This retuning compensates for changes in operating factors such as temperature, speed, load, and altitude.

The distinctions between open-loop and closed-loop operation are fundamental to any engine control system. The basic difference is simple: The computer either does not (open loop) or does (closed loop) respond to a feedback signal. The specific applications can become quite subtle and complex, however. The next section summarizes seven basic engine operating modes and how the computer responds to each. Then the following paragraphs summarize open- and closed-loop computer operations.

BASIC ENGINE OPERATING MODES

- **Engine Crank:** The air-fuel ratio is fixed between 2:1 and 12:1 depending on temperature. The oxygen sensor is cold and is not inputting a usable signal to the PCM.
- **Engine Warm-up:** The air-fuel ratio is fixed between 2:1 and 15:1, depending on start-up temperature and temperature changes as the engine warms. The oxygen sensor is still usually not hot enough to generate an accurate signal.
- **Open Loop:** The air-fuel ratio is fixed between 2:1 and 15:1, depending on start-up temperature and engine temperature changes as the engine warms. The oxygen sensor may be signaling but is ignored by the PCM. The criteria needed to enable closed loop may be the oxygen sensor voltage varying, the coolant temperature reaching a predetermined value, or a time period of engine run time.
- **Closed Loop:** The air-fuel ratio should be about 14:7:1 depending on the oxygen sensor input. The oxygen sensor signal is now being used by the PCM as a feedback signal.
- **Hard Acceleration:** The air-fuel ratio will vary to the rich side depending on driver demands. The oxygen sensor is still signaling the PCM but is ignored by the PCM until acceleration is complete.
- **Deceleration:** The air-fuel ratio will vary to the lean side. The oxygen sensor signal is ignored by the PCM until the deceleration is complete.
- **Idle:** The air-fuel ratio may be either slightly rich or lean depending on the calibration. In some systems the oxygen sensor signal may be ignored depending on the calibration.

Open-Loop Operating Mode

When a vehicle is first started, the control system is in open-loop mode. This means:

- The sensors provide information to the computer in the form of voltage signals.
- The computer compares the signals to its stored program and makes a decision.

- The computer transmits a voltage signal to its output drivers to implement the decision.
- The output drivers react to the signal by opening or closing the ground circuit of one or more output actuators.
- The actuators operate in a specific manner without providing any feedback to the computer.

For an engine computer to determine whether the air-fuel mixture is correct, it requires an oxygen sensor feedback signal. If the computer ignores this signal, it relies on its stored program and signals from other sensors to make its decision. In this case, suppose the engine coolant sensor tells the computer that the engine is cold. The throttle position sensor also signals that engine speed is increasing. Based on this information, the computer tells the fuel control actuator to enrich the mixture. It does this under certain specified conditions:

- During a cold start or hot restart
- Under low vacuum conditions
- At wide-open throttle or under full load, regardless of engine speed
- During idle or deceleration conditions (some systems/models)
- When brakes are applied (some systems/ models)

Closed-Loop Operating Mode

Once the computer switches system operation into a closed-loop mode, it responds to feedback signals provided by sensors or actuators. These feedback signals tell the computer whether the output is insufficient, optimum, or excessive. In other words, feedback signals regulate or manage the output control.

In a closed-loop mode, the computer responds to the oxygen sensor signal. If the sensor measures an excessively rich mixture, it signals the computer. The computer then directs the fuel control actuator to lean the mixture. If the actuator leans the mixture too much, the oxygen sensor informs the computer, which then directs the fuel control actuator to again enrich the mixture. This ongoing process occurs many times per second.

Modified-Loop Operating Mode

Some engines also operate in a modified-loop mode at part-throttle cruising speed. When this occurs, the computer provides an air-fuel ratio that is slightly leaner than 14.7 to 1 as a way to achieve better fuel economy, while still maintaining driveability without increasing emissions. Under other conditions, the oxygen sensor may cool off during idle. If this happens, its feedback signal may not be reliable. To compensate, the computer controls spark timing and idle speed in the closed-loop mode, but revert to the open-loop mode for control of fuel metering according to pre-programmed values.

AIR-FUEL RATIO, TIMING, AND EGR EFFECTS ON OPERATION

Consider the effects of air-fuel ratio, timing, and exhaust gas recirculation (EGR) operation on overall engine operation. Most engine systems recirculate a measured amount of exhaust gas into the intake air-fuel mixture. Exhaust gas recirculation was introduced in the early 1970s to reduce NO_x emissions by diluting the intake air-fuel charge (displacing oxygen) and thus lowering combustion temperatures. EGR also is an effective detonation control method, which allows spark timing to be maintained at optimum advance for performance and economy. Engineers achieve maximum engine efficiency, economy, and low emissions with computerized control of air-fuel ratios, spark timing, and EGR.

Air-Fuel Ratio—A Review

Assuming fixed timing and engine speed, variations in air-fuel ratio have a dramatic effect on pollutants, figure 8-2. When the ratio is lower, or richer, than 14.7 to 1, hydrocarbon (HC) and carbon monoxide (CO) emissions are high, but oxides of nitrogen (NO_x) emissions are low. Torque is greatest at ratios between 12 and 16 to 1. Above 14.7 to 1, CO and HC decrease, but the increase in cylinder temperatures creates greater NO_x. Above 16 to 1, torque decreases, as does NO_x, while HC increases again due to a lean misfire condition. Because the mixture is too lean to support normal ignition and combustion, raw gasoline goes right through the engine and out the exhaust pipe.

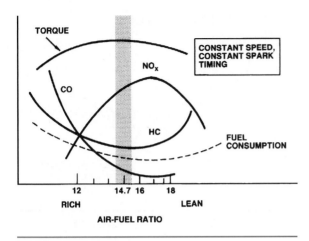

Figure 8-2. Changes in air-fuel ratio produce these fuel consumption, emission, and torque curves.

Stoichiometric Ratio

The stoichiometric or ideal air-fuel ratio is 14.7 to 1. This ratio gives the most efficient combination of air and fuel during combustion. This ratio is a compromise between maximum power and economy, and it delivers the optimum combination of performance and mileage.

Emission control is also optimum at this ratio if a three-way oxidation-reduction catalytic converter is used, figure 8-3. As the mixture gets richer, HC and CO conversion efficiency lowers. With leaner mixtures NO_x conversion efficiency is also lowered. As figure 8-3 shows, the range of the catalytic converter's efficiency is very narrow— between 14.65 and 14.75 to 1. A fuel system without feedback control cannot maintain this narrow range.

Fuel Ratio Control

Two types of fuel control actuators are used with carburetors:

- A solenoid or stepper motor mounted on or in the carburetor to directly control the fuel-metering rods or the air bleeds, or both
- A remote-mounted, solenoid-actuated vacuum valve to regulate vacuum diaphragms controlling the fuel-metering rods and air bleeds.

The computer sends a pulsed voltage signal to the control device, varying the ratio of on-time to off-time according to the signals received from the engine coolant temperature sensor, O_2S, and other sensors. As the percentage of on-time increases or decreases, the mixture becomes leaner or richer.

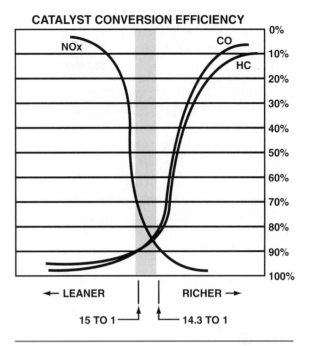

Figure 8-3. Three-way catalysts work properly only in a narrow air-fuel ratio range.

Fuel-injected engines may use a single throttle body with one or two fuel injectors intermittently spraying fuel into the intake manifold. Port fuel-injected engines rely on a single injector for each cylinder, spraying fuel directly behind the intake valve. Some port-injected systems inject fuel in ignition firing sequence just before each respective intake valve opens. This provides greater torque, even better fuel economy, and lower emissions.

With fuel-injection systems, the computer controls the air-fuel ratio by pulsing the injectors. Where engine rpm determines the switching *rate,* the computer varies the length of injector on-time (pulse width) to establish the air-fuel ratio, figure 8-4. As the computer receives data from its inputs, it increases the pulse width to supply more fuel for situations such as cold running, heavy loads, or fast acceleration. In a similar manner, it shortens the pulse width to lean the mixture for situations such as idling, cruising, or decelerating.

High engine cylinder temperatures, lean mixtures, or overadvanced ignition timing can lead to detonation of the fuel in the combustion chamber. Such conditions are intolerable and lead to engine destruction. The advantage of computerized engine control is the conditions leading to detonation are avoided, or at least minimized.

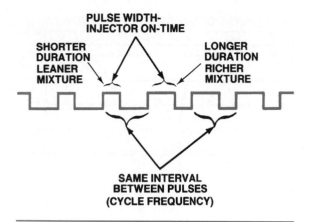

Figure 8-4. Fuel injector pulse width determines the air-fuel ratio.

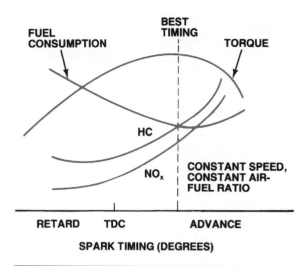

Figure 8-5. Changes in ignition timing produce these fuel consumption, emission, and torque curves.

Ignition Timing

Assuming a fixed air-fuel ratio and engine speed, variations in ignition timing may have a dramatic effect on fuel consumption and pollutants, figure 8-5. When timing is at top dead center (TDC) or slightly retarded, emissions are low and fuel consumption is high. As timing is advanced, fuel consumption decreases but emissions increase. Engine computers are programmed to calculate the best timing for any combination of air-fuel ratio and engine speed to prevent excessive emissions, or other problems like pinging, detonation, or misfiring.

Exhaust Gas Recirculation

Exhaust gas recirculation is the most efficient way to reduce NO_x emissions without adversely affecting fuel economy, driveability, and HC emission control. The recirculation of exhaust gases lowers the combustion temperature by displacing a portion of the burnable mixture. This lowers effective displacement of the cylinders and NO_x emissions drop off sharply when EGR is introduced into the air-fuel mixture. However, excessive reliance on EGR leads to an increase in both HC emissions and fuel consumption and lower engine output. Again, the engine computer is programmed to calculate the percentage of EGR, which delivers the best compromise between NO_x control, HC emissions, and fuel economy without detonation problems.

Computer Integration

One of the primary values of a computer is its ability to integrate the operation of two or more individual, single-function systems to form a larger, more precise multiple-function system. For example, centrifugal and vacuum advance mechanisms can control spark timing relative to engine speed and load. We know that fuel metering through a carburetor is controlled by airflow, and that manifold or ported vacuum can manage basic EGR flow. Integrating such independent systems through a computer provides faster, more precise regulation of each system, and allows the computer to calculate the overall effect of changing several variables at the same time.

The trend in computer integration is in the direction of smarter engine and powertrain control computers with faster microprocessors and larger memories as a way of handling multiple functions and systems within a single unit. One example of this approach is the integrated control of engines and electronic automatic transmissions. Early electronic transmissions were controlled by their own or transmission control module (TCM), which interfaced with the engine control computer to determine torque converter clutch application and shift scheduling. Late-model engine and transmission control modules are usually integrated into one module, the **powertrain control module,** or **PCM.**

Engine Mapping

Every computer needs instructions to do its job. These instructions are written as a computer program. The program for an engine control module, or computer, consists of several elements:

Ignition map: Electronically optimized

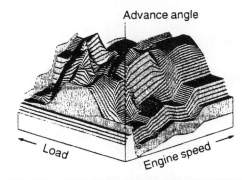

Figure 8-6. A typical spark advance "map" produced by the engine-mapping process. (Reproduced with permission from the copyright owner Robert Bosch GmbH. Further reproductions strictly prohibited.)

Figure 8-7. A typical GM PROM.

- The mathematical instructions that tell the computer how to process, or "compute," the information it receives
- The information that pertains to *fixed* vehicle values, such as vehicle weight, the number of cylinders, engine compression ratio, transmission type and gear ratios, firing order, and emission control devices
- The data that pertains to *variable* vehicle values such as coolant temperature, engine rpm, vehicle speed, intake airflow or manifold absolute pressure (MAP), fuel flow, fuel temperature, ignition timing, and others.

Since the mathematical instructions and vehicle values are constant values, they are fixed and are placed into computer memory. To place the variable values into memory, it is necessary to simulate the vehicle and its system in operation. Vehicle manufacturers use a computer to calculate all the possible variable conditions for any given system. This process of system simulation is called **engine mapping** and provides the control program for the individual onboard computer.

By operating a vehicle on a dynamometer and manually adjusting the variable factors such as speed, load, and spark timing, it is possible to determine the optimum output settings for the best driveability, economy, and emission control. Engine mapping creates a three-dimensional performance graph, figure 8-6, which applies to a specific vehicle and powertrain combination.

The vehicle information mapped in this manner is stored along with the mathematical instructions in a computer chip called programmable read-only memory (PROM), figure 8-7. This is

installed in the computer for that particular model. The computer refers to information mapped and stored in its PROM and compares it with incoming data from many sensors. It then makes decisions on how to manage the output data and, through system actuators, adjusts the engine and powertrain systems under its control accordingly.

Mapping allows a manufacturer to use one basic computer for many models; the unique PROM individualizes the computer for a particular model. Also, if a driveability problem can be resolved by a change in the program, the manufacturer can release a revised PROM to supersede the earlier part. Some PROMs are made so they can be erased when exposed to ultraviolet light and then reprogrammed. These are called erasable programmable read-only memories (EPROMs).

Most recently, electronically erasable programmable read-only memory (EEPROM) chips are being used. Instead of being erased by exposure to UV light, they are electronically erased and reprogrammed by computer. This makes updates both quick and economical.

Some vehicle manufacturers have used a replaceable PROM, which plugs into the computer, figure 8-8. Some computers have used a larger "calibration module," which contains the system PROM, figure 8-9. The PROM or calibration unit could be removed from the defective unit and installed in a replacement computer, should the originally installed computer fail. A module with the incorrect PROM causes all kinds of vehicle driveability problems and subsequent diagnostic technician headaches. Today, however, EPROMS are widely used. Should a change of operating instructions be required, the EPROM is reprogrammed onboard by the dealer via MODEM from the factory. This saves considerable time, expense, and customer inconvenience.

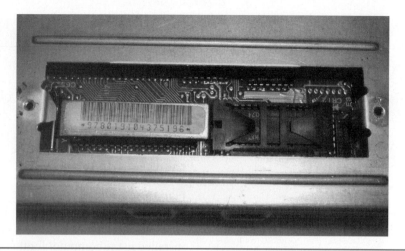

Figure 8-8. A GM engine control module (ECM) with a replaceable PROM.

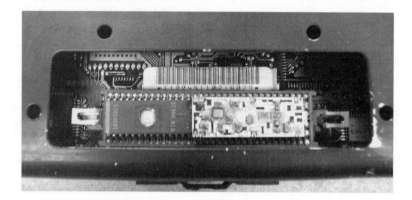

Figure 8-9. Some computers use a memory calibration unit instead of a single PROM chip.

Adaptive Memory, Integrator, and Block Learn

Engine control systems may be programmed to learn from their own experience through **adaptive memory.** Adaptive memory, or an adaptive learning strategy, allows the computer to adjust its memory for computing open-loop operation. Once the system is operating in closed loop, the computer compares its open-loop calculated air-fuel ratios against the average limit cycle values in closed loop. If there is a substantial difference, the computer updates its data and corrects its memory.

Adaptive learning strategies also compensate for production variations and gradual wear in system components. When a controlled value sent to the computer is not within the original design parameters, the adaptive learning strategy modifies the computer program to accept the new value and restore proper operation of the system. Such modifications are stored in RAM and remain in memory when the ignition is turned off, but not

when the battery is disconnected. When battery power is restored, vehicle driveability may be unsatisfactory until the computer relearns the parameters, which generally requires driving the vehicle several miles under varying conditions.

Vehicle driveability may also be affected when a malfunctioning sensor is replaced in a system with adaptive learning strategies. Since the computer modified its original program to compensate for the out-of-tolerance or unreliable sensor signals, the signals received from the new sensor are in conflict with the computer's modified program. This means driveability remains unsatisfactory until the computer relearns the necessary parameters.

Computer programs increasingly incorporate backup or fail-safe modes, as well as adaptive learning strategies. The backup or fail-safe modes compensate for such things as sensor failure. If a sensor does fail, the computer takes its required signal from another source, or from its memory. For example, fuel pump operation in an electronic fuel-injection system may normally be controlled by a pump relay,

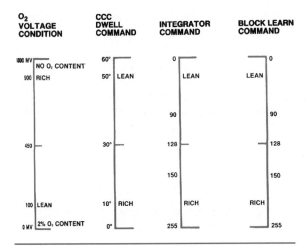

Figure 8-10. The relationship between O_2 sensor voltage, mixture control solenoid dwell, and the integrator and block learn functions for fuel-injected engines. (Courtesy of General Motors Corporation)

which relies on the presence of a tachometer signal. A tach signal failure would normally shut down the fuel pump relay and disable the engine, but the computer may compensate for a lost tach input signal by taking its engine-run signal from the oil pressure switch. The process is so sophisticated that few people would notice a difference in vehicle operation. The only clues to the failure would be the lighting of the malfunction indicator lamp (MIL) and the storage of a fault code in the PCM's memory.

Computers used with earlier GM fuel-injected engines in the 1970s and 1980s incorporated a pair of functions called **integrator** and **block learn.** On OBD II vehicles, these functions became known as short-term fuel trim and long-term fuel trim. Short- and long-term fuel trim functions are responsible for making both temporary and "permanent," minor adjustments to the air-fuel ratio of a fuel-injected engine, much like the MC (mixture-control) solenoid does on carbureted engines, figure 8-10.

Short-term and long-term fuel trim values affect injector on-time. Short-term fuel trim is a method of temporarily changing the fuel delivery and it functions only in closed loop. The PCM program contains a base fuel calculation in memory. The fuel trim function interfaces with this calculation, causing the PCM to add or subtract fuel from the base calculation according to the oxygen sensor feedback signal.

Fuel trim information may be monitored with a scan tool by connecting the scan tool to the serial data transmission line through the data link connector (DLC). Fuel trim values range as a number between 0 and 255. Since the average fuel trim value is one-half the maximum, or 128, the scan tool reads the base fuel calculation as the number 128. The fuel trim function reads the oxygen sensor output voltage, adding or subtracting fuel as required to maintain the 14.7 to 1 ratio. A fuel trim reading of 128 is neutral, which means the oxygen sensor is telling the PCM a 14.7 to 1 air-fuel ratio burned in the cylinders. If the scan tool shows a higher number, the PCM adds fuel to the mixture. If the number is less than 128, fuel is subtracted to lean the mixture.

This corrective short-term fuel trim action is effective only on a short-term basis allowing the computer to make constant corrections in fuel metering. Such corrections may be necessary, for example, when temporary barometric changes occur such as when driving from low altitude across a high mountain pass and back to low altitude within an hour or two.

Long-term fuel trim represents the long-term effect of short-term corrections, although the corrections are not as great as those made short term. The older GM term *block-learn* takes its name from a division of the operating range of the engine for any given combination of rpm and load into 16 cells or "blocks," figure 8-11. Fuel delivery is based on the value stored in memory in the block corresponding to a given operating range. As with short-term fuel trim, the number 128 is a base fuel calculation. A value of 128 means no correction is made to the value stored in that cell, or block. Long-term fuel trim starts out at 128, but shifts or adjusts as required, until the oxygen sensor reads the results as a 14.7 to 1 air-fuel mixture burning in the cylinders. Long-term fuel trim brings short-term fuel trim back to 128. In other words, long-term fuel trim monitors short-term fuel trim. If short-term maintains an increase or decrease, such as with a leaking vacuum line or other "permanent" change, long-term makes a correction in the same direction in order to bring short-term fuel trim gradually back to a stoichiometric value of 128.

Both short-term and long-term fuel trim do exactly what their names imply: Trim the fuel delivery for momentary, or more permanent changes, in operating conditions. Fuel trim is one form of adaptive memory designed to allow the computer to fine-tune the fuel-injection system. Both short- and long-term fuel trim have predetermined limits called "clamping" that vary according to engine design. If long-term fuel trim reaches the limits of its control without correcting the condition, the

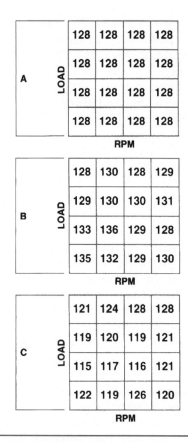

Figure 8-11. The block-learn function divides engine operating range into 16 cells or blocks according to load and speed conditions. In A, all cells are running at the desired 14.7 to 1 ratio under all engine load and speed conditions. B shows typical readings during compensation for a slightly lean condition. C shows typical readings during compensation for a slightly rich condition. (Courtesy of General Motors Corporation)

short-term fuel trim also reaches its limits. At this point, the engine starts to run poorly. Therefore, clamping is necessary to prevent driveability problems that might result under certain conditions. For example, stop-and-go traffic on a crowded roadway could last long enough for long-term corrections to take place. Were this to occur, once the traffic cleared and the vehicle returned to cruise speeds, a driveability problem would result.

Short- and long-term fuel trim adaptability has its limitations. Gross vacuum leaks, loss of engine compression, or other serious conditions, which cause out-of-range oxygen sensor readings, cannot be corrected by fuel trim alone. In such cases, the MIL illuminates, indicating the need for engine repairs. Short- and long-term fuel trim depend on KAM (keep-alive memory)

to store the adaptive values. Thus, disconnecting the battery erases fuel trim information from RAM. With a loss of this memory, the driver may notice different engine operating characteristics of the vehicle after a battery has been disconnected and reconnected. Corrective values must be relearned and stored. This could require one-half hour or more of driving the vehicle after the repair.

FULL-FUNCTION CONTROL SYSTEMS

The earliest electronic control systems first appeared in the 1977 model year and affected a single system: either ignition timing or fuel metering. These early systems gradually expanded to full-function systems controlling two or more engine functions. In doing so, they shared many components and operational principles.

The early, full-function control systems had one or more of the following characteristics; later-model systems have them all:

- The computer controls timing electronically instead of relying on distributor vacuum and centrifugal advance mechanisms.
- The computer keeps air-fuel ratio as close as possible to the stoichiometric value.
- The computer controls fuel metering by operating a carburetor mixture control solenoid or by pulsing fuel injectors according to data received from various sensors.
- Engine operation is divided into open- and closed-loop operational modes, with some systems also providing a modified closed-loop mode.
- A separate "limp-in mode," or "limited operational strategy," is provided when a serious system malfunction occurs, allowing the vehicle to be driven in for service.
- The computer controls EGR, air injection switching, and vapor canister purging.
- The computer can set a number of trouble codes and has a self-diagnostic capability.
- The computer controls automatic transmission shifting and torque converter lockup.
- The computer has the ability to monitor its sensors and their inputs, and actuators.
- Onboard diagnostic generation two (OBD II) systems periodically perform "self-tests" of each system for proper fault detection capability.

CONTROL SYSTEM DEVELOPMENT

Engine systems have developed considerably since first appearing several decades ago. Modern computer control systems, however, are undergoing an even more rapid development into highly sophisticated electronic systems composed of many microprocessors, or computers, which manage many, if not all, operational and passenger convenience systems in a vehicle. The body computer module (BCM) concept, which appeared on some 1986 GM luxury cars, was the first generation of these "total control" systems, and their development has been more rapid than the progress seen with engine control systems.

COMMON COMPONENTS

As a review, all engine control systems use a microprocessor computer, an ever larger variety of sensors, and numerous actuators. The sensors feed data to the computer in the form of voltage signals. The computer processes the sensor data according to its internal program and then signals the actuators to exercise the desired control over the subsystems requiring adjustment.

System Sensors

As explained in Chapter 6, a sensor is an input device used to change temperature, motion, light, pressure, and other forms of energy into voltage signals a computer can read. Input sensors tell the computer what is happening in many areas of vehicle operation at any given moment. Typical sensors used as computer inputs are shown in figure 8-12.

Basic types of sensors were dealt with at length in Chapter 6. Here is a quick review of their outstanding characteristics before looking at specific sensor applications:

- **Switch:** the simplest form of sensor, it signals an on or off condition
- **Timer:** used to delay a signal for a predetermined length of time to prevent the computer from compensating for momentary conditions that do not significantly affect engine operation
- **Transformer:** contains a movable core that varies its position between input and output windings to produce a voltage signal

- **Generator:** may be a magnetic pulse generator or a galvanic battery; these do not require a reference voltage but generate their own signal voltage.

■ Electronic Mufflers

Some vehicles, such as the Mitsubishi 3000GT, have two stage mufflers that are computer controlled. The muffler is staged for quiet mode for city driving, and a more flow-through low restriction mode for open road driving. The computer determines when sound baffles in the muffler are bypassed for a freer exhaust. The change is based on inputs from sensors such as the vehicle speed sensor. The result: better performance once you've hit the city limits and want to "open her up."

A considerably more advanced step in muffler design is based on acoustic cancellation of sound waves. Although it sounds like a joke, the concept is sound (no pun) and actually pretty simple: It is commonly known that two speakers operating out of phase in your car can cancel out sound waves, especially bass response. Electronic mufflers use this principle by using sensors and microphones to pick up sound pressure waves in the exhaust pipe. A computer analyzes the sound waves, and causes a mirror image pattern of sound pulses to be produced by a set of speakers mounted near the exhaust outlet. Since these computer-generated sounds are 180 degrees out of phase with the engine-produced sounds, the sounds from the speakers help nullify or cancel out the engine's exhaust noise. Since the system is computer controlled, it can be tailored to work over any rpm range or load condition.

With the better sound control that this system affords, the mufflers themselves can be less restrictive, resulting in greater performance and fuel economy. With a little tuning of this system, muffler engineers may even be able to give a car a tailored exhaust note. Better yet, the driver could select the desired exhaust sound, using onboard computerized exhaust sound control logic.

Vane Airflow Sensors

Some fuel injection systems use a **vane-type air flow sensor (VAF)**, figure 8-13, positioned between the air filter and the intake manifold. The air flow sensor monitors the volume of air entering the intake manifold. Most VAFs include a thermistor to sense intake air temperature, figure 8-14. The vane swings open with increased airflow, and the position of the vane is "read" by a potentiometer connected

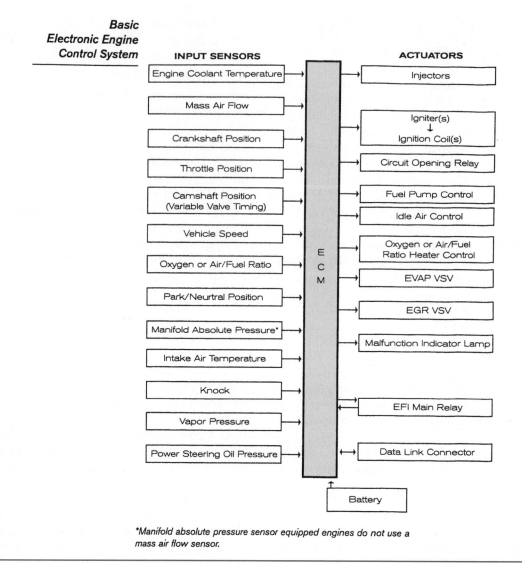

Manifold absolute pressure sensor equipped engines do not use a mass air flow sensor.

Figure 8-12. This chart shows a basic electronic engine control system. Sensors provide the needed data. The ECM will send the appropriate signal to the actuators. (Provided courtesy of Toyota Motor Sales U.S.A., Inc.)

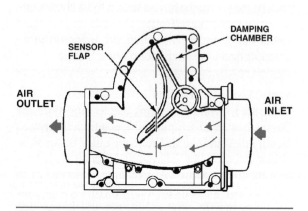

Figure 8-13. A vane-type air flow sensor.

to the vane, figure 8-14. The potentiometer returns a variable voltage signal proportional to intake air volume to the computer. Earlier VAFs included a microswitch to turn on the fuel pump relay once the vane moved from its rest position.

Mass Air Flow Sensors

A thermal measurement device, the mass air flow sensor is installed between the air intake and throttle body to determine the molecular mass of combustible air entering the engine, figure 8-15. It performs a similar function as a vane-type air flow sensor, but measures mass rather than air volume. Two basic types of mass air flow (MAF) sensors are used: a heated platinum wire or a heated thin-film semiconductor. Current pro-

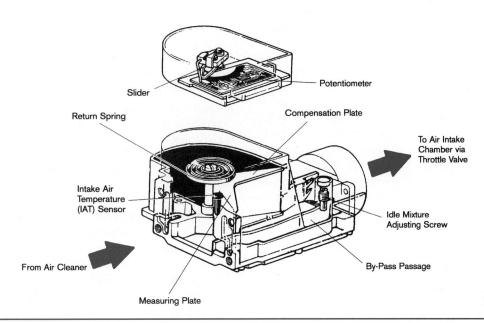

Slider

Potentiometer

Return Spring

Compensation Plate

To Air Intake
Chamber via
Throttle Valve

Intake Air
Temperature
(IAT) Sensor

Idle Mixture
Adjusting Screw

From Air Cleaner

By-Pass Passage

Measuring Plate

Figure 8-14. Vane air flow meter. (Provided courtesy of Toyota Motor Sales U.S.A., Inc.)

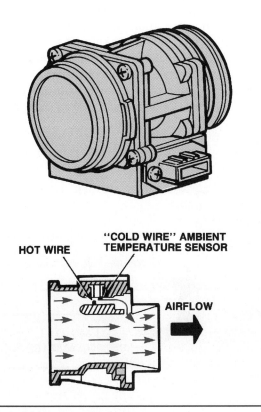

HOT WIRE

**"COLD WIRE" AMBIENT
TEMPERATURE SENSOR**

AIRFLOW

Figure 8-15. Operation of the hot wire MAF sensor.
(Courtesy of Ford Motor Company)

vided by the MAF module heats the wire or film to 100 degrees above ambient air temperature. The wire or film is cooled by incoming airflow. As the heated wire or film cools, the module must increase current to keep the sensor at the specified temperature. As the wire or film again heats up, current decreases once more. The module measures current changes, and converts them into voltage signals sent to the engine control computer for use in determining injector pulse width. Some MAFs, such as Bosch, send an analog signal; others, like Hitachi and Delco, emit a variable frequency digital square wave signal. Figure 8-16 shows the components of a typical heated thin-film MAF, and figure 8-15 shows operation of the heated wire design. Some MAFs initiate a "burnoff" sequence after engine shutdown to eliminate impurities that may have collected on the heated sensor wire.

Manifold Pressure and Vacuum

These sensors keep the computer informed about air density and engine load, allowing it to adjust fuel metering accordingly. Their input, along with an rpm reference signal from the distributor, also are used by the computer to adjust timing and EGR flow relative to load. These so-called speed-density fuel-injection systems rely on a manifold absolute pressure (MAP) sensor as the primary input in calculating mass airflow rate, figure 8-17. They may be piezoresistive devices, transformers, or potentiometers operated by an aneroid bellows or a vacuum diaphragm.

Temperature Sensors

A number of temperature sensors are used on modern engines: The two most important to fuel trim and timing are the engine coolant temperature (ECT) and **intake air temperature (IAT).** If the computer is only interested in whether coolant or air temperature is above a stated point, a simple bimetal switch is used. However, when the

computer requires information about a temperature range, a thermistor is used.

The thermistor-type sensor which is usually an NTC, refer to Chapter 6, is used to track coolant temperature. The sensor is threaded into a coolant passage where its sensing element is immersed in coolant, figure 8-18.

An intake air temperature sensor is similar in construction to the coolant temperature sensor, but provides a response to air temperature changes. It may be located in the air intake, figure 8-19 to measure only air temperature, or in the intake manifold where it measures the air-fuel mixture temperature.

Throttle Position Switches and Sensors

A throttle position switch is a simple on/off micro switch used to indicate wide-open throttle or idle position, sending either a voltage high or low signal. The switch used on some Chrysler carburetors or on Bosch throttle bodies is an example of this type.

A potentiometer also can be used as a throttle position (TP) sensor to indicate the exact position and speed of throttle movement. The TP sensor is usually a rotary potentiometer, figure 8-20 or a linear type, depending upon its application. A rotary throttle position sensor is used on fuel-injection throttle body assemblies. A linear TP sensor generally is used with carburetors. The two types differ primarily in how they work, but both send the computer a variable analog dc signal proportional to the angle of the throttle plate opening. The rotary potentiometer moves on an axis with the throttle shaft; the linear potentiometer uses a plunger that rides on a throttle shaft cam.

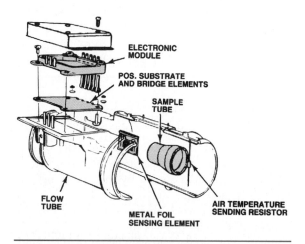

Figure 8-16. Components of a heated film MAF sensor. (Courtesy of General Motors Corporation)

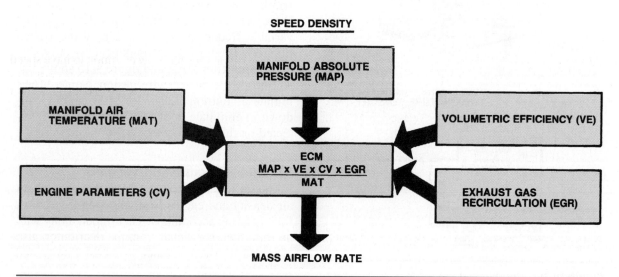

Figure 8-17. A speed-density fuel-injection system uses manifold absolute pressure and temperature to calculate mass airflow rate. (Courtesy of General Motors Corporation)

Figure 8-18. The engine coolant temperature (ECT) sensor screws into a coolant passage so that the tip of the sensor is immersed in the coolant stream.

Intake Air
Temperature
Sensor

Figure 8-19. The intake air temperature (IAT) sensor may screw into the intake manifold or be located in or near the air cleaner housing.

Figure 8-20. Most throttle position (TP) sensors are rotary potentiometers that are attached to the throttle body plates.

Ignition Timing, Crankshaft Position, and Engine Speed Sensors

Hall Effect switches, or sensors, send digital signals the computer uses to control ignition timing, fuel metering, EGR, automatic transmission converter lockup and shift patterns, and more. When a Hall Effect sensor is used for rpm reference or ignition timing, it is sometimes mounted in a distributor housing. For crankshaft position (CKP) sensing, a timing disc or interrupter ring is mounted on the harmonic balancer of the engine, figure 8-21. Some vehicles use the Hall Effect switch for camshaft position (CMP) sensing; the switch may be on the end of the camshaft.

Magnetic pulse generators also provide engine speed and crankshaft position information. They may be located in the distributor at the front or rear of the engine block, figure 8-22, or in the side of the block when the reluctor is cast as part of the crankshaft. This sensor produces an analog ac signal that is converted to digital information inside the PCM.

Vehicle Speed Sensors

A magnetic pulse generator or an optical sensor is typically used to provide the computer with vehicle speed information. This information is used by the computer in a variety of ways, including speedometer displays, torque converter lockup, and transmission shift scheduling. The pulse generator operates by breaking a magnetic field with a toothed reluctor wheel, thus creating an analog AC output. In the optical sensor, figure 8-23, a spinning blade passes through a light emitting diode (LED) beam. Each time the blade cuts through the beam, it reflects light back to a phototransistor. This creates a digital signal, which is then amplified and sent to the PCM.

As vehicle systems have become more complex, they are equipped with multiple speed sensors. For example, it is now common to have speed sensors in the transmission to signal converter input and output speed. Output shaft speed is also measured. The PCM then compares transmission input and output shaft speeds to determine line pressure, transmission shift patterns, and torque converter clutch apply pressure, in addition to calculating turbine speed, gear ratios, and torque converter clutch slippage for diagnostic purposes.

EGR Sensors

A potentiometer may be connected to the top of the EGR valve stem to inform the computer of EGR position or EGR flow rate, figure 8-24. This information is used by the computer to control

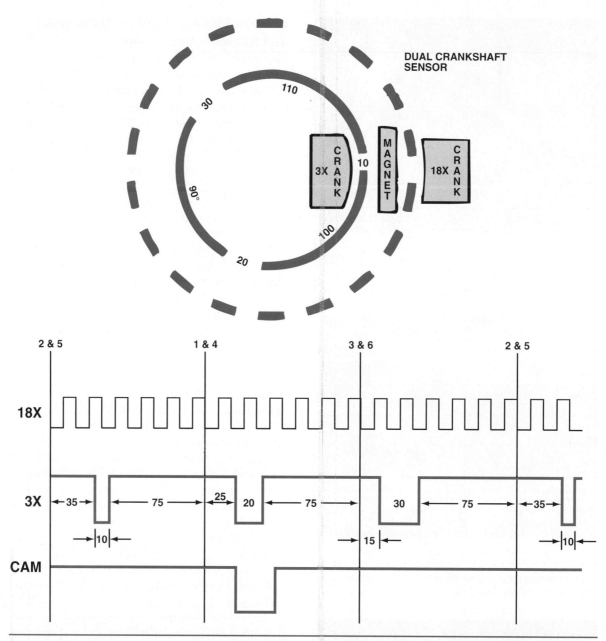

Figure 8-21. Rotation of the interrupter blades between the permanent magnet and Hall Effect switch creates a varying voltage signal. (Courtesy of General Motors Corporation)

timing, fuel metering, and as feedback to control EGR valve operation. General Motors has almost made the transition to 100 percent linear EGR. Linear EGR includes a position sensor.

Air conditioning Sensors

An air conditioning compressor adds to the engine load when the compressor clutch is engaged. To allow the computer to make the necessary adjustments to compensate for the increased load, a simple on/off switch tells the computer whether the compressor clutch is engaged or disengaged. On some vehicles, a magnetic speed sensor provides feedback to the computer when compressor clutch engagement has occurred. Air conditioning system pressure is also monitored on some systems.

Detonation Sensors

Detonation sensors generally use a piezoresistive crystal, which changes resistance whenever pressure is applied, figure 8-25. A reference voltage

Figure 8-22. Rotation of a crankshaft timing disc through a sensor field tells the computer the cylinder position and provides a triggering signal to fire the proper coil.

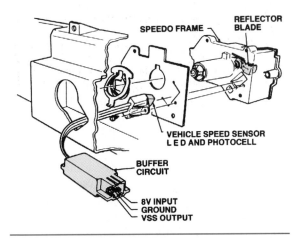

Figure 8-23. An LED and a photocell in the speedometer act as a speed sensor. (Courtesy of General Motors Corporation)

Figure 8-24. This Ford EGR valve includes a potentiometer that indicates the position of the EGR pintle.

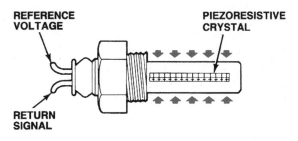

NO DETONATION—EQUAL PRESSURE

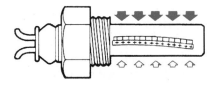

DETONATION—UNEQUAL PRESSURE

Figure 8-25. A piezoresistive crystal changes its resistance when pressure is applied. (Courtesy of Ford Motor Company)

from the computer is applied to one terminal. Engine vibrations apply pressure to the crystal, which changes the resistance of the circuit through the sensor. This alters the amount of voltage returned to the computer through the second terminal. The return signal is electronically filtered to isolate detonation frequencies from other generated signals. If detonation frequencies are detected, timing is retarded or turbo boost pressure is wastegated.

Pre- and Post-catalytic Converter Oxygen Sensors

With OBD II vehicles, oxygen content of the exhaust is detected both before and after catalytic treatment of the exhaust stream. Not only does this detect catalytic converter failures, but also allows the management system to test the converter for efficiency during the normal drive cycle. If a malfunction is detected and adversely affects emissions, an MIL illuminates and a trouble code is set in the computer. Some vehicles have dual exhausts, and thus have four oxygen sensors, two for each bank of cylinders. On OBD II, there is a control heater circuit with PWM.

SYSTEM ACTUATORS

As previously discussed, an actuator is an output device that changes the computer's voltage signal into a mechanical action. Most engine control actuators are solenoids, although stepper motors are used in some applications.

A solenoid is an electromechanical device that uses magnetism to move an iron core. The core provides mechanical motion to some other system part. A solenoid thus changes electrical voltage and current into mechanical movement. Solenoids operate fuel injectors, carburetor metering rods and air bleeds, vacuum control valves, transmission hydraulic circuits, and other devices in an engine control system. There are two types of solenoids used in automotive systems:

- On/off
- Pulse width modulated

All solenoids are digital devices, which can only be switched on or off. However, by regulating the amount of time the induction coil is switched on in relation to the amount of time that it is switched off, the PCM can vary the movement of the core. This type of solenoid control is known as pulse width modulation.

Solenoid Operation

Most solenoids used in a computer control system are grounded through the computer. This allows the computer to control voltage to the solenoid without having to interrupt the power supply. The solenoid is either energized for any length of time the computer desires, or pulsed on and off at either fixed or variable rates per second.

Pulse width modulation and duty cycle were covered in Chapter 6 and also in this chapter with fuel injector solenoids. Pulse width, or the length of time that a solenoid remains energized, is not a concern with solenoids energized, indefinitely. However, if a solenoid is required to pulse on and off rapidly, pulse width and duty cycle become factors. Figure 8-26 illustrates the relationship between pulse width, duty cycle, and on-time.

As discussed earlier, a computer-controlled fuel system operating in open-loop ignores the O_2S input. The PCM considers input from the ECT and the VAF, MAF, or MAP sensors when sending a signal to the fuel solenoid(s). After certain conditions are met, the system switches into closed-loop operation. Along with the inputs mentioned above, the fuel solenoid on-time is varied by the computer according to oxygen sensor input to maintain the 14.7 to 1 air-fuel ratio as closely as possible.

Fuel-Metering Actuators

The stepper motor is a DC motor moving in approximately 100 to 120 incremental steps as it travels. The stepper motor functions as an analog rotary actuator responding to digital signals. Stepper motors are primarily used for idle speed control and idle mixture control.

In earlier computer-controlled carbureted vehicles, a stepper motor was used to maintain a 14.7 to 1 air-fuel ratio at the midpoint of its operational range. In TBI and port fuel-injected vehicles, a stepper motor typically controls idle air bypass around the closed throttle plate to control engine idle speed under varying load conditions, such as when accessories like the AC are switched on or off.

All fuel control solenoids work on the duty-cycle principle, but they manage air-fuel ratios in

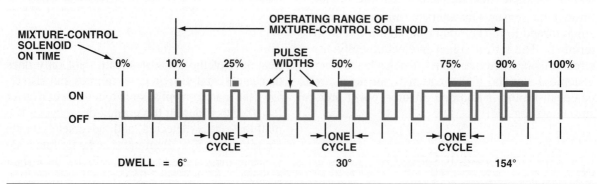

Figure 8-26. The solenoid on-time can be altered to control fuel metering with pulse width modulation. While the duty cycle remains constant, pulse width may vary. (Courtesy of General Motors Corporation)

different ways. In earlier applications, the mixture control solenoid used in Rochester carburetors, figure 8-27, was an integral part of the carburetor. It operated both a metering rod in the main jet and a rod controlling an idle air bleed passage.

In throttle-body applications, either one or two injectors feed the intake manifold with fuel.

Figure 8-27. A mixture control solenoid used in Rochester carburetors.

These injectors have a higher flow rate than port fuel injectors because they feed more than one cylinder with fuel. With port fuel-injected systems, one injector sprays fuel behind its own intake valve. Each cylinder has its own injector. In many systems, these injectors fire in "banks" or groups, one for each bank of the V configuration engine. In sequential fuel-injection systems, the injectors spray fuel in ignition timing sequence. Regardless of system type, fuel injectors are pulse width modulated and operate at varying hertz depending on rpm.

EGR Actuators

Many engine control computer systems that manage EGR flow use solenoids to regulate the amount of vacuum applied to the EGR valve. Although systems differ, figure 8-28 shows a typical system.

Some newer applications use a digital EGR valve, which is entirely electronic in operation and does not use vacuum. The valve consists of two or more computer-controlled solenoids, which operate separate valves to open or close EGR ports, figure 8-29. All share a common power supply but are grounded individually at the computer's output interface. The EGR valve's ports may differ in size according to engine requirements for EGR flow, but each valve is either completely open or completely closed. They can be operated separately or together as required to provide the required flow of EGR.

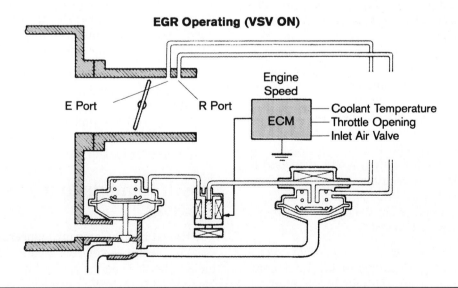

Figure 8-28. EGR cut-off operation. (Provided courtesy of Toyota Motor Sales U.S.A., Inc.)

Figure 8-29. A digital EGR valve consisting of three computer-controlled solenoids.

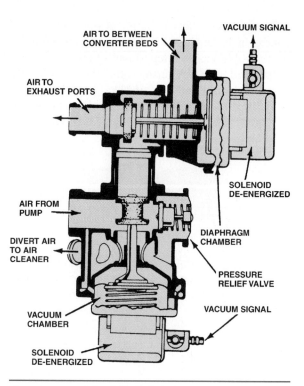

Figure 8-30. The Delco air switching valve uses two solenoids. (Courtesy of Delphi Technologies, Inc.)

Other System Actuators

In addition to controlling ignition timing and fuel metering, integrated electronic engine systems control numerous other functions:

- Torque converter lockup and/or transmission shifting—solenoid valves in the transmission or transaxle hydraulic circuits respond to computer signals based on vehicle speed and engine load sensors
- Air injection switching—one or more solenoids operate a valve in the vacuum line to the air switching or air control valve, figure 8-30
- Vapor canister purge—a solenoid installed in the canister-to-throttle body or intake manifold vacuum line opens and closes as directed by the computer
- Alternator regulation—control partial or full-charge output to limit engine load and emissions on cold starts
- Traction control—limit torque delivered to the wheels by retarding timing or limiting fuel delivery
- Vehicle speed or cruise control functions—in concert with the BCM

On many vehicles today, the powertrain control system is connected with the BCM, which commands and controls overall vehicle operation.

HISTORY OF ENGINE CONTROL SYSTEMS

Electronic engine controls appeared on the automotive scene with the 1977 models. The early control systems regulated only a single function: either ignition timing or fuel metering. However, they were rapidly expanded to control both systems, as well as numerous other engine functions. These included:

- Open- and closed-loop operation
- Electronic ignition timing control
- Fuel-metering control
- Air injection switching
- Automatic transmission or transaxle torque converter lockup

Bosch Lambda and Motronic Systems

The Robert Bosch Company pioneered fuel injection and electronic controls used in European vehicles. Bosch D-Jetronic was introduced on the United States model Volkswagen Type III in 1968.

The Lambda-Sond, or oxygen sensor system, from Bosch was combined with K-Jetronic fuel injection systems on the 1977 and later Volvo models sold in the United States. This was the first electronically controlled fuel-metering system using a three-way catalytic converter and oxygen sensor. The original Lambda-Sond system controlled only fuel metering.

In K-Jetronic systems, an input signal from the oxygen sensor results in an output signal to the K-Jetronic timing valve. This valve varies injection control pressure, which in turn regulates fuel supplied to the continuously spraying injection nozzles.

The Bosch Motronic or digital motor electronics (DME) system added ignition timing and electronic spark control to the fuel-metering control of the Lambda-Sond system. In addition to the three-way converter and oxygen sensor, the DME system receives input signals from crankshaft speed and position sensors, and a magnetic pulse generator in the distributor. Bosch Motronic systems first used vane-type airflow (VAF) sensors, and were known as L-Jetronic systems. This allowed the computer to:

- Adjust injector pulse width for air-fuel ratio control
- Adjust ignition timing for combined speed and load conditions
- Shut off injection completely during closed-throttle deceleration

Later Bosch systems replaced the VAF with a MAF sensor for better emission control.

Chrysler

Chrysler introduced its electronic lean-burn (ELB) spark timing control system in 1976 on some 6.6-liter V-8 engines. The system is based on a special carburetor that provides air-fuel ratios as lean as 18 to 1 and a modified electronic ignition controlled by an analog spark control computer attached to the air cleaner housing. Two printed circuit boards inside the computer contain the spark control circuitry.

In 1977, ELB became available on all Chrysler V-8 engines and the centrifugal advance mechanism was eliminated. A second-generation ELB design was used on 5.2-liter V-8 engines. The start pickup in the distributor was dropped and the computer was redesigned with the circuitry fitting onto a single board.

The second-generation system was adopted on all V-8s in 1978, and a new ELB version was in-troduced on the Omni and Horizon 4-cylinder engine. The 4-cylinder system uses a Hall Effect distributor instead of a magnetic pickup. It has variable dwell to control primary current and does not use a ballast resistor.

With the introduction of three-way catalytic converters, the ELB system was modified in 1979 to work with revised carburetors providing air-fuel ratios closer to 14.7 to 1. This third-generation system was renamed electronic spark control (ESC) and appeared on some 6-cylinder inline engines with an oxygen sensor and a feedback carburetor.

The 1980 model year was a transitional one for Chrysler. All California engines received ESC with feedback fuel control. The 5.9-liter V-8 and Canadian 5.2-liter 4-barrel V-8 continued to use ESC without the feedback system. All other engines reverted to basic electronic ignition with mechanical and vacuum advance mechanisms. Detonation sensors were introduced on some 1980 ESC systems, but the biggest change was the switch from an analog to a digital computer.

Chrysler introduced the modular control system (MCS) in late 1983 on throttle-body fuel-injected 4-cylinder engines. In 1984, its use was expanded to turbocharged port-injected engines. The modular control system regulates vehicle functions using two separate modules whose functions are similar to the two circuit boards in the original ELB computer. The logic module handles all of the low-current tasks within the system, including receiving the inputs and making control decisions. A replaceable PROM is mounted in the logic module housing, and a self-test program is provided to aid in system diagnosis. The logic module is mounted inside the vehicle to avoid underhood electrical interference.

The power module handles the high-current tasks and is located on the left front fender. It looks similar to the spark control computer used in 4-cylinder ESC systems. The power module contains the regulated power supply for the entire control system, along with the switching controls for the ignition coil, fuel injectors, and auto-shutdown (ASD) relay. The ASD supplies power to the coil, the fuel pump relay, and the power module when it detects a distributor cranking signal.

The modular control system (MCS) was replaced by the single-module engine control (SMEC) computer on some 1987 4-cylinder and V-6 engines, figure 8-31. The SMEC brought the two circuit boards used in MCS logic and power modules under one housing. Its advanced

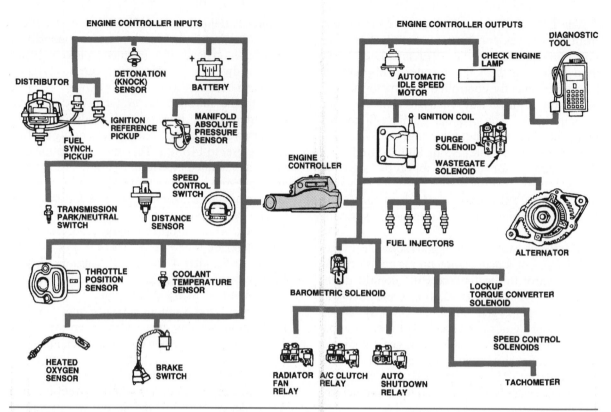

Figure 8-31. Chrysler's single-module engine control (SMEC) system; multipoint injection version shown. (Courtesy of DaimlerChrysler Corporation)

microprocessor was smaller, faster, and more powerful than the earlier MCS microprocessor. The module also has electrically erasable memory (EEPROM) that can be programmed in the assembly plant. The SMEC processed instructions twice as fast as the older MCS system to accommodate the V-6 engine, as the additional cylinders require the computer to process 50 percent more information per engine revolution than a 4-cylinder engine.

Further refinement of Chrysler's engine computer came in 1989 with the introduction of the single-board engine controller (SBEC) and the SBEC II in 1992, figure 8-32. The circuitry in the SBEC and SBEC II is simplified considerably from that used by previous engine controllers.

As shown in figure 8-32, the SBEC engine controller accepts a larger number of input signals and produces more output signals than earlier controllers. In addition to the controller's increased number of functions, its enlarged memory capacity incorporates expanded diagnostic capabilities.

As with other manufacturers, all Chrysler computer control systems are designed with an emergency "limp-in" mode. In case of a system failure, the computer reverts to a fixed set of operating val-

ues. This allows the vehicle to be driven to a shop for repair. In the above Chrysler systems however, if the failure is in the start pickup or the coil triggering circuitry, the engine will not start.

Ford

Ford introduced its feedback electronic engine control system on some 1978 2.3-liter, 4-cylinder engines. The system contains a three-way catalyst and conventional oxidation catalyst (TWC-COC) converter, an oxygen sensor, a vacuum control solenoid, and an analog computer. Its control was limited to fuel metering. In 1980, a digital computer replaced the analog unit and the system was renamed the microprocessor control unit (MCU) fuel feedback system. The major change in early applications is the addition of self-diagnostics, but later designs have expanded capabilities, including control over idle speedup, canister purge, and detonation spark control. They might be considered complete engine control systems except they lack continuous spark timing control.

Ford also introduced its first-generation electronic engine control (EEC-I) system on the 1978

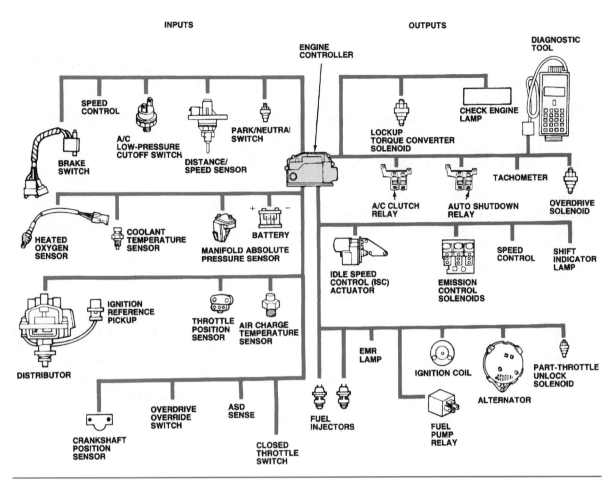

Figure 8-32. Chrysler's single-board engine control (SBEC) system; multipoint injection version shown. (Courtesy of DaimlerChrysler Corporation)

Lincoln Versailles. This system controls spark timing, EGR flow, and air injection. A digital microprocessor electronic control assembly (ECA) installed in the passenger compartment receives signals from various sensors. It then determines the best spark timing, EGR flow rate, and air injection operation and sends signals to the appropriate actuator devices.

All 1978–79 California EEC-I systems used a variation of the "blue-grommet" Dura-Spark II ignition. The 1979 Federal EEC-I system has a "yellow-grommet" dual-mode Dura-Spark II module. Although these modules appear similar to their non-EEC counterparts, they are controlled through the ECA and therefore cannot be tested with the same procedures.

The ignition switching signal on 1978 EEC-I systems is provided by a sensor at the rear of the engine block, which detects four raised ridges on a magnetic pulse ring mounted to the end of the crankshaft. In 1979, the pickup and pulse ring were moved to the front of the engine immediately be-

hind the vibration damper. This design was used on the later EEC-II and EEC-III systems as well.

Late in 1979, Ford's second-generation electronic engine control (EEC-II) system appeared on some 5.8-liter V-8 engines. An EEC-II system has added electronic controls for vapor canister purging and air injection switching. In addition, dual three-way converters were used with a feedback carburetor for precise air-fuel mixture control.

Ford's third-generation, or EEC-III, system appeared in 1980 and was available in two versions through 1984. The EEC-III/FBC incorporates a feedback carburetor similar to the EEC-II system; EEC-III/CFI has a throttle-body-type central fuel injection system. All EEC-III systems use the Dura-Spark III ignition module. EEC-III is the first Ford computer engine control system to have a self-test program.

Ford's EEC-IV was introduced in 1983 and incorporates the thick-film integrated (TFI) ignition system. The two-microchip EEC-IV microprocessor is much more powerful than the four- or five-microchip ECAs used with earlier EEC systems. EEC-IV has increased both memory and the ability to handle almost one million computations per second. Unlike earlier EEC systems, the EEC-IV's calibration assembly is located inside the ECA and cannot be replaced separately. All EEC-IV systems have an improved self-test capability with trouble codes stored for readout at a later date.

Ford made continuing and significant improvements in the processing speed, memory and diagnostic capabilities of its EEC-IV microprocessor without changing the basic designation of the system.

Ford introduced EEC-V with OBD II equipped vehicles. As with other manufacturers' vehicles, EEC-V systems include not only self diagnostics, but also override systems to protect onboard emissions components.

Modern Ford vehicles, along with other makes, are said to have more computer power than early space exploration modules. Later-model Ford vehicles equipped with electronic automatic transmissions have the transmission control functions integrated in the PCM. All Ford EEC systems have a limited operating strategy (LOS) mode in case of a failure within the system. The exact nature of the LOS varies from one system to another, but generally, the timing is fixed, and other ECA outputs are rendered inoperable.

Modern Ford onboard systems are designed to communicate with one another to improve not only emissions, but also overall vehicle performance, fuel economy, and safety.

General Motors

General Motors' first spark timing control was offered on 1977–78 Oldsmobile Toronados. The 1977 system was called microprocessed sensing and automatic regulation (MISAR). The MISAR is a basic spark timing system only. A rotating disc and stationary sensor on the front of the engine replace the pickup coil and trigger wheel of the distributor. Except for this change, the 1977 MISAR system uses a standard HEI distributor with a basic four-terminal ignition module.

The MISAR system was modified in 1978 and renamed electronic spark timing (EST). The crankshaft-mounted disc and stationary sensor were dropped and a conventional pickup coil and trigger wheel were again fitted in the HEI distributor. The ignition module is a special three-terminal design not interchangeable with any other HEI system.

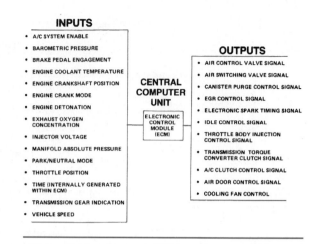

Figure 8-33. Typical CCC system sensors and controls.

In mid-1979, GM introduced the computer-controlled catalytic converter (C-4) system. At first, the C-4 system was purely a fuel control system used with three-way converters. But in 1980, Buick V-6 engines with C-4 were also fitted with an EST system. Although this is the same name applied to the MISAR system just discussed, the two systems are not the same. A C-4 system with EST has a single electronic control module (ECM) that regulates both fuel delivery and spark timing. It was the first complete computer engine control system General Motors used. The C-4 system was further upgraded in 1981 with EST in almost all applications, and additional control capabilities were added. The expanded system, figure 8-33, was renamed computer command control (CCC or C-3).

In 1986, GM began to update the CCC system through the introduction of a new high-speed ECM on certain vehicles. Systems using the new ECM often are called "P-4" systems. The high-speed ECM was smaller than previous models, but had more functional capabilities. It operated at twice the speed of previous ECMs, was capable of 600,000 commands per second, and contained fewer processor chips and internal connections. Service procedures for the unit allow repair and reprogramming by replacing several different integrated circuits in the controller housing.

The ECMs used on vehicles with electronic automatic transmissions had the transmission control circuitry integrated in the memory and calibration (MEM-CAL) unit, and were redesignated as powertrain control modules (PCMs). All C-4 and C-3 systems had self-diagnostic capabilities. The newer the system, the more comprehensive the diagnostic capabilities were. Several GM models provided diagnostic readouts accessible through instrument panel displays. Later systems provided such information only through the Tech I or similar scan tool.

ONBOARD DIAGNOSTICS (OBD) SYSTEMS

During the 1980s, manufacturers began to install full-function control systems capable of alerting the driver of a malfunction and allowing the technician to retrieve codes that identify circuit faults. Ideally, these early diagnostic systems were meant to reduce emissions and speed up vehicle repair.

The automotive industry calls these systems onboard diagnostics (OBD). In 1985, the California Air Resources Board (CARB) developed the first regulation requiring vehicle manufacturers selling vehicles in the state of California to install OBD. Called OBD Generation I (OBD I), it included the following requirements:

- An instrument panel warning lamp capable of alerting the driver of certain control system failures, now called a **malfunction indicator lamp (MIL)**
- The ability to record and transmit **diagnostic trouble codes (DTCs)** for emission-related failures
- Electronic system monitoring of the O_2S, EGR valve, and evaporative purge (EVAP) solenoid

The OBD I standards apply to all vehicles sold in California beginning with the 1988 model year. Although not U.S. EPA-required, during this time most vehicle manufacturers also equipped vehicles sold outside California with OBD I.

Representing only a beginning, these initial regulations failed to meet many expectations. By failing to monitor catalytic converter efficiency, evaporative system leaks, and the presence of engine misfire, OBD I did not do enough to lower automotive emissions. In addition, the OBD I monitoring circuits lacked sufficient sensitivity. By the time an engine with a lit OBD I MIL found its way into a repair shop, the vehicle would already have emitted an excessive amount of pollutants.

Aside from the OBD I's lack of emission-reduction effectiveness, another problem existed. Vehicle manufacturers implemented OBD I rules as they saw fit, resulting in a vast array of servicing tools and systems. Rather than simplifying the job of locating and repairing a failure, the aftermarket technician faced a tangled network of procedures often requiring the use of expensive, special test equipment and dealer-proprietary information.

It soon became apparent that more stringent measures were needed if the ultimate goal, reduced automotive emission levels, was to be achieved. This led the CARB to develop OBD Generation II (OBD II).

OBD II Objectives

Generally, the CARB defines an OBD II equipped vehicle as having the following abilities:

- Detect component degradation or a faulty emissions-related system preventing compliance with federal emission standards
- Alert the driver of needed emission-related repair or maintenance
- Use standardized DTCs, and accept a generic scan tool

These requirements apply to all 1996 and later-model light-duty vehicles; since 1997, these requirements apply to all light-duty trucks. Adopted by the U.S. EPA for use on a national level through the 1997 model year, new CARB-based federal standards took effect in 1998.

Based on these requirements, OBD II is the most significant automotive emissions-related development since the introduction of the catalytic converter during the mid-1970s. Also, to meet the first requirement listed above, engineers substantially increased OBD II's complexity. For example, OBD II must be able to detect minute increases in emission levels. The MIL must light when catalytic converter degradation allows increases in hydrocarbons (HCs) greater than 0.4 gram/mile (g/m) above the 0.6 g/m federal test procedure (FTP) standards. Other similarly strict emission-level requirements stipulate when the MIL must light after the engine misfires or the O_2S degrades.

OBD II Operation

An OBD II vehicle tests its components and communicates these test results without the driver's or technician's help. To use OBD II, it is important to understand the types of OBD II tests, test results, and communication methods.

The OBD II's PCM-held information manager, the software program in charge of testing and indicating test results, is called the **Diagnostic Executive.** The sections that follow also describe this software program's functions.

OBD II diagnostic tests

Depending on the OBD II system, the Diagnostic Executive (usually referred to as the Executive) performs tests on up to seven emission systems using dedicated emission monitors. In addition, an eighth emission monitor, called the comprehensive component monitor (CCM), tests components not included in the other seven monitors.

Each monitor conducts tests during a key-on, engine run, key-off cycle when certain operating conditions called **enabling criteria** are met. A driving cycle that includes an emission monitor test is called a **trip.** The criteria may include information such as elapsed time since start-up, engine speed, throttle position, and vehicle speed. Many tests require the vehicle to be "warmed-up" as part of the enabling criteria. Vehicle manufacturers may define warmed-up differently. For example, on most Ford vehicles this term means the engine temperature reached a minimum of 158°F (70°C) and rose at least 36°F (20°C) over the course of a key-on, engine run, key-off cycle.

A variety of factors influences whether the Executive must delay a test, figure 8-34. These are split into three groups:

* Pending tests—The Executive runs some secondary tests only after certain primary tests have passed. These delayed secondary tests are considered pending.
* Conflicting tests—Sometimes, different tests use the same circuits or components. In this case, the Executive requires each test to finish before allowing another to begin.
* Suspended tests—Each of the emission tests and monitors is prioritized. The Executive may suspend a test so one of higher priority may run.

Generally, the Executive runs three types of tests. Passive tests simply monitor a system or component without affecting its operation. Active tests, conducted when a passive test fails, require the monitor to produce a test signal so it can evaluate the response. Unlike the active tests, the third type, intrusive tests, do affect engine performance and

---

CATALYTIC CONVERTER EMISSION MONITOR

ENABLING CRITERIA

* ENGINE COOLANT TEMPERATURE GREATER THAN 170°F
* VEHICLE SPEED GREATER THAN 20 MPH FOR MORE THAN 2 MINUTES
* OPEN THROTTLE
* CLOSED LOOP OPERATION
* RPM BETWEEN 1,248 AND 1,952 (AUTO), OR BETWEEN 1,248 AND 2,400 (MANUAL)
* MAP VOLTAGE BETWEEN 1.50 AND 2.60

PENDING STATUS CONDITIONS

* MISFIRE DTC
* O2S MONITOR DTC
* UPSTREAM O2S HEATER DTC
* DOWNSTREAM O2S HEATER DTC
* FUEL SYSTEM RICH DTC
* FUEL SYSTEM LEAN DTC
* VEHICLE IS IN THE LIMP-MODE DUE TO MAP,TPS, OR TEMPERATURE DTC
* UPSTREAM O2S SENSOR RATIONALITY DTC
* DOWNSTREAM O2S SENSOR RATIONALITY DTC

CONFLICT STATUS CONDITIONS

* EGR MONITOR IS IN PROGRESS
* FUEL SYSTEM RICH INTRUSIVE TEST IS IN PROGRESS
* PURGE MONITOR IS IN PROGRESS
* TIME SINCE START IS LESS THAN 60 SECONDS
* ONE TRIP MISFIRE MATURING CODE
* ONE TRIP O2S MONITOR MATURING CODE
* ONE TRIP UPSTREAM O2S HEATER MATURING CODE
* ONE TRIP DOWNSTREAM O2S HEATER MATURING CODE
* ONE TRIP FUEL SYSTEM RICH MATURING CODE
* ONE TRIP FUEL SYSTEM LEAN MATURING CODE

SUSPENSION STATUS CONDITIONS

RESULTS OF THE MONITOR ARE NOT RECORDED UNTIL THE O2S MONITOR PASSES

Figure 8-34. This table specifies the conditions under which a Chrysler OBD II catalytic converter emission monitor operates.

emissions levels. The Executive performs these tests after both the passive and active tests have failed.

On completion of a test, the monitor reports a pass or fail to the Diagnostic Executive. For most monitors, the Executive does not set a DTC and light the MIL until the test fails during two consecutive trips. The following monitor descriptions include not only when that particular monitor lights the MIL and sets a DTC, but also a brief overview describing their operation and function.

The comprehensive component monitor (CCM) checks PCM input and output signals for malfunctions affecting any component or circuit not already reviewed by another monitor. Depending on the circuit or system, the CCM conducts tests for open or short circuits and for out-of-range values. Two additional tests for "rationality," or inputs, and "functionality," or outputs, compare voltage signals from one sensor or actuator with those of another. For example, a rationality test on a speed-density system TP sensor first examines TP sensor voltage and then compares it to MAP sensor voltage. Because a wider throttle opening decreases engine vacuum, and TP and MAP sensor voltages increase as the throttle opens and engine vacuum drops, the CCM can verify that the TP sensor works properly.

Depending on the system, the CCM may monitor the following components: MAF, IAT, ECT, TP, CMP, and CKP sensors; the fuel pump, IAC motor, and torque converter clutch (TCC). Typically, the Executive lights the MIL after the CCM detects the same emissions-related fault during two consecutive trips.

In addition to the CCM, the Diagnostic Executive relies on dedicated monitors. These are designed to detect deteriorated systems allowing emissions levels to exceed 1.5 times the FTP standard. These monitors evaluate the following systems and components:

- Catalytic converter
- O_2S
- Misfire detection
- System ability to correct the air-fuel ratio (fuel system monitor)
- EVAP system

The Catalytic Converter Monitor
An OBD II catalytic converter monitor uses upstream and downstream O_2S signals to *infer* catalytic efficiency, figure 8-35. When the engine burns a lean air-fuel mixture, higher amounts of oxygen flow through the exhaust into the converter. The catalyst materials absorb this oxygen for the oxidation process, thereby removing it from the exhaust stream. If a converter cannot absorb enough oxygen, oxidation does not occur. Engineers established a correlation between the amount of oxygen absorbed and converter efficiency.

The OBD II system monitors how much oxygen the catalyst retains. A voltage waveform from the downstream O_2S of a good catalyst should have few or no crosscounts, figure 8-36. A voltage waveform from the downstream O_2S of a degraded catalyst shows many crosscounts. In other words, the closer the activity of the downstream O_2S matches that of the upstream O_2S, the greater the degree of converter degradation. In operation, the OBD II monitor compares crosscounts between the two exhaust oxygen sensors. Depending on the system, the mon-

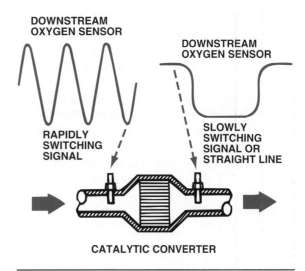

Figure 8-35. The OBD II catalytic monitor compares the signals of the upstream and downstream O_2 sensors to determine converter efficiency.

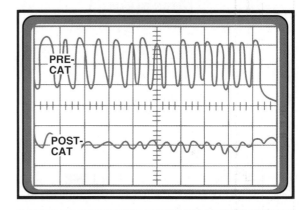

Figure 8-36. The waveform of an O_2S downstream from a properly functioning converter shows few, if any, crosscounts.

itor may run statistical tests or wait until two consecutive failed trips, or both, before setting a DTC.

The Exhaust Oxygen Sensor Monitor
The second monitor, the O_2S monitor, may run several tests, depending on whether it is testing the up or downstream O_2S. This monitor checks O_2S for proper heater circuits and PCM reference signals. Precatalyst O_2S tests include a check for high and low threshold voltage and switching frequency. The switching frequency test counts the number of times the signal voltage crosses the midpoint of 450 millivolts during a specific time and compares it to previously stored values. In addition, the monitor measures the time needed for a lean-to-rich and

a rich-to-lean transition, figure 8-37. Usually, the lean-to-rich transition requires less time. Again, the monitor compares the average response times with a stored calibrated threshold value.

For the downstream O_2S, whose voltage signal fluctuates little, if at all, the O_2S monitor runs two tests. During rich running conditions, the monitor checks whether the O_2S signal waveform is fixed lower. Likewise, while running a lean mixture, the monitor checks for an O_2S signal fixed higher. For both O_2S sensors, the Diagnostic Executive only lights the MIL following two failed consecutive trips.

The Misfire Monitor

Another critical monitor detects poor cylinder combustion which causes engine misfire. The un-

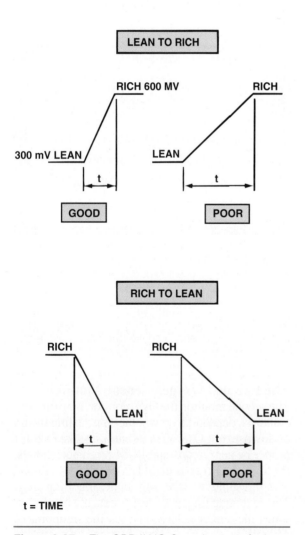

t = TIME

Figure 8-37. The OBD II HO$_2$S monitor tests for lean-to-rich and rich-to-lean transition response times. The Executive lights the MIL should the HO$_2$S monitor take too long and fail these tests.

derlying problem may be a lack of compression, proper fuel metering, or inadequate spark. Misfiring cylinders cause excess HCs to enter the converter. This accelerates converter degradation. In addition, misfire overloads the converter, causing an increase of HC emissions from the tailpipe.

Whenever a cylinder misfires, combustion pressure drops momentarily, slowing the piston. This retarded piston movement also slows the crankshaft. Using this principle, engineers are able to use CKP sensors to detect engine misfire. Misfire causes interruptions in the even spacing of the CKP sensor's waveform, alerting the monitor of a misfire. By comparing the CKP and CMP sensor signals, the monitor is also able to determine which cylinder misfired, figure 8-38.

Excessive driveline vibration caused by rough roads can mimic conditions that the monitor would recognize as a misfire. To prevent false DTCs and aid the technician in isolating a problem, the monitor maintains a cylinder-specific misfire counter. Each cylinder counter tallies the number of misfires occurred during the past 200 and 1000 revolutions. Every time the misfire monitor reports a failure, the Diagnostic Executive reviews all the cylinder misfire counters. The Executive sets a DTC only if one or more of the counters has a significantly greater amount of misfire counts. In effect, by using cylinder counters, the Executive decreases the number of false DTCs by eliminating most of the "background noise" caused by normal driveline vibrations.

The misfire monitor recognizes two types of misfire: those that damage the catalytic converter and those that cause exhaust emissions to exceed emission standards by 1.5 times. The Executive immediately sets a DTC and *causes the MIL to flash* when the monitor detects misfire in more than 15 percent of the cylinder firing opportunities during a 200-crank-revolution segment. In OBD II language, this is a Type "A" misfire.

A Type "B" misfire occurs after two trips in which the monitor detects a misfire in 2 percent of the cylinder firing opportunities during a 1000-crank-revolution segment. In either event, the Executive lights the MIL continuously, and sets one or more DTCs.

The Fuel System Monitor

The fourth monitor, the fuel system monitor, detects when the short-term and long-term FT are no longer able to compensate for operating conditions leading to an excessively rich or lean air-fuel mixture. The adaptive learning strategies discussed earlier in this chapter basically describe how this monitor functions. Unlike previous air-fuel mixture correction strategies measuring the correction in counts or dwell, the OBD II fuel system monitor measures it in percentages, figure 8-39. The Executive lights the MIL after detecting a fault on two consecutive trips.

The Evaporative Systems Monitor

The EVAP system monitor tests for purge volume and leaks. As mentioned earlier, the EVAP system collects and stores the HC vapors emitted from the fuel tank in a charcoal canister. Most applications purge the charcoal canister by venting the vapors into the intake manifold during cruise. To do this, the PCM typically opens a solenoid-operated purge valve installed in the purge line leading to the intake manifold.

A typical EVAP monitor first closes off the system to atmospheric pressure and opens the purge valve, figure 8-40. A fuel tank pressure sensor then monitors the rate at which vacuum increases in the system. The monitor uses this information

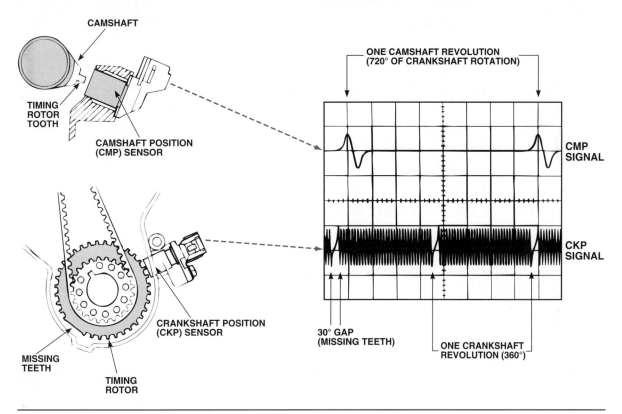

Figure 8-38. The OBD II misfire monitor uses position sensors. In this application, each peak on the crankshaft position signals 10 degrees of rotation. A misfire momentarily slows the crankshaft and interrupts the waveform.

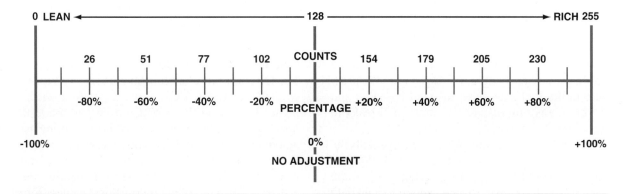

Figure 8-39. This chart shows GM fuel trim equivalencies between OBD I and OBD II systems. (Courtesy of General Motors Corporation)

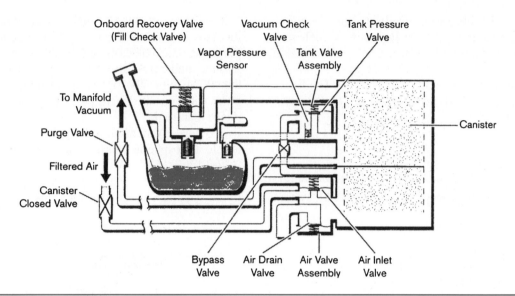

Figure 8-40. An enhanced EVAP system. (Provided courtesy of Toyota Motor Sales U.S.A., Inc.)

to determine the purge volume flow rate. To test for leaks, the EVAP monitor closes the purge valve, creating a completely closed system. The fuel tank pressure sensor then monitors the leak-down rate. If the rate exceeds previously stored values, a leak greater than or equal to the OBD II standard of 0.040 in. (1.0 mm) exists. After two consecutive failed trips testing either purge volume or the presence of a leak, the Executive lights the MIL and sets a DTC.

Natural Vacuum Leak Detection (NVLD)

Current California Air Resource Board (CARB) requirements are that the evaporative system monitor be able to detect a leak equivalent to a 0.020 in. (0.5 mm) hole. Earlier systems only had to be able to detect a 0.040 in. (1 mm) hole. A new method of detecting small leaks is the natural vacuum leak detection method. This method employs the laws of physics that state that the pressure in a sealed vessel will change if the temperature of the gas in the vessel changes. By monitoring the pressure in the fuel tank when the vent is sealed and the engine is shut off, detection of small leaks in the evaporative system is possible.

After the engine is shut off, a vent valve seals the canister vent. As the fuel in the tank cools down, a vacuum should be created in the tank. When the vacuum in the system reaches about 1 inch H_2O (0.25 KPA), a vacuum switch closes. When the

PCM detects the vacuum switch closed, it records a "pass" for the small leak monitor test when the vehicle is started. This test will only run after the vehicle has been driven and then shut off, the fuel level is not more than 85 percent and certain ambient temperature conditions are met. This test could require a week before a fault code will set.

If the small leak test is inconclusive, the system may then run a large leak test. The large leak test will run after the vehicle has been started. The ambient temperature must be between 40° to 90° F and fuel level under 85 percent. The vent valve is sealed and the purge solenoid is turned on. This causes a vacuum to be created in the evaporative system. The vacuum is allowed to build up past the diagnostic switch closing point (1 inch H_2O). After the switch closes the purge valve is deactivated and the time of the pressure decay in the tank is monitored. The time that it takes for the diagnostic switch to open is recorded. If the pressure decay occurs too quickly, a code will set for medium or large leaks.

The EGR System Monitor

The OBD II EGR emissions monitor uses a variety of methods to test EGR flow, depending on the manufacturer and the application. All of the tests open and close the EGR valve, while measuring the amount of change in a test sensor's voltage signal. Typically, the monitor opens the EGR valve and watches MAP and fuel trim. After comparing the test-sensor signal voltage with stored values from look-up tables correlating with exhaust-gas flow, the monitor calculates EGR system efficiency. If the EGR efficiency level does not meet a predetermined standard after two consecutive trips, the Executive lights the MIL and sets one or more DTCs.

The Air Injection System Monitor

Not all manufacturers presently include the seventh monitor, called the secondary air injection (AIR) system monitor. It usually performs a variety of continuity and functionality tests on the bypass and diverter valves. If the air injection system uses an electronic pump, the monitor tests that component as well. Most importantly, the AIR monitor performs a functional test of the amount of air flowing into the exhaust stream from the AIR pump. Almost all vehicle manufacturers use the upstream O_2S to detect the excess oxygen from the AIR system. Of course, the Executive places the AIR monitor in a pending status until all of the O_2S monitor tests have passed. Like most of the other emission monitors, the Executive lights the MIL and sets a DTC after failure of two consecutive trips.

OBD II Diagnostic Test Results

Having discussed all of the OBD II emission monitors, this section now examines how the OBD II Diagnostic Executive indicates test results to the technician. One of the goals of OBD II is to standardize electronic diagnosis so the same inexpensive scan tool can be used to test any compliant vehicle that comes into the shop. To aid the technician, the Society of Automotive Engineers (SAE) established OBD II guidelines—J1930—that provide:

- A universal **data link connector (DLC)** with dedicated pin assignments
- A standard location for the DLC, visible under the dash on the driver's side
- Vehicle identification automatically transmitted to the scan tool
- A standard list of DTCs used by all vehicle manufacturers
- The ability to record a "snapshot" of operating conditions when a fault occurs
- The ability to clear stored codes from vehicle memory with the scan tool
- A series of "flags" to alert the technician of a vehicle's readiness to take an inspection and maintenance (I/M) test
- A glossary of standard terms, acronyms, and definitions used for system components.

The Data Link Connector

The DLC, a 16-pin connector, provides a compatible connection with any generic scan tool for accessing the diagnostic data stream, figure 8-41. The female half of the connector is on the vehicle and the male end is on the scan tool cable. Pins are arranged in two rows of eight, numbered one to eight and nine to sixteen, figure 8-42. The connector is "D"-shaped and keyed so the two halves mate only one way.

Figure 8-41. The standard OBD II 16-pin data link connector.

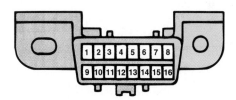

PIN NO.	ASSIGNMENTS
1.	MANUFACTURER'S DISCRETION
2.	BUS + LINE, SAE J1850
3.	MANUFACTURER'S DISCRETION
4.	CHASSIS GROUND
5.	SIGNAL GROUND
6.	CAN HIGH
7.	K LINE, ISO 9141
8.	MANUFACTURER'S DISCRETION
9.	MANUFACTURER'S DISCRETION
10.	BUS – LINE, SAE J1850
11.	MANUFACTURER'S DISCRETION
12.	MANUFACTURER'S DISCRETION
13.	MANUFACTURER'S DISCRETION
14.	CAN LOW
15.	L LINE, ISO 9141
16.	VEHICLE BATTERY POSITIVE

Figure 8-42. OBD II 16-pin DLC terminal assignments.

Seven of the sixteen pin positions have mandatory assignments. The individual manufacturer may assign uses for the remaining nine pins. Most systems use only five of the seven dedicated pins. Terminals seven and fifteen are for systems transmitting data conforming to the European standards established by the International Standards Organization (ISO) regulation 1941-2. This is an alternate communications network to the **bus link**, terminals 2 and 10, defined by SAE J1850 recommended practices.

Diagnostic Trouble Codes and the Malfunction Indicator Lamp

Once the Diagnostic Executive communicates through the DLC with the scan tool, it displays DTCs. Under OBD II guidelines, DTCs appear in a five-character, alphanumeric format. The first character, a letter, defines the system where the code was set. The second character, a number, reveals whether it is an SAE or a manufacturer-defined code. The remaining three characters, all numbers, describe the nature of the malfunction, figure 8-43.

The first character of the DTC currently has four letters assigned for system recognition. These are "B" for body, "C" for chassis, "P" for powertrain, and "U" for undefined. Undefined codes are reserved for future assignment by SAE. Universal powertrain codes are defined and used on current OBD II systems.

The second character of the DTC is either a "0," "1," "2," or "3." A zero in this position indicates a malfunction defined and controlled by the SAE; the number one is a code defined by the manufacturer. The numerals two and three are designated for future use and both are reserved for SAE assignment in powertrain codes. With body and chassis codes, the digit two is manufacturer controlled, and three is reserved for SAE.

The third character of a powertrain DTC indicates the system where the fault occurred. Both "1" and "2" designate fuel or air metering problems. A "3" in this position indicates an ignition malfunction or engine misfire. Number "4" is assigned to auxiliary emission control systems. Problems in the vehicle speed or idle control system set a number "5" in the DTC. Number "6" is used for computer or output circuit faults. Transmission control problems are indicated by either a "7" or "8," while "9" and "0" are reserved by SAE for future use.

The remaining two digits of the DTC show the exact condition triggering the code. SAE assigns different sensors, actuators, and circuits to specific blocks of numbers. The lowest numeral of the block indicates a general malfunction in the monitored circuit. This is the "generic" DTC. Ascending numbers in the block provide more specific information, such as low or high circuit voltage, slow response, or an out-of-range signal; these are known as enhanced OBD II DTCs. The system does not allow duplicate codes. If the system has the capability, an enhanced code takes precedence over a generic code.

With regard to the emission monitors, OBD II has divided DTCs into four categories:

* Type A—These indicate the presence of an emissions-related failure causing the Executive to light the MIL after only one trip. A Type "A" misfire would be an example.

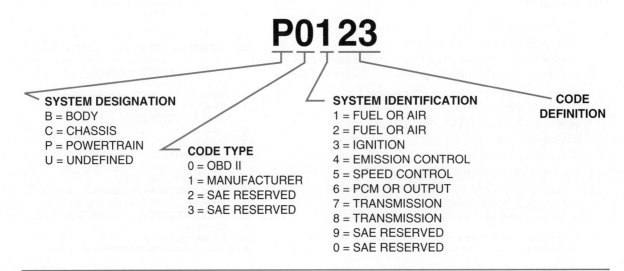

Figure 8-43. OBD-II five-character powertrain DTC structure.

- Type B—These indicate the presence of an emissions-related fault causing the Executive to light the MIL after a failure on two trips. This category includes many of the DTCs set when faults occur during each of two consecutive trips. The misfire monitor can generate two kinds of Type B DTCs: one occurring after two consecutive trips, and the other after two non-consecutive trips. Each time the monitor detects a misfire, it stores engine load, speed, and coolant temperature information. If the misfire occurs during just one trip, the monitor waits until the next misfire, again stores the three items and compares them with one another. When the two sets of data indicate similar operating conditions, the Executive sets a DTC and lights the MIL.
- Type C—These indicate the presence of a failure, not emissions-related, that causes the Executive to light a service lamp after only one trip.
- Type D—These indicate the presence of a failure, not emissions-related, that causes the Executive to light a service lamp after two consecutive trips.

When the Executive lights the MIL, it always stores a DTC. To clear a Type "A" or "B" DTC, the system in which the fault occurred must pass the test during 40 consecutive warm-up cycles conducted after the MIL is turned off. For most monitors, the Executive does not turn off the MIL until the fault-finding emissions monitor reports passes for three consecutive trips. MILs set by a fuel trim or misfire-related DTC require special handling. The three consecutive trips must occur during engine speed, load, and temperature conditions similar to those that originally caused the fault.

The Freeze-frame Record

Another type of OBD II test result indicator, the **freeze-frame record,** provides a wealth of diagnostic information to the technician. Whenever the MIL is lit by the Executive, it stores certain engine operating conditions into the freeze-frame record. The information may include:

- The reported DTC
- Air-fuel ratio
- Mass airflow rate
- Fuel trim, short- and long-term
- Engine speed
- Engine load

- Engine coolant temperature
- Vehicle speed
- MAP/BARO sensor signal voltages
- Fuel injector base pulse width
- Loop status

The Executive allows only operating conditions from the first occurrence of a fault to be recorded into the freeze-frame record. In the presence of multiple codes, the Executive usually allows only parameters for the first DTC to be recorded. However, if one of these multiple codes includes a fuel trim or misfire fault, the Executive allows the freeze-frame record to be overwritten. Erasure of a DTC causes the erasure of its accompanying freeze-frame record.

Fail Records

In addition to freeze-frame records, which are required on OBD II systems, many manufacturers provide additional records of failures to aid technicians in diagnosing fault codes. While these are not required by OBD II, they are an enhancement to freeze-frame records, which have limited abilities. These additional records may be called fail records or some other name.

Fail records record when any DTC is set whether it is emissions related or not. A fail record can store for Type A, B, C, or D codes. Many PCM's can store up to five or eight fail records. These fail records may store more operating parameters than are allowed by freeze-frame records. Unlike freeze-frame, fail records can overwrite existing fail records when the memory is full. The oldest fail record is overwritten by the latest failure that occurs.

Fail records are a valuable diagnositc tool for technicians. The technician can use fail records to determine what the operating conditions were when the DTC was set. This can be particularly helpful when trying to repair intermittent conditions that cause DTCs to set.

Inspection and Drive Cycle Maintenance Readiness Test Flags

I/M readiness test flags provide the technician with the final type of test result indicator. The OBD II emission monitors perform their diagnostic tests once during each trip. At the completion of each test, regardless of a pass or fail, the Executive sets a **flag.** Usually, normal driving after a period of time causes all of the flags to be set; in some instances, however,

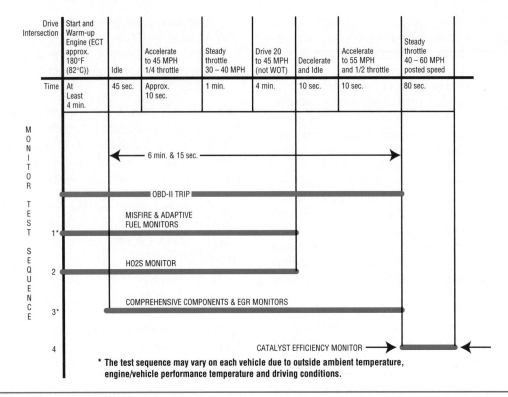

Drive Intersection	Start and Warm-up Engine (ECT approx. 180°F (82°C))	Idle	Accelerate to 45 MPH 1/4 throttle	Steady throttle 30 – 40 MPH	Drive 20 to 45 MPH (not WOT)	Decelerate and Idle	Accelerate to 55 MPH and 1/2 throttle	Steady throttle 40 – 60 MPH posted speed
Time	At Least 4 min.	45 sec.	Approx. 10 sec.	1 min.	4 min.	10 sec.	10 sec.	80 sec.

Figure 8-44. Driving an OBD II vehicle through its I/M Readiness Drive Cycle Test allows the emission monitors to run all of their tests.

the technician must drive the vehicle through a manufacturer-specific drive cycle to set all of the flags, figure 8-44. These drive cycles vary the time spent driving during idle, acceleration, steady throttle, and deceleration. Most require the vehicle to be started while cold.

The Technician and OBD II

After checking an OBD II vehicle's MIL and I/M readiness inspection flags, and evaluating the DTCs and freeze-frame record, you can usually form a diagnosis. Some manufacturers then have the technician perform an OBD II system check. After completing emission-related repairs, it is important to verify their effectiveness. Use a scan tool to clear the set DTCs and perform an I/M readiness drive cycle test. Check whether all of the I/M flags have set, and then see if the Executive has reported additional DTCs. If none appear, the repair is successful.

Control System Development

As covered in this chapter, engine systems have developed considerably since they first ap-

peared in the late 1970s. Computer control systems, however, are undergoing an even more rapid development. Highly sophisticated electronic networks composed of many microprocessors manage all operational and convenience systems in a vehicle. The body computer module (BCM) concept, which appeared on some 1986 GM luxury cars, was the first generation of these "total control" systems. Single "supercomputers" with enough memory and speed to integrate all computer-controlled systems on the vehicle into one "super program" are also prevalent. Communications between systems or modules are increasingly multiplexed and transmitted by fiber optics.

These trends underscore the fact that when a particular vehicle arrives for service, you will have to know the system's components, understand its operation, and use the correct procedures and specifications. As described in Chapter 1 of the *Shop Manual,* this information is available in service manuals from the manufacturer or from independent after-market publishers. It is impossible for technicians to carry all of this information in their heads.

■ Multiplexing is the encoding of signals transmitted from a master module such as the BCM. Signals are received and decoded by the appropriate slave module. All other modules online listen in, in the event they are commanded to fulfill a function. This is akin to the old telephone system of a master switchboard with many telephones tied to it. Multiplexing assists in reducing bulk and weight of wire bundles and promotes fail-safe communications between systems or subsystems. Use of fiber optics reduces line noise (improved signal-to-noise ratio) for more reliable communications between modules, along with reduced manufacturing cost and onboard weight.

Remember, the laws of electricity, physics, and chemistry have no manufacturer's trademark on them. They remain the same, and all of the systems we have studied operate on the same principles. Knowing the basics is the starting point. With the many vehicles and systems in the marketplace, a solid understanding of the basics ensures the technician's ability to survive in the service bay. The ability to read and digest written and computer-based information and the ability to operate and interpret data from electronic diagnostic equipment are mandatory skills if the technician is to remain competitive in the automotive service business.

SUMMARY

All computers must perform four basic functions: input, processing, storage, and output. Engine and powertrain control computers use various sensors to receive input data. This data is compared to lookup tables in the computer memory. Some data may be stored in memory for future use. The computer output takes the form of voltage signals sent to actuators.

A control system operates in an open-loop mode until specific parameters or conditions are met; then the computer switches into the closed-loop mode. In open loop, the computer ignores oxygen feedback signals and functions with a predetermined set of values. Once the system switches into closed loop, the computer acts on feedback signals and constantly "retunes" the engine while it is running.

Computers have the ability to adapt their operating strategies to account for changes of operating system conditions, including engine and operating system wear, and sensor and actuator degradation over time.

System sensors generally measure analog variables. Sensor output voltage signals are digitized by the computer, which compares the input signals to stored information in its look-up tables. The PCM processes the data and stores it momentarily or permanently for later use. Once the data is processed, it is sent as output signals to actuators, which perform work. Actuators change computer output signals into electromechanical motion.

Sensors are generally switches, variable resistors, transformers, or generators. Actuators are usually solenoids, relays, or stepper motors.

Computer-controlled engine systems began as a way of managing fuel metering for better mileage and emission control. Manufacturers used electronic ignition with fuel-metering systems to form the basis for an engine management system. Systems evolved into those with self-diagnostic capabilities and the ability to control many more functions, and became far more efficient and powerful than their predecessors. Modern control systems not only detect errors, but also perform periodic self-tests under specific conditions. They also store "snapshots" of data stream information for later retrieval by the service technician using a scan tool.

While early electronic systems controlled either ignition timing or fuel metering, later full-function systems control both. To further decrease emission levels, and to facilitate repair, the CARB developed the OBD I and OBD II regulatory guidelines.

An OBD II system detects system degradation and uses a standardized DLC, DTCs, and accepts a generic scan tool. This unique system, required of most light-duty vehicles since 1996, uses a Diagnostic Executive to manage the emissions monitors performing tests and to report test results. This sophisticated process takes place independently of the technician. Up to seven dedicated emission monitors test the following systems, components, and faults: the catalytic converter, O_2S, misfire detection, air-fuel ratio compensation, and the EVAP, EGR, and AIR systems. The OBD II system uses DTCs, the MIL, a freeze-frame record, and I/M Readiness Drive Cycle Test flags to indicate diagnostic information to the technician.

Review Questions

Choose the letter that represents the best possible answer to the following questions:

1. The first domestic engine control system was Chrysler's electronic lean-burn. It controls:
   a. Ignition timing
   b. Fuel metering
   c. Both a and b
   d. Neither a nor b

2. Current Ford engine control systems are called:
   a. Electronic feedback fuel control (EFC)
   b. Microprocessor control unit (MCU)
   c. Electronic engine control-V (EEC-V)
   d. Computerized emission control (CEC)

3. NO_x emissions may be reduced by:
   a. Retarding ignition timing
   b. Lowering the engine's compression ratio
   c. Recirculating exhaust gases
   d. All of the above

4. An engine coolant temperature sensor:
   a. Receives reference voltage from the computer
   b. Is a variable resistor
   c. Provides the computer with a digital signal
   d. Contains a piezoelectric crystal

5. Technician A says an exhaust gas oxygen sensor (O_2S) with two wires is grounded to the exhaust manifold.
   Technician B says the computer ignores an O_2S sensor in closed-loop operation. Who is right?
   a. Technician A only
   b. Technician B only
   c. Both a and b
   d. Neither a nor b

6. Technician A says a throttle position sensor (TPS) may be either a rotary or a linear potentiometer.
   Technician B says a TPS may be a simple on/off switch. Who is right?
   a. Technician A only
   b. Technician B only
   c. Both a and b
   d. Neither a nor b

7. Technician A says the output signal from one computer can act as an input signal for another computer.
   Technician B says late-model computers have self-diagnostic capabilities and can store trouble codes. Who is right?
   a. Technician A only
   b. Technician B only
   c. Both a and b
   d. Neither a nor b

8. Technician A says an inoperative exhaust gas recirculation system can increase NO_x.
   Technician B says an inoperative exhaust gas recirculation system can result in pinging or even detonation. Who is right?
   a. Technician A only
   b. Technician B only
   c. Both a and b
   d. Neither a nor b

9. The first electrically controlled fuel metering system with a three-way converter and O_2S was the:
   a. C-4 system
   b. Lambda-Sond system
   c. EEC system
   d. None of these

10. An OBD I system typically includes:
    a. Short-term fuel trim capability
    b. Adaptive learn capability
    c. Self-diagnostic capability
    d. All of the above

11. Which of the following is NOT true of a fuel-injected engine?:
    a. The computer controls the air-fuel ratio by switching injectors on or off
    b. The pulse width is increased to supply more fuel
    c. To lean the mixture, the computer opens the air bleeds
    d. Engine speed determines the injector switching rate

12. What is the most efficient way to reduce NO_x without adversely affecting fuel economy, driveability, and HC emissions?
    a. Engine mapping
    b. Retarded ignition timing
    c. Lower compression ratios
    d. Exhaust gas recirculation

13. Integrator and Block Learn refer to:
    a. Self-diagnostic capability
    b. Adaptive learn capability
    c. Engine mapping
    d. None of the above

14. Ford's EEC-V system integrates the following features:
    a. Standardized DTCs (fault codes)
    b. Standardized 16-pin DLC (diagnostic connector)
    c. Self-test capability
    d. All of the above

15. An OBD I self-test program performs which of these tests?
    a. Tests each sensor for proper operation
    b. Tests each input circuit for proper operation
    c. Compares incoming data with data in memory
    d. Both b and c

16. An OBD II self-test program performs which of these?
    a. Checks the fuel EVAP system for integrity
    b. Checks the fault detection system for proper operation
    c. Monitors engine operation for misfires
    d. All of the above

17. Technician A says OBD II equipped vehicles have a standardized DLC. Technician B says OBD II equipped vehicles may have as many as 4 O_2S. Who is right?
    a. Technician A only
    b. Technician B only
    c. Both a and b
    d. Neither a nor b

18. Technician A says MAF sensors may use either a hot wire or heated film. Technician B says automatic transmission operation is controlled by the PCM or TCM in many late-model vehicles. Who is right?
    a. Technician A only
    b. Technician B only
    c. Both a and b
    d. Neither a nor b

19. Technician A says the BCM may be a master computer over PCM, EBCM and other computers. Technician B says fiber optics are now used onboard in the interest of saving weight. Who is correct?
    a. Technician A only
    b. Technician B only
    c. Both a and b
    d. Neither a nor b

20. With OBD II equipped vehicles, the increased use of computer power has meant:
    a. Safer automobiles
    b. Cleaner automobiles
    c. Better-performing automobiles
    d. All of the above

21. Today's well-equipped, well-trained technician must be able to retrieve and interpret:
    a. Digital storage oscilloscope patterns
    b. Freeze-frame data stored onboard
    c. Multi-function scanner readouts
    d. All of the above

22. An OBD II equipped vehicle has just completed a trip. This means:
    a. The vehicle completed a warm-up cycle
    b. The vehicle completed a key-on, engine run, key-off cycle
    c. The vehicle completed a key-on, engine run, key-off cycle where an emissions monitor reported a pass for each test to the Diagnostic Executive
    d. The vehicle completed a key-on, engine run, key-off cycle where an emissions monitor reported a pass or fail for each test to the Diagnostic Executive

23. What can cause the MIL to flash on an OBD II equipped vehicle?
    a. A type "A" misfire
    b. A type "B" misfire
    c. Catalytic converter malfunction
    d. O_2S malfunction

24. Technician A Says OBD II DTCs cover only engine and emission related faults. Technician B says OBD II DTCs may be designated by the vehicle manufacturer, or follow an SAE standard. Who is right?
    a. Technician A only
    b. Technician B only
    c. Both a and b
    d. Neither a nor b

25. For most monitors, how long does it take for the Executive to light the MIL, and then to clear DTCs?
    a. Two consecutive trips; 40 consecutive warm-up cycles
    b. Four consecutive trips; 30 consecutive warm-up cycles
    c. Two consecutive trips; 30 consecutive warm-up cycles
    d. Four consecutive trips; 40 consecutive warm-up cycles

9

Gasoline Fuel-Injection Systems

OBJECTIVES

After completion and review of this chapter, you will be able to:

- Describe the difference between throttle body and port fuel-injection systems.
- List five advantages of fuel-injection systems.
- List three air intake sensors.
- List major fuel system components of a typical port fuel-injection system.
- Explain how a fuel pressure regulator operates.
- Define the term *speed density*.

KEY TERMS

continuous injection
  system (CIS)
duty cycle
electronic fuel
  injection (EFI)
engine coolant
  temperature (ECT)
fuel injection
idle air control
  (IAC) motor
idle speed control
  (ISC) motor

multipoint (port)
  injection
negative temperature
  coefficient (NTC)
on-time
  (pulse width)
speed-density
throttle body
  injection (TBI)

INTRODUCTION

A carburetor is a mechanical device that is neither totally accurate, nor particularly fast in responding to changing engine requirements. Adding electronic feedback fuel mixture control improves a carburetor's fuel-metering capabilities under some circumstances, but most of the work is still done mechanically by many jets, passages, and air bleeds. In recent years, feedback controls and other emission-related devices have resulted in very complex carburetors that are extremely expensive to repair or replace.

The intake manifold is also a mechanical device that, when teamed with a carburetor, results in less than ideal air-fuel control. Because of a carburetor's limitations, the manifold must locate it centrally over the intake ports (V-engines), or next to the intake ports (inline engines), while remaining within the space limitations under the hood. The manifold runners have to be kept as short as possible to minimize fuel delivery lag, and there cannot be any low points where fuel might puddle. These restrictions severely limit the amount of manifold tuning possible, and even the best designs still have problems with fuel condensing on cold manifold walls.

The solution to most of the problems posed by a carbureted fuel system is **fuel injection.** Fuel injection provides precise mixture control over all speed ranges and under all operating conditions. Its fuel-delivery components are simpler and often less expensive than those of a feedback carburetor system. Some designs allow a wider range of manifold designs.

Fuel injection may be mechanically or electronically controlled. Electronically controlled fuel injection, more commonly called **electronic fuel injection (EFI),** offers highly reliable and accurate fuel delivery to meet today's stringent standards for fuel economy, emission control, and good driveability. This chapter covers:

- Advantages of fuel injection over carburetion
- Differences among various fuel-injection systems
- Fundamentals of electronic fuel injection
- Subsystems and components of typical fuel-injection systems now in use

FUEL-INJECTION OPERATING REQUIREMENTS

The major difference between a carbureted and an injected fuel system is the method of fuel delivery. In a carbureted system, the carburetor mixes air and fuel. In a fuel-injection system, one or more injectors meter and spray atomized fuel into the intake air stream.

Electronic fuel-injection systems use the same principle of pressure differential as a carbureted system, but in a slightly different way. As discussed in Chapter 8, it is the difference in pressure between the inside and the outside of the engine that forces fuel out of the carburetor fuel bowl. A higher airflow through the carburetor creates a higher venturi vacuum to draw more fuel from the fuel bowl. In EFI systems, the amount of air entering the engine is measured by an air flow volume sensor or a mass air flow meter. The computer calculates the amount of air in each cylinder by the amount of air entering the engine and the engine rpm. On speed-density systems, the computer calculates the amount of air in each cylinder using manifold pressure and engine rpm. The amount of air in each cylinder is the major factor in determining the amount of fuel needed. Other sensors provide information to modify the fuel requirements.

The computer makes it possible for the fuel injectors to do all the functions of the main systems used in a carburetor. However, throttle response in an injected system is more rapid than in a carbureted system because the fuel is under pressure at all times. The computer calculates the changing conditions of incoming air and responds at the injector much more quickly and precisely than a carburetor can react under similar conditions.

Carburetors must break up liquid gasoline into a fine mist, change the liquid into a vapor, and attempt to distribute the vapor evenly to all cylinders. Fuel injectors do the same. A fuel injector delivers atomized fuel into the air stream where it is instantly vaporized. In **throttle body injection (TBI)** systems, this occurs above the throttle at about the same point as in a carbureted system, figure 9-1. With **multipoint (port) injection** systems, the injectors are mounted in the intake manifold near the cylinder head and injection occurs as close as possible to the intake valve, figure 9-2. Both systems are discussed in greater detail later in the chapter. To review basic engine air-fuel requirments:

- When an engine starts, low airflow and manifold vacuum, combined with a cold engine and poor fuel vaporization, require a rich air-fuel ratio.
- At idle, low airflow and high manifold vacuum, combined with low carburetor vacuum and poor vaporization, still require a slightly rich air-fuel ratio.
- At low speed, the air-fuel ratio becomes progressively leaner as engine speed, airflow, and carburetor vaccum increase.
- At cruising speed, air-fuel ratios become the leanest for best economy with light load and high, constant venturi vacuum and airflow.
- For extra power during acceleration or heavy load operation, the engines needs a richer air-fuel ratio; this requirement is combined with low vacuum and airflow on acceleration or with low vacuum and high airflow at wide-open throttle.

A carburetor satisfies all of these requirements as its systems respond to changes in air pressure and

Figure 9-1. Air and fuel combine in a TBI unit at the same point as in a carburetor.

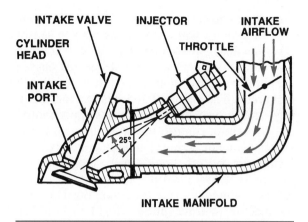

Figure 9-2. In a multipoint system, the air and fuel combine at the intake valve. (Courtesy of General Motors Corporation)

airflow. A fuel-injection system does the same things. All of the sensors and other devices in an injection system respond to the same operating conditions that a carburetor does, but an injection system responds faster and more precisely for a better combination of performance, economy, and emission control.

ADVANTAGES OF FUEL INJECTION

Fuel injection is not a new development. The first gasoline fuel-injection systems can be traced back to 1912 and the Robert Bosch Company in Germany. Bosch's early work in fuel injection concentrated on aircraft applications, but introduced automotive systems in the 1930s. A decade later, two Americans named Hilborn and Enderle developed injection systems for use on racing engines. Chevrolet, Pontiac, and Chrysler offered production mechanical fuel-injection systems during the late 1950s and early 1960s. Volkswagen pioneered widespread use of electronic fuel injection in the United States in 1968. The system, known as D-Jetronic, was made by Bosch based on patents purchased earlier from Bendix. Among domestic manufacturers, Cadillac made electronic fuel injection standard on its 1976 Seville model using a combination of Bosch D-Jetronic and Bendix components. Other automobile manufacturers soon followed suit.

What really made fuel injection a practical alternative to the carburetor was the development of reliable solid-state electronic compounds dur-

ing the 1970s. Automobile manufacturers were quick to apply these advances in electronics to fuel injection. The result was a more efficient and dependable system of fuel delivery.

Electronic fuel-injection systems offer several major advantages over carburetors:

- Injectors can precisely match fuel delivery to engine requirements under all load and speed conditions. This reduces fuel consumption with no loss of engine performance.
- Since intake air and fuel are mixed at the intake port, maintaining a uniform mixture temperature is easier with port fuel-injection systems than with carburetors. There is no need for manifold heat valves.
- The manifold in a port fuel-injection system carries only air, so there is no problem of the air and fuel separating.
- Exhaust emissions are lowered by maintaining a precise air-fuel ratio according to engine requirements. The improved air-fuel flow in fuel-injection systems also helps to reduce emissions.
- Continuing improvements in electronic fuel-injection design have allowed some automobile manufacturers to eliminate other emission control systems such as heated intake air, injection, and EGR valves.

AIR-FUEL MIXTURE CONTROL

Almost all of the electronic fuel-injection systems used on domestic vehicles in recent years are integrated into complete electronic engine control systems with exhaust gas oxygen sensor (O_2S) feedback for accurate fuel mixture control. The elctronic controls of a fuel-injected engine are similar to those of an engine with a feedback carburetor. The main difference is that instead of regulating a vacuum solenoid, a stepper motor, or a mixture control solenoid, the computer switches one or more solenoid-operated fuel injectors on and off. Engine rpm determines the rate of switching, and the computer varies the length of time the injectors remain open, known as injector **on-time (pulse width)** to establish the air-fuel ratio, figure 9-3. Injector on-time, pulse width, and **duty cycle** are explained in more detail later in this chapter.

Variable pulse width takes the place of all the different carburetor metering circuits. As the computer receives data from its inputs, it can lengthen the pulse width to supply additional fuel for cold running (choke), heavy loads (power enrichment), fast acceleration (accelerator pump), or several other situations. Similarly, it can shorten the pulse width to lean the mixture at idle (idle circuit), under cruise (main circuit), or during deceleration. On such systems, the O_2 sensor provides fine tuning of the mixture under most operating conditions.

By weight, fuel makes up only about one-fifteenth of the air-fuel mixture. The computer is mainly concerned with regulating the delivery, but to do this accurately, it must know the amount, or volume, of air entering the engine. Because airflow and air pressure cannot act directly on the fuel as they do in a carburetor, fuel-injection systems use three basic kinds of intake air sensors:

- A manifold absolute pressure sensor
- A volume airflow meter or sensor
- A mass air flow meter or sensor

Manifold pressure, airflow velocity, and the molecular mass of the intake air all directly affect the weight of intake air. (Remember that air-fuel ratios are expressions of air and fuel weight.) The computer has a set of values (similar to a graph) stored in memory. These values relate manifold pressure, airflow, or air mass to engine speed and the required fuel entering for any combination of conditions. Once the computer knows the amount of air entering the engine, as well as other conditions, it adjusts injector pulse width (or on-time) to achieve the required air-fuel ratio.

Some injection systems use combinations of the three basic kinds of sensors listed earlier. More about these sensors appears later in this chapter.

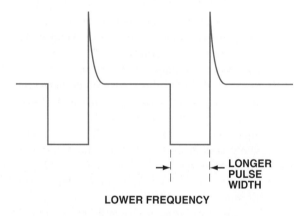

LONGER PULSE WIDTH

LOWER FREQUENCY

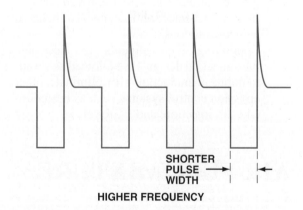

SHORTER PULSE WIDTH

HIGHER FREQUENCY

Figure 9-3. Fuel injector pulse width determines the air-fuel ratio.

TYPES OF FUEL-INJECTION SYSTEMS

There are three general ways to categorize modern fuel-injection systems:

- Mechanical or electronic injection
- Throttle body or multipoint (port or manifold) injection
- Continuous or intermittent injection

However, the distinctions between the various types of fuel-injection systems are not simple and clear-cut. Later mechanical systems make use of some electronic components, and electronic systems may share some of the features of port and TBI in a single system. One example is Chrysler's continuous-flow TBI system introduced in 1981 and used exclusively on V-8 engines in the Imperial. It broke the rules of operation about to be discussed. It had two injectors mounted in the throttle body. They responded to varying pressure from a control pump and delivered fuel continuously. A unique airflow sensor in the air-cleaner inlet measured the volume of air passing into a system. The entire operation was controlled electronically.

Mechanical or Electronic Injection

A mechanical fuel-injection system delivers gasoline by using fuel pressure to open the injector nozzle or valve. Mechanical injection systems generally deliver fuel continuously when the engine is running. The Bosch K-Jetronic is a typical example of a **continuous injection system (CIS).**

An electronic fuel-injection (EFI) system generally uses one or more solenoid-operated injectors to spray fuel in timed pulses, either into the intake manifold or near the intake port. EFI systems inject fuel intermittently, rather than continuously. All domestic manufacturers use some form of EFI.

Throttle Body or Port Fuel Injection

In the past, the most common type of EFI system was throttle body injection (TBI), figure 9-4. This design was something of a halfway measure be-

tween feedback carburetion and port fuel injection. TBI generally is classified as a single-point, pulse-width, modulated (variable on-time) injection system. TBI, as its name implies, has one or two injectors in a carburetor-like assembly mounted on a traditional intake manifold. Fuel is kept at the correct pressure by a regulator built into the throttle body, figure 9-4.

The computer controls injector pulsing in one of two ways: synchronized or nonsynchronized. If the system uses a synchronized mode, the injector pulses once for each distributor reference pulse. When dual injectors are used in a synchronized system, the injectors pulse alternately. In a nonsynchronized system, the injectors are pulsed once during a given period (which varies according to calibration) completely independent of distributor reference pulses.

A TBI system had certain advantages: It provided improved fuel metering over a carburetor, was easier to service, and was less expensive to manufacture. Its disadvantages were primarily related to the manifold: Fuel distribution was unequal and a cold manifold still caused fuel to condense and puddle. To compensate for this placement, some systems used two differently calibrated injectors. This resulted in a different amount of fuel being sprayed by each injector. Also, a TBI unit, like a carburetor, must be mounted above the combustion chamber level. This generally prevented the use of tuned intake manifold designs.

Multipoint, or port, fuel injection, figure 9-5, is older than TBI and because of its many advantages

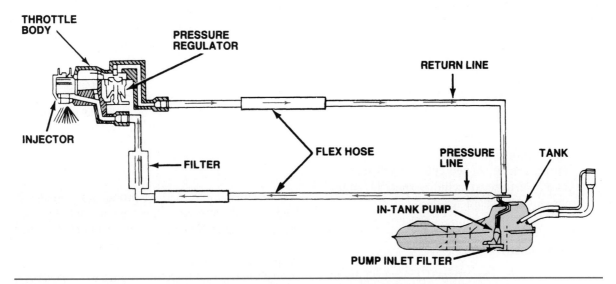

Figure 9-4. A typical throttle body injection system. (Courtesy of General Motors Corporation)

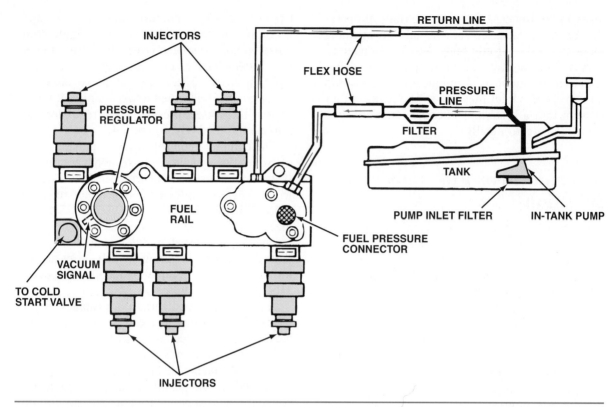

Figure 9-5. A typical 6-cylinder multipoint injection system. (Courtesy of General Motors Corporation)

will probably continue to be the system of choice in the future. Port fuel system have one injector for each engine cylinder. The injectors are mounted in the intake manifold near the cylinder head where they can inject fuel as close as possible to the intake valve, figure 9-2.

The advantages of this design also are related to characterisitics of intake manifolds:

- Fuel distribution is equal to all cylinders because each cylinder has its own injector, figure 9-6.
- The fuel is injected almost directly into the combustion chamber, so there is no chance for it to condense on the walls of a cold intake manifold.
- Because the manifold does not have to carry fuel or properly position a carburetor or TBI unit, it can be shaped and sized to tune the intake airflow to achieve specific engine performance characteristics.

Continuous Injection

A continuous injection system (CIS) constantly injects fuel whenever the engine is running. The most

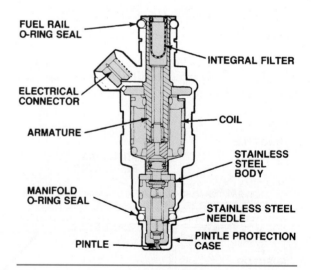

Figure 9-6. A multipoint (port) electronic fuel injector.

common example used on production vehicles is the Bosch K-Jetronic, a mechanical continuous injection system (K stands for *Kontinuierlich,* which means continuous). Many manufacturers refer to K-Jetronic simply as CIS.

The individual injectors, figure 9-7, operate on the opposing forces of fuel pressure and the

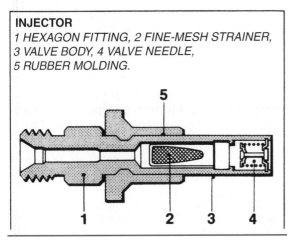

INJECTOR
1 HEXAGON FITTING, 2 FINE-MESH STRAINER,
3 VALVE BODY, 4 VALVE NEEDLE,
5 RUBBER MOLDING.

Figure 9-7. A Bosch K-Jetronic mechanical fuel injector. (Reproduced with permission from the copyright owner Robert Bosch GmbH. Further reproductions strictly prohibited.)

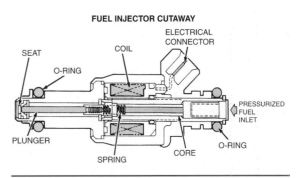

FUEL INJECTOR CUTAWAY

Figure 9-8. Solenoid actions intermittently open the EFI nozzles. (Courtesy of DaimlerChrysler Corporation)

spring-loaded valve in the injector tip. When inlet fuel pressure reaches about 49 psi (338 kPa), it overcomes spring pressure and forces the injector open. Each injector then delivers fuel continuously to the port near the intake valve. A cloud of fuel collects or "waits" at the valve to mix with incoming air. When the intake valve opens, the air-fuel mixture enters the cylinder.

In the K-Jetronic system, fuel pressure controls injector opening and the amount of fuel that is injected. Fuel pressure varies continually during different engine operating conditions. Under heavy load, for example, fuel pressure may reach 70 psi (483 kPa) or more. This forces the injector farther open to admit more fuel.

In the K-Jetronic system, fuel pressure opens the injector. The amount of fuel delivered is determined by the fuel distributor, which routes fuel to each individual injector. The amount of fuel delivered is determined by the airflow sensor plate moving the control plunger. As the sensor plate opens, it also increases a small slit to allow more fuel to each injector. The sensor plate is opened by airflow, and is restricted by control pressure. Control pressure varies with temperature to allow the sensor plate to open farther when cold for a richer mixture.

Injector spring force and fuel pressure cause the injector valve to vibarte or buzz to atomize the fuel. When the engine is shut off, fuel pressure at the injector drops below injector-tip spring pressure, and the injectors close. Residual fuel pressure is retained in both the supply and the injector lines to ensure a ready fuel supply when the engine is restarted, and to prevent hot restart problems.

Electronic Injection

To understand fuel injection as part of a complete engine control system, it is important to understand the operation and control of EFI injectors. An EFI injector is simply a specialized solenoid, figure 9-8. It has an armature winding to create a magnetic field, and a needle (pintle), a disc, or a ball valve. A spring holds the needle, disc, or ball closed against the valve seat, and when energized, the armature winding pulls open the valve when it receives a current pulse from the powertrain control module (PCM). When the solenoid is energized, it unseats the valve to inject fuel.

The injector always opens the same distance, and the fuel pressure is maintained at a controlled value by the pressure regulator. The amount of fuel delivered by the injector depends on the amount of time (on-time) that the nozzle is open. This is the injector pulse width—the on-time in milliseconds that the nozzle is open.

The PCM commands a variety of pulse widths to supply the amount of fuel that an engine needs at any specific moment. A long pulse width delivers more fuel; a short pulse width delivers less fuel.

Injector pulse width relates to another important concept that is used to test injection systems. This is the injector duty cycle. A duty cycle is a fixed number of on-off cycles in a specific amount of time. A duty cycle solenoid will turn on and off a fixed number of times per second. The amount of on-time and off-time will change within each on-off cycle, but there will always be the same number of cycles per second, figure 9-9. A common automotive example of a duty cycle solenoid is the carburetor mixture control solenoid. It has a computer-controlled, fixed frequency of 10 cycles per second. During each cycle (a tenth of a second) the solenoid has both

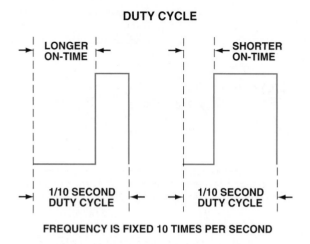

DUTY CYCLE

LONGER ON-TIME

SHORTER ON-TIME

1/10 SECOND DUTY CYCLE

1/10 SECOND DUTY CYCLE

FREQUENCY IS FIXED 10 TIMES PER SECOND

Figure 9-9. Within the duty cycle, as the percentage of on-time is increased or decreased, the mixture is made leaner or richer.

on- and off-time. By controlling the amount of time the solenoid is on during each cycle, the computer controls the amount of fuel flow. The amount of time energized is expressed as "on-time" and is usually measured by dwell. The on-time can also be measured on a voltage or duty-cycle scale in percent. In most cases an increase in on-time leans the mixture.

Various Sequence Combinations

Electronic fuel-injection systems use a solenoid-operated injector, figure 9-8, to spray atomized fuel in timed pulses into the manifold or near the intake valve. Injectors may be sequenced and fired in one of several ways, but their duty cycle and pulse width are determined and controlled by the engine computer.

Port systems have an injector for each cylinder, but they do not all fire the injectors in the same way. Domestic systems use one of five ways to trigger the injectors:

• Grouped single-fire
• Grouped double-fire
• Simultaneous double-fire
• Alternating synchronous double-fire
• Sequential

The first four of these combinations are typical of a pulsed injection system, since the injectors are opened and closed at regular intervals. This differs from a continuous injection system, which provides an uninterrupted flow of atomized fuel when the injectors are open. The pulsed system controls fuel flow by varying the length of time during which the injectors are open. Sequential injection is similar, but is called a timed injection system. The injector operates the same as in a pulsed system, but its pulse is timed to inject fuel just as the intake valve for each cylinder opens.

Grouped Single-Fire
Based on the Bosch D-Jetronic system, injectors in this system are split into two groups. The groups are fired alternately with one group firing each engine revolution. Only one injector pulse is used for each combustion stroke in a four-stroke cycle. Early Cadillac V-8 systems used this design and while it worked reasonably well, it was not as precise as newer designs. Since only two injectors can be fired relatively close to the point where the intake valve is about to open, the fuel charge for the remaining six cylinders must stand in the intake manifold for varying times. Since a new charge is released only once every two crankshaft revolutions, it is necessary to wait this long before any change can be made in the air-fuel mixture.

Grouped Double-Fire
This system again divides the injectors into two equal-size groups. The groups fire alternately; each group fires once each crankshaft revolution, or twice per four-stroke cycle. This means the air-fuel mixture changes can be made sooner than with a single-fire system.

Simultaneous Double-Fire
This design fires all of the injectors at the same time once every engine revolution: two pulses per four-stroke cycle. Many port fuel-injection systems on 4-cylinder engines use this pattern of injector firing. It is easier for engineers to program this system and it can make relatively quick adjustments in the air-fuel ratio, but it still requires the intake charge to wait in the manifold for varying lengths of time.

Alternating Synchronous Double-Fire
In this design, used on some GM Quad 4 engines, the four injectors are divided into two pairs: 1 and

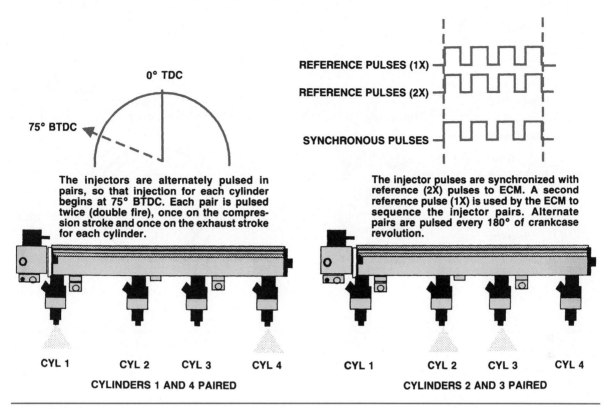

Figure 9-10. The Quad 4 alternating synchronous double-fire (ASDF) 2 × 2 firing diagram. (Courtesy of General Motors Corporation)

4, and 2 and 3. Alternate pairs are triggered every 180 degrees of crankshaft revolution, with each pair fired twice per four-stroke combustion cycle, figure 9-10.

Sequential

Sequential firing of the injectors according to engine firing order is the most accurate and desirable method or regulating port fuel injection. However, it is also the most complex and expensive to design and manufacture. In this system, the injectors are timed and pulsed individually, much like the spark plugs are sequentially operated in firing order of the engine. Each cylinder receives one charge every two crankshaft revolutions, just before the intake valve opens. This means that the mixture is never static in the intake manifold and mixture adjustments can be made almost instantaneously between the firing of one injector and the next. A camshaft sensor signal or a special distributor reference pulse informs the PCM when the No. 1 cylinder is on its compression stroke. If the sensor fails or the reference pulse is interrupted, some injection systems shut down, while others revert to pulsing the injectors simultaneously.

COMMON SUBSYSTEMS AND COMPONENTS

Regardless of the type, all fuel-injection systems have three basic subsystems:

- Fuel delivery
- Air control
- Electronic fuel trim using auxiliary sensors or actuators

Fuel Delivery System

The fuel delivery system consists of an electric fuel pump, a filter, a pressure regulator, one or more TBI injectors (conversely, a fuel rail with individual port injectors), and the necessary connecting fuel feed and return lines.

Fuel Pump

An electric fuel pump provides constant and uniform fuel pressure at the injectors. Most systems use a positive-displacement vane, turbine, or roller

Fuel Pump

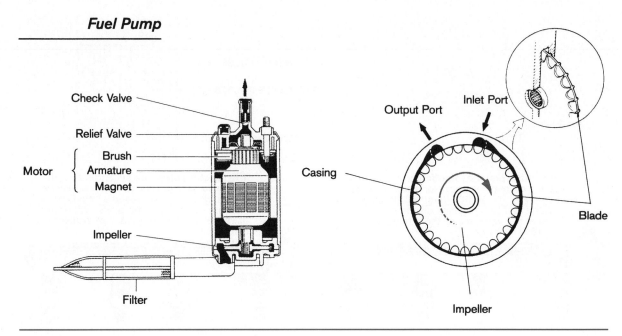

Figure 9-11. Fuel-injection system uses an electronic in-tank fuel pump. (Provided courtesy of Toyota Motor Sales U.S.A., Inc.)

pump, figure 9-11. Fuel-injected vehicles generally use one of the following pump applications:

- High pressure, in-tank
- Low pressure, in-tank
- Low pressure, in-tank and high pressure, inline
- High pressure, inline

In-tank fuel pumps may be separate units, but often are combined with the fuel gauge sender assembly, figure 9-12. The pump supplies more fuel than the system requires with the pressure regulator controlling volume and pressure. An internal relief valve protects the pump from excessive pressure if the filter or fuel lines become restricted.

Fuel Filter

Clean fuel is extremely important in a fuel-injection system because of the small orifice in the injector tip through which the fuel must pass. Most injection systems use three filters. The first is a fuel strainer or filter of woven plastic attached to the fuel pump inlet, figure 9-12. This prevents contamination from entering the fuel line and separates water from the fuel. The filter is self-cleaning and requires no maintenance. If a fuel restriction occurs at this point, the tank contains too much sediment or moisture and should be removed and cleaned.

Additional filtration is provided by a high-capacity inline filter, figure 9-13, to remove con-

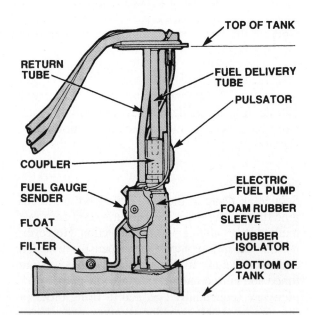

Figure 9-12. This in-tank roller vane fuel pump is combined with the fuel gauge sender unit. (Courtesy of General Motors Corporation)

tamination larger than 10 to 20 microns (0.0025 to 0.0050 inch). This filter generally is mounted on a frame rail under the vehicle, but sometimes it may be mounted in the engine compartment.

The final line of defense against fuel contamination is in the fuel injector itself. Each injector contains its own inlet filter screen, figure 9-6, to prevent contamination from reaching the tip. Some

Figure 9-13. Fuel-injection systems use large disposable filter canisters to ensure proper filtration. They may be the threaded type, which require an O-ring, or the quick disconnect type.

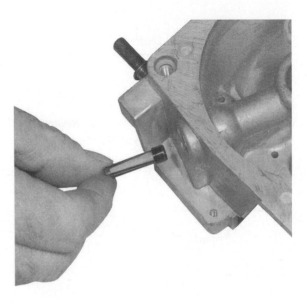

Figure 9-14. A filter screen may be installed in the fuel inlet of some throttle bodies.

TBI units have a similar filter screen installed in the fuel inlet fitting, figure 9-14.

Fuel Lines

A supply line carries fuel from the pump through the filter to the TBI unit or fuel rail. In fuel-return type systems, a return line carries excess fuel back to the tank. This allows fuel to be continuously circulated from the tank to the injectors and back to the tank, figure 9-15. The system maintains constant fuel pressure drop and volume while minimizing fuel heating and vapor lock. Most injection-system fuel lines are made of steel tubing, although Ford vehicles and some imports also use nylon fuel-line tubing. Flexible hose may be used in some low-pressure return lines.

Pressure Regulator

The pressure regulator and fuel pump work together to maintain the required pressure drop at the injector tips. The fuel pressure regulator typically consists of a spring-loaded, diaphragm-operated valve in a metal housing.

Fuel pressure regulators on fuel-return EFI systems are installed on the return (downstream) side of the injectors at the end of the fuel rail, or are built into or mounted upon the throttle body housing. Downstream regulation minimizes fuel pressure pulsations caused by pressure drop across the injectors as the nozzles open. It also ensures positive fuel pressure at the injectors at all times and holds residual pressure in the lines when the engine is off. On returnless systems, the regulator is located back at the tank with the fuel filter.

In order for excess fuel (about 80 to 90 percent of the fuel delivered) to return to the tank, fuel pressure must overcome spring pressure on the spring-loaded diaphragm to uncover the return line to the tank, figure 9-16. This happens when system pressure exceeds operating requirements. On some systems, the spring side of the diaphragm is exposed to inlet venturi vacuum to fine-tune fuel delivery. With TBI, the regulator is close to the injector tip, so the regulator senses essentially the same air pressure as the injector.

The pressure regulator used in a port fuel-injection system, figure 9-17, has an intake manifold vacuum line connection on the regulator vacuum chamber. This allows fuel pressure to be

■ The Importance of Constant Pressure Drop

Constant pressure drop across the injectors is important to ensure that the fuel sprays properly into the intake manifold regardless of manifold pressure. A manifold vacuum of 20 inches equates to approximately 10 psi absolute. A turbo boost of 11 pounds above 14.7 atmospheric equates to an absolute manifold pressure of almost 26 psi. 26 pounds full boost, minus 11 pounds at idle, equals a variation of pressure in the intake of about 15 pounds. The fuel rail pressure must be adjusted accordingly by the pressure regulator if fuel quantity injected (given a fixed on-time) is to always be the same.

Return Fuel Delivery System

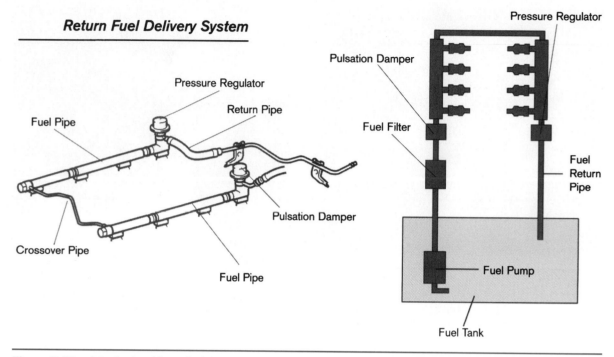

Figure 9-15. A typical multiport fuel delivery system. (Provided courtesy of Toyota Motor Sales U.S.A., Inc.)

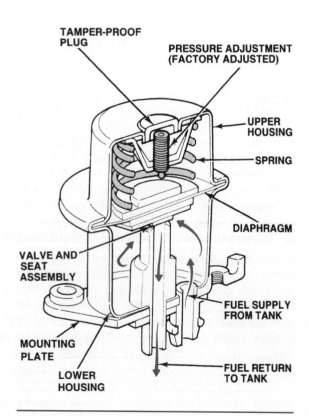

Figure 9-16. The pressure regulator commonly used with a TBI system functions by fuel pressure alone. (Courtesy of Ford Motor Company)

modulated by a combination of spring pressure and manifold vacuum acting on the diaphragm.

In both TBI and port systems, the regulator shuts off the return line when the fuel pump is not running. This maintains pressure at the injectors for easy restarting after hot soak as well as reduced vapor lock.

On the Bosch K-Jetronic mechanical injection systems without electronic control, there are actually two pressure regulators: the primary system pressure regulator, and the control pressure regulator. The system pressure regulator is located on the side of the fuel distributor and connects to the fuel return line. Spring force operates on one side of the regulator with pump pressure on the other side to modulate the return line opening.

Bosch K-Jetronic systems vary control pressure to tailor fuel delivery for different operational requirements with the control pressure regulator, which is sometimes called the warmup regulator. Primary system pressure is typically around 75 psi (517 kPa) and may go as high as 100 psi (690 kPa).

In later Bosch CIS-E mechanical-injection system designs with electronic control, a current-controlled *electrohydraulic pressure actuator* replaces the control pressure regulator. This device is built into the fuel distributor, figure 9-18, and is operated by the computer. A separate spring-loaded, diaphragm-operated pressure regulator governs pri-

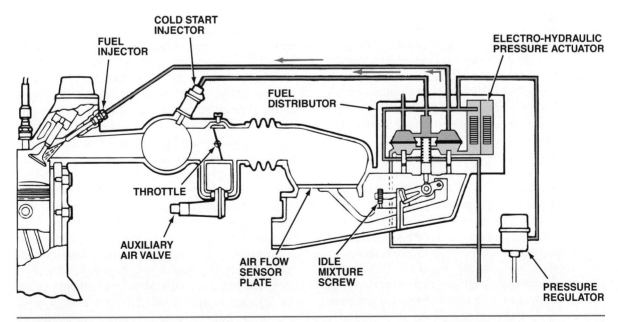

Vacuum Modulated Pressure Regulator

At Zero Vacuum

36 or 41 PSI

Fuel Pressure

36 or 41 PSI

A

B

Atmospheric Pressure

Intake manifold Pressure

Spring

Intake Chamber

Diaphragm

Check Valve

From Delivery Pipe

To Fuel Return

Intake Manifold Pressure	Low	High
Effective Spring Tension	Small	Large
Fuel Pressure	Low	High
Injection Volume	Same	Same

Figure 9-17. The pressure regulator commonly used with a multiport system works on a combination of fuel pressure and manifold vacuum engine control systems. (Provided courtesy of Toyota Motor Sales U.S.A., Inc.)

FUEL INJECTOR

COLD START INJECTOR

ELECTRO-HYDRAULIC PRESSURE ACTUATOR

FUEL DISTRIBUTOR

THROTTLE

AUXILIARY AIR VALVE

AIR FLOW SENSOR PLATE

IDLE MIXTURE SCREW

PRESSURE REGULATOR

Figure 9-18. An electrohydraulic pressure actuator is used instead of a control pressure regulator on CIS-E late-model electronic systems. (Courtesy of Volkswagen of America)

mary system pressure. This eliminates the need for a pressure relief valve in the fuel distributor.

Port fuel-injection systems generally operate with pressures at the injector of about 30 to 55 psi (207 to 379 kPa), while TBI systems work with injector pressures of about 10 to 20 psi (69 to 138 kPa). The difference in system pressures results from the difference in how the systems operate. Remember that an injection system requires only enough pressure to move the fuel through the injector and help atomize it. Since injectors in a TBI system inject the fuel into the airflow at the manifold inlet (above the throttle), there is more time for atomization in the manifold before the air-fuel charge reaches the intake valve. This allows TBI injectors to work at lower pressures than injectors used in a port system.

Mechanical-Injection Nozzles

The most common type of mechanical injection is the Bosch K-Jetronic system. Two types of injectors are used. Nonelectronic (CIS) systems use an injector with a spring-loaded tip to seal the nozzle against fuel pressure, figure 9-7. When pressure is greater than a specified level (of about 49 psi or 310 kPa), it forces the nozzle pintle to lift and nozzle opens. The fuel distributor controls injection volume by movement of the control plunger, affected by opposing forces of the airflow sensor lift and control pressure resistance to movement acting on top of the plunger. K-Jetronic nozzles inject fuel constantly as long as the pump is operating to supply pressure, and the airflow sensor plate is lifted by airflow. Bosch K-Jetronic nozzles are described in detail in the Continuous Injection section, earlier in this chapter.

Electronically controlled K-E-Jetronic systems use an air-shrouded injector. These injectors work in the same way as those used with nonelectronic systems, but design changes have made them more efficient. Air flows in through a cylinder head passage, passes between the injector and plastic shroud surrounding it, and exits close to the injectors tip. This improves atomization of the fuel as it leaves the injector, reducing fuel condensation in the manifold. A second-generation air-shrouded injector, figure 9-19, uses a circlip to retain the seal ring. A separate plastic injector shroud and fluted air directional shield on the injector tip improve airflow around the injector tip. These changes further improve fuel atomization.

■ Engine Modifications Can't Do the Whole Job

Since the automobile was discovered to be the biggest source of air pollution, many changes in engine design have been made to "clean it up." These engine modifications have reduced exhaust emissions, but it is impossible to eliminate the major cause of HC emissions simply by changing the engine design. This is because the major cause of HC emissions is the effect of the "quench area" on combustion.

The quench area is the inner surface of the combustion chamber. When the ignition flame front passes through the combustion chamber, it burns the fuel charge as it goes until the quench area is reached. This is a thin layer between 0.002 and 0.10 inch (0.05 and 0.25 mm) thick at the edge of the combustion chamber. When the flame front reaches the quench area, it is snuffed out because the quench area is so close to the cylinder head water jacket that the temperature there is too low for combustion to continue. Consequently, hydrocarbons within the quench area do not burn. They are ejected from the cylinder on every exhaust stroke along with the exhaust gases formed by combustion and enter the atmosphere as pollutants.

Electronic Injectors

EFI systems use solenoid-operated injectors. Figure 9-20 shows typical TBI injectors; figure 9-21 shows the two types of bottom-feed, port fuel injectors currently in use. This electromagnetic device contains an armature and a spring-loaded needle valve or ball valve assembly. When the computer energizes the solenoid, voltage is applied to the solenoid coil until the current reaches a specified reference level (usually about five amperes). This permits a quick pull-in of the armature during turn-on. The armature is pulled off of its seat against spring force, allowing fuel to flow through the inlet filter screen to the spray nozzle, where it is sprayed in a conical pattern. When current reaches the reference level, it is regulated at a specified value (usually about one ampere) until the injector is turned off. The low energy level during the holding state prevents overheating of the solenoid coil. The injector opens the same amount each time it is energized, so the amount of fuel injected depends on the length of time the in-

A — CUTAWAY VIEW

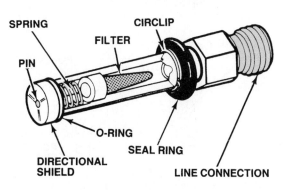

SPRING
CIRCLIP
FILTER
PIN
O-RING
SEAL RING
DIRECTIONAL
SHIELD
LINE CONNECTION

B — EXTERNAL VIEW

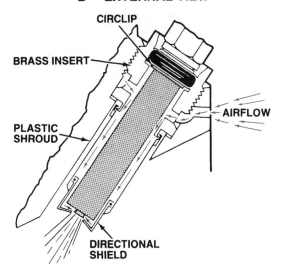

CIRCLIP
BRASS INSERT
AIRFLOW
PLASTIC
SHROUD
DIRECTIONAL
SHIELD

Figure 9-19. Air-shrouded injectors are similar to the K-Jetronic design, but flow air past the nozzle to improve atomization. (Courtesy of Volkswagen of America)

jector remains open. The injector duty cycle and the operating pulse width are explained earlier in this chapter.

Most injectors use a needle or pintle valve, figure 9-21A, for spray control. A diffuser positioned below the valve seat atomizes the fuel in a spray pattern of about 25 degrees. This design, however, has the disadvantage of allowing varnish deposits to gather on the pintle and seat, causing the injector eventually to malfunction unless it is cleaned regularly. Bosch reduced this deposit problem by extending the length of the protective cap or "chimney" on the injector tip. Varnish problems have also been reduced by better-quality fuel additives.

Early injectors also suffered from rust damage caused by water in the fuel, or varnish deposits as a result of hot soaking, a condition caused by the combination of short trips and high underhood temperatures. Bosch, for example, combated the rust problem with its so-called X-90 stainless steel injector, which all but eliminated the rusting problems.

Rochester Products took a different approach to the varnish buildup problem with the introduction of its Multec injector in 1987. The Multec injector uses a stainless steel ball and seat valve, figure 9-21B, instead of a needle valve. To assure a positive seal, the ball and seat are polished to a near-mirror finish. Spray pattern control is provided by a recessed director plate containing six machined holes. Since the Multec fuel-metering area does not extend as deeply into the intake port as a pintle-type injector does, it is not as exposed to deposit-forming intake gases, figure 9-22.

By machining four holes in the director plate at an angle, Rochester created a dual-spray Multec injector for use with 3.4-liter engines. By angling the director hole plates, the injector sprays fuel more directly at the intake valves, figure 9-23. This further atomizes and vaporizes the fuel before it enters the combustion chamber.

Lucas released another design of port injector similar to those described above, but uses a disc-type armature to control the opening and closing of the fuel spray area.

PFI injectors typically are a top-feed design in which fuel enters the top of the injector and passes through its entire length to keep it cool before being injected. Recent changes in some manifold and fuel rail designs have led to the use of a Multec bottom-feed injector in GM PFI systems, figure 9-24, similar to that used in TBI systems.

A further refinement in Multec PFI injector design appeared on some 1993 GM engines. The new injector has a stamped spray tip with a larger bore to deliver an improved fuel spray pattern, figure 9-25. The stamped spray tip injectors are slightly shorter in length and have a plastic collar installed behind the lower O-ring for proper positioning in the manifold.

Ford introduced two basic designs of deposit-resistant injectors on some 1990 engines. The design manufactured by Bosch uses a four-hole director/metering plate similar to that used by the Rochester Multec injectors. The design manufactured by Nippondenso uses an internal upstream orifice in the adjusting tube. It also has a redesigned

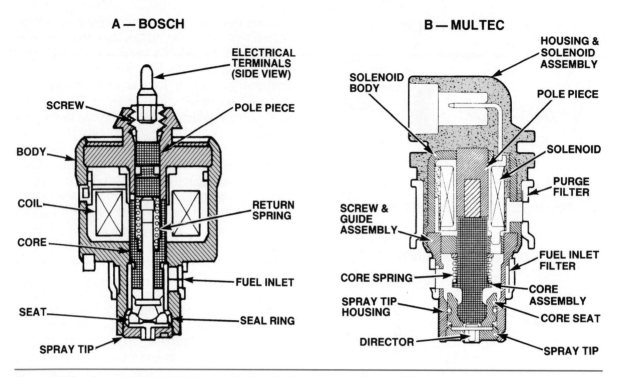

Figure 9-20. Bosch (left) and Rochester Multec (right) injectors used in a TBI unit. (Courtesy of General Motors Corporation)

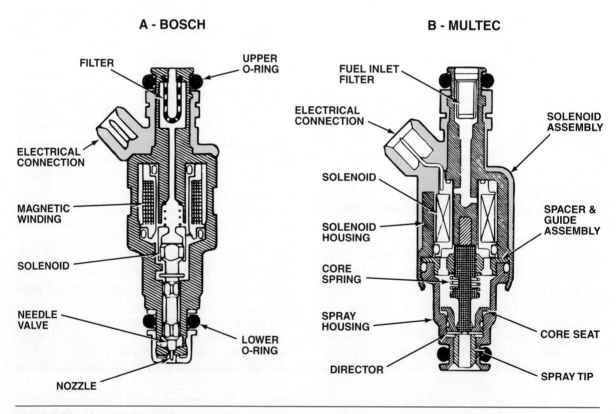

Figure 9-21. Bosch (left) and Rochester Multec (right) injectors used in multipoint systems. (Courtesy of General Motors Corporation)

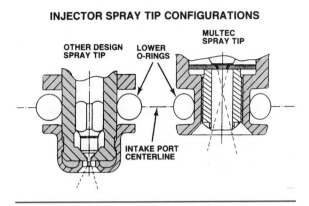

Figure 9-22. Since the Multec spray tip is not as exposed to hot intake gases as other injector designs, fewer deposits are formed to clog the injector. (Courtesy of General Motors Corporation)

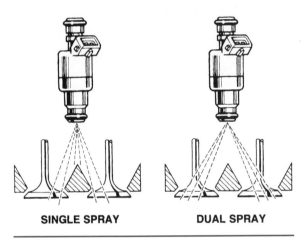

Figure 9-23. A comparison of single versus dual spray injectors. (Courtesy of General Motors Corporation)

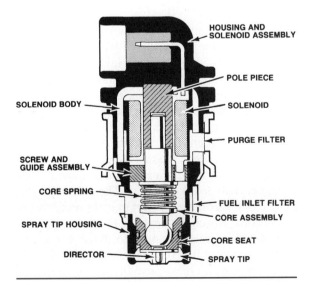

Figure 9-24. Cross section of a bottom-feed port (BFP) injector. (Courtesy of General Motors Corporation)

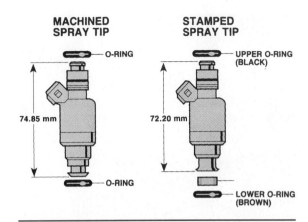

Figure 9-25. A comparison of machined spray tip and stamped spray tip injectors. (Courtesy of General Motors Corporation)

pintle/seat containing a wider tip opening that tolerates deposit buildup without affecting injector performance.

Air Control System

As discussed in Chapter 3, the difference between fuel bowl air pressure and carburetor barrel air pressure controls fuel metering in a carbureted system. A fuel-injection system needs the same kind of control. However, since the fuel is not introduced to the airflow until it passes through the injectors, there must be some way to measure intake air volume and meter the fuel accordingly. All spark-ignited engines use a throttle to control the volume of air. Those with fuel injection must measure intake air volume, usually using one of the following methods:

- An airflow meter or sensor to measure air volume
- A manifold pressure sensor to measure pressure
- An air mass air flow sensor to measure air weight (mass)

Throttle

The throttle works exactly the same in both a carbureted and a fuel-injected system. It is connected to the accelerator linkage, and regulates the amount of air taken into the engine. With a TBI system, the throttle is mounted downstream of the injectors but above the intake manifold, figure 9-1. The injectors spray directly into the throttle bore. Since the fuel

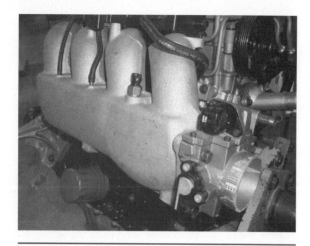

Figure 9-26. As can be seen in this photo, the horizontal throttle plates control the airflow into the intake manifold runners.

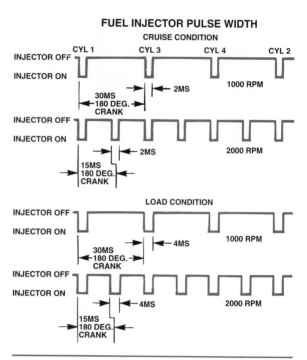

Figure 9-27. Fuel flow timing and pulse width. (Courtesy of DaimlerChrysler Corporation)

is introduced to the air above the throttle plate, the throttle regulates the amount of air-fuel mixture that enters the manifold.

On port injection systems, the throttle is mounted in the air intake throttle body upstream of the port injectors and controls only airflow, figure 9-26. A sensor determines the intake air volume. It commonly does this in one of three ways: by measuring airflow speed, manifold pressure, or the molecular mass (weight in grams/second) of the air drawn into the engine. Each of these methods directly relates to the amount of fuel required from the injectors.

Electronic Control System with Auxiliary Sensors and Actuators

A fuel-injection system does the same jobs as a carburetor. A fuel-injection system handles enrichment and mixture control by using information from electronic sensors and actuators.

Computer Control

An electronic fuel-injection system uses engine sensors and auxiliary metering devices to perform the required enrichment and mixture control functions. Sensors provide input data to the engine computer. After processing the data, the computer sends output signals to actuators. In an EFI system, solenoid-operated fuel injectors are one type of actuator used.

Remember that the injector opens the same amount each time it is energized, so the amount of

fuel injected depends on the length of time the injector is energized. Time duration is called pulse width, and is measured in milliseconds. The computer varies the injector pulse width according to the amount of fuel required under any operating condition, figure 9-27. Each on/off sequence of an injector is a *full* cycle. The pulse width is a portion, or percentage, of that full cycle.

Cold Starting

A cold engine requires a richer air-fuel mixture for starting. Port fuel-injection (PFI) systems sometimes use a separate cold-start injector to provide additional fuel during the crank mode. During engine cranking, the individual injectors send fuel into the cylinder ports. At the same time, the cold-start injector provides the additional fuel by injecting it into a central area of the manifold or into a separate passage in the manifold, figure 9-28. In either location, the extra fuel is distributed more or less equally to all cylinders. The colder the engine, the longer the cold-start injector remains on.

On some Bosch systems, the cold-start injector is also pulsed intermittently to assist hot starts. The evaporating fuel in the intake manifold cools the incoming air charge, making it more dense.

In older systems, cold-start injector duration is regulated by a thermal-time switch in the ther-

Non-ECM Controlled Cold Start Timer Circuit

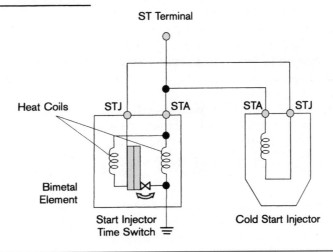

Figure 9-28. The cold-start injector provides extra fuel when cranking. (Provided courtesy of Toyota Motor Sales U.S.A., Inc.)

mostat housing or cylinder head. This switch grounds the cold-start injector circuit when the engine is cranked at a coolant temperature below a specified value. A heating element inside the switch also is activated during cranking and starts to heat a bimetallic strip in the ground circuit. Once the bimetallic strip reaches the specified temperature, it opens the ground circuit and shuts off ground for the cold-start injector.

Idle Control

Port fuel-injection systems generally use an auxiliary air bypass or auxiliary air regulator, figure 9-29 to do the same job as the fast idle cam in a carburetor. This air bypass or regulator provides needed additonal airflow, and thus more fuel. The engine needs more power when cold to maintain its normal idle speed to overcome the increased friction from cold lubricating oil. It does this by opening an intake air passage to let more air into the engine. The system is calibrated to maintain engine idle speed at a specified value regardless of engine temperature.

Some PFI and TBI systems use an **idle air control (IAC) motor** to regulate idle bypass air. The IAC is computer-controlled, and is either a solenoid-operated valve or a stepper motor, figure 9-29, that regulates the airflow around the throttle.

Other systems, like the earlier versions of the Bosch D, K, and L-Jetronic systems, use an auxiliary air valve that contains a heated bimetallic

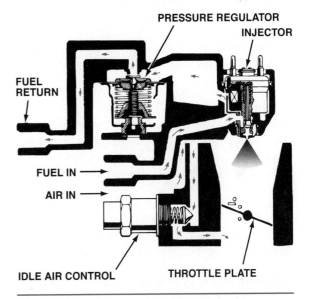

Figure 9-29. The idle air control controls the idle bypass air. (Courtesy of General Motors Corporation)

strip. When the engine is first started, air passes through the gate valve and current is applied to the bimetallic strip. As the engine warms up, the bimetallic strip starts to deflect and gradually closes off the gate valve passage to shut off the additional airflow. By the time the engine has fully warmed up, the gate valve is completely closed. Idle speed is now determined by the bypass adjustment built into the throttle housing.

Idle airflow in a TBI system travels through a passage around the throttle and is controlled by a stepper motor. In some applications, an externally mounted permanent magnet motor called the **idle speed control (ISC) motor** mechanically advances the throttle linkage to advance the throttle opening.

Idle fuel metering on a Bosch K-Jetronic system is controlled by a mechanical adjustment. Electronic fuel-injection systems may control idle fuel metering either by a predetermined program in the computer or from signals provided by a manifold pressure sensor. With systems using a vane airflow sensor, an idle mixture screw allows adjustment of bypass air past the moveable vane.

Acceleration and Load Enrichment

Airflow and manifold pressure change rapidly when the throttle is opened. With a mechanical injection system, these changes cause an increase in the volume of fuel flow. In an electronic injection system, signals from the VAF or the MAP or MAF sensors to the computer result in an increase in injector pulse width. The throttle position sensor movement also assists in a momentary enrichment requirement being signaled to the computer.

Engine Speed and Crankshaft Position

Electronic injection systems use engine speed and crankshaft position to time the injection pulses and determine the pulse width. Speed and position signals can be provided either by a magnetic sensor in the engine or by Hall Effect switches or pulse generators in the distributor, at the crankshaft or camshaft.

Engine-mounted sensors generally are positioned in the bell housing, block, or vibration dampener area. These magnetic pulse generators react to the changes in magnetic reluctance caused by flywheel grooves or teeth, figure 9-30. On some GM engines with a distributorless ignition system, the sensor is in the engine block and uses a special reluctor or wheel on the crankshaft to generate the signal, figure 9-31. Distributor signals generally are provided by the magnetic pulse generator, a Hall Effect switch, figure 9-32, or optical sensor, although a separate signal generator or switching device can be used.

As mentioned previously, sequential port systems require a number 1 cylinder signal to time the injector firing order. This signal can be provided by the same sensor that indicates crankshaft position or engine speed. Some engines use a separate sensor for the sequential timing signal, but it works

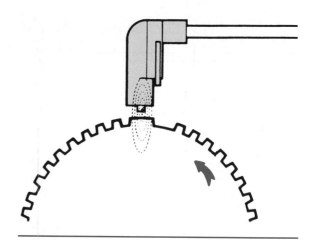

Figure 9-30. A typical crankshaft sensor that reads crankshaft position from the flywheel teeth.

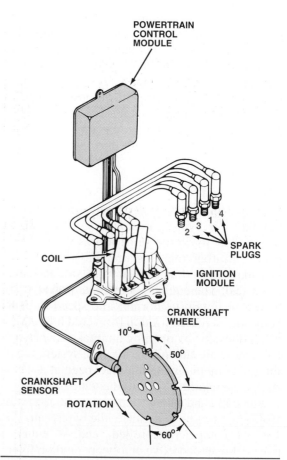

Figure 9-31. Distributorless ignitions use a sensor and notched wheel or reluctor on the crankshaft to inform the computer of crankshaft position. (Courtesy of General Motors Corporation)

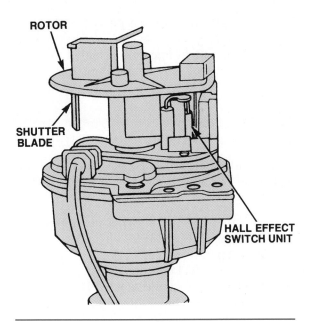

Figure 9-33. Most throttle position sensors (TPS) are a rotary-type potentiometer that attaches directly to the throttle plates.

Figure 9-32. A Hall Effect switch in the distributor also can be used to determine crankshaft position. (Courtesy of DaimlerChrysler Corporation)

on the same principle as other position or speed sensors. Without the number 1 cylinder signal, a sequential injection system either will not run, or will revert to a programmed pulsed firing sequence that allows the car to be driven in for repair.

■ **Poor Driving Means Poor Mileage**

Manufacturers have succeeded in producing cars that use less fuel and are more efficient. Still, the driving patterns of individual drivers can make a big difference in how much fuel the car will use. Here are some helpful points to remember:

- Stopping and restarting a car engine consumes less fuel than idling for one minute.
- A minor tune-up can increase mileage by about 10 percent.
- One failed spark plug in an 8-cylinder engine can cut mileage by 12 percent and increase HC emission by as much as 300 percent.
- Turning off the air conditioner can improve gas mileage by up to 10 percent.

Throttle Position

A sensor monitors throttle position and rate of change for the computer. Throttle position sensors (TPS) may be a simple on/off switch, or a potentiometer that sends a variable voltage signal to the computer, figure 9-33. By interpreting the 0.4-to 4.5-volt signal, the computer determines when the engine is at:

- Closed throttle (during idle and deceleration) ~0.5 volts
- Part throttle (normal operation) ~2-3 volts
- Wide-open throttle (acceleration and full-power operation) ~4.5 volts

Proper TPS adjustment is critical to ensure that fuel enrichment and fuel cut-off occur at the appropriate times. Otherwise, driveability problems such as stalling, surging, and hesitation will occur. Some throttle position sensors can be adjusted; others are nonadjustable and must be replaced if out of specifications.

Air and Coolant Temperature

The computer must be kept aware of intake air temperature and engine coolant temperature. Intake air temperature is used as a density corrector for calculating fuel flow and proportioning cold enrichment fuel flow. The **engine coolant temperature (ECT)** sensor is the single most important input for fuel trim calculations, and is a major input for fuel delivery when the engine is cranking. Intake air temperature (IAT) input is also used to help determine the correct air-fuel ratio according to the engine's operating temperature. This information is provided to the computer by the coolant and aif temperature sensors.

Most temperature sensors are thermistors or variable resistors, figure 9-34. Usually, this type of sensor lowers its resistance as the sensor is heated. Because resistance lowers with temperature

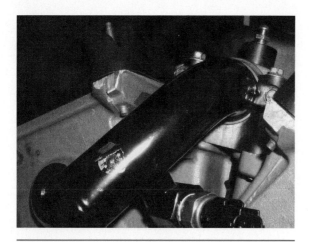

Figure 9-34. The engine coolant temperature (ECT) sensor is usually screwed directly into a coolant passage so that it is immersed in the coolant stream.

increase, it is often referred to as having a **negative temperature coefficient (NTC)**. The reference voltage applied by the computer to this type of resistor is thus altered according to temperature, because of the varying voltage drop across the NTC thermistor. This data is integrated with data from other sensors by the computer, and used to change the fuel trim by adjusting injector pulse width or fuel pressure.

On late-model GM and Chrysler cars, for greater ECT measuring accuracy, midway through the warmup sequence of the engine, an additional resistance is placed in the circuit causing the signal to suddenly jump to a higher level, only to gradually drop once more as the engine continues to warm up. Thus, a graph of ECT circuit resistance as the engine warms up appears to have two distinct slopes instead of the normal one.

SPECIFIC SYSTEMS

Fuel-injection systems have replaced the carburetor. The summary descriptions that follow provide the basic features of these systems and can help you recognize the different devices, and how they do similar tasks.

Late-Model Port Fuel-Injection Systems

The port fuel-injection systems used on domestic and many Japanese vehicles are based on

Bosch designs and contain many components manufactured by or under license from Bosch. All have individual injectors installed in some form of fuel rail and mounted in the intake manifold.

General Motors

General Motors introduced port fuel injection on some 1984 models and now uses five different types:

- Port fuel injection (PFI) or multipoint fuel injection (MFI or MPFI)
- Tuned port injection (TPI)
- Sequential fuel injection (SFI)
- Central port fuel injection (CPI)
- Central sequential port fuel injection (CSFI)

PFI and MFI or MPFI systems, figure 9-35, are simultaneous double-fire systems in which all injectors fire once during each engine revolution. In other words, two injections of fuel are mixed with incoming air for each combustion cycle. Some Quad 4 engines, however, use a variation called alternating synchronous double-fire (ASDF) injection. Two of the four injectors are triggered every 180 degrees of crankshaft revolution with each pair (1 and 4, 2 and 3) fired twice per combustion cycle.

TPI systems function in the same manner as PFI and MFI systems, but the intake air plenum runners are individually tuned to provide the best airflow for each cylinder. This system generally is used on high-performance V-8 engines in Corvettes, Camaros, and Firebirds.

SFI systems are essentially the same design as PFI and MFI, but the injectors are triggered in firing order sequence and timed to the opening of the intake valves. SFI engines are used with a distributorless ignition system. This type of ignition system requires sequential fuel injection so that the "waste-spark" that fires in the cylinder on an exhaust stroke does not ignite any air-fuel mixture.

A cross between port fuel injection and throttle body injection, CPI was introduced on 1992 4.3-liter V-6 M-van truck engines, figure 9-36. The CPI assembly consists of a single fuel injector, a pressure regulator, and six poppet nozzle assemblies with nozzle tubes, figure 9-37. The CSFI system has six injectors in place of the CPI's one.

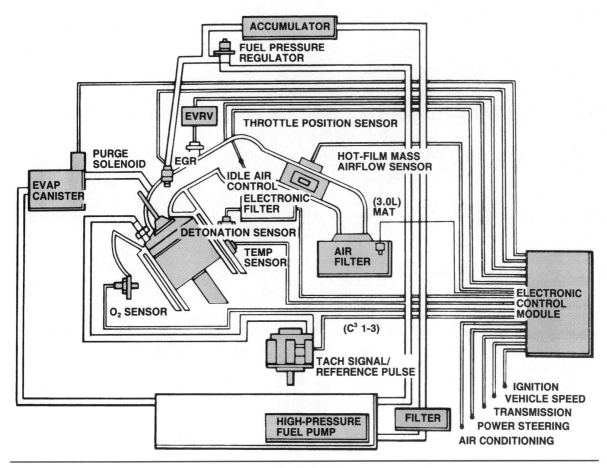

Figure 9-35. A typical GM simultaneous double-fire injection system. (Courtesy of General Motors Corporation)

Figure 9-36. The GM 4.3 L central point fuel injection (CPI) system.

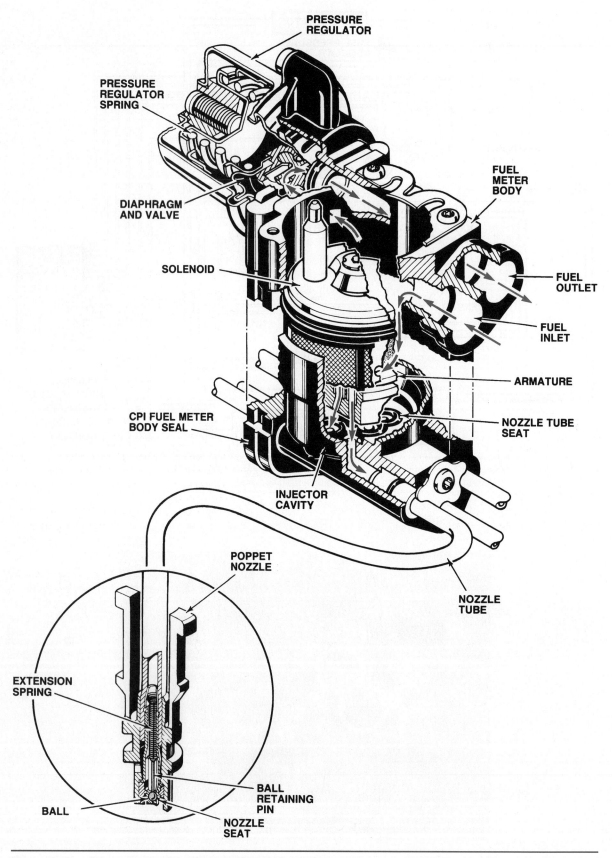

Figure 9-37. Cross section of the CPI injection assembly. (Courtesy of General Motors Corporation)

■ Bosch Jetronic Systems

A mechanical continuous injection system (CIS), figure 9-38, the Bosch K-Jetronic system was used on many European vehicles in the early 1970s. Air volume is measured by sensing airflow; fuel delivery pressure is regulated relative to air volume and activates the individual injectors. A second-generation injection system, K-Jetronic with Lambda (oxygen sensor), features a closed-loop system using a computer-driven frequency valve to accurately adjust fuel pressure in the fuel distributor. A third-generation system, KE-Jetronic, uses additional sensors to provide increased electronic control capabilities. A milliampere-driven electrohydraulic pressure actuator within the fuel distributor controls fuel pressure (quantity) according to signals from the system computer.

The Bosch D-Jetronic system (D stands for *Druck,* or pressure), figure 9-39, is an electronically controlled injection system used on 1968–1973 Volkswagens and other European vehicles. Its solenoid-operated injectors operate in alternating banks in intermittent pulses; fuel is measured relative to engine speed and manifold air pressure.

The Bosch L-Jetronic system (L stands for *Luft,* or air) was introduced in 1974 and uses individual solenoid-operated injectors that operate intermittently. It is similar to the mechanical K-Jetronic system in that fuel metering is controlled relative to airflow, but the vane-type airflow sensor is connected to an electronic module that also regulates fuel metering relative to tempera-

ture and engine speed, figure 9-40. The L-Jetronic system sometimes is called airflow-controlled (AFC) injection. The system was revised in the 1980s to measure air volume more precisely by replacing the vane-type airflow sensor with a hot-wire mass air flow sensor (MAF). Closed-loop and digital processing enhanced the L-Jetronic's capability to meet industry demands. A variation of L-Jetronic that also incorporates ignition mapping is called Bosch Motronic.

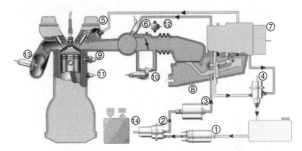

1. ELECTRIC FUEL PUMP	8. AIR-FLOW SENSOR
2. FUEL ACCUMULATOR	9. THERMO-TIME SWITCH
3. FUEL FILTER	10. AUXILIARY-AIR REGULATOR
4. SYSTEM PRESSURE REGULATOR	11. ENGINE-TEMPERATURE SENSOR
5. INJECTOR	12. THROTTLE-VALVE SWITCH
6. COLD-START VALVE	13. LAMBDA SENSOR
7. FUEL DISTRIBUTOR	14. ECU

Figure 9-39. KE-Jetronic is similar to the K-Jetronic, with a mechanical hydraulic injection system. However, it was an additional electronic control unit which increases flexibility and enables other performance enhancements. (Reproduced with permission from the copyright owner Robert Bosch GmbH. Further reproductions strictly prohibited)

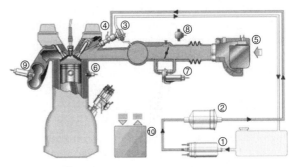

1. ELECTRIC FUEL PUMP	6. THERMO-TIME SWITCH
2. FUEL FILTER	7. AUXILIARY-AIR REGULATOR
3. FUEL PRESSURE REGULATOR	8. THROTTLE-VALVE SWITCH
4. INJECTOR	9. LAMBDA SENSOR
5. AIR-FLOW SENSOR	10. ECU

Figure 9-40. L-Jetronic. This is an electronically controlled fuel injection system which injects fuel intermittently into the intake manifold. It combines the advantages of direct air-flow sensing and electronic precision-without the need for any form of drive. (Reproduced with permission from the copyright owner Robert Bosch GmbH. Further reproductions strictly prohibited)

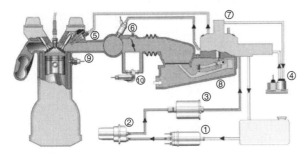

1. ELECTRIC FUEL PUMP	6. ELECTRIC START VALVE
2. FUEL ACCUMULATOR	7. FUEL DISTRIBUTOR
3. FUEL FILTER	8. AIR-FLOW SENSOR
4. WARM-UP REGULATOR	9. THERMO-TIME SWITCH
5. INJECTOR	10. AUXILIARY-AIR REGULATOR

Figure 9-38. K-Jetronic. This is a mechanical fuel injection system which constantly measures the fuel in accordance with the amount of air being drawn by the engine. (Reproduced with permission from the copyright owner Robert Bosch GmbH. Further reproductions strictly prohibited)

When the injector is energized, its armature lifts off of the six fuel tube seats and pressurized fuel flows through the nozzle tubes to each poppet nozzle. The increased pressure causes each poppet nozzle ball to also lift from its seat, allowing fuel to flow from the nozzle. This hybrid injection system combines the single injector of a TBI system with the equalized fuel distribution of a PFI system. It eliminates the individual fuel rail while allowing more efficient manifold tuning than is otherwise possible with a TBI system.

Some GM port fuel-injection systems use a mass air flow (MAF) sensor. The MAF sensor is installed in the air intake path and messures the mass or density of the incoming air. Through the MAF sensor signals, the PCM is able to determine the air temperature, density, and humidity to calculate the "mass" of the incoming air in grams per second. GM uses Delco, Bosch, and Hitachi mass air flow sensors to provide digital signals to the PCM. They operate as two types: heated film and hot wire.

Both types operate on the same principle: The resistance of a conductor varies with temperature. The film or wire conductor is kept at a constant calibrated temperature by the PCM. The greater the volume of air passing the heated conductor, the more heat the air carries away with it, and the more current required from the PCM to maintain the calibrated temperature. The PCM translates this current requirement into a voltage or frequency signal to determine the amount of airflow.

Other GM port fuel-injection systems rely on a **speed-density** air measurement system instead of mass airflow. A speed-density system uses manifold absolute pressure and temperature along with an engine mapping program in the computer to calculate mass air flow rate, figure 9-41. This type of system is sensitive to engine and EGR variations.

From 1987 through the early 1990s, the trend in fuel rail design for GM V-6 and V-8 port systems was toward dual, or split, fuel rails. Each bank of the engine has its own fuel distribution rail and injectors, figure 9-42. As a further refinement, some of these split-rail systems have biased injectors. The computer controls each bank differently, and the injectors for right and left banks may have different flow rates for identical duty cycles and pulse widths. For example, with a pulse width of five milliseconds, the left bank injectors may flow 1.5 cc, and the right bank injectors may flow 1.6 cc. Biased injectors are color coded for identification. Many engines with split-rail injection also have two O_2 sensors to allow the engine computer

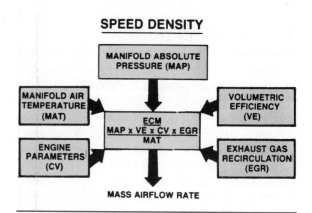

Figure 9-41. A General Motors speed-density fuel-injection system takes several inputs into consideration to determine mass air flow rate. (Courtesy of General Motors Corporation)

to control fuel metering independently for each bank of the engine.

The split-rail design now is being challenged by lower intake manifold assemblies that have a longitudinal fuel passage as an integral part of the casting, with intersecting injector bores in each runner, figure 9-43. This design totally eliminates the fuel rail. The injectors are held in place by a retaining plate and the fuel feed and return lines connect directly to the manifold housing.

Ford

Ford refers to its port fuel-injection systems as Electronic Fuel Injection (EFI). Originally based on the Bosch L-Jetronic design, Ford EFI systems for four-cylinder engines use simultaneous double-fire injector control. Ford introduced its first port EFI system on the 1.6-liter engine in 1983 Escort, Lynx, and EXP/LN7 models. The same basic system continued in use on some later 1.6- and 1.9-liter engines.

Two new EFI systems were introduced in 1986. Taurus and Sable models with 3.0-liter V-6 engines use a port simultaneous-injection system similar to the one used on 4-cylinder engines. Starting in 1986, Ford introduced a new speed-density, sequential electronic fuel-injection (SEFI) system. In the SEFI system, the EEC-IV computer operates the injectors in firing order sequence. Each injector fires once every two crankshaft revolutions as the intake valve opens.

Ford EFI systems on 4-cylinder engines originally used a vane-type airflow meter to measure intake airflow. Similar to the Bosch L-Jetronic, the airflow meter is installed ahead of the throttle

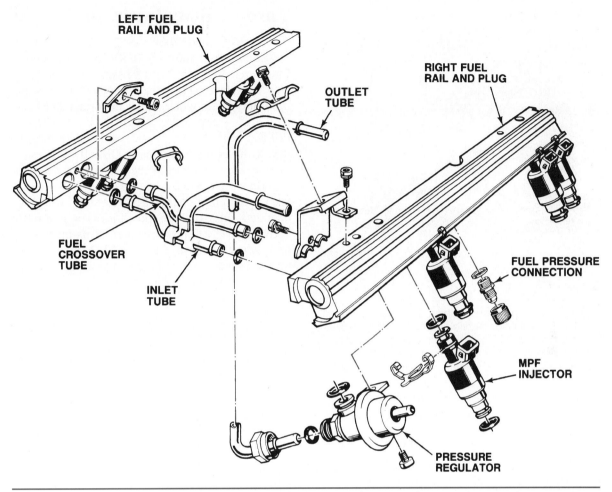

Figure 9-42. A typical GM dual fuel rail design used on the 5.7-liter engine. The individual rails are connected by fuel inlet, outlet, and crossover tubes. (Courtesy of General Motors Corporation)

body in the air intake system. The vane moves a potentiometer sensor that sends a voltage signal to the EEC computer proportionate to intake airflow. When Ford introduced SEFI to its 4-cylinder engines, the vane airflow meter was replaced by a mass air flow sensor.

Chrysler

A grouped double-fire port fuel-injection system, figure 9-44, is used on some 1984 and later 2.2-liter 4-cylinder engines. This system is similar to a Bosch D-Jetronic system and uses Bosch-designed fuel injectors in combination with Chrysler-engineered electronics. The basic Chrysler 4-cylinder injection systems are speed-density designs.

Starting in 1989, Chrysler gradually converted its turbocharged 4-cylinder engines, all V-6 and most V-8 engines to sequential multiport fuel injection (SMFI). These are either speed-density or mass air flow designs. Figure 9-45 shows a typical SMFI system as used on 3.3- and 3.8-liter V-6 engines.

Imports

The major Japanese automobile manufacturers use port fuel-injection systems derived from the Bosch L-Jetronic or AFC system. Figure 9-46 shows common components used in both Nissan and Toyota EFI systems. Nissan introduced its system in 1975, with Toyota following in 1979. These early Japanese systems control only injector duration. Systems introduced in 1981 (Nissan) and 1983 (Toyota) are fully integrated into the engine management system, controlling idle speed, EGR, and ignition timing in relation to fuel injection.

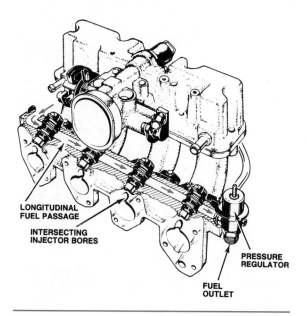

Figure 9-43. The fuel inlet and fuel passage are cast into the lower intake manifold in GM's integrated air flow system (IAFS). The design requires the use of bottom feed port injectors. (Courtesy of General Motors Corporation)

Throttle Body Injection Systems

TBI systems made their first appearance on domestic engines in the early 1980s. Once accounting for more than half of the EFI systems in use, they now are relegated primarily to economy cars and light-duty trucks where cost is important. (TBI systems are less expensive to build than PFI systems.) PFI and SFI systems are the dominant designs today, with SFI predominant on OBD II vehicles.

With the exception of the Chrysler V-8 system introduced on 1981 Imperials, all TBI systems use similar components. Single or dual solenoid-operated injectors are positioned in a throttle body assembly that owes much of its inspiration to a carburetor base, figure 9-47. A MAP sensor measures intake air and the computer controls injection volume proportionately.

General Motors

All GM TBI systems are manufactured by Rochester Division of General Motors and con-

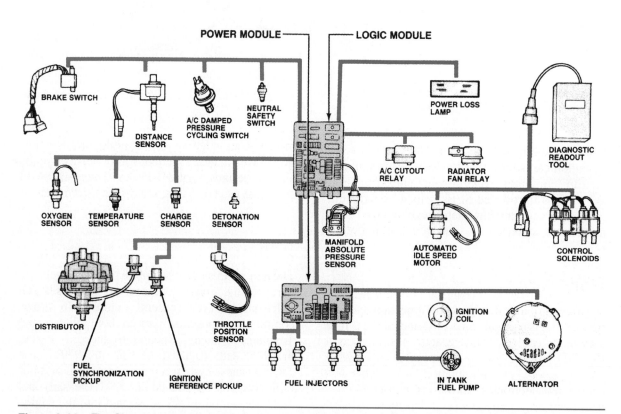

Figure 9-44. The Chrysler grouped double-fire injection system used on 2.2-liter engines. (Courtesy of DaimlerChrysler Corporation)

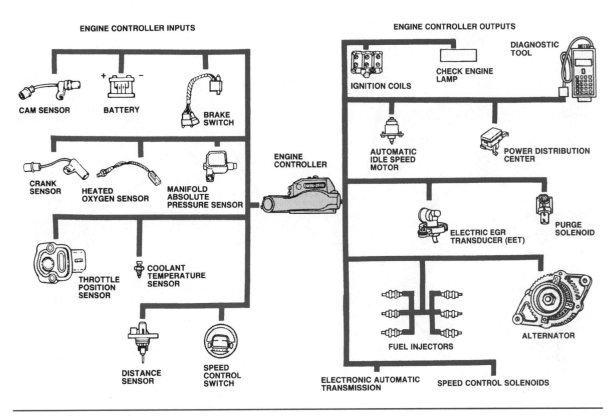

Figure 9-45. The Chrysler sequential multiport fuel-injection system on 3.3- and 3.8-liter engines. (Courtesy of DaimlerChrysler Corporation)

tain one or two injectors in the throttle body casting. There are several basic models in use:

- Models 100, 200, and 220 are two-barrel, two-injector assemblies. The Model 220 was introduced in 1985 and is the only one of the three currently in use. Some Model 220 TBI units use injectors that are calibrated to flow fuel at a different rate in each throttle bore.
- Models 300, 500, and 700 are one-barrel, one-injector assemblies. With the exception of minor fuel and airflow differences, the Models 300 and 500 are the same. The Model 700 was introduced on 4-cylinder engines as a replacement for both the 300 and 500.
- Model 400 is a unique crossfire assembly of two one-barrel, one-injector throttle bodies mounted on a single intake manifold, figure 9-48. This system was used on some Corvette, Camaro, and Firebird high-performance V-8 engines. Like the Chrysler V-8 TBI system, the GM crossfire fuel-

injection (CFI) system proved to be a technological dead end.

All GM throttle bodies use an idle speed control (ISC) motor or idle air control (IAC) valve, a throttle position sensor, and a fuel pressure regulator, figure 9-49. The regulator controls fuel pressure by opening and closing the fuel return line.

Ford

Through 1992, Ford called all of its throttle body injection systems central fuel injection (CFI) because the fuel is injected at a central point. Starting with 1993 models, it adopted the TBI designation used by other manufacturers. The high-pressure, dual injector fuel charging assembly (now called a throttle body), has been used on V-6 and V-8 engines. A low-pressure, single injector unit was introduced in 1985 for use on 2.3- and 2.5-liter 4-cylinder engines. Like GM throttle bodies, the Ford TBI units also use throttle position sensors, idle speed control motors, and pressure regulators.

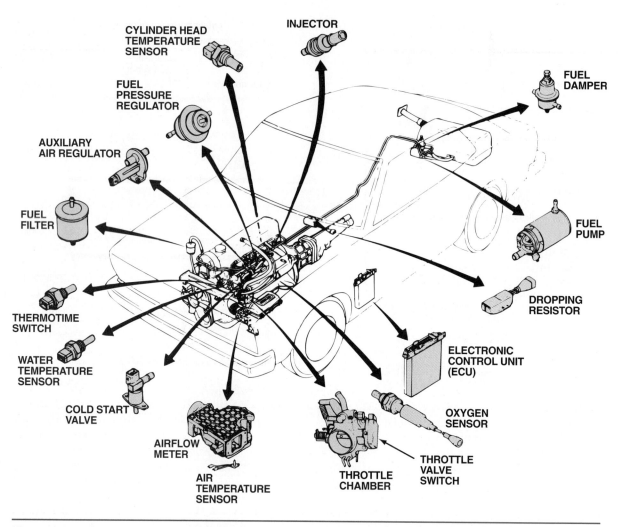

Figure 9-46. Common components used in early Nissan and Toyota port fuel-injection systems.

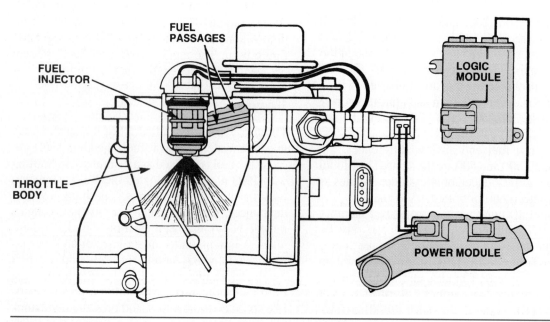

Figure 9-47. The throttle body assembly in a TBI system looks much like a carburetor, but contains only an injector, idle air control, and pressure regulator. (Courtesy of DaimlerChrysler Corporation)

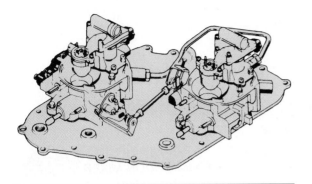

Figure 9-48. The Model 400 crossfire injection (CFI) assembly consists of two TBI units mounted on a common manifold. (Courtesy of General Motors Corporation)

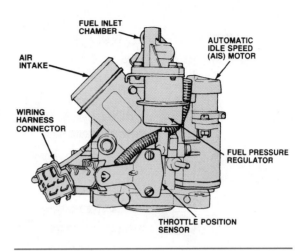

Figure 9-50. The Chrysler-Holley-Bendix high-pressure TBI unit. (Courtesy of DaimlerChrysler Corporation)

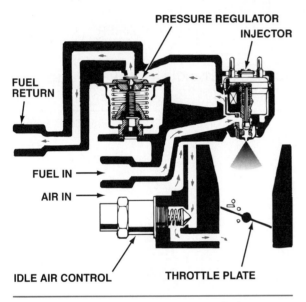

Figure 9-49. Throttle body injection fuel pressure regulation. (Courtesy of General Motors Corporation)

Chrysler

Chrysler's first TBI system was a continuous flow system used on its V-8 engine in 1981–83 Imperials. A special sensor in the air cleaner snorkel measures air intake volume. Radial vanes swirl the air and a U-shaped pressure probe in the vortex of the swirling air determines the amount of air entering the engine.

Most 1981 and later nonturbocharged Chrysler-built 4-cylinder engines use a single injector TBI system. The original system used through 1985 had a Holley-designed throttle body, figure 9-50, with an injector supplied by Bendix. The TBI system was revised in 1986, with the newer version using a low-pressure injector designed by Bosch. The ball-type injector used through 1988 was replaced by a pintle-type injector in 1989 and later TBI units.

Imports

Most import vehicles equipped with TBI use systems that are quite similar in design and operation to those used on domestic vehicles. Less typical is the TBI system used on turbocharged Japanese vehicles sold by Mitsubishi (and Chrysler Corporation) in the United States, figure 9-51. This system measures intake airflow rate based on Karman's vortex theory (a well-known theory dealing with aerodynamics and ultrasound waves). The system uses a Karman vortex airflow sensor in the air cleaner housing, figure 9-52, to measure the rate of intake airflow. Two injectors installed in the air intake throttle body are fired in an alternating sequence. A swirl nozzle design atomizes the fuel and "jiggles" the fuel spray, figure 9-53. This system is considerably more complex, figure 9-54, than the other TBI systems that have been shown.

TRENDS

With greater sophistication under the hood, methods of monitoring fuel-injection system performance have also become more high tech in nature. A technician can routinely monitor electronic waveforms or signatures using a lab oscilloscope or graphing multimeter (GMM). Injector waveforms, which represent the voltage variations as the injector pulses, can easily be read on the vehicle. Likewise, intake manifold pressure waves, fuel line pressure pulses, idle air control motor waveforms, and other signatures can produce data that, in the hands of an experienced technician,

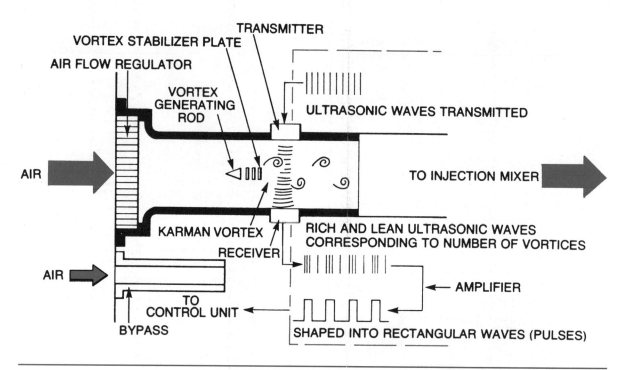

Figure 9-51. Operation of the Mitsubishi-Chrysler TBI system that measures airflow rate according to Karman's vortex theory. (Courtesy of DaimlerChrysler Corporation)

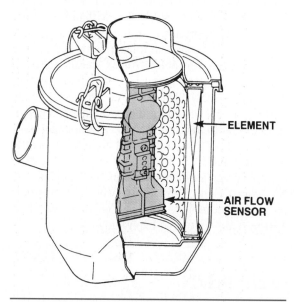

Figure 9-52. The air flow sensor is mounted inside the air-cleaner filter element. (Courtesy of DaimlerChrysler Corporation)

may reveal component weaknesses or faults even before the component actually fails.

Injector Patterns

Injectors are solenoids and thus offer inductive reactance to the flow of current. For example, as an inductor coil charges, a distinctive waveform

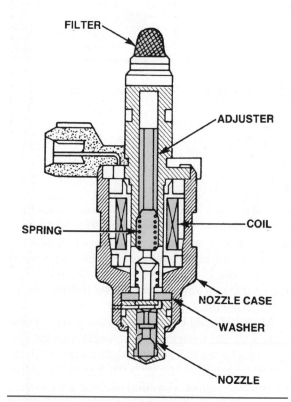

Figure 9-53. The Mitsubishi-Chrysler TBI injector uses a swirl nozzle for better fuel atomization, but operates like any other solenoid-actuated injector. (Courtesy of DaimlerChrysler Corporation)

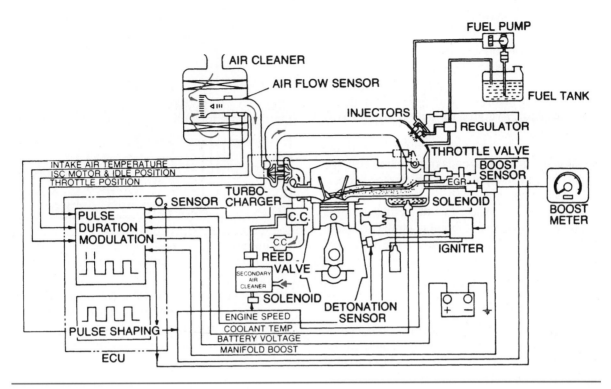

Figure 9-54. The Mitsubishi-Chrysler ECI system components. (Courtesy of DaimlerChrysler Corporation)

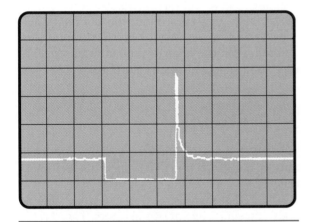

Figure 9-55. Typical fuel-injector voltage pattern. Note the prominent spike.

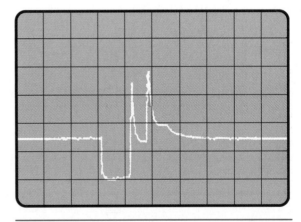

Figure 9-56. The typical peak and hold injector pattern shows two spikes as the current increase is stopped.

is displayed on the scope as the resistance to current flow rises with the buildup of magnetism in the coil.

When the injector's coil is discharged, an even more prominent voltage spike is seen, figure 9-55. The injector's voltage spike may reach as high as 70 volts; the negative current flow helps to pull the injector pintle closed much faster and without bounce than would otherwise be possible if the spike were "clamped."

Some injectors are current controlled to avoid coil overheating. One method is to open with full current, then reduce the current very quickly to just "hold" the injector pintle off of its seat. These are called peak and hold circuits, and they offer a distinctive two voltage spike pattern, figure 9-56.

Changes to known-good patterns can be caused by carbon or varnish buildup, corrosion or erosion of the pintle or tip of the injector, as well as electrical problems such as shorted windings.

IAC and ISC Motor Patterns

These motor or solenoid components have several types of patterns, depending on the type of control. An amperage-controlled IAC has a typical "sawtooth" pattern, figure 9-57. Many others are controlled by varying the duty cycle on PWM systems, figure 9-58.

Stepper motors have an on/off signal. If the signal is taken on both ends of the same coil, the signals will switch polarity as the motor is operated. As one end goes high, the other end will go low, figure 9-59. Changes to the normal patterns can reveal dirt or carbon buildup on the armature that would alter the reaction time of the motor and affect its signature.

Intake Pressure

Normal intake manifold pressure waves created by opening and closing valves can be monitored on the scope by using a pressure transducer tapped into the manifold. Far more accurate than the tradi-

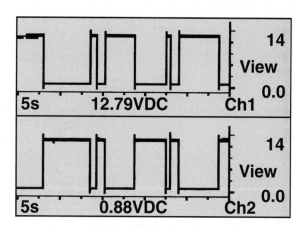

Figure 9-59. Dual trace of a stepper motor shows switching of polarity. (Courtesy of Snap-on Vantage®)

tional vacuum gauge, the scope can reveal not only abnormal patterns caused by vacuum leaks, burned valves, abnormal valve or ignition timing, but also a whole host of other more subtle problems.

Fuel Line Pressure

Abnormal scope patterns in fuel line or fuel rail pressures can reveal injector faults, pressure regulator faults, and even imminent fuel pump failures from clogging, faulty brushes/commutator and other faults. Again, a pressure transducer is used to input signals to the labscope.

SUMMARY

Two basic types of injection systems are used: throttle body injection (TBI) and port fuel injection (PFI).

The TBI system uses a carburetor-like throttle body containing one or two injectors. This throttle body is mounted on the intake manifold in the same position as a carburetor. Fuel is injected above the throttle plate and mixed with incoming air.

In a PFI system, individual injectors are installed in the intake manifold at a point close to the intake valve where they inject the fuel to mix with the air as the valve opens. Intake air passes through an air intake throttle body containing a butterfly valve or throttle plate and travels through the intake manifold where it meets the incoming fuel charge at the intake valve. The difference in injector location permits more advanced manifold designs in port systems to aid in fuel distribution.

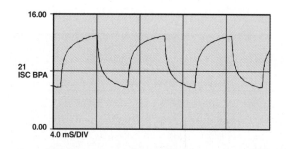

Figure 9-57. Idle air control (IAC) and idle speed control (ISC) motor produce a sawtooth pattern.

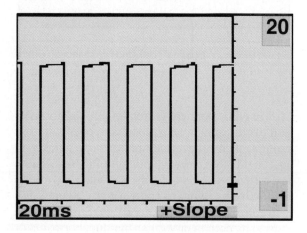

Figure 9-58. The duty cycle on a PWM system shows regular waveforms. (Courtesy of Snap-on Vantage®)

Fuel injection has numerous advantages over carburetion. It can match fuel delivery to engine requirements under all load and speed conditions, maintain an even mixture temperature, provide more efficient fuel distribution, and reduce emissions while improving driveability.

Electronic fuel injection (EFI) is far more common than mechanical injection systems such as the Bosch K-Jetronic. EFI integrates the injection system into a complete engine management system, which includes control of EGR, ignition timing, canister purging, and various other functions.

The EFI computer (PCM) uses various sensors to gather data on engine operation. Intake air volume is determined by measuring airflow speed, manifold pressure, or the mass of the air drawn into the engine. This data is added to information about throttle position, idle speed, engine coolant and air temperature, crankshaft position, and other operating conditions. After processing the data, the computer signals the solenoid-operated injectors when to open and close. Injectors may be fired in various combinations, but all use the principle of pulse width modulation, in which the computer varies the percentage of on-time in each full on/off cycle according to engine requirements. In this way, the amount of fuel injected can be varied instantly to accommodate changing conditions.

Review Questions

Choose the letter that represents the best possible answer to the following questions:

1. Fuel-injection systems can lower emissions:
   a. By matching the air-fuel ratio to engine requirements
   b. Only at high speeds
   c. By using the intake manifold to vaporize fuel
   d. By matching engine speed to load conditions

2. All of the following are common kinds of gasoline fuel-injection systems in use today *except:*
   a. Multipoint injection
   b. Throttle body injection
   c. Direct cylinder injection
   d. Continuous injection

3. Technician A says Bosch K-Jetronic systems are mechanical injection systems that use an air sensor plate and fuel distributor.
   Technician B says Bosch K-Jetronic systems are intermittent injection systems. Who is right?
   a. A only
   b. B only
   c. Both A and B
   d. Neither A nor B

4. Bosch L-Jetronic and similar systems measure airflow with a:
   a. BMAP sensor
   b. Hot-wire air mass sensor
   c. Vane-type airflow sensor
   d. Air charge temperature sensor

5. Technician A says TBI systems are electronic, intermittent injection systems.
   Technician B says TBI systems require fuel pressure regulators. Who is right?
   a. A only
   b. B only
   c. Both A and B
   d. Neither A nor B

6. Technician A says fuel pressure regulators always are on the upstream (inlet) side of the injectors in an EFI system.
   Technician B says mechanical-injection systems do not require a pressure regulator. Who is right?
   a. A only
   b. B only
   c. Both A and B
   d. Neither A nor B

7. Technician A says the throttle in a multipoint gasoline fuel-injection system is located between the injectors and the manifold.
   Technician B says the throttle in an EFI system operates the same as a throttle on a carbureted engine. Who is right?
   a. A only
   b. B only
   c. Both A and B
   d. Neither A nor B

8. Mechanical fuel injectors are operated:
   a. Continuously by fuel pressure
   b. Intermittently by the computer
   c. Alternately by air pressure
   d. Sequentially by firing order

9. Modern fuel-injection systems are based on work begun by:
   a. Rochester Products Division of GM
   b. Ford Motor Company
   c. Hitachi
   d. Robert Bosch GmbH

10. A cold-start injector is a:
    a. Mechanical continues injector
    b. Solenoid-operated auxiliary injector
    c. Mechanical intermittent injector
    d. Dual-fire electronic actuator

11. Multec injectors used in a multipoint injection system differ from those produced by Robert Bosch in that they use a:
    a. Pintle valve
    b. Swirl nozzle
    c. Pin and spring
    d. Ball-seat valve with director plate

12. All of the following are filter locations in a fuel-injection system *except:*
    a. In the pressure regulator
    b. At the fuel pump
    c. In the fuel line
    d. In the fuel injectors

13. Technician A says an EFI system can use an on/off throttle position switch.
    Technician B says a throttle position sensor is a variable resistor. Who is right?
    a. A only
    b. B only
    c. Both A and B
    d. Neither A nor B

14. Technician A says Bosch K-Jetronic fuel-injection systems operate at higher fuel pressures than other commonly used systems.
    Technician B says fuel metering in the K-Jetronic system relies on a balance of forces between incoming air and control pressure exerted on the control plunger. Who is correct?
    a. A only
    b. B only
    c. Both A and B
    d. Neither A nor B

15. The cold-start injector on Bosch L- and K-Jetronic systems is controlled by:
    a. The ignition circuit
    b. The thermo-time switch
    c. The auxiliary air valve
    d. The PCM

16. The fuel pump relay in port fuel-injection systems may be controlled by all of these *except:*
    a. A micro switch in the VAF
    b. The ignition tachometer signal
    c. The engine coolant temperature sensor
    d. The starting circuit

17. Fuel injectors in pulsed fuel-injection systems:
    a. Have an on-time measured in milliseconds
    b. Match frequency to engine speed
    c. Can be diagnosed with a lab-type oscilloscope
    d. All of the above

18. Technician A says "peak and hold" refers to a method of controlling current to the fuel injector.
    Technician B says the waveform of a "peak and hold" circuit contains two inductive spikes. Who is correct?
    a. A only
    b. B only
    c. Both A and B
    d. Neither A nor B

19. Technician A says the voltage spike from an injector can be as high as 70 volts.
    Technician B says the voltage spike helps to pull the injector closed quickly. Who is correct?
    a. A only
    b. B only
    c. Both A and B
    d. Neither A nor B

10

Supercharging and Turbocharging

OBJECTIVES

Upon completion and review of this chapter, you will be able to:

- Describe the effects of heat and pressure on air density.
- Explain the difference between a turbocharger and a supercharger.
- Identify a turbocharger and supercharger and their components on a vehicle.
- Explain the purpose of the wastegate control.

KEY TERMS

boost	turbocharger
exhaust scavenging	turbo lag
intercooler	wastegate
normally aspirated	water injection
supercharging	

INTRODUCTION

Engines with carburetors rely on atmospheric pressure to push an air-fuel mixture into the combustion chamber vacuum created by the downstroke of a piston. The mixture is then compressed before ignition to increase the force of the burning, expanding gases. The greater the mixture compression, the greater the power resulting from combustion. In this chapter, we will study three ways to increase mixture compression.

ENGINE COMPRESSION

One way in which mixture compression can be increased is through the use of a high compression ratio. During the late 1960s, high-performance car engines used compression ratios as great as 11 to 1. However, compression ratios fell to the range of 8 to 1 or 8.5 to 1 during the 1970s because of increasing emission control requirements and the tendency of higher compression engines to emit too much NO_x. Lower compression ratios also were necessary due to the greatly reduced lead content in the gasoline. The use of electronic engine management systems, however, has made it possible to raise compression ratios once again to the 9 to 1 or 10 to 1 range.

There are two major benefits to the use of high compression ratios:

- They increase volumetric efficiency because the piston displaces a larger percentage of the total cylinder volume on each intake stroke.
- They increase thermal efficiency because they raise compression temperatures, resulting in hotter, more complete combustion.

Just as high compression ratios have advantages, however, they also offer major disadvantages:

- Because the compression ratio remains unchanged throughout the engine's operating range, combustion temperature and pressure also are high and cause increased emissions during deceleration, idle, and part-throttle operation.
- A high compression ratio increases NO_x emissions.
- High compression ratios require the use of high-octane gasoline with effective anti-knock additives.

Lead used to be the most effective antiknock additive, but it is poisonous, produces harmful emissions, and destroys catalytic converters, so it is not used anymore.

THE BENEFITS OF AIR-FUEL MIXTURE COMPRESSION

The amount of force an air-fuel charge produces when it is ignited is largely a function of the charge density. Density is the mass of a substance in a given amount of space, figure 10-1. The greater the density of an air-fuel charge forced into a cylinder, the greater the force it produces when ignited, and the greater the engine power.

Atmospheric pressure is really a poor choice to push air through the intake system, as it has a predetermined limit to its efficiency. An engine that uses atmospheric pressure for intake is called a **normally aspirated** engine. A better way to increase air density in the cylinder is to use a pump. When air is pumped into the engine, however, two variables that are present even with a normally aspirated engine assume a far more important role in determining density: pressure and temperature.

When air is pumped into the cylinder, the combustion chamber receives an increase of air pressure known as **boost.** This boost in air pressure generally is measured in pounds per square inch (psi). While boost pressure increases air density, friction heats air in motion and causes an increase in temperature. This increase in temperature works in the opposite direction, decreasing air density. Because of these and other variables, an increase in pressure does not always result in greater air density, figure 10-2.

SUPERCHARGING

Another way to achieve an increase in mixture compression is called **supercharging.** This method uses a pump to pack a denser air-fuel charge into the engine's cylinders. Since the density of the air-fuel charge is greater, so is its weight—and power is directly related to the weight of an air-fuel charge consumed within a given time period. The result is similar to that of a high compression ratio, but the effect can be controlled during idle and deceleration to avoid high emissions.

Supercharging Principles

As you recall from your study of how an automotive engine breathes in Chapter 3 of this manual, air is drawn into a normally aspirated engine by atmospheric pressure forcing it into the low-pressure area of the intake manifold. The low pressure or vacuum in the manifold results from the reciprocating motion of the pistons. When a piston moves downward during its intake stroke, it creates an empty space, or vacuum, in the cylinder. Although atmospheric pressure pushes air to fill up as much of this empty space as possible, it has a difficult path to travel. The air must pass through the air filter, the carburetor or throttle body, the manifold, and the intake port before entering the cylinder. Bends and restrictions in this pathway limit the amount of air reaching the cylinder before the intake valve closes, figure 10-3. Because of this, the engine's volumetric efficiency is considerably less than 100 percent. Volumetric efficiency, as we have seen, is a percentage measurement of the volume of air drawn into a running engine compared to the maximum amount the engine could draw in (total displacement volume). While a stock engine averages approximately 70 to 80 percent volumetric efficiency, racing engines with the

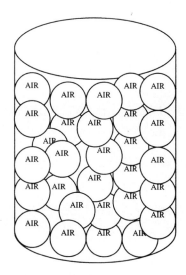

High Density Air

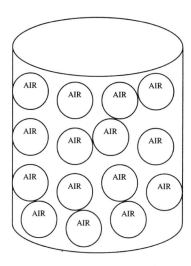

Low Density Air

Figure 10-1.   The more air that can be packed in a cylinder, the greater the density of the air charge.

proper intake and exhaust tuning often can exceed 100 percent volumetric efficiency at their power curve peak.

The volumetric efficiency of any normally aspirated engine is related to the density of the air drawn into it. Since atmospheric pressure and air density are greatest at (or below) sea level, both pressure and density normally decrease as altitude above sea level increases. For example, atmospheric pressure at sea level is about 14.7 psi (101 kPa); at higher elevations, atmospheric pressure may be only 8 or 9 psi (55 or 62 kPa), figure 10-4. Therefore, the volumetric efficiency

of any engine will be greater at sea level than at an altitude above sea level.

Pumping air into the intake system under pressure forces it through the bends and restrictions at a greater speed than it would travel under normal atmospheric pressure, allowing more air to enter the intake port before it closes, figure 10-5. By increasing the engine's air intake in this manner, more fuel can be mixed with the air while still maintaining the same air-fuel ratio. The denser the air-fuel charge entering the engine during its intake stroke, the greater the potential energy released during combustion. In addition

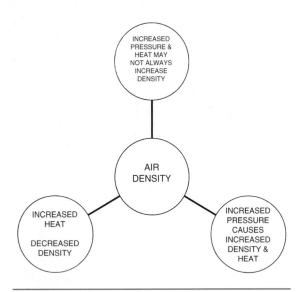

Figure 10-2. The effects of heat and pressure on density.

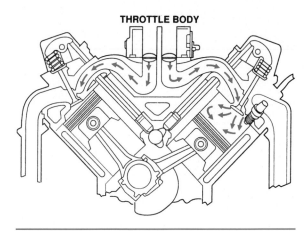

Figure 10-3. Normal engine airflow. (Courtesy of General Motors Corporation)

to the increased power resulting from combustion, there are several other advantages of supercharging an engine:

- It increases the air-fuel charge density to provide high-compression pressure when power is required, but allows the engine to run on lower pressures when additional power is not required.
- The pumped air pushes the remaining exhaust from the combustion chamber during intake and exhaust valve overlap.
- The engine burns more of the air-fuel charge, lowering emissions.
- The forced airflow and removal of hot exhaust gases lowers the temperature of the cylinder head, pistons, and valves, and helps extend the life of the engine.

A supercharger pressurizes air to greater than atmospheric pressure. The pressurization above atmospheric pressure, or boost, can be measured in the same way as atmospheric pressure, which we studied in Chapter 8. Atmospheric pressure drops as altitude increases, but boost pressure remains the same. If a supercharger develops 12 psi (83 kPa) boost at sea level, it will develop the same amount at a 5,000-foot altitude.

As we saw earlier, when air is both compressed and heated, its density may increase, decrease, or not change at all, figure 10-2. When a supercharger compresses the air it pumps into an engine, the friction caused by turbulence also heats the air. If the intake temperature becomes too

great, the overheated air-fuel charge can cause premature detonation in the combustion chamber and result in engine damage. Superchargers generally use two devices to prevent overheating the air-fuel charge and premature detonation.

To prevent the air-fuel charge from overheating, it must be cooled down before reaching the combustion chamber. This is the function of an **intercooler,** or heat exchanger, installed between the supercharger and the engine. The intercooler may use air or engine coolant as a cooling medium. Air-to-liquid intercoolers are heavier and more complicated than the air-to-air type. How much the air or air-fuel charge cools as it passes through the intercooler depends both on the temperature of the cooling medium and its flow rate. By cooling the air-fuel charge, an intercooler increases its density, allowing a greater quantity to enter the engine. Intercooling also lowers the thermal loading on engine components. However, premature detonation still can occur even when an intercooler is used.

To prevent premature detonation, boost pressure is limited to a predetermined amount through use of a spring-loaded blowoff valve, figure 10-6. The bypass actuator generally is installed between the supercharger and the engine to bleed off part of the air-fuel charge. If the maximum allowable boost pressure is reached, it overcomes the blowoff valve spring pressure and vents back to the supercharger. Superchargers that compress only air, however, can be vented to the atmosphere.

Superchargers

In basic concept, a supercharger is nothing more than an air pump mechanically driven by the engine

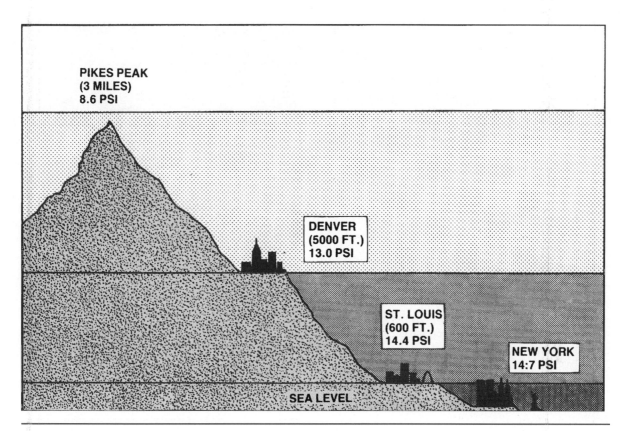

Figure 10-4. Air is a substance and has weight, which causes atmospheric pressure. Atmospheric pressure decreases with increases in altitude.

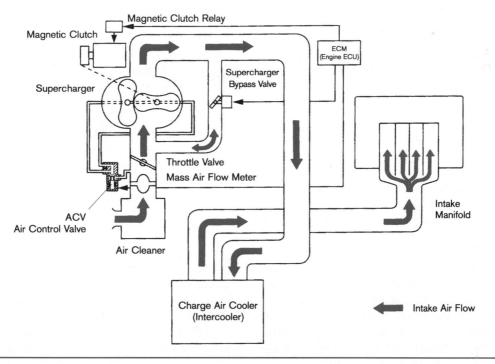

Figure 10-5. Air intake operation in a supercharged or turbocharged engine. (Provided courtesy of Toyota Motor Sales U.S.A., Inc.)

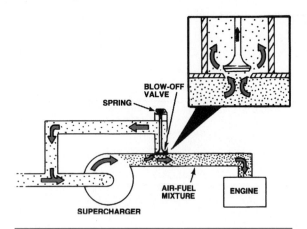

Figure 10-6. Blowoff valve operation. (Courtesy of Ford Motor Company)

itself. Gears, shafts, chains, or belts from the crankshaft can be used to turn the pump, figure 10-7. This means that the air pump or supercharger pumps air in direct relation to engine speed.

There are two general types of superchargers:

- *Positive displacement*—pumps the same volume of air on each cycle regardless of engine speed.
- *Variable displacement*—the volume of air pumped varies with engine speed.

Positive displacement pumps include the reciprocating, the lobe (Roots), and the vane designs, figure 10-8. These operate by drawing in a large column of air, compressing it into a small area and then forcing it through an outlet at high pressure. Unlike a turbocharger, the positive displacement supercharger provides substantial boost even when the engine is at idle.

The lobe-type, or Roots blower, figure 10-9, is the most common positive displacement supercharger in use. It was first developed in 1864 as a device to separate wheat from chaff but was applied to automotive engines around the turn of the century. Roots blowers are used on many high-performance and racing engines, but also are found on two-stroke diesel engines, where they improve air intake and **exhaust scavenging** instead of acting as primary superchargers. Vane-type superchargers are not commonly used.

Variable displacement superchargers include the centrifugal, axial flow, and pressure-wave designs. The centrifugal pump is the most commonly used and most efficient design. It does not heat the air as much as a lobe-type, and the resulting airflow is much smoother. As the impeller turns at high speed, air pulled into the center of the impeller is speeded up and then thrown outward from the

Figure 10-7. Most superchargers used today are belt driven.

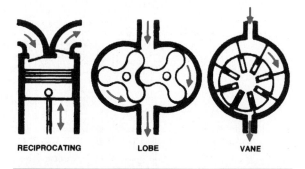

Figure 10-8. These are three common positive displacement pump designs.

blades by centrifugal force. Air moved to the perimeter of the pump housing is forced through an outlet. Since centrifugal force increases as speed increases, a centrifugal supercharger will draw in more air and create a higher boost pressure as its speed increases. Its pumping output increases roughly as a square of the engine speed. When engine speed is doubled, the centrifugal supercharger provides four times as much boost pressure. If engine speed is tripled, the supercharger delivers nine times as much boost pressure.

Superchargers are installed to pump air directly into the intake manifold on diesel and fuel-injected

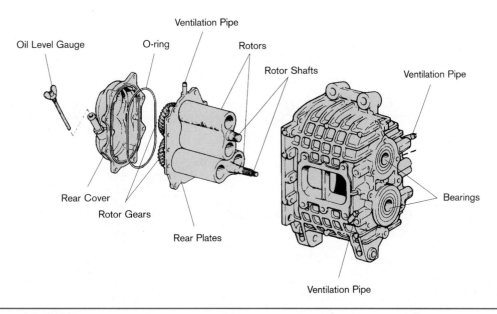

Oil Level Gauge   O-ring   Ventilation Pipe   Rotors   Rotor Shafts   Ventilation Pipe   Bearings   Rear Cover   Rotor Gears   Rear Plates   Ventilation Pipe

Figure 10-9. The lobe (Roots) supercharger is the most commonly used positive displacement pump used as a supercharger. (Provided courtesy of Toyota Motor Sales U.S.A., Inc.)

engines. With carbureted engines, two systems are possible: downstream or upstream. The downstream system is the most popular, with the supercharger mounted between the carburetor and the intake manifold. This permits the supercharger to draw the air-fuel charge from the carburetor and then pump it into the intake manifold. This system helps atomize the fuel particles by mixing the air-fuel charge as it passes through the supercharger.

■ A Challenge to the Roots Blower

One of the most successful non-Roots blower designs to emerge from the drawing board and onto actual production cars, the G-Lader blower has appeared in recent years on some Volkswagen Corrado models. The G-Lader takes its name from the housing, which is shaped like the letter "G," and the German word for charger—*lader*. This spiral channel unit was derived from a French design patented in 1905, and is both lighter and quieter than a traditional Roots blower.

Concentric spiral ramps in both sides of a rotor mesh with similar ramps cast in the split casing. The rotor moves around an eccentric shaft instead of spinning on its axis, as in most other supercharger designs. As air is drawn into the casing, it is compressed by squeezing it through the spiral and then forced through a cluster of ports in the center of the casing before passing into the engine. Boost pressure is slightly less than 12 psi, yet the G-Lader uses approximately 17 horsepower under maximum throttle and rpm.

The upstream system locates the supercharger between the air filter and carburetor. In this arrangement, the carburetor must be encased in a pressure box because the supercharger blows air through the carburetor into the intake manifold. Since a carburetor requires a pressure drop at the venturi in order to draw fuel from the bowl, the main jets will not deliver fuel into the intake air stream unless the air around the carburetor is pressurized.

The mechanical drive methods that operate a supercharger all have inherent limitations. They require up to 20 percent of the engine's power even when the supercharger is not in use. They also need a high overdrive ratio to obtain a speed great enough to provide the desired boost. Since the drive belt or gear speed must be quite high, the mechanical components are subjected to heavy wear.

Supercharging and the 1990s

Although carmakers experimented with a variety of superchargers on limited production cars during the 1980s, none appeared on domestic vehicles until 1989, when Ford introduced a 3.8-liter V-6 in the Thunderbird Super Coupe with a Roots-type blower manufactured by Eaton Corporation. This is Ford's first venture into supercharging since the late 1950s, when it also offered a belt-driven centrifugal supercharger.

As used by Ford, the intercooled Eaton blower develops 12 psi maximum boost at low manifold vacuum. When closed, a vacuum-operated bypass

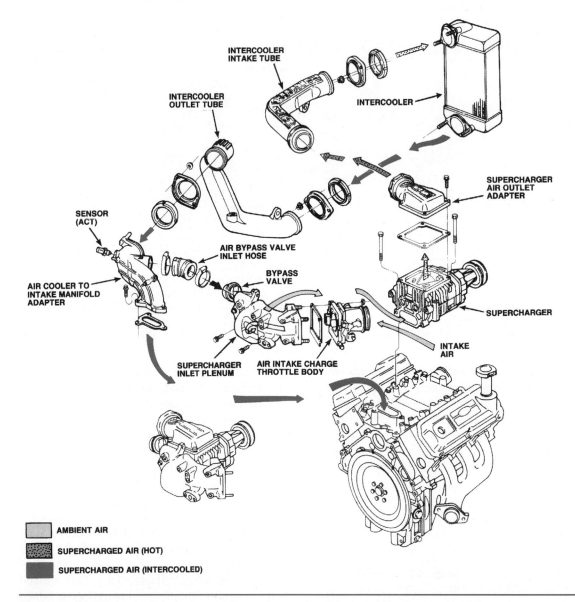

Figure 10-10. The air inlet system and supercharger flow pattern of the Ford 3.8-liter V-6 as used on the Thunderbird Super Coupe. (Courtesy of Ford Motor Company)

valve sends the full flow of air from the supercharger directly to the intake manifold. The butterfly valve starts to open at about 3 in. Hg vacuum to bleed boost pressure back to the blower inlet; at 7 in. Hg vacuum, the valve is fully open. The airflow pattern is shown in figure 10-10.

General Motors introduced a similar system without an intercooler on some 1992 3800 V-6 engines. In this design, intake air enters at the rear of the blower and exits directly into the intake manifold plenum from the bottom of the blower. The system also uses a vacuum-operated bypass valve, figure 10-11, and a PCM-controlled boost solenoid to regulate induction boost pressure during rapid deceleration, under high engine load, or in reverse gear. Boost system operation is shown in figures 10-12 and 10-13. Maximum boost pressure can range from 7 psi to 11 psi, but is maintained at about 8 psi on applications with the 4T60E transaxle due to the transaxle's inability to deal with high torque.

It is too early to tell if superchargers will find the same favor with buyers that turbochargers have, but their reappearance on production cars may make this the decade of the supercharger, as the 1980s were for turbochargers.

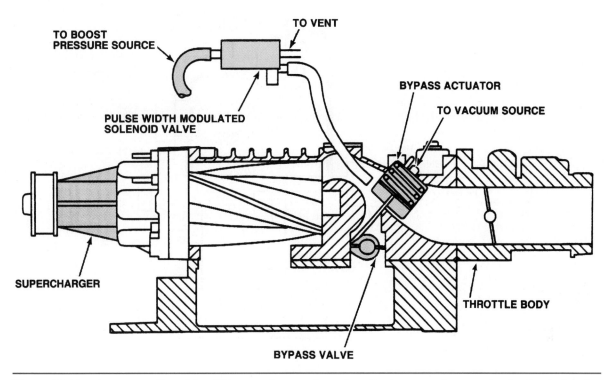

Figure 10-11. A cross section of the GM supercharger shows how the bypass valve operates to divert air to the blower inlet and limit boost pressure. The Ford system is similar. (Courtesy of General Motors Corporation)

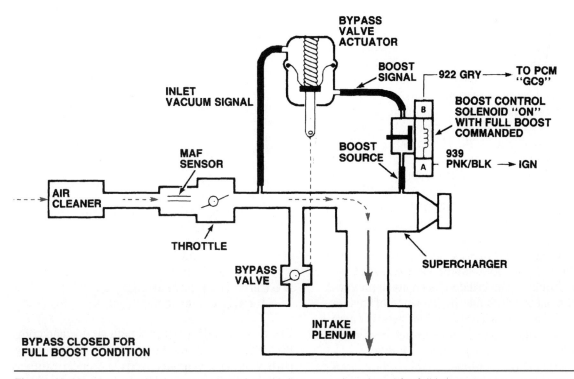

Figure 10-12. Boost control system operation with bypass valve closed for full boost. (Courtesy of General Motors Corporation)

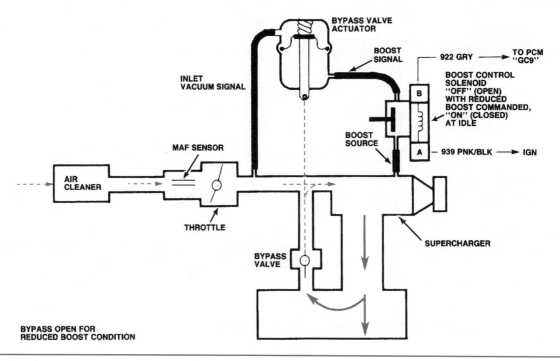

Figure 10-13. Boost control system operation with bypass valve open for reduced boost. (Courtesy of General Motors Corporation)

TURBOCHARGERS

The major disadvantage of a supercharger is its reliance on engine power to drive the unit. In some installations, as much as 20 percent of the engine's power is used by a mechanical supercharger. However, by connecting a centrifugal supercharger to a turbine drive wheel and installing it in the exhaust path, the lost engine horsepower is regained to perform other work and the combustion heat energy lost in the engine exhaust (as much as 40 to 50 percent) can be harnessed to do useful work. This is the concept of a **turbocharger.**

The turbocharger's main advantage over a mechanically driven supercharger is that the turbocharger does not drain power from the engine. In a normally aspirated engine, about half of the heat energy contained in the fuel goes out the exhaust system, figure 10-14. Another 25 percent is lost through radiator cooling. Only about 25 percent is actually converted to mechanical power. A mechanically driven pump uses some of this mechanical output, but a turbocharger gets its energy from the exhaust gases, converting more of the fuel's heat energy into mechanical energy.

A turbocharger turbine, figure 10-15, looks much like a typical centrifugal pump used for supercharging. Hot exhaust gases flow from the com-

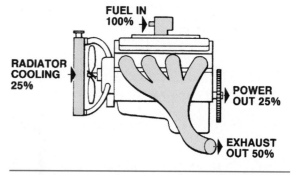

Figure 10-14. A turbine uses some of the heat energy that normally would be wasted.

bustion chamber to the turbine wheel. The gases are heated and expanded as they leave the engine. It is not the speed of force of the exhaust gases that forces the turbine wheel to turn, as is commonly thought, but the expansion of hot gases against the turbine wheel's blades.

Turbocharger Design and Operation

The modern turbocharger is both simple and compact, with few moving parts. Because its moving parts work at very high speeds and under extreme heat, however, a turbocharger must be manufactured to very precise tolerances. A turbocharger con-

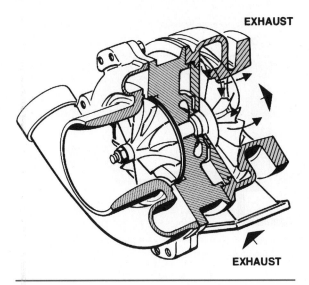

Figure 10-15. A turbine wheel is turned by the expansion of gases against its blades.

sists of two chambers connected by a center housing. The two chambers contain a turbine wheel and a compressor wheel connected by a shaft, which passes through the center housing, figure 10-16.

To take full advantage of the exhaust heat that provides the rotating force, a turbocharger must be positioned as close as possible to the exhaust manifold. This allows the hot exhaust to pass directly into the unit with a minimum of heat loss. As exhaust gas enters the turbocharger, it rotates the turbine blades. The turbine wheel and compressor wheel are on the same shaft so that they turn at the same speed. Rotation of the compressor wheel draws air in through a central inlet and centrifugal force pumps it through an outlet at the edge of the housing. A pair of bearings in the center housing

supports the turbine and compressor wheel shaft, figure 10-17, is lubricated by engine oil.

Both the turbine and compressor wheels must operate with extremely close clearances to minimize possible leakage around their blades. Any leakage around the turbine blades causes a dissipation of the heat energy required for compressor rotation. Leakage around the compressor blades prevents the turbocharger from developing its full boost pressure.

When the engine is started and runs at low speed, both exhaust heat and pressure are low and the turbine runs at a low speed (approximately 1,000 rpm). Because the compressor does not turn fast enough to develop boost pressure, air simply passes through it and the engine works like any normally aspirated engine. As the engine runs faster or load increases, both exhaust heat and flow increase, causing the turbine and compressor wheels to rotate faster. Since there is no brake and very little rotating resistance on the turbocharger shaft, the turbine and compressor wheels accelerate as the exhaust heat energy increases. When an engine is running at full power, the typical turbocharger rotates at speeds between 100,000 and 150,000 rpm.

Engine deceleration from full power to idle requires only a second or two because of its internal friction, pumping resistance, and drive train load. The turbocharger, however, has no such load on its shaft, and is already turning many times faster than the engine at top speed. As a result, it can take as much as a minute or more after the engine has returned to idle speed before the turbocharger also has returned to idle. If the engine is decelerated to idle and then shut off immediately, engine lubrication stops flowing to the center housing bearings while the

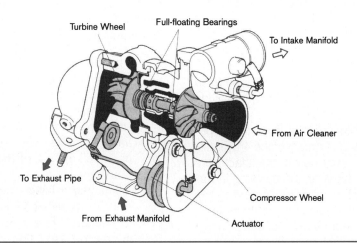

Figure 10-16. The components of a typical turbocharger. (Provided courtesy of Toyota Motor Sales U.S.A., Inc.)

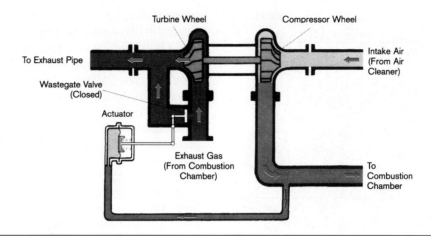

Figure 10-17. Basic operation of a turbocharger. (Provided courtesy of Toyota Motor Sales U.S.A., Inc.)

■ Switching the Emphasis on Turbos

Some drivers may think of turbochargers as exotic trappings for souped-up street cars and racers. That's natural when you consider that automotive engineers normally bring a turbo into play high up on the engine's performance curve to deliver maximum power at top rpm. The result is additional peak power that has won many races, and further convinced the general driving public that turbine-operated superchargers are not for them.

The reappearance of turbocharging as a factory option approached the subject from the opposite viewpoint. Engineers wanted 6-cylinder fuel economy with V-8 performance and the turbocharger was the way to get it. Since its appearance on 1978 domestic engines, the turbocharger has been used for low- and medium-speed passing, and acceleration from about 1,200 rpm up. As engine speed climbs, the wastegate begins to open, preventing an overload of the engine. This means that the turbocharger will be used for only about 5 percent of the time the engine is operating.

The performance-oriented 1980s brought turbocharging into favor with consumers. When used with 4-cylinder engines, turbochargers provide the power to make them perform like V-6 and small V-8 engines, while retaining good fuel economy. By the mid-1980s, 7 percent of the 4-cylinder engines produced by domestic manufacturers were turbocharged. However, the upswing in turbocharger popularity has been challenged by the revival in popularity of V-6 and small V-8 engines, as well as the recent appearance of superchargers on some Ford and GM engines.

turbocharger is still spinning at thousands of rpm. The oil in the center housing is then subjected to extreme heat and can gradually "coke" or oxidize. If this happens, a new turbocharger is in order.

The high rotating speeds and extremely close clearances of the turbine and compressor wheels in their housings require equally critical bearing clearances. The bearings must keep radial clearances of 0.003–0.006 inch (0.08–0.15 mm). Axial clearance (endplay) must be maintained at 0.001–0.003 inch (0.025–0.08 mm). Constant lubrication with clean oil is very important to bearings that must maintain these very critical clearances at extreme speeds. Such close clearances and high rotational speeds equal instant failure if a small dirt particle or other contamination enters the exhaust or intake housings.

The turbocharger is a simple but extremely precise device in which heat plays a critical role. If properly maintained, the turbocharger also is a trouble-free device but to prevent problems, three conditions must be met:

- The turbocharger bearings must be constantly lubricated with clean engine oil—turbocharged engines should have regular oil changes at half the time or mileage intervals specified for non-turbocharged engines.
- Dirt particles and other contamination must be kept out of the intake and exhaust housings.
- Whenever a basic engine bearing (crankshaft or camshaft) has been damaged, the turbocharger must be flushed with clean engine oil after the bearing has been replaced.

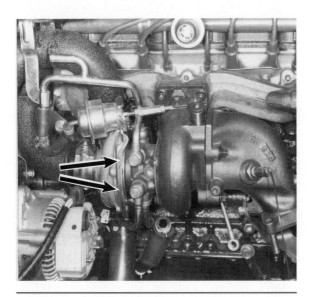

Figure 10-18. Water-cooled turbochargers induct engine coolant into passages in the center housing (arrows) to cool the center bearings.

- If the turbocharger is damaged, the engine oil must be drained and flushed and the oil filter replaced as part of the repair procedure.

Late-model turbochargers all have liquid-cooled center bearings to prevent heat damage. In a liquid-cooled turbocharger, figure 10-18, engine coolant is circulated through passages cast in the center housing to draw off the excess heat. This allows the bearings to run cooler and minimizes the probability of oil coking when the engine is shut down.

Turbocharger Size and Response Time

As we have seen, a time lag exists between an increase in engine speed and the turbocharger's ability to overcome inertia and spin up to speed as the exhaust gas flow increases. This delay between acceleration and turbo boost is called **turbo lag.** Like any material, moving exhaust gas has inertia. Inertia also is present in the turbine and compressor wheels, as well as the intake airflow. Unlike a supercharger, the turbocharger cannot supply an adequate amount of boost at low speed.

Turbocharger response time is directly related to the size of the turbine and compressor wheels. Small wheels accelerate rapidly; large wheels accelerate slowly. While small wheels would seem to have an advantage over larger ones, they may not have enough airflow capacity for an engine. To minimize turbo lag, the intake and exhaust breathing capacities of an engine must be matched to the exhaust and intake airflow capabilities of the turbocharger.

The location of the turbocharger on the engine is another factor influencing response time. The most efficient arrangement is to locate the turbine outlet close to the intake manifold, as shown in the operational cycle for a fuel-injected engine.

Engineers continue to explore new ways to improve turbocharger response. One method is to reduce the weight of turbine and compressor wheels by using lightweight ceramic rotating parts. Another is to control the flow of inlet gas to the turbine through the use of a variable-nozzle turbine. In this design, a curved flap is used at the turbine inlet. When the flap is closed, exhaust gas entering the turbine is deflected to strike the turbine at almost a 90-degree angle, producing a quicker response at low speed. As engine speed increases, the flap starts to open. This reduces exhaust backpressure and increases the amount of exhaust gas to keep the turbine spinning at high speed. The flap is controlled by a computer-operated vacuum solenoid. Chrysler used a variable-nozzle turbocharger on some 1990 models, but the cost and complexity of this design have not made it a popular choice.

Turbocharger Installation

Like superchargers, turbochargers can be installed either on the air intake side (upstream) of the carburetor or fuel injectors or on the exhaust side (downstream) of the carburetor. In an upstream installation, the turbocharger compresses and delivers a denser air charge to the carburetor or injectors. When installed downstream, the air-fuel mixture is compressed and delivered to the cylinders. Both types of installations are used with turbocharger applications.

Downstream Installations

Turbochargers generally are installed downstream with carbureted engines, as are superchargers (this is called a draw-through turbocharger). The carburetor usually is located on the intake side of the turbocharger, although it may be positioned on the outlet side. Positioning the carburetor at the compressor inlet allows the turbocharger to increase airflow and pressure drop through the carburetor. This provides the required air-fuel mixture, which is compressed and sent to the cylinders through the manifold, figure 10-19.

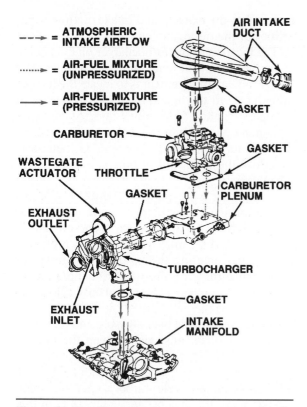

Figure 10-19. The turbocharger compresses the air-fuel mixture when the carburetor is installed at the compressor inlet. (Courtesy of General Motors Corporation)

Positioning the carburetor at the compressor inlet simplifies air-fuel control, and the carburetor does not have to be modified to withstand boost pressure. If the carburetor is located at the compressor outlet, it must be calibrated to meter fuel correctly under both atmospheric and above-atmospheric conditions, and pressurized to withstand boost pressure without leaking fuel. In addition, the fuel system must also move the fuel at a higher pressure to overcome the boost pressure present at the carburetor. These factors all make air-fuel ratio control difficult when the carburetor is located at the turbocharger outlet.

There are, however, minor drawbacks that must be overcome with carburetor placement at the compressor inlet:

- The carburetor is located away from the intake manifold, which may cause a slight hesitation when the throttle is opened rapidly.
- Throttle and choke linkages must be more complex in design and operation.
- Preheating the air-fuel mixture on a cold engine is more difficult.

- A special seal must be installed between the center housing and compressor to prevent the air-fuel mixture from entering the engine lubrication system.
- Fuel separation from the compressed mixture may occur before it reaches the manifold.

Upstream Installations

Turbochargers are installed upstream with fuel-injected engines (this is called a blow-through turbocharger). The unit is positioned at the manifold air intake and compresses only intake air. This reduces fuel delivery time and increases the amount of turbine energy available. Most turbocharged engines with fuel injection use a multipoint system, with an individual injector at each cylinder. Just before the compressed air charge enters the cylinder, fuel is injected into it. The throttle plate installed between the turbocharger and injectors regulates airflow.

Mitsubishi uses a throttle body injection (TBI) system on some engines in which the turbocharger delivers compressed air to the injectors in the TBI unit. The throttle plate, however, is located between the injectors and the intake manifold, figure 10-20. In this system, the throttle plate regulates the intake volume of compressed air-fuel mixture.

Twin Turbo Installations

Twin turbochargers with individual intercoolers and electronic controls are found on some expensive high-performance cars. Chrysler and Mitsubishi have used a complicated twin turbocharger system since 1991. In this type of installation, small and lightweight turbochargers are used and each is driven independently by the exhaust from one cylinder bank of a V-type engine. Because the turbochargers are small, they can respond quicker at low engine speeds with a combined volume sufficient to deliver the required boost throughout the engine's operating range. Each turbocharger **wastegate** is computer-controlled through a pulse-width-modulated (PWM) solenoid to bleed off boost pressure when necessary. Mazda introduced a twin turbocharger system on its 1993 RX-7 that provides boost throughout the engine's power range by using a small turbocharger for low-speed boost, and a larger one for use at higher engine speeds. Turbocharger engagement and disengagement, as well as wastegate operation are computer-controlled.

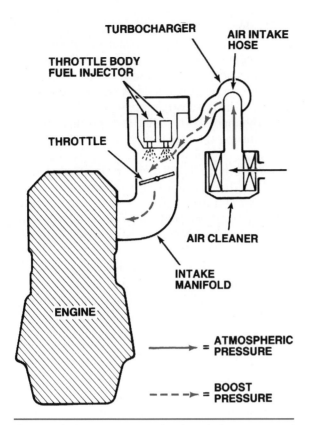

Figure 10-20. The throttle is placed downstream of the turbocharger and fuel injectors in this Mitsubishi TBI system. (Courtesy of Mitsubishi Motors Corporation)

TURBOCHARGER CONTROLS

You cannot simply bolt a turbocharger onto an engine and automatically pick up free power. As we have seen, a turbocharger increases both compression pressure and loads on an engine's moving parts, as well as increasing underhood temperatures. Most late-model engines are strong enough to withstand the higher compression pressures without major modifications. Many turbocharged engines, however, use oil coolers and heavy-duty radiators to deal with higher engine and underhood temperatures.

Three factors must be controlled because of the high pressures and temperatures created by a turbocharger:

- Boost pressure
- Air-fuel mixture temperature
- Detonation

These factors are interrelated. Higher pressures (boost) raise intake mixture temperatures, which raises combustion temperatures. Higher combustion temperatures and pressures can combine to cause detonation.

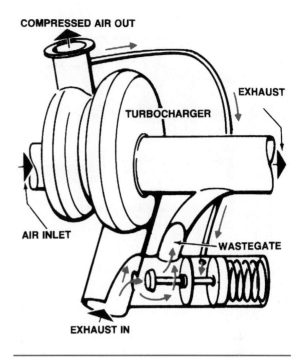

Figure 10-21. The wastegate reacts to intake manifold pressure and controls the amount of exhaust gas reaching the turbine wheel.

Boost Pressure Control

Boost increases steadily as turbocharger rotation increases. Boost pressure must be limited to the maximum the engine can withstand without detonation or serious engine damage. This can be done either by limiting the amount of exhaust gas reaching the turbine, or by venting off some of the compressed air-fuel charge before it reaches the combustion chamber. This is done most efficiently by an exhaust wastegate or a blowoff value.

The wastegate is a bypass valve at the exhaust inlet to the turbine. It allows all of the exhaust into the turbine, or it can route part of the exhaust past the turbine to the exhaust system, figure 10-21. The wastegate is operated by a vacuum-actuated diaphragm exposed to compressor outlet pressure. When pressure rises to a predetermined level, the diaphragm moves a linkage rod, figure 10-22, to open the wastegate. This diverts some or all of the exhaust to the turbine outlet, limiting the maximum turbine and compressor speed.

A pressure hose connects the actuator to the compressor outlet, figure 10-22. During deceleration, pressure rises at this point because the compressor is working against the closed throttle of an engine that requires little airflow. This causes the actuator to open the wastegate, eliminating overboost during a closed-throttle condition.

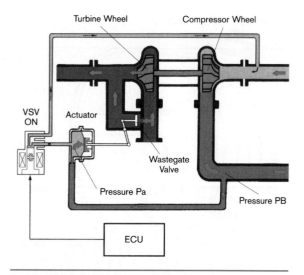

Figure 10-22. The ECU controls the VSV solenoid, which controls vacuum to the wastegate valve. (Provided courtesy of Toyota Motor Sales U.S.A, Inc.)

The location of a turbocharger on the engine does not affect wastegate control of boost pressure. The turbocharger can be installed either upstream or downstream from the fuel injectors or carburetor. Location does not affect, nor is it affected by, intake air-fuel mixture.

Turbochargers installed on production engines during the 1970s were limited to 6–8 psi (41–55 kPa) boost pressure. During the early 1980s, turbocharging was refined and integrated with electronic engine management systems. Many turbocharger systems now operate at boost pressures of 10–14 psi (69–97 kPa). All turbochargers with wastegate control operate as described above, but some can be electronically controlled to modulate boost according to the octane content of the fuel being used.

Another form of boost pressure control is called a blowoff valve. As you learned when you

■ The Next Generation of Turbochargers

Integrating the turbocharger into electronic engine management systems has brought it a well-deserved reputation for reliability and performance. To this point, all production turbocharger applications have been based on the centrifugal pump. The next generation of turbochargers will see numerous design variations, the first of which is Nissan's N2-VN or variable nozzle (scroll) turbo.

A movable curved flap in the turbo housing changes the throat area to vary its output according to engine requirements. The flap can move 27 degrees in stepless increments. Flap position is determined by a vacuum-operated diaphragm controlled by a pressure-controlled modulator and the engine computer.

With the flap closed, the flow of exhaust gas that enters the turbine housing is speeded up, increasing its pressure and rotating the turbine rapidly at low-speed, low-load conditions. At high-speed operation, the flap opens fully and the reduction in exhaust resistance aids in filling the cylinders.

The design provides a normal boost up to 15 psi (103 kPa). However, if operating conditions will permit it without engine damage, the computer will allow boost up to 17 psi (117 kPa) for short periods, such as overtaking another vehicle.

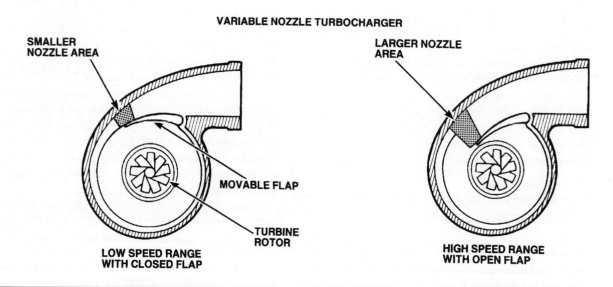

studied superchargers, this is a large, spring-loaded relief valve that opens whenever pressure exceeds the desired maximum, figure 10-6. The blowoff or bypass valve generally is a poppet or flapper design with a damper attached to prevent fluttering that could damage the valve. When a blowoff or bypass valve is used with a turbocharger, its location in the system differs according to whether the turbocharger is upstream or downstream.

A blowoff or bypass valve also can be used to control exhaust emissions instead of boost pressure by operating it through intake manifold vacuum. During idle, closed-throttle deceleration, and choked operation, vacuum in the intake manifold causes a diaphragm to move and open the valve. The engine operates as a normally aspirated unit, avoiding the excessive emissions that turbocharger compression would cause. Some blowoff or bypass valves are operated by vacuum routed to one side of the diaphragm and boost pressure to the other.

Intake Mixture Cooling

As turbocharging increases intake air pressure, the temperature of the air also increases. The higher temperature results in two unwanted effects:

- It reduces the air-fuel charge density, working against the boost pressure.
- It makes the air-fuel charge prone to premature detonation as it enters the cylinders.

When detonation combines with the high pressures involved in a turbocharged engine, it can burn a piston, bend a piston connecting rod, or do other damage to internal components. There are two ways to cool the compressed mixture before it reaches the combustion chamber: water injection and intercooling.

Water injection is rarely used on factory-installed turbocharger installations, but it is common on after-market systems and generally used on carbureted engines when the turbocharger is downstream from the carburetor. At maximum boost, an electric or vacuum-operated pump in-

jects a fine spray of water into the carburetor inlet. The water vaporizes with the gasoline and has no effect on combustion, but cools the intake charge and leaves the engine as water vapor in the exhaust.

An intercooler is nothing more than a heat exchanger, figure 10-23, as you saw when you studied superchargers. Compressed air from the turbocharger is sent through the intercooler, which transfers heat to the ambient airflow much like a radiator. This is called an air-to-air intercooler. Air-to-liquid heat exchangers also have been used as intercoolers, but they are heavier and more complicated than the air-to-air type.

Intercoolers work best on fuel-injected systems or with carbureted systems in which the carburetor is downstream from the turbocharger. Since only air passes through the intercooler, fuel cannot separate from the mixture as it cools.

Spark Timing and Detonation

Excessive pressure and heating of the air-fuel mixture are not the only causes of detonation in a turbocharged engine. Spark timing that is excessively advanced will ignite the dense air-fuel charge too quickly. This causes the charge to burn unevenly and maximum cylinder power develops too early. When a turbocharged engine is under maximum boost, ignition timing must be retarded from maximum advance.

Retarding ignition through boost pressure was adequate for basic turbocharger systems used in the 1970s, but modern turbocharged engines with electronic controls use a detonation or knock sensor for precise control. The sensor is a piezoelectric crystal mounted in the engine block or intake manifold that generates a voltage or varies the return signal of a reference voltage from the computer when engine knock occurs. The sensor detects detonation as physical vibration within the engine and instantly signals the computer to retard timing. Timing is retarded in 2- to 4-degree increments until either a maximum retard setting is reached or the detonation stops. The computer then gradually restores the proper spark advance required for optimum engine operation.

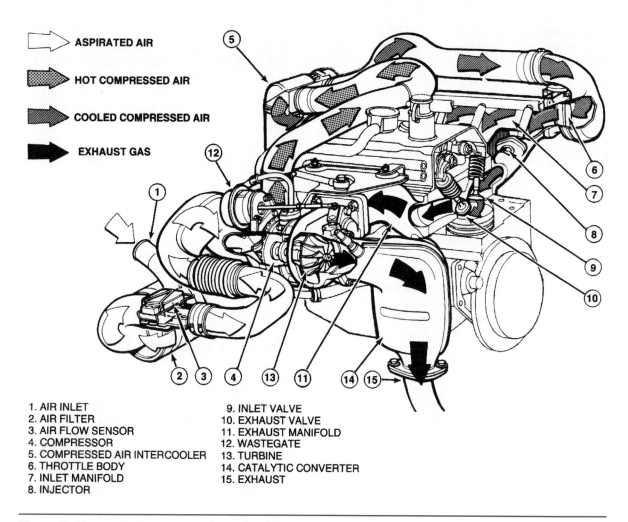

⇨ ASPIRATED AIR

⇨ HOT COMPRESSED AIR

⇨ COOLED COMPRESSED AIR

⇨ EXHAUST GAS

1. AIR INLET
2. AIR FILTER
3. AIR FLOW SENSOR
4. COMPRESSOR
5. COMPRESSED AIR INTERCOOLER
6. THROTTLE BODY
7. INLET MANIFOLD
8. INJECTOR
9. INLET VALVE
10. EXHAUST VALVE
11. EXHAUST MANIFOLD
12. WASTEGATE
13. TURBINE
14. CATALYTIC CONVERTER
15. EXHAUST

Figure 10-23.   The Renault version of an air-to-air intercooler.

SUMMARY

Both supercharging and turbocharging are proven technologies useful in mixture compression that increase combustion power without increasing exhaust emissions. A supercharger is an air pump mechanically driven by the engine in direct relationship to engine speed. A turbocharger is an air pump driven by exhaust gases. Except for their power source, superchargers and turbochargers do essentially the same job in increasing the volumetric efficiency of an engine. Both increase intake air pressure above atmospheric pressure, resulting in a denser-air fuel charge. Supercharging, however, can use up to 20 percent of the engine's power even when not in use. Turbochargers increase engine power only on demand when additional power is required. One major disadvantage of a turbocharger is "turbo lag," the time interval between increasing

engine speed and the turbocharger's ability to overcome inertia and spin up to speed.

Boost pressure, air-fuel mixture temperature, and detonation all are problems caused by pressurizing the intake air-fuel charge and must be controlled to prevent high exhaust emissions and possible engine damage. Boost pressure is controlled through use of a wastegate (bypass valve) or blowoff valve in the exhaust inlet to the turbine. The wastegate is vacuum-operated; the blowoff valve is a spring-loaded device. Mixture temperature can be controlled by injecting a fine water spray into the intake charge, or by using an intercooler (heat exchanger) to remove heat from the air charge before it mixes with the fuel. Detonation can be controlled by retarding ignition timing mechanically through a vacuum diaphragm or electronically through a detonation sensor and the engine computer.

Review Questions

Choose the letter that represents the best possible answer to the following questions:

1. Supercharging delivers the air-fuel mixture to the cylinder at:
   a. Lower than atmospheric pressure
   b. Atmospheric pressure
   c. Higher than atmospheric pressure
   d. Three times atmospheric pressure

2. Which is *not* true of superchargers?
   a. They cannot operate at very low rpm
   b. They are mechanically driven
   c. There are two types
   d. Mechanical superchargers consume a lot of engine power

3. Positive displacement pumps:
   a. Contain impellers
   b. Pump the same volume of air each revolution
   c. Increase air pressure by decelerating it
   d. Are highly efficient

4. A turbine wheel turns because of:
   a. Centrifugal force
   b. Exhaust gas speed
   c. Manifold vacuum
   d. Exhaust gas expansion

5. Of the heat energy contained in gasoline:
   a. 50 percent is converted to mechanical power
   b. 50 percent is lost to cooling
   c. 25 percent is converted to engine power
   d. 50 percent goes out the exhaust system

6. Despite the high compression pressures turbochargers achieve, they are okay for emission-controlled engines because:
   a. They have fixed compression ratios
   b. Turbocharger boost can be varied to meet engine needs
   c. They are placed between the air cleaner and carburetor
   d. They are placed between the carburetor and engine

7. Which is *not* a method of controlling a turbocharger system?
   a. Changing the amount of boost
   b. Cooling the compressed mixture
   c. Readjusting the carburetor idle
   d. Altering spark timing

8. The wastegate controls boost by controlling:
   a. Exhaust gas flow
   b. Compressed air
   c. Air-fuel mixture
   d. Exhaust emissions

9. Retarding spark timing controls detonation by:
   a. Cooling the compressed mixture
   b. Cooling the exhaust manifold
   c. Reducing compression pressure
   d. Lowering peak temperature of combustion

10. Technician A says turbochargers are constant-displacement pumps.
    Technician B says Roots blowers are variable-displacement pumps. Who is right?
    a. A only
    b. B only
    c. Both A and B
    d. Neither A nor B

11. Boost control can be limited by a wastegate or by:
    a. An intercooler
    b. A blowoff valve
    c. A detonation sensor
    d. Manifold vacuum

12. Engine exhaust is used to drive a turbocharger:
    a. Turbine
    b. Wastegate
    c. Compressor
    d. Intercooler

13. The momentary hesitation between throttle opening and boost delivery by the turbocharger is called:
    a. Underboost lag
    b. Turbo lag
    c. Ignition lag
    d. Compression lag

14. Technician A says a turbocharger installed upstream from the fuel source compresses only air.
    Technician B says a turbocharger installed downstream from the fuel source compresses air and fuel. Who is right?
    a. A only
    b. B only
    c. Both A and B
    d. Neither A nor B

15. Technician A says energy to drive a turbocharger comes from exhaust heat. Technician B says energy to drive the turbocharger comes from exhaust pressure. Who is right?
    a. A only
    b. B only
    c. Both A and B
    d. Neither A nor B

16. Which is *not* a benefit of high compression ratios?
    a. Increased thermal efficiency
    b. Increased volumetric efficiency
    c. Decreased NO_x formation
    d. Increased compression pressure and temperature

17. A heat exchanger used to control intake air charge temperature is called a(n):
    a. Wastegate
    b. Blowoff valve
    c. Air charge temperature valve
    d. Intercooler

18. Technician A says except for their power source, superchargers and turbochargers do essentially the same job. Technician B says reducing turbocharger size and weight is one way of minimizing turbo lag. Who is right?
    a. A only
    b. B only
    c. Both A and B
    d. Neither A nor B

11

Variable, Flexible, and Bi-Fuel Systems

OBJECTIVES

Upon completion and review of this chapter, you will be able to:

- Describe the difference between a bi-fuel vehicle and a variable fuel vehicle.
- List some dangerous characteristics of methanol.
- Compare the characteristics of gasoline to CNG.
- Have knowledge of and describe the safety precautions when working on a flexible fuel vehicle.
- Describe the pressure stages of a CNG vehicle.

KEY TERMS

fuel compensation sensor
methanol

M85
remote injector driver

INTRODUCTION

All domestic manufacturers have designed and produced variable fuel vehicles (VFVs) or flexible fuel (FF) vehicles. VFVs/FFs can operate on unleaded gasoline, or on a new fuel called **M85,** a blend of 85 percent methanol and 15 percent unleaded gasoline. In fact, they can operate equally well on any blend of **methanol** and gasoline between 0/100 percent and 85/15 percent.

GM uses the term VFV, while Ford calls its version an FF system. To avoid confusion throughout this chapter, we will use the GM VFV terminology when referring to those vehicles designed to run on an M85 methanol-blend fuel.

The use of M85 or other methanol blends in the vehicle's fuel does not change the basic operation of the fuel-injection system, but it does require a considerable redesign of system components for compatibility. In this chapter, we will learn about some of the changes required, as well as necessary precautions to be observed when servicing a VFV fuel system.

VARIABLE OR FLEXIBLE FUEL VEHICLES

Vehicles capable of running on fuels other than gasoline will play an increasing role in meeting the emission requirements established by the federal government, as well as regulations promulgated by individual state governments. California, for example, has legislated that a specified percentage of vehicles sold within the state must, by the year 2000, meet emission standards so stringent that most of the vehicles sold in 1995 could not meet them. California's eventual goal is to eliminate vehicle emissions altogether.

During the late 1980s, the most feasible approach seemed to be the conversion of the internal combustion fuel system to one that will burn a combination of gasoline and methanol, figure 11-1. Methanol was a leading choice as a gasoline substitute for many reasons:

- Performance
- Cost
- Availability
- Ease of transition
- Reduced emissions

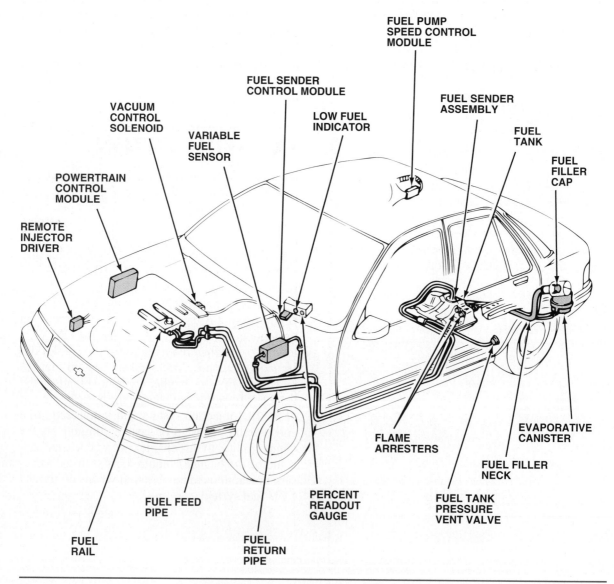

Figure 11-1. GM's methanol variable fuel system used with the Chevrolet Lumina. (Courtesy of General Motors Corporation)

Although there are many political factors involved concerning the final approach taken by the domestic automotive industry, both the fuel and cars to use it gradually are becoming available.

General Motors introduced an M85 methanol-blend system on the 3.1-liter V-6 MFI engine used in some 1991 and later Chevrolet Lumina models, figure 11-1. Ford brought forth an M85 system on some 1992 and later Taurus, Sable, and Econoline models. Chrysler also has its own variable fuel version, but we will look only at the GM Lumina and Ford Econoline systems as representative examples.

Servicing the variable fuel system, however, will require additional knowledge on the part of the technician, since its components differ considerably from those used with gasoline fuel systems.

M85 AND METHANOL-BLENDED FUELS

One major difference between methanol and gasoline is the nature of methanol and its properties. Methanol can be manufactured from wood, coal, or natural gas and contains about 60 percent of the energy present in regular unleaded gasoline, figure 11-2. However, methanol is *highly poisonous* and *cannot* be made nonpoisonous. It is so dangerous that it can cause headaches, blindness, or even death. Unlike gasoline, methanol can be absorbed through your skin, and swallowing only a mouthful can result in death. For this reason, VFVs use either a check ball or screen in the fuel filler neck to prevent the siphoning of M85 by mouth.

Methanol is also colorless and clear (like water) with a faint alcohol odor, and breathing its vapors over a period of time can cause blindness or death. For these reasons, the following precautions should be observed whenever you are working around a VFV fuel system:

- Avoid methanol leaks, spills, and splashing.
  a. If any of the fuel gets on your skin, wash the affected parts immediately.
  b. If the fuel spills on your clothes, change them at once and then immediately wash any areas of skin the fuel might have touched. Let the clothes air-dry, then machine wash to remove methanol contamination.
  c. If the fuel spills on a painted surface, flush the surface with water and air-dry. If wiped off with a cloth, the paint surface can be permanently damaged.
- Always wear methanol-resistant gloves, such as nitrile gloves, and wraparound safety glasses or goggles when working on a VFV fuel system.
  a. Do not use methanol fuel for cleaning parts. This would expose your skin to the substance; it is absorbed rapidly with a

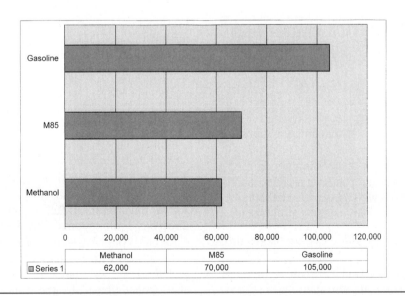

	Methanol	M85	Gasoline
▦ Series 1	62,000	70,000	105,000

Figure 11-2. This chart compares the energy content of methanol, M85, and gasoline.

slight cooling effect, like the evaporation of rubbing alcohol.

b. Like gasoline, methanol is highly flammable and burns violently. If methanol splashes on you and something ignites it, you could be burned severely. Do not smoke when working on a methanol fuel system, and do not work near any source of sparks or an open flame.

c. While the exhaust of an engine burning gasoline contains carbon monoxide, an odorless poison, the exhaust of an engine burning methanol contains formaldehyde, which gives off a sharp odor. Exposure to the exhaust fumes from a VFV, however, will cause a severe burning of the eyes, nose, and throat. As with carbon monoxide, exposure to formaldehyde can cause death.

d. If you remove the fuel filler cap on a variable fuel vehicle too quickly, fuel can spray out on you. This generally happens when the tank is nearly full, and is most likely to occur during hot weather. To avoid the possibility of a fuel spray, turn the fuel filler cap slowly, and wait for any hissing noise to stop before removing the cap completely.

e. In case of a methanol fire, use a B, C, BC, or ABC dry chemical extinguisher or an ARF-(alcohol-resistant foam) type extinguisher. Do not use water on a methanol-blend fire, as it causes gasoline in the blend to rise to the surface. Because gasoline burns hotter than methanol, spraying water on the fuel will make the situation worse, not better.

Symptoms of Methanol Exposure

Individuals react differently to methanol exposure. While some may experience symptoms immediately, others may not notice anything wrong for 10 to 15 hours after exposure, causing them to not associate the symptoms with methanol. Methanol exposure symptoms generally occur in three stages:

1. Headache, nausea, giddiness, stomach pains, chills, or muscle weakness.
2. A 10- to 15-hour period free of any symptoms.
3. Failing eyesight, intense nausea, dizziness, difficulty breathing, and intense headaches.

METHANOL AND AUTOMOTIVE ENGINES

Methanol is a highly corrosive agent when it comes in contact with materials such as aluminum, rubber, zinc, and low-grade steels. When used as a fuel, methanol causes corrosion problems with exhaust valves and seats, piston compression rings, and engine oil. Exhaust valves, valve seats, and compression rings can be made of stainless steel or chrome plated to reduce corrosion problems, but engine oil circulates throughout the engine. When standard engine oil is used, methanol tends to degrade it quickly. For this reason, M85-fueled engines require the use of a specific oil formulation, which should be changed at regular 3,000-mile intervals.

If an approved VFV oil is not available, any SAE 10W-30 SG engine oil can be used in an emergency, but only for a short duration. Prolonged use of standard engine oil in an engine running on M85 or other methanol blends can result in engine damage. If a standard engine oil has to be used, it should be changed as soon as possible. The oil filter also should be changed at the same time to prevent contaminating the VFV oil.

The ignition coil used with M85-fueled Luminas differs from those used with gasoline-fueled engines. The M85 coil must provide more current, a faster rise time, and a shorter duration spark. Using a standard coil on a VFV engine will result in cold-start problems because methanol vapor pressure at temperatures below about 50°F (10°C) is not great enough to produce an easily flammable air-fuel mixture. Ford M85 applications use a spark plug with a wider side electrode and a nontapered center electrode for increased heat transfer.

COMMON SUBSYSTEMS AND COMPONENTS

As we have seen, all fuel injection systems have three basic subsystems. The VFV air intake and electronic control systems are essentially the same as those used with a gasoline-powered vehicle. The fuel delivery and metering system, figure 11-3, also operates basically the same as in a gasoline system, with simultaneous injection used on the

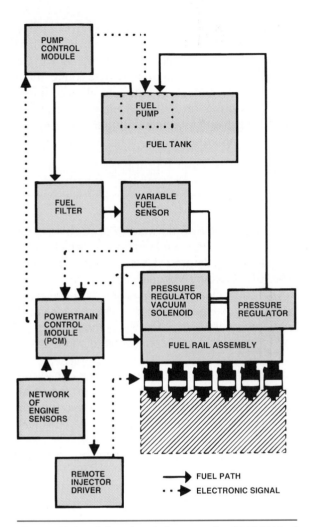

Figure 11-3. A functional diagram of a GM VFV fuel system. (Courtesy of General Motors Corporation)

GM 3.1-liter V-6 and sequential injection on Ford's 4.9-liter inline 6-cylinder.

There are, however, numerous changes in the components required to deal with the methanol content in the fuel. Methanol is very corrosive to aluminum, rubber, zinc, and low-grade steel. It is less corrosive to stainless steel, nylon, teflon, and chromium. All components that come in contact with methanol-blended fuel are made of methanol-resistant materials, or treated with methanol-tolerant coatings and paints.

GENERAL MOTORS VFV FUEL SYSTEM

The Chevrolet Lumina VFV fuel system, figure 11-1, consists of an in-tank electric fuel pump and fuel pump speed controller, flame arresters, an inline fil-

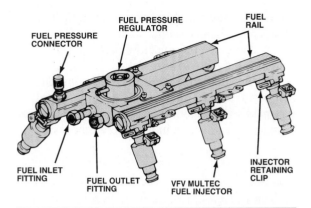

Figure 11-4. The GM R620 fuel rail has a replaceable pressure regulator assembly, but is otherwise the same as fuel rails used with gasoline fuel systems. (Courtesy of General Motors Corporation)

ter, a variable fuel sensor, a fuel rail with pressure regulator and six injectors, and the necessary connecting fuel distribution hoses and lines. We will examine the fuel system, starting from the fuel rail and working back to the fuel tank.

Fuel Rail, Injectors, and Hoses

With two exceptions, the basic fuel rail design, figure 11-4, is the same as that used with gasoline fuel systems:

- The aluminum fuel rail is yellow anodized to prevent methanol corrosion, as well as to provide identification as a VFV fuel rail.
- The replaceable fuel pressure regulator diaphragm is made from a methanol-resistant material.

The fuel rail is equipped with low-impedance Multec top-feed fuel injectors. Injector design is the same as that used with gasoline systems, but the coil bobbin, coil wire insulation, and the encapsulated wire/bobbin assembly are manufactured with alcohol-resistant materials. However, since methanol contains only about one-half the energy per unit as gasoline, the VFV injectors deliver a higher fuel output (3.84 g/s) than gasoline injectors (1.95 g/s) when wide open. They also have a faster reaction time and a longer-duration wide-open pulse signal than gasoline injectors. Injectors used with 1992 and later models have an operating resistance of 1 ohm, reduced from the 2-ohm injectors used with 1991 models. All VFV injectors draw more current than gasoline injectors. Because this current draw creates more heat than the powertrain control module (PCM) can dissipate, a **remote injector driver** controlled by the PCM is located on the right front fender. The driver energizes the injectors and dissipates the heat through its own heat sinks.

The fuel feed and return hoses in the engine compartment are braided hoses made of teflon wrapped with a stainless steel mesh and encased in nylon sheaths at points where the hoses bend.

This construction protects the hoses from engine heat as well as methanol corrosion. If standard flexible hoses are used as replacements for the braided hoses, methanol will deteriorate them within a few minutes, resulting in plugged injectors. Fuel hoses on 1991 models use O-rings in the fittings; quick-connect fittings are used on 1992 and later models.

■ The Future of Propane as an Alternative Fuel

As a result of the two energy crises during the 1970s, intense effort was devoted to developing a replacement fuel for gasoline. Ford Motor Company led the way in translating this effort into vehicles that would use such fuels. One result was the propane-fueled, 2.3-liter inline 4-cylinder engine that briefly graced 1982 Ford Granada/Mercury Cougar and 1983 Ford LTD/Mercury Marquis cars.

Propane, or liquified natural gas (LNG), is a liquid form of the same clean-burning natural gas used in the home. Liquified by chilling to $-258°F$ ($-161°C$), it is stored in thermos-type containers. Extensive testing of propane in automobiles showed that exhaust pollutants were practically eliminated. Since many drivers of motor homes were accustomed to using propane gas, it was thought that cars could operate on the same fuel.

The Ford system included a unique air cleaner, a propane carburetor, a fuel lock, and a converter/regulator assembly—all in the engine compartment. Twin propane tanks mounted beneath the trunk floor provided the necessary fuel. The only external sign of the vehicle's power source was a small "PROPANE" emblem on each front fender.

Propane is a flammable substance like gasoline, but is a vapor at normal temperatures and barometric pressures. The Ford system had two relief valves to vent excessive pressure resulting from high ambient temperatures. If a leak developed or the system vented through the relief valves, the propane immediately vaporized and expanded to about 270 times its liquid volume. Since it is heavier than air, propane settles in low spots and gradually dissipates. This created the possibility of a dangerous fire hazard.

For this reason, there were many restrictions placed on propane-fueled vehicles:

- Do not vent fuel unnecessarily.
- Do not drain the fuel tanks.
- Do not use a drying oven when refinishing the paint.
- Do not weld near the fuel system tanks or components.
- Do not service the vehicle near electrical equipment, such as motors or switches, that may discharge sparks.
- Do not store or service the vehicle over a confined area, such as a lube pit, where vapors might accumulate.

Technicians were not thrilled with such restrictions because the numerous safety precautions involved with propane vehicles interfered with their normal shop operation. Furthermore, these vehicles did not prove popular with the driving public, and the majority of these Ford cars ended up as fleet sales to companies interested in fuel conservation.

The lack of consumer response to Ford's effort was attributed to the difficulty in refueling the propane tanks and a general unavailability of the fuel in many areas. The propane-powered cars required a greater-than-normal amount of care on the part of drivers and died a quick death in the marketplace.

Gasoline prices started to stabilize at about the same time as the propane-fueled Fords were made available to the public. Most drivers were not willing to cope with the particular problems presented by this alternative fuel source. Thus, the propane-powered Fords passed into the pages of automotive history, much as the Chrysler turbine-powered vehicles had a decade earlier.

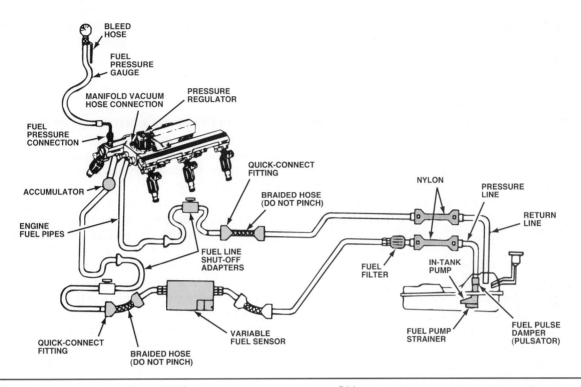

Figure 11-5. A 1992 and later VFV fuel system pressure test on a GM system. (Courtesy of General Motors Corporation)

The 1991 fuel system pressure of 41 to 47 psi (284 to 325 kPa) was raised on 1992 and later models to a range of 48 to 55 psi (333 to 376 kPa). The higher system pressure reduces vapor in the fuel lines and minimizes hot-starting problems. Refer to figure 11-5 (1992 and later) when performing a fuel system pressure test.

Variable Fuel Sensor

Because the methanol/gasoline ratio can vary from one tankful of fuel to another, the PCM must be able to determine the percentage of methanol in the fuel at all times to control ignition timing and fuel delivery. The variable fuel sensor (VFS) consists of a variable coaxial capacitor sensor, a temperature sensor, and an electronic module contained in a housing and installed in the fuel feed line near the brake booster, figure 11-6. The VFS is powered by voltage through the ignition switch and grounded through the chassis.

When the ignition is turned on, the PCM sends a 5-volt reference voltage to the VFS module. As fuel passes through the sensor housing, its dielectric properties are monitored by the variable capacitor. The voltage buildup in the capacitor is proportional to the percentage of methanol contained in the fuel.

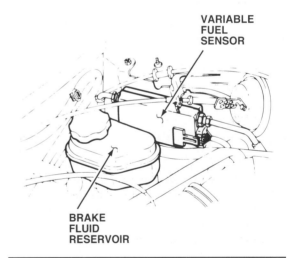

Figure 11-6. Location of the variable fuel sensor in GM's Lumina engine compartment.

Because the capacitance reading varies according to fuel temperature, the temperature sensor allows the PCM to compensate for any capacitor variance. The sensor signals to the VFS module are returned to the PCM as fuel density and fuel temperature signals. The 1991 sensor uses a separate wiring harness connector and has fuel inlet and outlet nuts to connect into the fuel line. To remove the sensor, the threaded hose fittings are disconnected at the sensor. The housing was redesigned for 1992 and later models with an integral wiring harness connector and a length of fuel tubing on either end containing quick-connect fittings. Although the fuel tubes are connected to the sensor housing by inlet and outlet nuts, the threaded hose fittings should not be disconnected to remove the sensor assembly. If sensor removal is required, disconnect it from the system at the quick-connect fittings. The entire unit is manufactured of methanol-tolerant materials.

If the variable coaxial capacitor in the VFS fails, it will cause a code to be set in PCM memory. With the 1991 sensor, a code 56 is stored; the 1992 and later sensors will cause the PCM to set a code 56 (voltage low), code 57 (voltage high), or code 58 (degraded fuel sensor). A temperature sensor failure will set a code 64 (temperature high) or code 65 (temperature low). The Tech 1 scan tool can be used to determine the percentage of alcohol present in the fuel.

The VFS is serviced by replacement only, and should be handled with care. If dropped or otherwise subjected to a shock, it must be replaced.

Inline Fuel Filter

The inline fuel filter used with the variable fuel system works like any other fuel filter, but is made from methanol-tolerant stainless steel. Filters designed for use with gasoline fuel systems should not be used as a replacement.

Flame Arresters

Two flame arrester devices made of corrugated stainless steel are installed in the filler neck and vent hoses where they connect to the fuel tank. The flame arresters act as filters to divide and quench any sparks or flames that originate outside the vehicle before they reach the fuel tank. They also serve as antisiphon devices to prevent accidental ingestion of fuel if someone should attempt to siphon the fuel tank.

Fuel Tank and Filler Cap

The size and shape of the fuel tank are identical to other Lumina tanks, but the VFV tank is made from stainless steel to resist the corrosive effects of methanol. The fuel tank filler tube contains a check ball to prevent siphoning fuel. The 1991 filler cap uses a pressure relief valve; 1992 and later filler caps contain a vacuum relief valve. A fuel tank

pressure vent valve is used on 1992 and later VFVs to vent excessive tank pressure to the atmosphere.

Fuel Pump

Like most other in-tank fuel pumps used on fuel-injected vehicles, the fuel pump is part of the fuel sending unit, figure 11-7. The sending unit and pump assembly also contain a pressure control and rollover valve.

Because the energy density of methanol is about 40 percent less than the energy density of gasoline, the high-pressure roller vane pump must run almost twice as fast as a comparable pump used with a gasoline fuel system if it is to deliver sufficient fuel for engine operation. For this reason, pump capacity has been nearly doubled. Except for the use of a check ball in the pump outlet line to prevent fuel backflow, the pump design is the same as that installed in the modular fuel assembly and used with gasoline fuel systems.

The fuel pump has a two-stage pumping action. The low-pressure turbine section uses an impeller with a staggered blade design to minimize pump noise and to separate vapor from the liquid

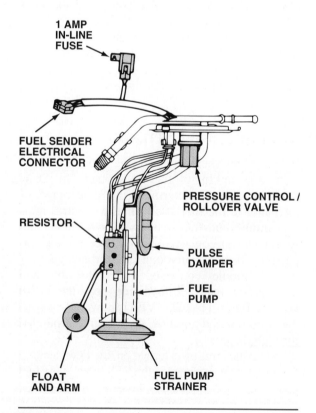

Figure 11-7. The GM variable fuel sender assembly. (Courtesy of General Motors Corporation)

fuel. It is designed to prevent vapor lock resulting from the higher vapor pressures common with methanol blends. The roller vane section creates the high pressure required for fuel injection. The end cap assembly contains an outlet check valve, a pressure relief valve, and a radio frequency interference (RFI) module. The brushes are a methanol-tolerant design, with coated springs, insulator disc, plated shunt wire, and soldered connections. All metal parts and wire insulation are treated to resist electro-galvanic corrosion, and all elastomers and other seals are designed to minimize swelling caused by methanol.

If the fuel pump is removed from the sender assembly for service or replacement, always install a new fuel pump strainer before reinstalling the unit in the fuel tank.

Fuel Pump Speed Controller

Fuel pump operational speed on 1991–92 systems is controlled by the PCM and a fuel pump speed controller (FPSC). To prolong pump life, the controller reduces its speed whenever a high fuel flow is not necessary, as when the vehicle is operating on 100 percent gasoline. When the ignition switch is turned on, the PCM energizes the controller for two seconds, allowing the fuel pump to pressurize the system. If ignition reference pulses are not received by the PCM within the two seconds, it deenergizes the controller to shut the pump off. If the FPSC malfunctions, the PCM stores a code 54 (low voltage) in memory. Use of the FPSC was found to be unnecessary and was eliminated on 1993 and later models, allowing the fuel pump to operate at a constant speed.

Fuel Sender Unit and Control Module

Specifically designed for use with variable fuels, the fuel sender unit, figure 11-8, uses a unique electrical circuit. A pulse-width-modulated (PWM) signal generated by the fuel sender control module is sent to a straight strip resistor on the sender assembly. The resistor modifies the PWM signal according to the position of the float and returns it to the control module, where it is processed and sent to the instrument panel gauge. A low fuel indicator lamp on the instrument panel is controlled by the PCM according to signals from the control module.

As with other components of a variable fuel system, the sender unit is designed to be

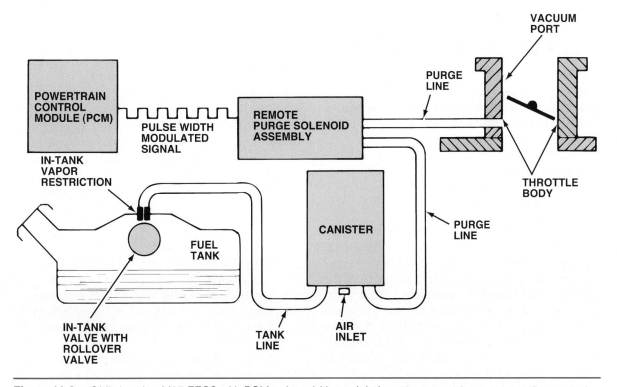

Figure 11-8. GM's Lumina M85 EECS with PCM pulse width modulation. (Courtesy of General Motors Corporation)

methanol-tolerant, with stainless steel used for all metal parts, fluorosilicone (orange) and Viton (black) seals, and an alcohol-resistant plastic float. All electrical connections have been modified to reduce positive-to-negative terminal current migration and potential shorting. A 1-amp fuse installed in the external sender harness provides circuit protection for the PWM circuit between the control module and strip resistor. If the fuse blows, the fuel gauge will go off-scale to the full side. The fuse is not serviced separately; if it blows, the entire sender assembly must be replaced.

When the sender unit is removed for sender or pump service, check the fuel pulse dampener or pulsator unit, figure 11-7 for degraded dampener seals. The 1991–92 dampeners use orange (fluorosilicone) seals; 1993 and later dampeners have black (Viton) seals. If the dampener seals have deteriorated, they may have contaminated the inline fuel filter. Install a new dampener and change the fuel filter.

The fuel pump strainer and inline fuel filter are more likely to clog in a VFV fuel system than in a gasoline fuel system. If the strainer requires replacement because of contamination, the inline filter should be replaced at the same time.

GENERAL MOTORS VFV EMISSION CONTROL SYSTEMS

The same emission control systems are used on both the gasoline and VFV versions of the 3.1-liter V-6 MFI engine. However, the evaporative emission control system (EVAP) and the catalytic converter required modifications.

Evaporative Emission Control System

The evaporative emission control system (EVAP) used with methanol fuel, figure 11-8, is essentially the same as that used with a gasoline-powered engine. When the engine is off, fuel vapors from the tank are sent through the combination tank pressure control valve (TPCV) and rollover valve to a charcoal canister for storage. When the engine is running, the vapors are purged from the canister by intake airflow and burned in the combustion chambers. Purging is controlled by the PCM.

The canister is made of methanol-tolerant materials and used in an inverted position. The canister

capacity is 2.3 quarts (2200 cc), or 0.7 quart (700 cc) greater than the canister used with the 189 cu in. (3.1-liter) gasoline-powered Luminas. A unique remote high-purge solenoid valve is controlled by a PWM signal from the PCM. A limited amount of purging takes place during idle or closed-throttle operation. As engine speed increases, so does the amount of purging. Ambient air enters the bottom of the canister through the inlet air tube, where it mixes with the stored vapors. When the purge solenoid is modulated by the PCM, the vapor-air mixture is drawn from the canister and passes through the air intake throttle body to the intake manifold for burning. If the purge solenoid malfunctions, the PCM will set a diagnostic code.

The canister is color-coded to prevent its use in a gasoline EVAP system, or vice versa. All GM canisters used with gasoline systems are black; the variable fuel canister is white, and is located underneath the rear of the vehicle.

Catalytic Converter

The catalyst in a regular production converter cannot cope with the formation of formaldehyde emissions that result from methanol combustion. For this reason, the catalyst used in a VFV converter has been recalibrated to make it effective in controlling the combustion by-products of methanol-blended fuels.

FORD FLEXIBLE FUEL SYSTEM

The Ford FF system used on the Econoline consists of an in-tank electric fuel pump and sending unit, an in-line filter, a VFS sensor, a fuel rail with pressure regulator and six injectors, and the necessary connecting fuel distribution hoses and lines. We will examine the differences in the fuel system components from those of the GM application, starting in the engine compartment.

- The same fuel rail is used on M85 and gasoline engines. The pressure regulator diaphragm is made from a methanol-resistant material. The fuel rail is equipped with high-resistance (13 to 16 ohm) methanol-resistant injectors.
- The injector fuel nozzles are larger than those on gasoline injectors to provide the higher flow capacity required when using a methanol blend.

- A cold-start injector (CSI) installed on top of the intake air throttle body near the idle air control (IAC) solenoid provides additional fuel for cold weather starting.
- Stainless steel and nylon fuel lines are used to prevent corrosion damage. Standard rubber hoses or steel fuel lines should not be used as replacements, as methanol deteriorates these materials and will cause plugged injectors.
- The fuel system pressure of 45 to 60 psi (310 to 415 kPa) reduces vapor in the fuel lines and minimizes hot starting problems.
- The fuel sensor used on Ford M85 fuel systems performs the same function as that used by GM, and operates in a similar manner. Unlike the GM system, however, its reading is translated by a control module into a linear display positioned between the steering column shroud and instrument panel.
- The in-line fuel filter works like other fuel filters, but is made of methanol-tolerant materials. Filters designed for use with gasoline fuel systems should not be used as a replacement.
- The size and shape of the fuel tank are the same as other Econoline tanks, but the steel M85 tank contains a methanol-resistant coating to resist the corrosive effects of methanol.
- The fuel tank filler tube uses a methanol-resistant coating and contains a screen to prevent siphoning fuel from the tank.
- The combination in-tank pump/sending unit assembly consists of nickel-plated metal parts. Composite parts of the assembly are made of materials that are compatible with methanol.

FORD FF EMISSION CONTROL SYSTEMS

The emission control systems used on gasoline and M85 versions of the 4.9-liter SFI engine are essentially the same. Like the GM VFV 3.1-liter V-6, however, the evaporative emission control system (EECS) and the catalytic converter are different.

Evaporative Emission Control System

The Ford M85 EVAP system functions in the same way as the EVAP system on a gasoline-powered vehicle, but there are two major differences in the components:

- Increased canister capacity
- A vapor management valve

To provide increased vapor capacity, four-carbon canisters are connected in pairs and tray-mounted to the outer side of the left frame rail. With the engine off, fuel vapors reach the canister assembly through vapor valves installed at each end of the tank. A vapor management valve (VMV) prevents vapor flow from the canister until the engine is started. Once the engine is running, vapor purge is controlled by Ford's Electronic Engine Control, Version IV (EEC-IV) module.

E-85 VEHICLES

E-85 vehicles are vehicles that can run on ethanol blends of up to 85 percent. As discussed in Chapter 3, ethanol is alcohol made from corn, grains, or other agricultural waste. Ethanol is a renewable energy source that can be domestically produced in many countries. Current policies encourage using greater amounts of ethanol as a way to reduce dependence on oil. While ethanol has commonly been used in gasoline blends of up to 10 percent to produce oxygenated fuels, it has not been available widespread in formulations of 85 percent. Production of E-85 vehicles in the United States began around 1998. By 2003 it is estimated that over 3 million vehicles are equipped to run on E-85 fuel.

General Motors, Ford, DaimlerChrysler, Mazda, and Honda are a few of the manufacturer's offering E-85-compatible vehicles. Generally with today's modern engine management systems used on vehicles, very few modifications have to be made to a vehicle designed to run on various ethanol blends. Most E-85 systems are transparent to the driver and require no action on the driver's part whether running on gasoline or 85 percent ethanol. E-85 vehicles have more robust fuel system parts designed to withstand the additional alcohol content, modified driveability programs that adjust fuel delivery and timing to compensate the various percentages of ethanol fuel, and a **fuel compensation sensor** that measures both the percentage of ethanol blend and the temperature of the fuel.

Table 11-1 shows a comparison of ethanol, methanol, and gasoline. As can be seen, ethanol has less BTU/lb content than gasoline. However, it does have a higher octane rating and a lower Reid vapor pressure than gasoline. Because it provides fewer miles per gallon than gasoline, either the fuel range of a vehicle will be reduced or

Table 11-1. Comparison of fuel properties.

Property	Methanol	Ethanol	Gasoline (87 Octane)	E85
Chemical formula	CH_3OH	C_2H_5OH	C_4 to C_{12} chains	*
Main constituents (% by weight)	38 C, 12 H, 50 O	52 C, 13 H, 35 O	85–88 C, 12–15 H	57 C, 13 H, 30 O
Octane (R+M)/2	100	98–100	86–94	96
Lower heating value (Btu/lb)	8,570	11,500	18,000–19,000	12,500
Gallon equivalent	1.8	1.5	1	1.4
Miles per gallon compared to gasoline	55%	70%	—	72%
Relative tank size to yield (driving range equivalent to gasoline)	Tank is 1.8 times larger	Tank is 1.5 times larger	1	Tank is 1.4 times larger
Reid vapor pressure (psi)	4.6	2.3	8–15	6–12
Ignition point				
Fuel in air (%)	7–36	3–19	1–8	*
Temperature (approx.) (°F)	800	850	495	*
Specific gravity (60/65°F)	0.796	0.794	0.72–0.78	0.78
Cold weather starting	Poor	Poor	Good	Good
Vehicle power	4% power increase	5% power increase	standard	3%–5% power increase
Stoichiometric air/fuel ratio (by weight)	6.45	9	14.7	10

Depends on percentage and type of the hydrocarbon fraction.

larger fuel tanks will need to be added to E-85 vehicles. While the energy content of ethanol is about 27 percent less than gasoline, it is estimated that actual fuel mileage reductions will range between 5 percent and 12 percent.

The benefits of E-85 vehicles are less pollution, less CO_2 production, and less dependence on oil. Ethanol-fueled vehicles generally produce the same pollutants as gasoline vehicles; however, they produce less CO and CO_2 emissions. In addition, evaporative emissions are lower from ethanol than gasoline. While CO_2 is not considered a pollutant, it does lead to global warming.

A Typical E-85 Vehicle

Most E-85 vehicles are very similar to non-E-85 vehicles. Fuel system components may be redesigned to withstand the effects of higher concentrations of ethanol. In addition, since the stoichiometric point for ethanol is 9:1 instead of 14.7:1 as for gasoline, the air-fuel mixture has to be adjusted for the percentage of ethanol present in the fuel tank. In order to determine this percentage of ethanol in the fuel tank, a fuel composition sensor is used. The fuel composition sensor is the only additional piece of hardware required on an E-85 vehicle.

Figure 11-9 shows an electrical schematic for an E-85 vehicle's fuel composition sensor. The fuel composition sensor provides both the ethanol percentage and the fuel temperature to the PCM. The PCM uses this information to adjust both the ignition timing and the quantity of fuel delivered to the engine. The fuel compensation sensor uses a microprocessor to measure both the ethanol percentage and the fuel temperature. This information is sent to the PCM on the signal circuit. The compensation sensor produces a square wave frequency and pulse-width signal. The normal frequency

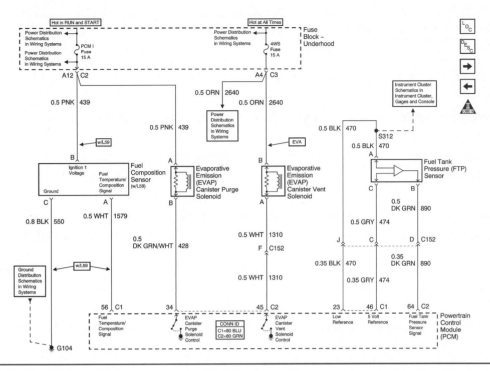

Figure 11-9. The fuel composition sensor provides the PCM with percentage of ethanol and fuel temperature. (Courtesy of General Motors Corporation)

range of the fuel composition sensor is 50 hertz, which represents 0 percent ethanol and 150 hertz, which represents 100 percent ethanol. The pulse width of the signal varies from 1 millisecond to 5 milliseconds. One millisecond would represent a fuel temperature of −40°F (−40°C) and 5 milliseconds would represent a fuel temperature of 257°F (125°C). Since the PCM knows both the fuel temperature and the ethanol percentage of the fuel, it can adjust fuel quantity and ignition timing for optimum performance and emissions.

BI-FUEL VEHICLES

Bi-fuel vehicles are vehicles that can run on more than one type of fuel. The Energy Policy Act of 1992 was established to encourage the development of vehicles that use alternate fuels. The goal is to reduce dependence on petroleum and lower emissions by using cleaner-burning fuels. No alternate fuels are mandated by the act. Clean fuels are any fuel that meets low emission standards, which could be any of the following:

- Methanol
- Natural gas (compressed natural gas CNG/ liquid natural gas LNG)
- Ethanol
- Reformulated gasoline
- Electricity
- Liquid petroleum gas (LPG)
- Reformulated diesel fuel
- Hydrogen

The Energy Policy Act applies to certain areas of the country and fleets with more than ten vehicles. This act requires a certain percentage of the fleet purchases to be alternate fuel vehicles.

A number of manufacturers are producing bi-fuel vehicles. These vehicles can operate on gasoline and some other type of fuel. Most often this other fuel is compressed natural gas (CNG). Another type of bi-fuel vehicle is the hybrid gasoline/ electric vehicle such as the Toyota Prius or the Honda Insight. As technology advances, more vehicle manufacturers will offer hybrid vehicles that use either electricity/gasoline or bi-fuel arrangements. The next section will cover a bi-fuel-equipped vehicle that uses gasoline and compressed natural gas.

Compressed natural gas is a by-product of petroleum exploration. It is made up of a blend of methane, propane, ethane, N-butane, carbon dioxide, and nitrogen. Once it is processed, it at least 93 percent methane. Natural gas is nontoxic, odorless, and colorless in its natural state. It is odorized during processing to allow for easy leak detection. Natural gas has a vapor density of 0.68 as compared to air, which is 1. Therefore, it is lighter than air and will rise when released into the air.

Table 11-2. Comparison of gasoline and natural gas.

Characteristic	Gasoline	Natural Gas
Chemical Symbol	C_4H_{10}-$C_{12}H_{24}$	CH_4
Vapor Density	3.5	0.6–0.7
Flammability Limits in Air	1.0–7.6%	5.0–15.0%
Flammability Ratio	5.4:1	3:1
Air-Fuel Ratio	14.7:1	16.5:1
Ignition Temperature	540°F – 800°F	1,200°F
Flame Speed	2.72 FPS	2.20 FPS
Octane Rating	86–93	115+
Fuel Quantity Measurement	Gallons	Gasoline Gallon Equivalent 1 GGE = 125 cu. Ft.
Energy Content (BTU per lb.)	18,400	20,500

In air the lower limit for flammability of natural gas is 5 percent, meaning it will not burn if there is less than 5 percent concentration of natural gas. The upper limit for flammability for natural gas is 15 percent. If there is more than 15 percent concentration of natural gas in air, it will not burn. The combustion characteristic of natural gas is different than gasoline. Since it is already a vapor, it does not need heat to vaporize before it will burn. This leads to improved cold start-up and lower emissions during cold operation. However, because it is already in a gaseous state, it does replace some of the air charge in the intake manifold. This leads to about a 10 percent reduction in engine power as compared to an engine operating on gasoline.

Table 11-2 shows a comparison between natural gas and gasoline. As can be seen from the table, natural gas has less energy content than gasoline; therefore, the fuel mileage when operating on natural gas will not be as great as with gasoline. Natural gas also burns slower than gasoline; therefore, the ignition timing must be advanced more when the vehicle operates on natural gas. This is not a problem with natural gas because its octane rating is much higher than gasoline, about 115 octane compared to 86–93 octane for gasoline. Gasoline requires between 540°F to 800°F ignition temperature to ignite, while natural gas requires an ignition temperature of 1200°F to ignite. The stoichiometric ratio is the point at which all the air and fuel is used or burned. For gasoline you have learned that the stoichiometric point for gasoline is 14.7:1. For natural gas the stoichiometric point is 16.5:1. This means that more air is required to burn one pound of natural gas than is required to burn one pound of gasoline.

A typical CNG-equipped vehicle is shown in figure 11-10. This vehicle is designed to run on CNG during most operating modes. As long as the CNG tank pressure is greater than 600 psi and engine coolant temperature is more than 10°F, the vehicle will normally start in the CNG mode and operate on compressed natural gas. If the vehicle cranks for longer than eight seconds without starting, it will switch to gasoline and start in that mode.

The CNG tank, when completely filled, has 3600 psi of pressure in the tank. When the ignition is turned on, the alternate fuel electronic control unit activates the high pressure lock-off, which allows high pressure gas to pass to the high pressure regulator. The high pressure regulator reduces the high pressure CNG to approximately 170 psi and sends it to the low pressure lock-off. The low pressure lock-off is also controlled by the alternate fuel electronic control unit and is activated at the same time that the high pressure lock-off is activated. From the low pressure lock-off the CNG is directed to the low pressure regulator. This is a two-stage regulator that first reduces the pressure to approximately 4 to 6 psi in the first stage and then to 4.5 to 7 inches of water in the second stage. From here, the low pressure gas is delivered to the gas mass sensor/mixture control valve. This valve controls the air-fuel mixture. The CNG gas distributor adapter then delivers the gas to the intake stream.

CNG vehicles are designed for fleet use that usually have their own refueling capabilities. One of the drawbacks to using CNG is the time that it

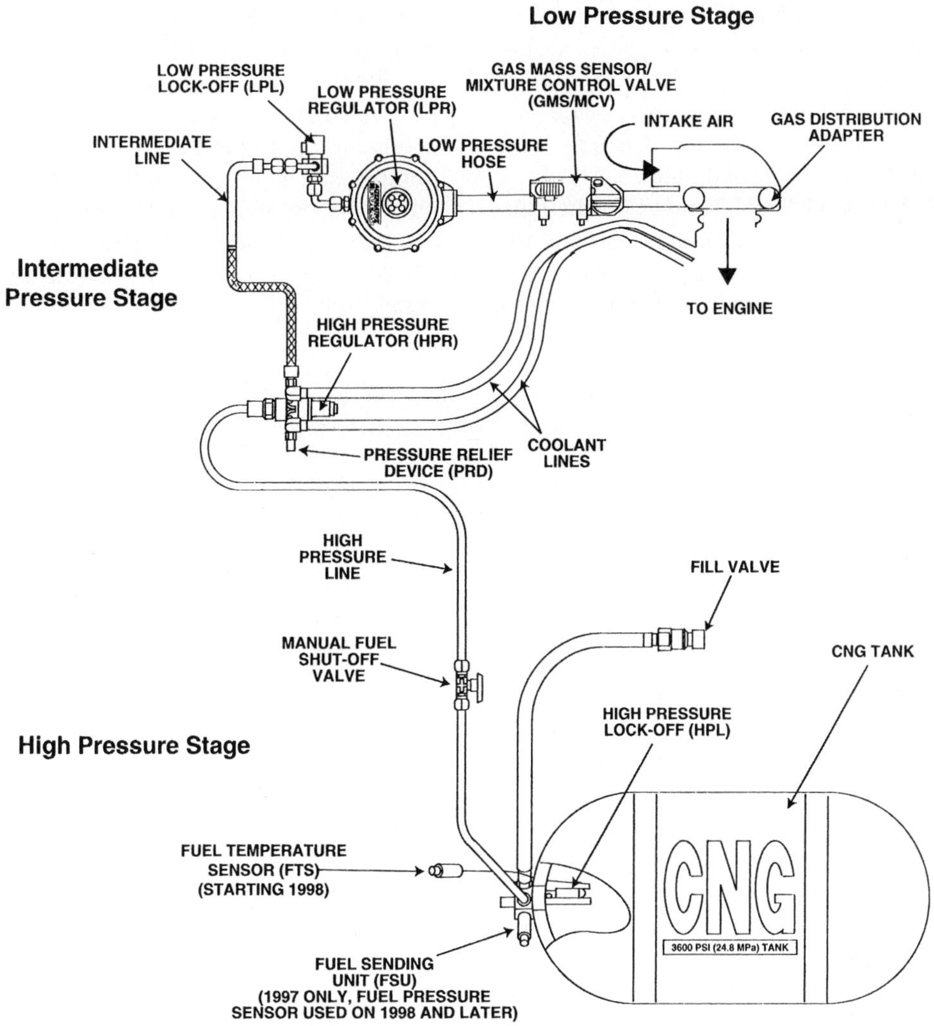

Low Pressure Stage

Figure 11-10. Typical gaseous fuel system. (Courtesy of General Motors Corporation)

takes to refuel a vehicle. The ideal method of refueling is the slow-fill method. The slow-filling method compresses the natural gas as the tank is being fueled. This method ensures that the tank will receive a full charge of CNG; however, this method can take 3 to 5 hours to accomplish. If more than one vehicle needs filling, the facility will need multiple CNG compressors to refuel the vehicles.

The fast-fill method uses CNG that is already compressed. However, as the CNG tank is filled rapidly, the internal temperature of the tank will rise, which causes a rise in tank pressure. Once the temperature drops in the CNG tank, the pressure drops, also resulting in an incomplete charge in the CNG tank. This refueling method may take only about 5 minutes; however, it will result in an incomplete charge to the CNG tank, thus reduced driving range.

SUMMARY

Variable fuel vehicles (VFVs) or flexible fuel (FF) vehicles can operate on unleaded gasoline or M85, a blend of 85 percent methanol and 15 percent unleaded gasoline. Using M85 does not change basic operation of the fuel injection system, but it does require component redesign for compatibility.

Methanol is highly poisonous and can cause headaches, blindness, or even death. While some people exposed to methanol poisoning may experience symptoms immediately, others may not notice anything wrong for 10 to 15 hours.

Methanol is highly corrosive to aluminum, rubber, zinc, and low-grade steels, and tends to degrade ordinary engine oil. Methanol-system exhaust valves, valve seats, and compression rings are made of stainless steel or chrome plated to reduce corrosion problems. Engines combusting methanol require the use of a specific oil formulation. All components that come in contact with methanol-blended fuel are made of methanol-resistant materials, or treated with methanol-tolerant coatings and paints.

During the early 1990s, GM and Ford had the most highly developed variable fuel systems for production cars. GM uses the term variable fuel vehicle (VFV), while Ford calls its version a flexible fuel (FF) system.

Bi-fuel vehicles are vehicles designed to run on more than one fuel. The vehicles may be hybrid gasoline/electric, gasoline/CNG, or other combinations of fuel systems. Many manufacturers are producing vehicles that are designed to run on compressed natural gas (CNG) as the primary fuel and gasoline as the secondary fuel system. Because of the lack of refueling stations, these are generally used by fleets with their own refueling capabilities.

Review Questions

Choose the letter that represents the best possible answer to the following questions:

1. The alternative fuel M85 consists of:
   a. 85 percent unleaded gasoline, 15 percent methanol
   b. 85 percent ethanol, 15 percent gasoline
   c. 85 percent methanol, 15 percent gasoline
   d. 85 percent unleaded gasoline, 15 percent ethanol

2. The energy content of methanol is approximately _____ that of gasoline:
   a. 60 percent
   b. 85 percent
   c. 25 percent
   d. 15 percent

3. Which of the following materials is *least* affected by methanol?
   a. Rubber
   b. Zinc
   c. Aluminum
   d. Teflon

4. Technician A says symptoms of exposure to methanol fuel occur immediately after exposure.
   Technician B says some symptoms may not occur for several hours after exposure. Who is right?
   a. A only
   b. B only
   c. Both A and B
   d. Neither A nor B

5. Technician A says the Lumina VFV fuel rail is yellow anodized to make it impervious to methanol.
   Technician B says the aluminum fuel rail is lined with Teflon. Who is right?
   a. A only
   b. B only
   c. Both A and B
   d. Neither A nor B

6. The GM variable fuel sensor contains a variable capacitor and electronic circuitry to determine the percentage of:
   a. Unleaded gasoline
   b. Methanol
   c. Water contamination
   d. None of these

7. The GM variable fuel sensor also senses fuel:
   a. Consumption
   b. Purity
   c. Temperature
   d. Flow

8. Technician A says a remote injector driver is used because GM VFV injectors draw more current than gasoline injectors.
   Technician B says VFV fuel system pressure on 1992 and later Luminas is 41 to 47 psi (284 to 325 kPa). Who is right?
   a. A only
   b. B only
   c. Both A and B
   d. Neither A nor B

9. Ford M85 fuel system pressure can be checked with a pressure tester that reads:
   a. 41 to 47 psi (284 to 325 kPa)
   b. 48 to 55 psi (333 to 376 kPa)
   c. 45 to 60 psi (310 to 415 kPa)
   d. 41 to 55 psi (284 to 376 kPa)

10. Technician A says the operating resistance of Ford FF injectors is 2 ohms.
    Technician B says their resistance is 1 ohm. Who is right?
    a. A only
    b. B only
    c. Both A and B
    d. Neither A nor B

11. Which of the following devices prevents siphoning fuel from the tank of a VFV Lumina?
    a. An arrester in the filler hose
    b. A check ball in the tank filler tube
    c. A fuel tank pressure vent valve
    d. All of the above

12. Technician A says regular production catalytic converters can be used with a VFV.
    Technician B says VFV converters must be recalibrated to reduce formaldehyde emissions. Who is right?
    a. A only
    b. B only
    c. Both A and B
    d. Neither A nor B

13. Technician A says the variable fuel sensor on a 1992 Lumina VFV is removed by disconnecting the threaded hose fittings. Technician B says the variable fuel sensor on a 1991 Lumina VFV is removed by using the quick-connect fittings. Who is right?
    a. A only
    b. B only
    c. Both A and B
    d. Neither A nor B

14. Which of the following statements regarding the oil used in M85-fueled engines is false?
    a. It should be changed every 3,000 miles (4828 km).
    b. It is specially formulated for use in methanol-fueled engines.
    c. The oil filter is manufactured of methanol-tolerant materials.
    d. Any SAE 10W-30 engine oil can be used in an emergency.

15. Bi-fuel vehicles are vehicles that are designed to run on:
    a. Gasoline/electricity
    b. Gasoline/CNG
    c. One fuel only
    d. Either A or B

16. Compressed natural gas (CNG) is comprised mostly of:
    a. Ethanol
    b. Methanol
    c. Methane
    d. Butane

17. CNG has a _____ octane rating than gasoline and burns _____ than gasoline.
    a. Lower, slower
    b. Higher, faster
    c. Higher, slower
    d. Lower, faster

18. What is the typical air-fuel ratio of CNG-equipped vehicles?
    a. 14.7:1
    b. 12.0:1
    c. 18.0:1
    d. 16.5:1

19. E-85 vehicles can run on any percentage of gasoline or ethanol up to 85 percent ethanol.
    a. True
    b. False

20. The fuel composition sensor measures:
    a. Ethanol percentage
    b. Fuel temperature
    c. Fuel density
    d. Both A and B

12

Emissions, Five Gas Theory, and I/M Programs

OBJECTIVES

Upon completion and review of this chapter, you will be able to:

- Name the five gases produced in an internal combustion engine.
- Identify which of the five gases are considered good gases versus pollutants.
- Have knowledge of various I/M testing programs that are in use.
- Describe the differences between constant volume sampling and percentage sampling.
- Describe the basic OBD II systems test.

KEY TERMS

ASM	NO_x
CO	sealed housing for
CO_2	evaporative
FTP	determination
HC	(SHED)
I/M	state implementation
I/M 240	plan (SIP)

INTRODUCTION

All internal combustion engines burn fuel comprised of mainly hydrocarbons (HC). The oxygen (O_2) used in the combustion process comes from the air. Air is comprised of approximately 78 percent nitrogen (N), 21 percent oxygen, and 1 percent other gases. If perfect combustion were to occur, the output from an internal combustion engine would be heat, water (H_2O), and carbon dioxide (CO_2). Unfortunately, perfect combustion is impossible to achieve even under laboratory conditions.

There are five primary gases that need to be analyzed in the exhaust stream from an internal combustion engine. These five gases are hydrocarbons (**HC**), carbon monoxide (**CO**), nitrogen oxides (**NO_x**), oxygen (O_2), and carbon dioxide (**CO_2**). Measuring and analyzing these gases is often referred to as the Five Gas Theory. As you learned in a previous chapter, HC, CO, and NO_x are the bad gases that lead to air pollution. O_2 and CO_2 are basically good gases that indicate engine efficiency. Measuring and analyzing these gases can indicate whether the engine is running efficiently and indicate whether the engine is emitting too many pollutants.

275

While emission control devices and systems have been used on automobiles since the early 1970s, air quality in many areas of the country have not always improved. A typical automobile is 96 percent cleaner today than a pre-emission equipped vehicle. Despite these improvements, due to more vehicles on the road and more miles being driven, pollution remains a threat to our health and the environment. It has been estimated that only between 10 and 30 percent of the vehicles on the road produce the majority of the pollution from mobile sources. These vehicles are usually vehicles in need of maintenance or repair. Consequently, most regions of the country with large population densities have implemented emission inspection maintenance (**I/M**) programs that are designed to detect vehicles that are the gross polluters and have them repaired or replaced. A number of different types of I/M programs are used throughout the country; however, their purpose is essentially the same.

THE COMBUSTION PROCESS

Figure 12-1 shows an example of perfect combustion. For perfect combustion to occur, a fuel comprised of HCs is needed. In addition, O_2 is needed to allow the fuel to burn. A heat source is needed to ignite the air-fuel mixture in the cylinder. Normally, this heat source is spark in a gasoline combustion engine or the heat of compression in a diesel engine. In a perfect world, combustion would occur and the only outputs would be water (H_2O), carbon dioxide (CO_2), and heat. This heat is transformed into mechanical energy in the internal combustion engine. Unfortunately, perfect

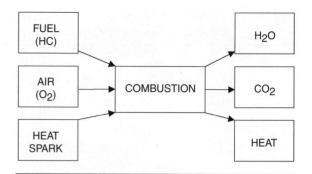

Figure 12-1. Perfect combustion.

combustion is not possible in an internal combustion engine. This is due to a variety of different factors, which will be explained next.

Figure 12-2 shows the result of normal combustion. In this process, we can see that we use air instead of oxygen in the combustion process. Air contains approximately 78 percent nitrogen. Nitrogen is considered an inert gas (a gas that doesn't normally combine with other gases); however, during the combustion process high heat causes some of the nitrogen to combine with varying amounts of oxygen to form compounds referred to as NO_xs. NO_xs are a pollutant that, when combined with sunlight and HCs, forms what we know as smog. Carbon monoxide (CO) is also produced during the combustion process due to the lack of oxygen. Whenever there is not enough oxygen left as the combustion process occurs, some CO is produced instead of CO_2. Any fuel that does not completely burn during the combustion process causes HCs to be produced in the exhaust stream. The purpose of the emission control devices that are used on internal combustion engines is to try to minimize the production of HC, CO, and NO_x or to eliminate these after they have been pro-

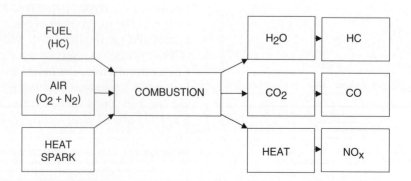

Figure 12-2. Normal combustion.

duced. Through careful engine design and accurate control of the air-fuel mixture, the production of these pollutants can be reduced to very low levels. The pollutants that are produced can then be eliminated through the use of additional emission control devices. While the production of these pollutants will probably never be completely eliminated in the internal combustion process, well-designed and well-maintained engines can achieve very low emission levels of these harmful pollutants.

THE FIVE GASES

Now that we know what gases are produced as a result of the combustion process, we will look at each of these gases individually and see how each of them is affected by air-fuel ratios. We will also discuss the causes for higher than normal production of these gases.

Hydrocarbons (HCs)

Hydrocarbon production is the result of incomplete combustion of fuel being burnt. Hydrocarbons combined with NO_xs contribute to ground-level smog and can be very harmful to the environment and to human health. Even in a well-designed and properly maintained engine, some hydrocarbons will be produced due to the quench effect.

Quenching is caused by the metal surfaces of the combustion chamber absorbing heat from the combustion process, figure 12-3. This creates cooler areas in the cylinder and causes the com-

bustion flame to extinguish. Some of the fuel does not burn; therefore, the HCs pass into the exhaust stream. Modern designed engines reduce this quench effect through combustion chamber design. Even so, some quenching will always occur in the combustion chamber.

Other causes for excessive HC production may be:

- Excessive lean air-fuel ratio
- Excessive rich air-fuel ratio
- Low compression
- Misfire
- Inaccurate spark timing

In figure 12-4 we can see the results of air-fuel ratio on the production of HC emissions. At the stoichiometric point of 14.7:1 HC production is at a fairly low level. HCs will continue to decline until an air-fuel ratio of about 16:1 is reached. If the air-fuel ratio goes leaner than 16:1, the HCs start to increase again. This is because the air-fuel ratio has now reached the lean misfire region. This means that because the amount of fuel in the cylinder is so low, it will not ignite or burn. There is too much air and not enough fuel to support combustion. While we could reduce the production of higher HCs by increasing the air-fuel ratio to about 16:1, we have to take into consideration the other pollutants, such as NO_x. This is why we strive to maintain the air-fuel ratio to 14.7:1, whenever possible.

You cannot determine if the fuel mixture is excessively rich or lean on a vehicle by just observing HCs. As can be seen in figure 12-3, high HCs can be the result of too lean of a mixture as well as too rich of a mixture.

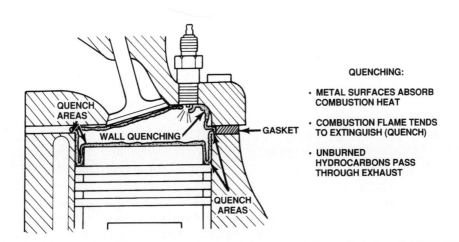

Figure 12-3. Quenching effect. (Courtesy of DaimlerChrysler Corporation)

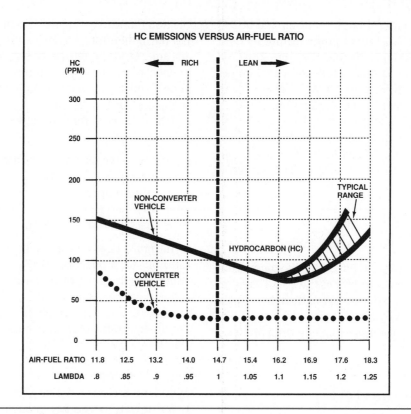

Figure 12-4. HC emissions versus air-fuel ratio. (Courtesy of DaimlerChrysler Corporation)

Carbon Monoxide (CO)

In a perfect combustion process, one atom of carbon would combine with two atoms of oxygen to form carbon dioxide (CO_2). However, whenever there is a lack of oxygen in the combustion process, some carbon atoms only combine with one atom of oxygen, resulting in the production of carbon monoxide (CO). CO is a colorless, odorless gas that is a very poisonous gas to humans and animals. When CO is breathed into the lungs, it enters the bloodstream and absorbs oxygen from the red blood cells. This can lead to the loss of consciousness and even death if too much CO is absorbed into the bloodstream.

High CO production in an engine is always the result of a rich air-fuel mixture. As can be seen in figure 12-5, CO is very high at air-fuel ratios of less than 14.7:1. As the air-fuel ratio goes leaner, the CO is reduced to almost zero. Past about 15.5:1 air-fuel ratio, the CO is as low

as it will ever get. Like HC, it is not at the lowest point at 14.7:1 air-fuel ratio; however, it is very low in comparison to richer air-fuel ratios. Thus, CO can be considered the rich indicator for the engine. Anytime it is higher than normal, it indicates that the engine is not getting enough air.

Oxides of Nitrogen (NO_x)

Oxides of nitrogen are pollutants that are formed by combining various amounts of oxygen with nitrogen. The nitrogen comprises about 78 percent of the air drawn into the engine. As mentioned previously, nitrogen is an inert gas that does not easily combine with other elements. However, under high temperature conditions above 2500°F (1371°C) nitrogen can combine with molecules of oxygen, resulting in the formation of nitric oxide (NO) or nitrogen dioxide (NO_2). This is why

---

WARNING: Always use exhaust extraction equipment when running an engine in the shop and ensure that the area is well ventilated. Even short periods of exposure to high CO levels can cause health concerns.

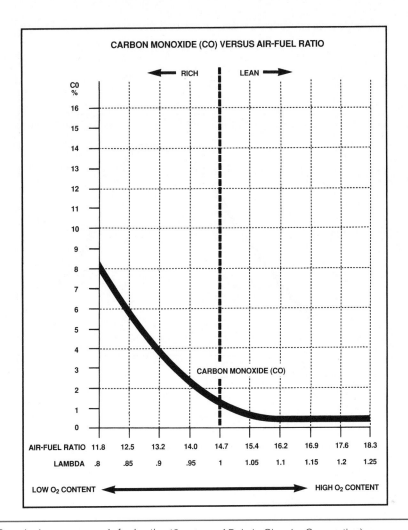

Figure 12-5. CO emissions versus air-fuel ratio. (Courtesy of DaimlerChrysler Corporation)

the term NO_x is used. The letter X represents an unknown quantity of oxygen molecules.

Smog, the brownish haze that can be seen above urban areas, is a combination of NO_x, HC, and ultraviolet radiation from the sun. This combination of pollutants causes eye irritation and respiratory problems.

As can be seen in figure 12-6, NO_x formation is low at very rich air-fuel ratios and increases as the fuel mixture gets leaner. NO_x production is at its highest in the range of 15 to 16:1 air-fuel ratio. Unfortunately, NO_x is at its highest within the air-fuel ratios where HC and CO are usually at their lower levels. This presents a problem when trying to control all three of these gases to their lowest level. If the mixture is rich, the NO_x decreases; however, HC and CO increases. If the mixture is leaner, the NO_x increases and HC may also increase.

Primarily, the formation of NO_xs is controlled by controlling the combustion temperature in the cylinder. Compression ratios have been lowered, timing is controlled more accurately with the engine control module and the exhaust gas recirculation (EGR) valve is used. If combustion temperatures can be kept under 2500°F (1371°C), NO_x formation is reduced. Even by controlling combustion temperatures, some NO_x formation still takes place during certain operating conditions. The other method used to control NO_xs is through the use of a three-way catalytic converter. The three-way converter has a reduction bed that separates the oxygen from the nitrogen. This type of converter has been used on most vehicles since 1981.

Carbon Dioxide (CO_2)

Carbon dioxide is not considered a pollutant; however, it is a gas that we measure when looking at exhaust emissions. The reason for this is because CO_2

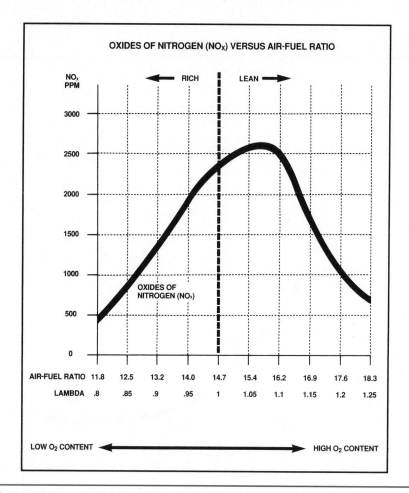

Figure 12-6. NO$_x$ emissions versus air-fuel ratio. (Courtesy of DaimlerChrysler Corporation)

is considered to be an efficiency gas. CO_2 is formed when oxygen in the combustion process combines with the carbon that is a component of the fuel. It also is produced in the catalytic converter when CO is oxidized by the converter. While not a pollutant, it is considered a greenhouse gas that contributes to global warming. Burning any fossil fuel contributes to the greenhouse effect. There are attempts around the word to limit the amount of greenhouse gases being produced. However, at this time there are no limits to the production of CO_2 from the internal combustion engine.

As can be seen in figure 12-7, CO_2 production is at its highest, 14 to 15 percent when the air-fuel ratio is at 14.7:1. This makes CO_2 an ideal indicator of the efficiency at which the engine is running. The higher the CO_2 readings, the better when it comes to determining the efficiency of the combustion process. CO_2 levels decrease whenever the air-fuel ratio is richer or leaner than the stoichiometric point of 14.7:1. Another factor

about CO_2 is that if we are sampling the exhaust gas after the converter, it and O_2 are the only gases that accurately reflect what is taking place in the engine. That is because HCs and COs are reduced by the action of the catalytic converter. Therefore, while they may be high leaving the engine, they may not show high indicators exiting the tailpipe. CO_2, on the other hand, indicates to us the efficiency of the combustion process.

Oxygen (O_2)

Oxygen is another gas that is used to indicate the efficiency of the combustion process. As can be seen in figure 12-8, whenever the air-fuel ratio is on the rich side, oxygen levels are very low. At about 14:1 air-fuel ratio, the oxygen starts to rise and continues to rise as the air-fuel ratio increases. At 14.7:1, the oxygen content should be about 1 to 1.5 percent. Oxygen levels above that

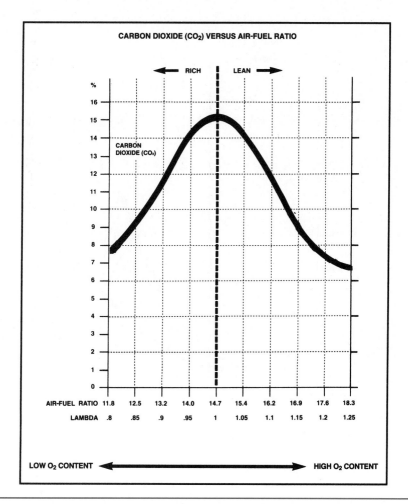

Figure 12-7. CO_2 versus air-fuel ratio. (Courtesy of DaimlerChrysler Corporation)

usually indicate that the exhaust mixture is getting too lean. Oxygen and CO should respond exactly opposite of each other. That is, when CO is high, O_2 will be low, and when CO is low, O_2 is higher.

Air-Fuel Ratio

As you can see by studying the previous air-fuel ratios and their effect on the various gases, maintaining the stoichiometric air-fuel ratio of 14.7:1 is very critical on a gasoline engine. This is the point where all three of the bad gases CO, HC, and NO_x are best controlled. While going richer will reduce the levels of NO_x production, the HCs and COs will increase. If we try to make the mixture leaner, HCs and COs will decrease; however, the NO_xs will increase.

As learned in the previous chapters on engine management, the primary purpose of the engine

management system is to control the air-fuel ratio to as close to 14.7:1 whenever it is possible. To achieve this precise control requires engine computers and fuel delivery systems that are very accurate. Carburetors were not accurate enough to achieve this level of fuel metering control under different operating conditions. This is why carburetors are no longer used.

Even with more precise fuel control, we cannot achieve the low emissions levels required on today's vehicles without the use of a catalytic converter. Since perfect combustion is impossible, the converter is used to reduce the pollutants to acceptable levels. However, for the converter to work properly we need accurate fuel control to start with. When all these systems are working correctly and are well maintained, the emissions levels are very low on a modern automobile. When systems fail or maintenance is ignored, the clean running automobile can quickly become

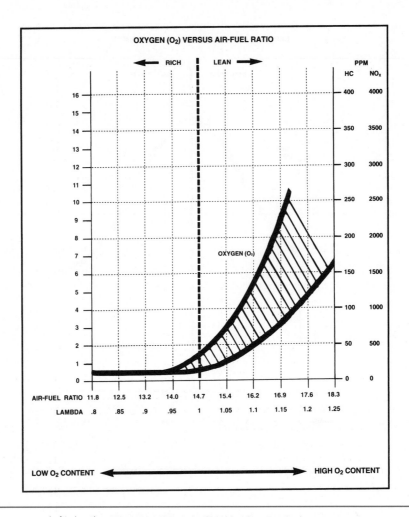

Figure 12-8. O_2 versus air-fuel ratio. (Courtesy of DaimlerChrysler Corporation)

one of the gross polluters. This is why there is a need for inspection/maintenance (I/M) programs to ensure that vehicles are properly maintained and meeting the emission standards as designed.

INSPECTION/ MAINTENANCE PROGRAMS (I/M)

ALIGN

Since approximately 10 to 30 percent of vehicles on the road today contribute 80 percent of the pollution from mobile sources, it is important to have I/M programs to identify these vehicles. Many states had I/M programs in place before the 1990 Clean Air Act Amendments. These programs were not particularly effective. In many states these were basic tailpipe tests that only identified the grossest polluters. In addition, many pro-

grams had such low waiver limits that there was no incentive to properly repair the vehicles that failed these tests. The Clean Air Act amendments require enhanced I/M programs in areas of the country with the worst air quality and the Northeast Ozone Transport region. The states must submit to the EPA a state implementation plan (SIP) for their programs. Each enhanced I/M program is required to include as a minimum the following items:

- Computerized emission analyzers
- Visual inspection of emission control items
- $450 minimum waiver limit (to be increased based on the inflation index)
- Remote on-road testing of one half of 1 percent of the vehicle population
- Registration denial for vehicles not passing an I/M test
- Denial of waiver for vehicles that are under warranty or that have been tampered with

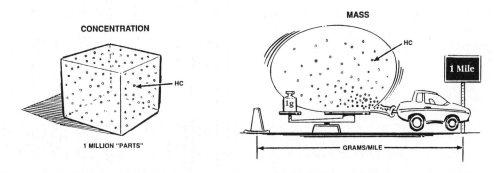

Figure 12-9. Concentration versus mass emission testing. (Courtesy of General Motors Corporation)

- Annual inspections
- OBD II systems check for 1996 and newer vehicles

Federal Test Procedure (FTP)

Before we look at the different type of I/M programs and test methods, we need to understand what the federal test procedure is. The FTP is the test used to certify all new vehicles before they can be sold. All preproduction vehicles must be tested using the FTP in order to certify their ability to meet and maintain emissions standards for the useful life of the vehicle. Once a vehicle meets these standards, it is certified by the EPA for sale.

The FTP test procedure is a loaded-mode test lasting for a total duration of 505 seconds and is designed to simulate an urban driving trip. A cold start-up representing a morning start and a hot start after a soak period is part of the test. This FTP driving cycle was designed to represent a typical drive commute in an urban setting. In addition to this drive cycle, a vehicle must undergo evaporative testing. Evaporative emissions are determined using the sealed housing for evaporative determination (SHED). The test measures the evaporative emissions from the vehicle after a heat-up period, representing a vehicle sitting in the sun. In addition, the vehicle is driven and then tested during the hot soak period.

It is important for you to understand the way that emissions are measured during the FTP. The vehicle is placed on a chassis dynamometer, which places a fixed load on the vehicle's wheels based on the weight of the vehicle. The vehicle is driven at different speeds and accelerates and decelerates during the 505-second drive cycle. The exhaust is sampled using a constant volume sampling method that collects and measures exhaust emissions of the vehicle during the drive cycle. This method samples a percentage of the volume of the exhaust stream. The end measurement of this procedure represents the mass of emissions of the various gases. This mass is measured in grams per mile (g/mi).

You need to understand the difference between measuring the mass of the emissions from the vehicle and measuring the percentage of the emissions in the exhaust stream, figure 12-9. Stand-alone and portable emission analyzers measure a small percentage of the exhaust emissions and display the result in percentage of CO, CO_2, and O_2. HCs are measured in parts per million (ppm). The constant volume sampling method, on the other hand, measures the weight of the emissions in the exhaust stream. This comparison is important to understand. If you compared a large vehicle equipped with a V-8 cylinder engine (whose CO emissions measured 1 percent CO) and a small 4-cylinder engine (whose CO emissions measured 1 percent), you would think that they are both equal in their emission levels. This is because we are measuring the CO in percentages. However, the V-8 engine would emit almost twice as many emissions as the 4-cylinder engine would. If we measured these emissions using the constant volume sampling method, the V-8 engine would produce more emissions than the 4-cylinder engine. It is important for you to realize that there is no direct comparison between the measurements obtained using constant volume sampling equipment and the measurements obtained when using stand-alone measuring equipment. The FTP is a much more stringent test of a vehicle's emissions than is any test type using equipment that measures percentages of exhaust gases. The federal emission standards for each model year vehicle are the same for that model regardless of what size engine the vehicle is equipped with. This is why larger V-8 engines often are equipped with more emission control devices than smaller 4- and 6-cylinder engines.

Types of I/M Test Programs

A variety of I/M testing programs have been implemented by the various states. These programs may be centralized testing programs or decentralized testing programs. Each state is free to develop a testing program suitable to their needs as long as they can demonstrate to the EPA that their plan will achieve the attainment levels set by the EPA. This approach has led to a variety of different testing programs. These test types may include any of the following:

- Visual tampering checks
- One-speed idle test
- Two-speed idle test
- Loaded mode test
- Acceleration simulation mode (ASM)
- I/M 240 tests
- Remote sensing
- Random roadside visual inspection
- OBD II testing

Visual Tampering Checks

Visual tampering checks may be part of an I/M testing program such as one-speed, two-speed tailpipe tests, ASM tests, or other testing programs. In addition, a visual tampering check may be implemented in some areas in place of any tailpipe testing programs. The visual inspection is designed to determine if the required emission control devices are present on a vehicle and if they appear to be connected properly and the right type for the vehicle. Visual tests do not determine if the components are actually working, simply that they are present and haven't been tapered with. The visual checks usually include the following items:

- Catalytic converter
- Fuel inlet restrictor
- Exhaust gas recirculation (EGR)
- Evaporative emission system
- Air injection reaction system (AIR)
- Positive crankcase ventilation (PCV)

If any of these systems are missing, not connected, or tampered with, the vehicle will fail the emissions test and will have to be repaired/replaced by the vehicle owner before the vehicle can pass the emissions test. Any cost associated with repairing or replacing these components may not be used toward the waiver amount required for the vehicle to receive a waiver.

One-Speed and Two-Speed Idle Test

The one-speed and two-speed idle test measures the exhaust emissions from the tailpipe of the vehicle at idle and/or at 2500 rpm. This uses stand-alone exhaust gas sampling equipment that measures the emissions in percentages. Each state chooses the standards the vehicle has to meet in order to pass the test. These standards vary by model year and the type of vehicle being tested such as automobiles and light-duty trucks up to 6000 pounds, or vehicles heavier than 6000 pounds up to 9000 pounds. Typically, for a newer vehicle from 1994 and up the standards may be 1 percent CO and 100 to 220 ppm HC. This type of test will measure HC, CO, CO_2, and O_2. Included with this test may also be the visual anti-tampering check and a gas cap pressure test. This test will detect vehicles with badly worn engines or fuel delivery systems that are badly out of calibration. Unfortunately, this test cannot test for NO_x emissions due to the fact that this is a no-load test. NO_x emissions are very low at idle and at 2500 rpm when there is no load placed on the engine. This test also cannot detect excessive HC or CO emissions that may be emitted by a vehicle when it is under heavy load such as when accelerating or going up a steep grade.

The advantage to using this type of testing is that the equipment is relatively cheap, around $15,000 to $20,000. This allows states to have a decentralized testing program because many facilities can afford the necessary equipment required to perform this test. Many states prefer decentralized testing programs due to political concerns and customer convenience.

Loaded Mode Test

The loaded mode test uses a dynamometer that places a "single weight" load on the vehicle, figure 12-10. The load applied to the vehicle varies with the speed of the vehicle. Typically, a 4-cylinder vehicle's speed would be 24 mph, a 6-

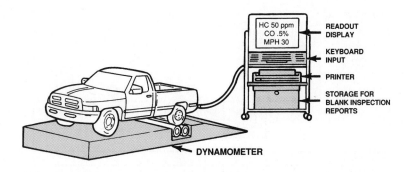

HC 50 ppm
CO .5%
MPH 30

READOUT
DISPLAY

KEYBOARD
INPUT

PRINTER

STORAGE FOR
BLANK INSPECTION
REPORTS

DYNAMOMETER

Figure 12-10. Equipment used in a loaded mode I/M test. (Courtesy of DaimlerChrysler Corporation)

cylinder vehicle's speed would be 30 mph, and a 8-cylinder vehicle's speed would be 34 mph. Conventional stand-alone sampling equipment is used to measure HC and CO emissions. This type of test is classified as a basic I/M test by the EPA.

Acceleration Simulation Mode (ASM)

The **ASM** type of test uses a dynamometer that applies a heavy load on the vehicle at a steady-state speed. The load applied to the vehicle is based on the acceleration rate on the second simulated hill of the FTP. This acceleration rate is 3.3 mph/sec. There are different ASM tests used by different states.

The ASM 50/15 test places a load of 50 percent on the vehicle at a steady 15 mph. This load represents 50 percent of the horsepower required to simulate the FTP acceleration rate of 3.3 mph/sec. This type of test produces relatively high levels of NO_x emissions; therefore, it is useful in detecting vehicles that are emitting excessive NO_x.

The ASM 25/25 test places a 25 percent load on the vehicle while it is driven at a steady 25 mph. This represents 25 percent of the load required to simulate the FTP acceleration rate of 3.3 mph/sec. Because this applies a smaller load on the vehicle at a higher speed, it will produce a higher level of HC and CO emissions than the ASM 50/15. NO_x emissions will tend to be lower with this type of test.

While ASM tests are effective in testing vehicles under load and a simulated driving condition, it is not as effective as the FTP test. This is because of the type of sampling equipment being used. Stand-alone percentage type of sampling equipment is used so the measurements are in percent CO and ppm HC. Unfortunately, the results achieved by this type of test cannot be correlated

to the FTP. The advantage of this type of test is that the equipment is much cheaper, about $50,000, as compared to the constant volume sampling equipment required by the FTP or I/M 240 test. This means that a state may have a decentralized program utilizing this type of equipment.

I/M 240 Test

The **I/M 240** test is the EPA's enhanced test. It is actually a portion of the 505 second FTP test used by the manufacturers to certify their new vehicles. The "240" stands for 240 seconds of drive time on a dynamometer. This is a loaded-mode transient test that uses constant volume sampling equipment to measure the exhaust emissions in mass just as is done during the FTP, figure 12-11. The I/M 240 test simulates the first two hills of the FTP drive cycle. Figure 12-12 shows the I/M 240 drive trace.

The I/M 240 test is a good representative test that closely duplicates the FTP. It measures the mass of emissions from the vehicle and calculates them in grams per mile. This test also measure NO_x emissions from the vehicle because the test produces enough load on the vehicle to create NO_xs. Some states have adopted the I/M 240 test for their enhanced I/M programs. The drawback to this type of testing is that the equipment is very expensive, approximately $170,000 per test lane. This requires that states that adopt this test method have centralized test facilities to perform the I/M testing. The other drawback to this type of test is repair validation. If a vehicle fails this test and needs to be repaired, it is difficult for a repair facility to validate their repair without sending the vehicle back to the test facility to be retested. This is due to the fact that repair facilities are using equipment that measures the exhaust in concentrations as mentioned earlier.

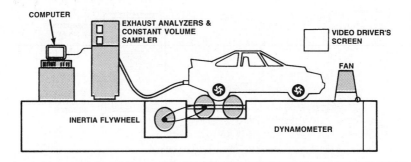

Figure 12-11. Test lane dynamometer. (Courtesy of General Motors Corporation)

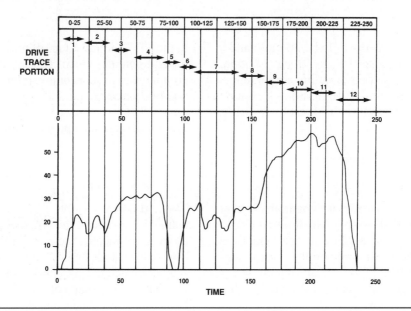

Figure 12-12. I/M 240 driving trace. (Courtesy of DaimlerChrysler Corporation)

Since there is no direct correlation between concentration readings and mass readings, it is difficult for a repair facility to determine positively that the vehicle has been repaired properly and will pass a retest at the I/M 240 test facility. This can cause some concern to customers and repair facilities in areas that have adopted the enhanced I/M 240 test.

OBD II Testing

In 1999 the EPA requested that states adopt OBD II systems testing for 1996 and newer vehicles. The OBD II system is designed to illuminate the MIL light and store trouble codes any time a malfunction exists that would cause the vehicle's emissions to exceed one and a half times the FTP limits. If the OBD II system is working correctly, the system should be able to detect a vehicle failure that would cause emissions to increase to an unacceptable

level. The EPA has determined that the OBD II system should detect emission failures of a vehicle even before that vehicle would fail an emissions test of the type that most states are employing. Furthermore, the EPA has determined that as the population of OBD II-equipped vehicles increases and the population of older non-OBD II-equipped vehicles decreases, tailpipe testing will no longer be necessary.

The OBD II testing program consists of a computer, figure 12-13, that can scan the vehicle's OBD II system using the DLC connector. The technician first performs a visual check of the vehicle's MIL light to determine if it is working correctly. Next, the computer is connected to the vehicle's DLC connector. The computer will scan the vehicle's OBD II system and determine if there are any codes stored that are commanding the MIL light on. In addition, it will scan the status of the readiness monitors and determine if they have all run and passed. If the readiness

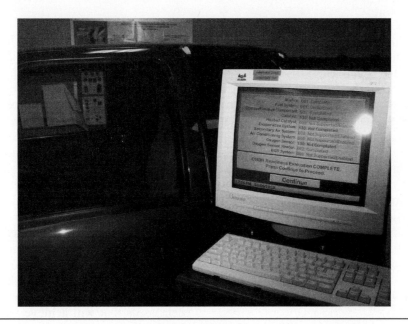

Figure 12-13. A computer is connected to the vehicle during the OBD II systems check.

monitors have all run and passed, it indicates that the OBD II system has tested all the components of the emissions control system. This indicates all of the vehicle's emissions components and systems are working correctly and the vehicle is meeting emissions standards as designed. An OBD II vehicle would fail this OBD II test if:

- The MIL light does not come on with the key on, engine off.
- The MIL is commanded on.
- A number (varies by state) of the readiness monitors have not been run.

If none of the above conditions are present, the vehicle will pass the emissions test. This type of test ensures that the vehicle's OBD II system is working correctly and that any needed maintenance is performed on the system before the vehicle can be registered.

Remote Sensing

The EPA requires that in high-enhanced areas states perform on-the-road testing of vehicles' emissions. The state must sample .5 percent of the vehicle population base in high-enhanced areas. This may be accomplished by using remote sensing devices, figure 12-14. This type of sensing may be equipment that projects an infrared light through the exhaust stream of a passing vehicle. The reflected beam can then be analyzed to deter-

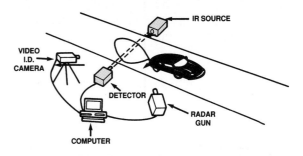

Figure 12-14. Remote sensing. (Courtesy of Daimler-Chrysler Corporation)

mine the pollutant levels coming from the vehicle. If a vehicle fails this type of test, the vehicle owner will receive notification in the mail that they have to take their vehicle to a test facility to have the vehicle's emissions tested.

Random Roadside Testing

Some states may implement random roadside testing that would usually involve visual checks of the vehicle's emission control devices to detect tampering. Obviously this method is not very popular as it can lead to traffic tie-ups and delays on the part of commuters.

As can be seen, states have much latitude in developing their I/M testing programs. Each state with high-enhanced designated areas has adopted a testing program that they feel best suits their needs. The type of testing program that is chosen is based on a variety of factors such as cost to the consumer, convenience, the number of high-enhanced areas, and the political climate. Test methods vary and the effectiveness of the program varies also. However, if the program utilized by a state does not show reductions in emissions, then the EPA can force the states to implement tougher programs or even restrictions to traffic volume in non-attainment areas. It is generally in the state's best interest to implement a program that will achieve the goal of reducing emissions in non-attainment areas to acceptable levels. The goal of any program should be to detect vehicles that are gross polluters and require these vehicles to be repaired or removed from the vehicle fleet.

SUMMARY

The harmful emissions from the internal combustion engines are hydrocarbons (HC), carbon monoxide (CO), and nitrogen oxides (NO_x). The other two gases that are measured are oxygen (O_2) and carbon dioxide (CO_2). Analyzing these five gases during the emission measuring process is referred to as the Five Gas Theory. CO is a rich indicator and indicates if the exhaust stream is too rich. High HCs can be caused by a mixture that is either too rich or too lean. NO_x emissions are always formed under high temperature conditions that exceed 2500°F (1371°C). To minimize the creation of all three of these gases, an engine needs to maintain the stoichiometric air-fuel ratio of 14.7:1 whenever possible.

I/M testing programs are needed in non-attainment areas to detect vehicles that are gross polluters. These programs employ a variety of test methods and may be centralized or decentralized programs. Emission test programs may be one- or two-speed tailpipe tests, IM 240 testing, ASM testing, visual anti-tampering checks, or OBD II testing. Each state designs a test method that best suits their needs that will achieve reductions in emission levels in non-attainment areas. As the population of older non-OBD II-equipped vehicles declines and the population of OBD II-equipped vehicles increases, states may discontinue tailpipe testing and incorporate OBD II testing.

Review Questions

Choose the letter that represents the best possible answer to the following questions:

1. Which of the following gases are considered pollutants?
   a. HC, CO_2, & NO_x
   b. CO, CO_2, & NO_x
   c. HC, CO, & NO_x
   d. O_2, NO_x, & CO_2

2. Which of the following gases could be used as a rich indicator?
   a. HC
   b. CO_2
   c. NO_x
   d. CO

3. Technician A says that if the vehicle is running too lean, the NO_x will increase. Technician B says that if the vehicle is running too lean, the HC may increase. Who is correct?
   a. Technician A
   b. Technician B
   c. Both A and B
   d. Neither A nor B

4. CO_2 is an indicator of the combustion:
   a. Leanness
   b. Richness
   c. Efficiency
   d. Process

5. High NO_xs are caused by excessive:
   a. Fuel
   b. Air
   c. Heat
   d. Cooling

6. The FTP is the test used by most states in their I/M programs.
   a. True
   b. False

7. Technician A says that the I/M 240 test is patterned after the FTP.
   Technician B says that the I/M 240 test uses stand-alone percentage concentration measuring equipment to record the emissions. Who is correct?
   a. Technician A
   b. Technician B
   c. Both A and B
   d. Neither A nor B

8. Visual tampering check usually involves which of the following items?
   a. Fuel inlet restrictor
   b. EGR
   c. Evaporative system
   d. All of the above

9. Technician A says that the OBD II testing program may eventually replace tailpipe testing.
   Technician B says that an OBD II system should be able to detect failures that would result in the vehicle's emissions exceeding 1.5 times the FTP. Who is correct?
   a. Technician A
   b. Technician B
   c. Both A and B
   d. Neither A nor B

10. Technician A says that the ASM test is a loaded-mode transient type of test and measures the pollutants in grams per mile. Technician B says that the ASM test is a more stringent test than the IM 240 type of test. Who is correct?
    a. Technician A
    b. Technician B
    c. Both A and B
    d. Neither A nor B

13

Positive
Crankcase
Ventilation,
Air-Injection
Systems,
Catalytic
Converters,
and EGR
Systems

OBJECTIVES

Upon completion and review of this chapter, you will be able to:

- Describe the purpose of the positive crankcase ventilation system.
- Identify PCV system components.
- Explain the purpose of the air-injection reaction system.
- Identify the components of a typical air-injection reaction system.
- Explain the operation of a catalytic converter.
- Explain the difference between a two-way and three-way catalytic converter.
- Describe the purpose of the EGR system.
- Have knowledge of the different types of vacuum-actuated EGR valves.
- Describe how an engine management system controls EGR solenoids.
- Describe how an OBD II system monitors the EGR system.

KEY TERMS

AIR	gulp valve
air injection	high-speed surge
backfire	inert
catalyst	light-off
catalytic converter	noble metals
diverter valve	OSC
(DPFE) sensor	oxidation
EGR valve position	PCV
sensor	pintle valve
exhaust gas	reduction
recirculation	substrate
(EGR)	TWC

INTRODUCTION

The process of combustion produces power in an internal-combustion engine. Under perfect conditions, combustion would completely consume the air-fuel mixture, leaving only harmless by-products, such as water vapor (H_2O) and carbon dioxide (CO_2). Combustion of the air-fuel mixture is never perfect, however, and at best is incomplete. By-products other than H_2O and CO_2 remain after the combustion process to go out the tailpipe. These by-products of incomplete combustion include carbon monoxide (CO), hydrocarbons (HC), and oxides of nitrogen (NO_x). This chapter covers the three most common automotive systems used to control combustion by-products, increase engine efficiency, and reduce emissions.

CRANKCASE VENTILATION

The problem of crankcase ventilation has existed since the beginning of the automobile. No piston ring, new or old, can provide a perfect seal between the piston and the cylinder wall. When an engine is running, the pressure of combustion forces the piston downward. This same pressure also forces gases and unburned fuel from the combustion chamber, past the piston rings, and into the crankcase. This process of gases leaking past the rings is called blowby, and the gases form crankcase vapors.

These combustion by-products, particularly unburned hydrocarbons, caused by blowby, figure 13-1, must be ventilated from the crankcase. However, the crankcase on modern engines cannot be vented directly to the atmosphere, because the hydrocarbon vapors add to air pollution. Positive crankcase ventilation (**PCV**) systems were developed to ventilate the crankcase and recirculate the vapors to the engine's induction system so they can be burned in the cylinders.

■ Draft Tube Ventilation

Before the 1960s, most vehicles used a road draft tube, figure 13-2, to ventilate the engine crankcase. This was nothing more than a tube connected to the engine crankcase that allowed vapors to pass into the air. With draft tube crankcase ventilation, fresh air enters through a vented oil filler cap and passes into the crankcase to mix with the vapors. Airflow past the road draft tube creates a vacuum that draws the crankcase vapors out into the atmosphere. Road draft ventilation has three drawbacks:

* At vehicle speeds below 25 mph, not enough vacuum is created to remove the vapors from the crankcase.
* At high vehicle speeds, high crankcase pressure may combine with low pressure at the end of the tube to draw oil from the engine.
* It releases the unburned hydrocarbons directly into the atmosphere, causing air pollution.

Closed PCV Systems

All new cars sold in the United States since 1968 have a Type 4 closed PCV system. The design of

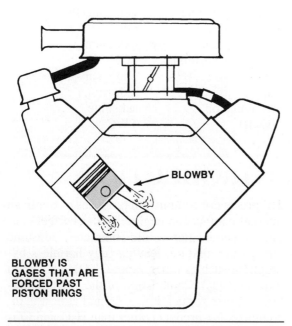

Figure 13-1. Piston rings do not provide a perfect seal. Combustion gases blow by the rings into the crankcase.

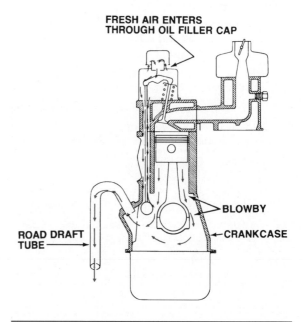

Figure 13-2. The road draft tube ventilates the crankcase to the atmosphere when the car is moving.

closed PCV systems is essentially the same, regardless of manufacturer. All use a PCV valve, calibrated orifice or separator, an air inlet filter, and connecting hoses. An oil-vapor or oil-water separator is used in some systems instead of a valve or orifice, particularly with turbocharged and fuel-injected engines. The oil-vapor separator lets oil condense and drain back into the crankcase. The oil-water separator accumulates moisture and prevents it from freezing during cold engine starts.

Type 4 System

Unlike open PCV systems, the Type 4 system uses a sealed oil filler cap (not vented to the atmosphere). Type 4 systems were required on all new California cars in 1964, all nationwide cars in 1968, and are still used on all new cars sold in the United States. System operation is shown in figures 13-3 and 13-4.

In this system, crankcase ventilation air may come from either the clean side (inside) or the dirty side (outside) of the carburetor air filter. When air is drawn from the clean side, the air cleaner filter acts as a PCV filter and a wire screen flame arrester is installed in the PCV air intake line to prevent a crankcase explosion if the engine backfires. When air is drawn from the dirty side of the air cleaner, a separate crankcase ventilation filter is used. Although generally a polyurethane foam type installed in the air cleaner, figure 13-5, the filter may also take the form of a wire gauze or mesh filter located in either the oil filler cap or the inlet air hose connection to the valve cover.

PCV Valves

The PCV valve in most systems is a one-way valve containing a spring-operated plunger, figure 13-6, that controls valve flow rate. Flow rate is established for each engine and a valve for a different engine should not be substituted. The flow rate is determined by the size of the plunger and the holes inside the valve. PCV valves usually are located in the valve cover or intake manifold.

The PCV valve regulates airflow through the crankcase under all driving conditions and speeds. When manifold vacuum is high (at idle, cruising, and light-load operation), the PCV

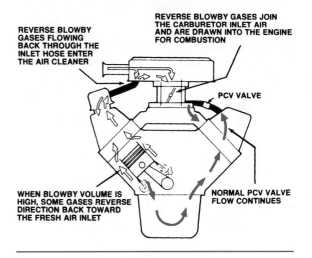

Figure 13-4. Closed PCV system operation under heavy load.

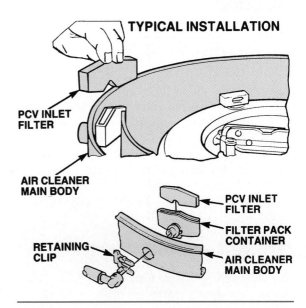

Figure 13-5. The PCV inlet filter on many vehicles is located in the air cleaner housing and is serviced as shown.

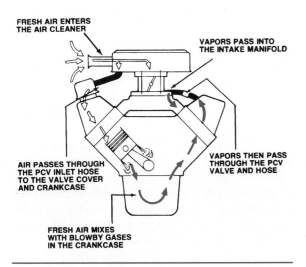

Figure 13-3. Closed PCV system operation under normal conditions.

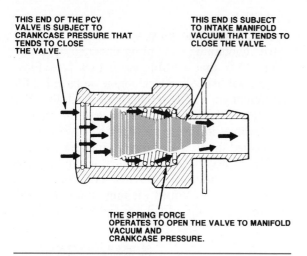

THIS END OF THE PCV VALVE IS SUBJECT TO CRANKCASE PRESSURE THAT TENDS TO CLOSE THE VALVE.

THIS END IS SUBJECT TO INTAKE MANIFOLD VACUUM THAT TENDS TO CLOSE THE VALVE.

THE SPRING FORCE OPERATES TO OPEN THE VALVE TO MANIFOLD VACUUM AND CRANKCASE PRESSURE.

Figure 13-6. Spring force, crankcase pressure, and manifold vacuum work together in regulating PCV valve flow rate.

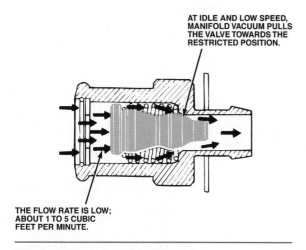

AT IDLE AND LOW SPEED, MANIFOLD VACUUM PULLS THE VALVE TOWARDS THE RESTRICTED POSITION.

THE FLOW RATE IS LOW; ABOUT 1 TO 5 CUBIC FEET PER MINUTE.

Figure 13-7. PCV valve airflow during idle, cruising, and light-load operation.

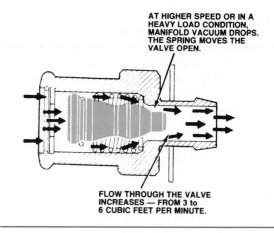

AT HIGHER SPEED OR IN A HEAVY LOAD CONDITION, MANIFOLD VACUUM DROPS. THE SPRING MOVES THE VALVE OPEN.

FLOW THROUGH THE VALVE INCREASES — FROM 3 to 6 CUBIC FEET PER MINUTE.

Figure 13-8. PCV valve airflow during acceleration and heavy-load operation.

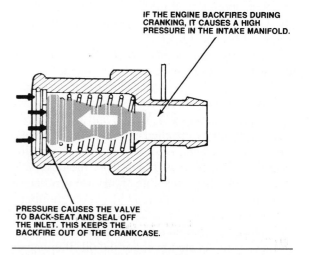

IF THE ENGINE BACKFIRES DURING CRANKING, IT CAUSES A HIGH PRESSURE IN THE INTAKE MANIFOLD.

PRESSURE CAUSES THE VALVE TO BACK-SEAT AND SEAL OFF THE INLET. THIS KEEPS THE BACKFIRE OUT OF THE CRANKCASE.

Figure 13-9. PCV valve operation in case of a backfire.

valve restricts the airflow, figure 13-7, to maintain a balanced air-fuel ratio. It also prevents high intake manifold vacuum from pulling oil out of the crankcase and into the intake manifold. Under high speed or heavy loads, the valve opens and allows maximum airflow, figure 13-8. If the engine backfires, the valve will close instantly, figure 13-9, to prevent a crankcase explosion.

Orifice-Controlled Systems

The closed PCV system used on some 4-cylinder engines contains a calibrated orifice instead of a PCV valve. The orifice may be located in the valve cover or intake manifold, or in a hose connected between the valve cover, air cleaner, and intake manifold.

While most orifice flow control systems work the same as a PCV valve system, they may not use fresh air scavenging of the crankcase. The carbureted 1981–85 Ford 1.6/1.9-liter Escort engine is a good example of this design, figure 13-10. Crankcase vapors are drawn into the intake manifold in calibrated amounts depending on manifold pressure and the orifice size. If vapor availability is low, as during idle, air is drawn in with the vapors. During off-idle operation, excess vapors are sent to the air cleaner.

A dual orifice valve was used on carbureted 1986 and later Ford 1.9-liter Escort engines to increase PCV flow during off-idle engine operation. At idle, PCV flow is controlled by a 0.050-inch (1.3-mm) orifice. As the engine moves off-idle, spark port vacuum pulls a spring-loaded valve off of its seat, allowing PCV flow to pass through a 0.090-inch (2.3-mm) orifice.

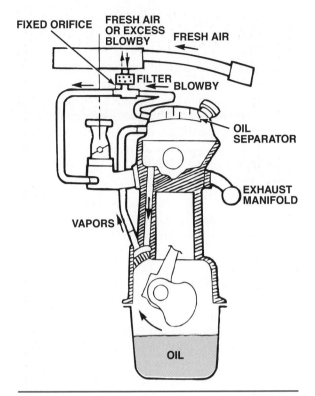

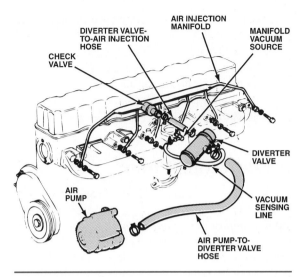

Figure 13-11. The basic components of an air-injection system.

Figure 13-10. The orifice flow control system used on 1981–85 Ford 1.6-liter and 1.9-liter Escort engines does not use fresh air scavenging of the crankcase. (Courtesy of Ford Motor Company)

Separator Systems

Turbocharged and many fuel-injected engines use an oil-vapor or oil-water separator and a calibrated orifice instead of a PCV valve. In the most common applications, the air intake throttle body acts as the source for crankcase ventilation vacuum and a calibrated orifice acts as the metering device.

PCV System Efficiency

When intake air flows freely, the PCV system functions properly, as long as the PCV valve or orifice is not clogged. Modern engine design includes the air and vapor flow as a calibrated part of the air-fuel mixture. In fact, some engines receive as much as 30 percent of their idle air through the PCV system. For this reason, a flow problem in the PCV system results in driveability problems.

A PCV system that is not properly vented, or scavenged, will cause oil dilution, sludge formation throughout the engine, and oil deposits in the air cleaner. Unlike driveability problems, these may not be noticed immediately, but their long-term effects are premature wear on moving engine parts and inefficient intake and idle circuit operation.

AIR INJECTION

Air injection was one of the first add-on devices used to help oxidize HC and CO exhaust emissions. An air-injection system, figure 13-11, sometimes called secondary air, or air-injection reactor (AIR), actually prolongs the combustion process in the exhaust. Combustion is an oxidation reaction, but usually an incomplete one because of an inherent lack of sufficient oxygen. Exhaust gas is hot enough to support continued burning as it leaves the cylinder, so additional air is supplied to the exhaust system as soon as the hot exhaust gases leave the cylinder. The air-injection system continues the oxidation (burning) of any HC and CO remaining in the exhaust manifold and the converter. As a result, the HC and CO combine with O_2 to form H_2O vapor and CO_2.

The development of sophisticated electronic fuel management and engine control systems has made it possible for many of the new and smaller engines equipped with electronic fuel injection to meet emission standards without air injection. With the increasing precision of electronic controls, air injection is sometimes limited to the first minutes of engine operation before it goes into closed loop.

Regardless of the various names that have been used by vehicle manufacturers for their AIR systems, all are relatively simple in design, function in the same way, and use the same basic components, figure 13-11.

- A belt-driven pump with inlet air filter (older models)
- An electric air pump (newer models)

- One or more air distribution manifolds and nozzles
- An anti-backfire valve
- One or more exhaust check valves
- Connecting hoses for air distribution
- Air management valves and solenoids on all newer applications

Air-Injection Pump

On vehicles manufactured up through the mid-1990s, the AIR pump, sometimes referred to as a *smog pump* or *thermactor pump,* figure 13-12, is mounted at the front of the engine and driven by a belt from the crankshaft pulley. It pulls fresh air in through an external filter and pumps the air under slight pressure to each exhaust port through connecting hoses or a manifold. Pumps used before 1968 contain three vanes and use a separate filter installed in the air inlet hose. Pumps used since 1968 contain two vanes and have an impeller-type, centrifugal air filter fan mounted on the pump rotor shaft. This is not a true filter, but cleans the air entering the pump by centrifugal force, figure 13-13. The relatively heavy dust particles in the air are forced in the opposite direction to the inlet air flow. The lighter air is then drawn into the pump by the impeller-type fan. Most pumps use a pressure relief valve, which opens at high engine speed. Pumps without a pressure relief valve use a diverter valve, explained later in this section.

Air Distribution Manifolds and Nozzles

Before the appearance of catalytic converters in 1975, the air-injection system sent air from the pump to a nozzle installed near each exhaust port in the cylinder head. This provided equal air injection for the exhaust from each cylinder and made it available at a point in the system where exhaust gases were the hottest.

Air is delivered to the engine's exhaust system in one of two ways:

- An external air manifold, or manifolds, distributes the air through injection tubes with stainless steel nozzles, figure 13-14. The nozzles are threaded into the cylinder heads or exhaust manifolds close to each exhaust valve. This method is used primarily with smaller engines.
- An internal air manifold distributes the air to the exhaust ports near each exhaust valve through passages cast in the cylinder head or the exhaust manifold. This method is used mainly with larger engines.

The addition of oxidation catalytic converters as an emission control device in 1975 increased the use of air injection. Air injection was used to improve converter efficiency by speeding up converter warm-up. When air is injected into the exhaust system upstream from the converter, it mixes with the exhaust gas. This raises the exhaust gas temperature, helping the converter to reach operating tempera-

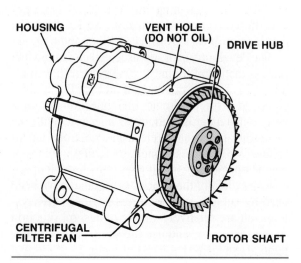

Figure 13-12. Late-model two-vane air pumps have an external centrifugal filter fan mounted on the front of the housing.

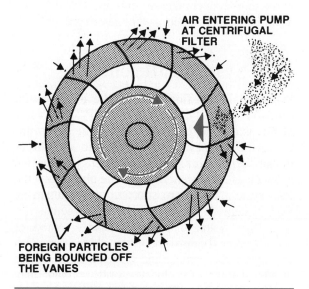

Figure 13-13. Centrifugal force removes dust and dirt from the inlet air.

Figure 13-14. External air manifolds are used with many air-injection systems.

ture quicker. When the exhaust gas enters the converter, the additional air in the mixture helps the converter oxidize HC and CO more completely.

■ It Wasn't Always So Simple

The early PCV systems caused a good deal of grief and engine troubles for vehicle manufacturers. Many garages, even franchised dealers, ignored the PCV systems on 1963–64 cars. They required a lot of care and cleaning, and they clogged quickly when ignored. Contaminants remained in the crankcase, and sludge and moisture formed. This clogged oil lines and prevented adequate engine lubrication. The result was disaster for the engine, and major overhauls on engines still under warranty were often required.

The situation reached a crisis point for one major manufacturer, which stopped using PCV on its cars from the spring of 1964 until early in 1965. Auto engineers were frustrated by the problems PCV systems were creating. The systems had been designed to be simple and require only a minimum amount of service. But mechanics in the field completely ignored the emission control device, and engines began to fail.

These problems resulted in a crash project by the manufacturers. While engineers worked overtime developing a "better" PCV system, manufacturers started a program to educate dealers, service technicians, and car owners. The so-called "self-cleaning" PCV valve was developed and began appearing on mid-1965 models. This second-generation PCV system is practically the same as the one in use today.

With the introduction of three-way converters, the air distribution pattern changed again. This type of converter requires additional air in the exhaust to bring it to operating temperature, but once it becomes operational, it is most efficient with less air in the exhaust. This brought about the use of air-switching devices, which will be explained later in this chapter.

Anti-backfire Valves

In a carbureted engine, during engine deceleration, high intake manifold vacuum enriches the air-fuel mixture. If the air pump supplies air to the exhaust manifold during deceleration, it will combine with excess unburned fuel in the exhaust. The result could be an engine backfire—a rapid combustion of the unburned gases that can destroy a muffler, or cause damage to the air cleaner or intake hoses. To prevent backfire, two different methods are used. One is to use a **gulp valve** to allow additional air into the intake manifold on deceleration to lean the mixture, while air continues to be directed to the exhaust manifold. The other method is to shut off air to the exhaust manifold and divert it to the atmosphere using a diverter valve. Both types of valves are called backfire suppression or anti-backfire valves.

Diverter Valve

The **diverter valve** is also known as a dump or bypass valve. The diverter valve uses a manifold-vacuum-operated diaphragm to redirect the airflow from the air pump. However, secondary air passes through the diverter valve continuously on its way to the air-injection manifold. During deceleration, manifold vacuum operates the valve diaphragm to divert, or dump, the air directly to the atmosphere, rather than into the air-injection manifold.

Some diverter valves vent the air to the engine air cleaner for muffling. Others vent it through a muffler and filter built into the valve. Diverter valves are used with vacuum solenoids, vacuum differential valves, vacuum vent valves, and idle valves to fine-tune air injection. Some diverter valves contain the air-injection system relief valve instead of the pump. Because diverter valves have no effect on the air-fuel mixture in the intake manifold, they are more trouble-free than gulp valves.

Exhaust Check Valves

All air-injection systems use one or more one-way check valves, figure 13-15, to protect the air pump

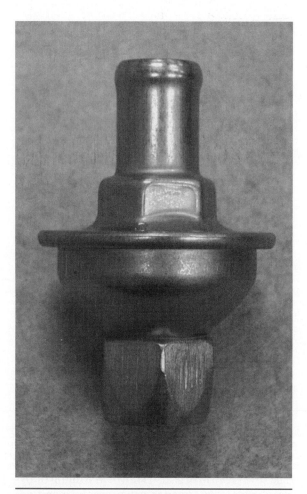

Figure 13-15. The check valve protects the system against reverse flow of exhaust gases.

and other components from reverse exhaust flow. A check valve contains a spring-type metallic disc or reed that closes the air line under exhaust back-pressure. Check valves are located between the air manifold and the gulp or diverter valve. If exhaust pressure exceeds injection pressure, or if the air pump fails, the check valve spring closes the valve to prevent reverse exhaust flow.

Air-Management Valves

Air management is a term used to describe the large variety of diverter, bypass, and air-control or switching valves that have been used by vehicle manufacturers to fine-tune air injection systems since about 1980. Figure 13-16 shows one such valve used by GM. Ford uses a similar valve that combines the divert and switching functions in one valve. Ford referred to the valve as a thermal AIR bypass/thermal AIR divert valve, or TAB/TAD. The air switching section sends air upstream to the exhaust ports during open-loop operation, such as immediately after start-up when the exhaust is rich with unburned fuel.

When the engine control system moves into the closed-loop mode, the valve switches airflow down-stream to a point between the catalytic converter beds. The divert, or air control, section of the valve protects the converter during wide-open-throttle operation and high temperatures.

A combination switch/relief valve used by Chrysler, figure 13-17, is controlled by a coolant vacuum switch cold open (CVSCO) or by a vacuum solenoid. On a cold start, air is injected as close as possible to the exhaust valves. When engine coolant temperature reaches the point where exhaust gas recirculation (EGR) begins, the CVSCO or vacuum solenoid shuts off the vacuum signal to the valve. This causes the valve to send most of the pump air downstream to the catalytic converter. The rest of the secondary air continues to reach the exhaust ports by passing through slots in the upstream valve seat.

Regardless of manufacturer or system, most traditional air-switching or control valves operate with:

- Manifold vacuum working on a vacuum diaphragm in the valve.
- Output pressure of the air pump working against vacuum or a spring in the vacuum chamber.
- One or more solenoids in the vacuum line opening or closing the vacuum supply to the diaphragm.

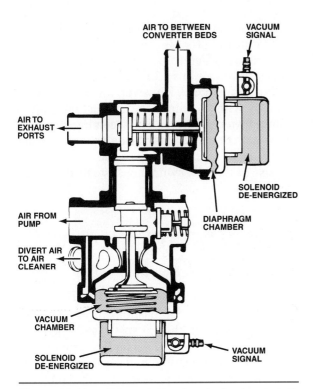

Figure 13-16. GM uses this electrically signaled diverter valve for switching and diverting tasks on air-injection systems used with CCC engines. (Courtesy of General Motors Corporation)

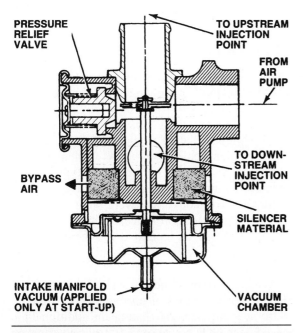

Figure 13-17. Cutaway of Chrysler's air switch/relief valve. (Courtesy of DaimlerChrysler Corporation)

With modern engine designs using rapidly heated oxygen sensors (HO$_2$S) and better catalytic converters, air injection is required only during the first few minutes of open-loop operation when the exhaust is relatively rich with HC and CO. In order to save valuable space under the hood, and to save horsepower, the belt-driven AIR pump has been replaced by an electrically driven pump, which operates for only 3 or 4 minutes after start-up.

■ PCV System Service

When a PCV system becomes restricted or clogged, the cause is usually an engine problem or the lack of proper maintenance. For example, scored cylinder walls or badly worn rings and pistons will allow too much blowby. Start-and-stop driving requires more frequent maintenance and causes PCV problems more quickly than highway driving, as will any condition allowing raw fuel to reach the crankcase. Using the wrong grade of oil, or not changing the crankcase oil at periodic intervals will also cause the ventilation system to clog.

When a PCV system begins to clog, the engine tends to stall, idle roughly, or overheat. As ventilation becomes more restricted, burned spark plugs or valves, bearing failure, or scuffed pistons can result. Also look for an oil-soaked distributor, or leaking from around valve covers or other gaskets. Do not overlook the PCV system while troubleshooting. A partly or completely clogged PCV valve, or one of the incorrect capacity, may well be the cause of poor engine performance. Although many PCV valves look alike, they are specifically calibrated for the engine on which they are installed and may not be interchanged. A plugged valve in the upper hose will allow the engine to pressurize, causing oil leaks and possible blown seals. A plugged PCV valve can cause water condensation in the crankcase and excess oil in the air filter.

Air Injection and Catalytic Converters

Introduction of the catalytic converter changed the air distribution needs of the engine. On earlier vehicles, the role of air injection shifted from that of being an oxidation device to a converter-assist device. Until more recent model vehicles with improved catalytic converters came along, air injection was used to improve converter efficiency and accelerate catalyst warm-up.

On these models, air can be injected in several different places:

- Near each exhaust port
- Into the exhaust manifold outlet

- Into the exhaust pipe ahead of the converter (upstream)
- Directly into the converter (downstream)

When air flows into any of these points, it mixes with the exhaust gases and continues the oxidation process. During initial engine operation, injected air increases exhaust temperature, oxidizing the exhaust in the exhaust manifold, which helps bring the converter to operating temperature more quickly.

■ Don't Oil the Air Pump!

No air-injection system is completely quiet. Pump noise usually increases in pitch as engine speed increases. If the drive belt is removed and the pump shaft turned by hand, it will squeak or chirp. Many who work on their own cars and even some technicians are not aware that air-injection pumps are permanently lubricated, and require no periodic maintenance.

Suppose the air pump is determined to be the source of the noise. It would seem that a few squirts of oil would silence it. See those small holes in the housing? While it is easy to mistake them for oiling points, these are actually vents. Don't oil them. More than a few pumps have failed because someone assumed that taking "good" care of the pump would make it last longer!

However, too much HC and CO in the exhaust of a cold engine (rich mixture) or an engine that idles for a long time can damage the catalysts or overheat the converter. Switching the injected air downstream from the exhaust ports to a point near or at the converter helps to dilute the HC and CO concentration in the exhaust. In newer vehicles, the engine-control computer monitors the duration of air injection as well as engine temperature to prevent catalyst overheating.

Computer-controlled engines also use one or more exhaust oxygen (O_2) sensors. Prior to electrically heated HO_2S sensors, these devices took a few minutes to reach operating temperature of about 600°F (315°C). Until the oxygen sensor came up to temperature, the engine ran in open loop with relatively high tailpipe emissions. Air injection in the exhaust stream during engine warm-up was used to help the sensor reach operating temperature more rapidly.

With the introduction of NO_x reduction converters (also called dual-bed, three-way converters, or TWC) in 1977, the problem of air distribution became even more complex. The oxidation process *adds* O_2 to HC and CO, but the reduction process *removes* O_2 from NO_x compounds. Therefore, a reduction converter requires *more* air in the exhaust to bring it to operating temperature. However, it works most efficiently with *less* air in the exhaust once up to operating temperature. This seeming paradox, along with the problems of catalyst damage and converter overheating, led to the development and use of the AIR switching systems described earlier.

CATALYTIC CONVERTERS

Like other emission controls, a catalytic converter modifies the combustion process. This device is installed in the exhaust system between the exhaust manifold and the muffler, and usually is positioned beneath the passenger compartment, figure 13-18. The location of the converter is important, since as much of the exhaust heat as possible must be retained for effective operation. The nearer it is to the engine, the better.

Oxidation and Reduction Reactions

A **catalyst** is a substance that promotes a chemical reaction but is not changed or affected by that reaction. An oxidation, or burning, reaction takes place when oxygen is added to an element or compound. Platinum and palladium are catalytic elements called noble metals that promote oxidation. If there is not enough oxygen in the exhaust, an air pump or aspirator valve provides extra air. Early converters were intended only to promote the oxidation process, which mixes the HC and CO with oxygen to render them harmless by forming water vapor, H_2O, and carbon dioxide, CO_2.

When emission standards were tightened to reduce levels of oxides of nitrogen, NO_x, the converter was engineered to handle the additional job of *reducing* NO_x. Since oxidation has no effect on NO_x, a separate reaction, called reduction (the removal of oxygen) is required. The reduction reaction changes oxides of nitrogen—NO_x—into nitrogen (N_2) and oxygen (O_2). The most common reduction catalysts are platinum and another noble metal, rhodium.

Figure 13-18. Catalytic converters are located near the engine. Most vehicles use one converter except when the vehicle is equipped with dual exhaust as shown in this figure.

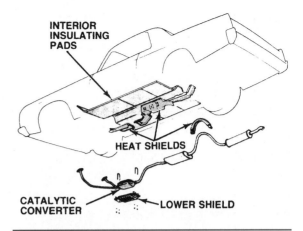

Figure 13-19. Monolithic converters require heat shielding. (Courtesy of DaimlerChrysler Corporation)

Converter Light-off

The catalytic converter does not work when cold; it must be heated to its light-off temperature of close to 500°F (260°C) before it starts working at 50 percent effectiveness. When fully effective, the converter reaches a temperature range of 900° to 1,600°F (482° to 871°C). In spite of the intense heat, however, catalytic reactions do not generate the flame and radiant heat associated with a simple burning reaction. Because of the extreme heat (almost as hot as combustion chamber temperatures), a converter remains hot long after the engine is shut off. Most vehicles use a series of heat shields, figure 13-19, to protect the passenger compartment, the automatic transmission, and other parts of the chassis—as well as the area beneath the vehicle—from excessive heat. More than once, vehicles have been incinerated because of the hot converter causing tall grass or dry leaves beneath the just-parked vehicle to ignite.

Converter Design

Platinum, palladium, rhodium, and in more recent years, cerium catalysts are used in converters. These are not pure metal, but thin deposits on a ceramic or aluminum oxide substrate. Three kinds of substrates are used:

- Tiny pellets or beads
- A large, porous ceramic block
- Formed metal

Each type of substrate material creates several thousand square yards or meters of catalyst surface area for contact with the exhaust gases as they flow through the converter. In converters using a ceramic (monolith) substrate, figure 13-20, a diffuser inside the converter shell produces a uniform flow of exhaust gases over the entire substrate surface. If the converter uses a pellet substrate, figure 13-21, the gas flows over the top and down through the substrate layers. The metal substrate consists of alternating layers of corrugated fins and formed metal plates that resemble a honeycomb pattern, figure 13-22. The substrate is contained within a round or oval shell made by welding two stamped pieces of aluminum or stainless steel together. These metals are used because of their ability to withstand the high temperatures of oxidation.

Pellet-type converters create a fair amount of exhaust restriction, but are less expensive to manufacture, and the pellets in some models can be replaced if they become contaminated. The ceramic substrate in monolithic converters is much less restrictive, but breaks more easily

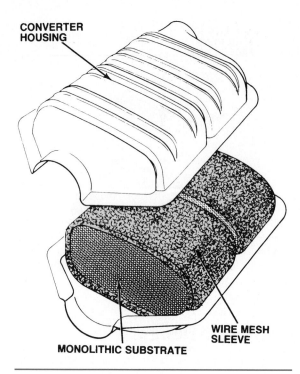

Figure 13-20. Typical catalytic converter with a monolithic substrate.

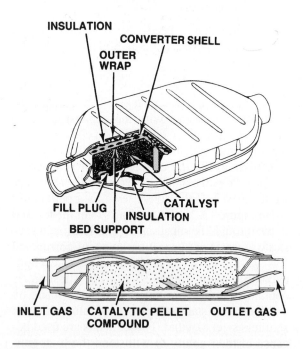

Figure 13-21. Typical catalytic converter with a pellet-type substrate.

when subject to shock or severe jolts, and is more expensive to manufacture. Monolithic converters can be serviced only as a unit; the entire component must be replaced.

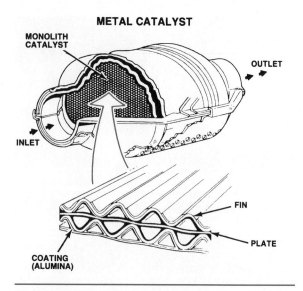

Figure 13-22. This cross-sectional view of a metal monolith substrate converter shows how the metal substrate is formed. (Courtesy of General Motors Corporation)

Converter Usage

A catalytic converter must be located as close as possible to the exhaust manifold to work effectively. The farther back the converter is positioned in the exhaust system, the more the gases cool before they reach the converter. Since positioning in the exhaust system affects the oxidation process, cars that use only an oxidation converter generally locate it underneath the front of the passenger compartment.

Some vehicles have used a small, quick heating oxidation converter called a pre-converter, pup, or mini-converter that connects directly to the exhaust manifold outlet. These have a small catalyst surface area close to the engine that heats up rapidly to start the oxidation process more quickly during cold engine warm-up. For this reason, they were often called light-off converters, or LOC. The oxidation reaction started in the LOC is completed by the larger main converter under the passenger compartment.

Three-way Converters (TWC)

A vehicle requiring both oxidation and reduction catalysts uses a two-stage, three-way converter, or **TWC.** The two catalysts can be housed in separate converters, or they can be located in separate chambers of the same housing. The reduction catalyst must be located ahead of the oxidation catalyst, since it produces oxygen that can be used in the oxidation process. A three-way converter may use a monolith substrate with

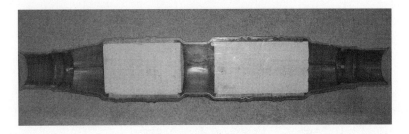

Figure 13-23. This three-way catalyst shows the two separate converter beds. One bed reduces NO_xs and the other bed oxidizes HCs and COs.

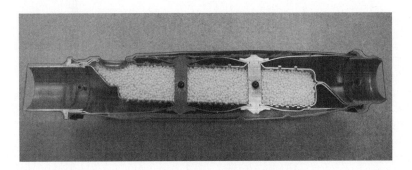

Figure 13-24. A three-way pellet-type catalytic converter construction.

the reduction catalyst located at the front of the converter, figure 13-23. A three-way converter with a pellet substrate has one placed above the other, with an air plenum separating the two catalysts, figure 13-24. Secondary air is supplied by the air pump to the converter. The reduction catalyst is positioned on top of the plenum.

Some three-way converters contain a hybrid reduction and oxidation catalyst at the front of the housing, and a second oxidation catalyst installed behind it. The hybrid catalyst works best to reduce NO_x when the CO level in the exhaust is between 0.8 and 1.5 percent. The second catalyst completes the oxidation process. Figure 13-25 shows the two-stage reaction in this type of converter.

Oxidation-reduction converters work most efficiently when the air-fuel mixture is maintained at the stoichiometric ratio of 14.7 to 1, figure 13-26. The most complete combustion of air and fuel occurs at this ratio, resulting in the least amount of harmful pollutants. HC and CO emissions are high at ratios richer than 14.7; NO_x emissions are greatest with ratios leaner than 14.7.

As covered in Chapters 8 and 9 of this manual, the three-way converter system is used with an air-fuel management feedback system or electronic engine controls. To help maintain the ratio at the ideal, or 14.7 to 1, ratio, the three-way converter system uses an exhaust oxygen sensor (O_2S). This

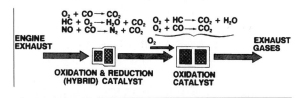

Figure 13-25. Catalytic oxidation and reduction reactions. (Courtesy of DaimlerChrysler Corporation)

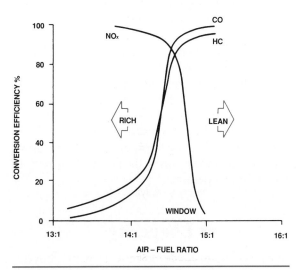

Figure 13-26. Characteristic conversion efficiencies of a three-way catalyst. (Courtesy of DaimlerChrysler Corporation)

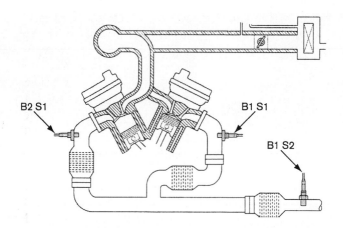

Figure 13-27. OBD II systems monitor converter efficiency by comparing data from the oxygen sensor ahead of the catalytic converter to the one behind. (Provided courtesy of Toyotoa Motor Sales U.S.A., Inc.)

sensor measures the amount of oxygen in the exhaust gas and sends a voltage signal to the engine computer. The computer then controls the feedback carburetor or fuel-injection system to keep the ratio as close as possible to 14.7 to 1.

With OBD II-equipped vehicles, catalytic converter performance is monitored by heated oxygen, HO_2 sensors both before and after the converter, figure 13-27. The converters used on these vehicles have what is known as OSC or *ALIGN* oxygen storage capacity. OSC is due mostly to the cerium coating in the catalyst rather than the precious metals used. When the TWC is operating as it should, the post-converter HO_2S is far less active than the pre-converter sensor. The converter stores, then releases the oxygen during its normal reduction and oxidation of the exhaust gases, smoothing out the variations in O_2 being released. Thus, if the converter is working properly, the post-HO_2S sensor is not exposed to the cycling that the pre-converter sensor sees, figure 13-28.

Where a cycling sensor voltage output is expected before the converter, because of the converter's action, the post-converter HO_2S should read a steady signal without much fluctuation. With the rapid light-off and more efficient converters used today, the air pump needs to supply only secondary air during the first few minutes of cold-engine operation.

Converter Longevity

Since converters have no moving parts, they require no periodic service. Under federal law, catalyst effectiveness is warranted for 80,000 miles (129,000 kilometers) or eight years. However, a

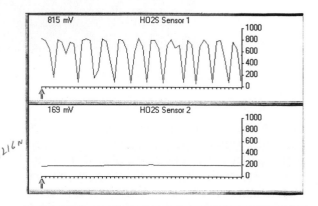

Figure 13-28. If the converter is working properly, there will be very little voltage fluctuation from the post-converter oxygen sensor. If the converter degrades, the rear oxygen signal will become more erratic.

catalyst will eventually wear out. When it does, the entire converter (monolith) or catalyst (pellet) must be replaced to maintain effective emission control.

It is possible, however, to damage a converter before the catalyst wears out. The three main causes of premature converter failure are:

- Contamination
- Excessive temperatures
- Improper air-fuel mixtures

When leaded gasoline is used in the fuel system, the lead will plate the catalyst and form a coating that prevents the exhaust gases from reaching the catalyst. A certain amount of lead in the exhaust will gradually burn off from the catalyst, allowing the converter to resume near-normal operation, but continued use of leaded fuel will eventually destroy the converter. Other substances that can destroy the converter include exhaust that contains excess motor oil, antifreeze, sulfur (from poor fuel), and various other chemical substances.

■ Catalytic Converter Odors

Although catalytic converters control HC and CO emissions, they also produce other undesirable emissions in small quantities. For example, most gasoline has a little bit of sulfur in it. This reacts with the water vapor inside a converter to produce hydrogen sulfide (H_2S). This toxic by-product has the distinct odor of rotten eggs. The smell is usually most noticeable while the engine is warming up, or during deceleration. Hydrogen sulfide, when combined with water vapor in the atmosphere, forms H_2SO_4 (sulfuric acid) and this acid rain destroys the landscape.

When the odor is very strong at normal operating temperatures, it may mean that the engine is out-of-tune and is running too rich. But the odor does not necessarily mean an incorrect mixture adjustment; changing brands of gasoline may help control the odor in some cases, since the amount of sulfur present in gasoline varies from one brand to another.

Although a converter operates at high temperature, it can be destroyed by excessive temperatures. This most often occurs either when too much unburned fuel enters the converter, or with excessively lean mixtures. Excessive temperatures may be caused by long idling periods on some vehicles, since more heat develops at those times than when driving at normal highway speeds. Severe high temperatures can cause the converter to melt down, leading to the internal parts breaking apart and either clogging the converter, or moving downstream to plug the muffler. In either case, the restricted exhaust flow severely reduces engine power.

Rich mixtures or raw fuel in the exhaust can be caused by engine misfiring, or an excessively rich air-fuel mixture resulting from a stuck choke pull-off, defective coolant temp sensor, or defective fuel injectors. Lean mixtures are commonly caused by intake manifold leaks. When either of these circumstances occurs, the converter can become a catalytic furnace, causing the damage described above.

To avoid excessive catalyst temperatures and the possibility of fuel vapors reaching the converter, follow these rules:

1. Use only unleaded gasoline in a vehicle equipped with a converter.
2. Do not try to start the engine on compression by pushing the vehicle. Use jumper cables instead.
3. Do not crank an engine for more than 40 seconds when it is flooded or firing intermittently.
4. Do not turn off the ignition switch when the car is in motion.
5. Do not disconnect a spark plug for more than 30 seconds to test the ignition.
6. Fix engine problems such as dieseling, misfiring, or stumbling from excess fuel that affect performance as soon as possible.

Exhaust gas recirculation (EGR) is an emission control that lowers the amount of nitrogen oxides (NO_x) formed during combustion. In the presence of sunlight, NO_x reacts with hydrocarbons in the atmosphere to form ozone (O_3) or photochemical smog, an air pollutant. This chapter discusses:

- The principles of EGR systems
- How the EGR system reduces NO_x and prevents detonation
- Various domestic EGR systems

NO_x FORMATION

Nitrogen N_2 and oxygen O_2 separate into N, N, and O, O atoms during the combustion process. These then bond to form NO_x (NO, NO_2). When combustion flame front temperatures exceed $2,500°F$ ($1,370°C$), NO_x formation increases dramatically. Because ignition timing affects peak combustion chamber temperature and ignition timing advance controls were economical to adapt in production, the first attempts during the early 1970s to meet NO_x control requirements were spark-timing advance control systems. Delayed spark advance produces less pressure and heat. This helps lower peak combustion chamber temperatures. In 1972, the EPA introduced new test procedures to determine NO_x levels. Auto manufacturers responded by developing more effective ways of controlling NO_x.

Controlling NO_x

The amounts of NO_x formed at temperatures below $2,500°F$ ($1,370°C$) can be controlled in the exhaust by a catalyst. To handle the amounts generated above $2,500°F$ ($1,370°C$), the following are some methods that have been used to lower NO_x formation:

- **Enrich the air-fuel mixture.** More fuel lowers the peak combustion temperature, but it raises hydrocarbon (HC) and carbon monoxide (CO) emissions. The reduction in fuel economy also makes this solution unattractive.

- **Lower the compression ratio.** This decreases NO_x levels somewhat but also reduces combustion efficiency. When the compression ratio becomes too low, HC and CO emissions rise.
- **Delay spark-timing advance.** Most manufacturers found they could meet the NO_x emission standards prior to 1972 by using devices to delay ignition spark advance. However, performance and fuel economy suffered. This method failed to cut emission levels enough to meet the 1972 standards.
- **Dilute the air-fuel mixture.** To lower emission levels further, engineers developed a system that introduces small amounts of **inert** exhaust gas into the engine intake. This lowers combustion temperatures by displacing some of the air and absorbs heat without contributing to the combustion process. Currently, this is one of the most efficient methods to meet NO_x emission level cut-points without significantly affecting engine performance, fuel economy, and other exhaust emissions. The EGR system routes small quantities (up to 14 percent) of exhaust gas from the engine's exhaust to the intake manifold.

EGR System Operation

Since small amounts of exhaust are all that is needed to lower peak combustion temperatures, the orifice that the exhaust passes through is small.

The level of NO_x emission changes according to engine speed, temperature, and load. At idle speed NO_x levels are low, so EGR operation is not required. During cold engine operation, heavy acceleration, or at wide-open throttle, NO_x formation is not critical because of ignition timing controls and richer air-fuel mixtures. In fact, EGR dilution of the air-fuel mixture during any of these operating conditions would cause engine performance problems. However, during light acceleration and at cruising speeds NO_x emission levels are high; this is when EGR systems work to lower NO_x formation.

The power output of early engines equipped and retrofitted with EGR systems was less than that of the same engine without an EGR, giving rise to the idea that exhaust gas recirculation automatically meant a reduction in power output.

During the early years of EGR, many drivers were convinced that they could increase engine performance by disconnecting the EGR system, a popular myth that still exists today. For many years, engineers have designed engines that take into account and require EGR dilution of the air-fuel mixture. The performance of these engines requires proper EGR functioning or driveability problems will occur.

The EGR System and Detonation

In addition to lowering NO_x levels, the EGR system also helps control detonation. Detonation, or ping, occurs when high pressure and heat cause the air-fuel mixture to ignite. This uncontrolled combustion can severely damage the engine.

Using an EGR system allows for greater ignition timing advance and for the advance to occur sooner without detonation problems. This increases power and efficiency.

EGR SYSTEM HISTORY

EGR systems first appeared on 1972 Chrysler corporation cars sold in California. The system used calibrated floor jets installed in the bottom of the intake manifold underneath the carburetor to provide an opening between the exhaust crossover passage and the intake manifold, figure 13-29. A combination of exhaust pressure and manifold vacuum caused exhaust gases to flow through a set of calibrated orifices or jets. This was the simplest of all EGR system designs, but proved to be the least efficient because the jets allowed exhaust gases to enter the intake manifold whenever the engine was running. Since EGR flow is not wanted with either a cold or idling engine, the result was rough engine idle and poor cold engine operation. After 1973, the floor jet EGR system became history.

Buick also introduced an EGR system on its 1972 engines. The Buick system controlled EGR flow through a spring-loaded, vacuum-operated poppet valve, figure 13-30. The vacuum EGR valve was adopted by other manufacturers and soon became the basic component of all EGR systems. Although later systems include devices to modulate the valve and some use taper-stem valves, the basic operation is still the same as the original.

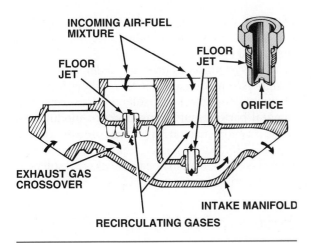

Figure 13-29. A Chrysler corporation floor jet EGR system.

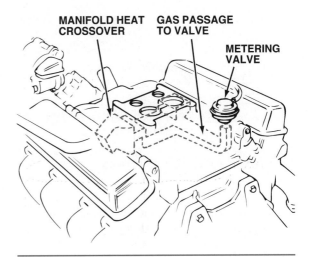

Figure 13-31. Passages supply exhaust gas and intake access for the EGR valve.

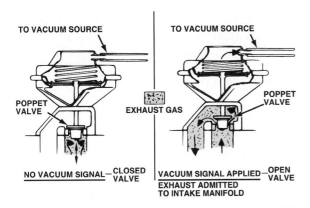

Figure 13-30. Typical single spring-loaded diaphragm poppet-type EGR valve.

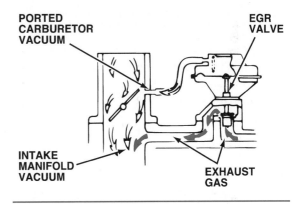

Figure 13-32. Ported vacuum opens the EGR above idle speed only during light or no-load conditions.

In many EGR systems, the valve mounts on the intake manifold and connects to the intake and exhaust systems through internal passages in the intake manifold, figure 13-31. In other systems, the valve may connect to the intake and/or exhaust systems through external steel tubing. The EGR valve is held in the closed position by the diaphragm spring. When vacuum is applied, the diaphragm overcomes the spring tension and opens the **pintle valve.** On some systems a vacuum line connects the diaphragm to a carburetor vacuum port located above the throttle. This EGR port is usually located at a different level than the spark port to control when the EGR opens. Using ported vacuum for actuating the diaphragm keeps the valve closed at idle, during deceleration, and during wide-open throttle since ported vacuum is weak during wide-open throttle. During light loads moderate acceleration, and cruising, ported vacuum operates the valve to provide EGR flow, figure 13-32.

Some early Ford and Chrysler EGR systems used venturi vacuum as a control signal during the 1970s. Since venturi vacuum is very weak, it was used only as a signal to a switch in a vacuum amplifier that had a manifold vacuum supply and reserve to actuate the EGR valve diaphragm. When the system was working properly, it produced a fairly good modulation of EGR flow proportional to engine speed. However, the system was abandoned in the early 1980s because more accurate and reliable electronic controls became available.

The basic vacuum-operated EGR valve system would not provide adequate control of EGR flow to accommodate engine temperature. Without some form of temperature control, EGR flow would occur during cold engine operation, affecting driveability. To prevent the EGR valve from opening while the engine is cold, a temperature-sensitive vacuum control valve is used. The valve may be installed in the radiator or in an engine

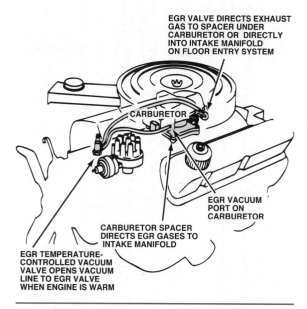

EGR VALVE DIRECTS EXHAUST
GAS TO SPACER UNDER
CARBURETOR OR DIRECTLY
INTO INTAKE MANIFOLD
ON FLOOR ENTRY SYSTEM

CARBURETOR

EGR VACUUM
PORT ON
CARBURETOR

CARBURETOR SPACER
DIRECTS EGR GASES TO
INTAKE MANIFOLD

EGR TEMPERATURE-
CONTROLLED VACUUM
VALVE OPENS VACUUM
LINE TO EGR VALVE
WHEN ENGINE IS WARM

Figure 13-33. A coolant temperature-controlled EGR system prevents EGR flow during cold engine operation.

water jacket to control the vacuum signal, figure 13-33. The following are some common names and acronyms for these devices:

- Ported vacuum switch (PVS)
- Coolant temperature override (CTO)
- Thermal vacuum switch (TVS)

When the engine's coolant is cold, the temperature valve is closed, preventing the vacuum from actuating the EGR valve. When the engine reaches a calibrated temperature, the valve opens and allows the vacuum to open the EGR valve, figure 13-34.

Since 1972, engineers have developed a variety of methods for modulating, or controlling, EGR valve operation relative to engine operating conditions. High or low ambient air temperature vacuum controls may be used to bleed off vacuum to weaken the EGR signal at either high or low temperatures. Chrysler has used an intake air temperature switch, time delay module, and an EGR solenoid to electrically delay EGR vacuum, figure 13-35. On this system, an intake charge temperature switch installed in the intake manifold turns the vacuum solenoid off when the intake charge air temperature is cold. Another method of EGR control uses a vacuum-bias valve to counteract the force of ported vacuum. It uses a second vacuum diaphragm to modulate the EGR valve at higher manifold vacuum conditions such as cruising speeds to eliminate a high-speed surge problem. A dual-diaphragm EGR valve, figure 13-36, uses

ALIGN

manifold vacuum to help the valve spring offset the carburetor vacuum under certain cruise and light-load conditions.

EGR systems used on later-model vehicles are designed to modulate EGR flow, through the use of computer-control and solenoids, external and integral backpressure transducers or modulators. More recently General Motors has used digital and linear EGRs that no longer require the use of vacuum. These digital and linear valves use a solenoid to open and spring tension to close the EGR valve.

Ported EGR Valves

Ported EGR valves use ported engine vacuum to operate the valve. Ported vacuum is vacuum that is taken above the throttle plates of the carburetor or the throttle body. This vacuum differs from manifold vacuum in that, when the throttle plates are closed, ported vacuum is not present. As the throttle plates open, ported vacuum starts to rise and continues to increase as the throttle plates are opened more. Thus, ported vacuum can serve as an indicator of the throttle plate position, which represents the load on the engine.

The vacuum supplied to a ported EGR valve is applied to the EGR valve diaphragm. When the ported vacuum signal reaches a predetermined level, it starts to overcome the calibrated spring tension that is against the diaphragm and the valve begins to open. When the throttle plates are closed, the ported vacuum drops and the spring tension acts against the diaphragm to close the pintle valve and stop exhaust gases from flowing back into the intake manifold.

Ported-type EGR valves are rarely used on newer vehicles due to the difficulty of accurately controlling EGR valve operation. To control NO_xs as close as possible and not cause driveability concerns, the EGR operation has to be controlled very precisely. More accurate control of vacuum-actuated EGR valves is achieved by using exhaust backpressure EGR valves and computer-controlled solenoids to control vacuum to the EGR valve.

Backpressure Transducer EGR Valves

Chrysler was the first to add an external backpressure transducer to control vacuum to the EGR valve on some 1973 engines, as a way of regulating the vacuum signal according to engine load. The transducer is installed between an exhaust passage and a

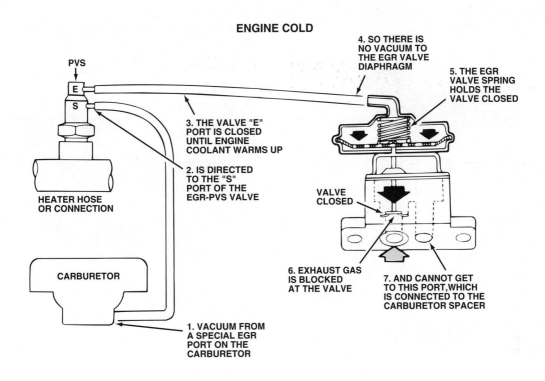

ENGINE COLD

PVS

4. SO THERE IS NO VACUUM TO THE EGR VALVE DIAPHRAGM

5. THE EGR VALVE SPRING HOLDS THE VALVE CLOSED

3. THE VALVE "E" PORT IS CLOSED UNTIL ENGINE COOLANT WARMS UP

2. IS DIRECTED TO THE "S" PORT OF THE EGR-PVS VALVE

HEATER HOSE OR CONNECTION

VALVE CLOSED

CARBURETOR

6. EXHAUST GAS IS BLOCKED AT THE VALVE

7. AND CANNOT GET TO THIS PORT, WHICH IS CONNECTED TO THE CARBURETOR SPACER

1. VACUUM FROM A SPECIAL EGR PORT ON THE CARBURETOR

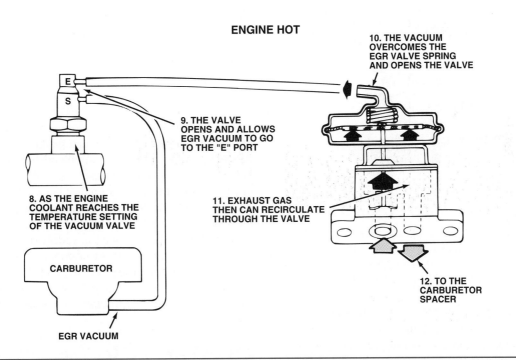

ENGINE HOT

10. THE VACUUM OVERCOMES THE EGR VALVE SPRING AND OPENS THE VALVE

9. THE VALVE OPENS AND ALLOWS EGR VACUUM TO GO TO THE "E" PORT

8. AS THE ENGINE COOLANT REACHES THE TEMPERATURE SETTING OF THE VACUUM VALVE

11. EXHAUST GAS THEN CAN RECIRCULATE THROUGH THE VALVE

CARBURETOR

12. TO THE CARBURETOR SPACER

EGR VACUUM

Figure 13-34. A comparison of a ported vacuum, coolant temperature-controlled EGR system operation under hot and cold engine conditions.

lower diaphragm chamber in the EGR valve, figure 13-37. When enough exhaust backpressure is present, it closes an air bleed opening in the EGR diaphragm, resulting in maximum EGR at high speed.

As engine speed decreases, backpressure drops and a spring opens the vacuum line bleed. Since this design incorporates the engine's load effect on vacuum and engine's speed effect on backpressure, it

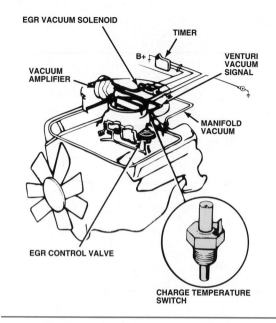

Figure 13-35. A time delay, temperature-sensitive, EGR system. (Courtesy of DaimlerChrysler Corporation)

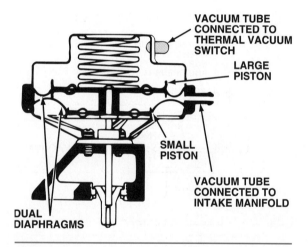

Figure 13-36. A dual-diaphragm EGR valve uses both manifold and ported vacuum to modulate flow.

causes a modulation of how much the EGR opens under different speed and load conditions.

By 1977, the external transducer had been replaced in many General Motors and Ford EGR systems by an integral (incorporated inside) transducer EGR valve, figure 13-38. This design is called a positive backpressure EGR valve. The valve will not open just by applying vacuum to the diaphragm. Exhaust backpressure entering the exhaust gas chamber must be strong enough to close the normally open internal air bleed (control valve) before the EGR valve will operate. When

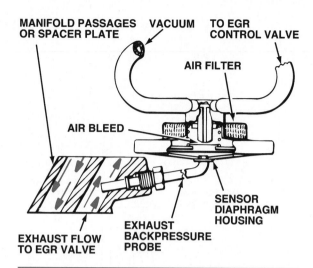

Figure 13-37. This external backpressure transducer uses exhaust backpressure to modulate EGR flow. (Courtesy of DaimlerChrysler Corporation)

the valve opens, pressure drops in the exhaust chamber and the control spring opens the air bleed, causing the valve to move back toward its closed position. This cycle repeats, opening and closing the valve pintle about 30 times per second. This in turn modulates the amount of EGR flow relative to the amount of vacuum and exhaust backpressure.

The negative backpressure EGR valve, figure 13-39, was introduced in 1979 for use with engines that have relatively little exhaust backpressure. In this design, the transducer air bleed is normally closed. As ported vacuum opens the EGR valve, a negative pressure signal from the vacuum in the intake manifold is modulated by the exhaust system pressure that travels up the inside of the EGR valve stem to the backside of the transducer diaphragm. When the pressure signal is low enough (high vacuum), it opens the air bleed and reduces the amount of EGR. Like the positive backpressure EGR valve, this modulating process goes on constantly.

COMPUTER-CONTROLLED EGR

Chrysler continued to use an external positive backpressure transducer with a single-diaphragm EGR valve, figure 13-40, on all normally aspirated engines until the mid 1980s. Chrysler's late-model transducer has an integral solenoid vacuum switch and is called an electronic EGR transducer

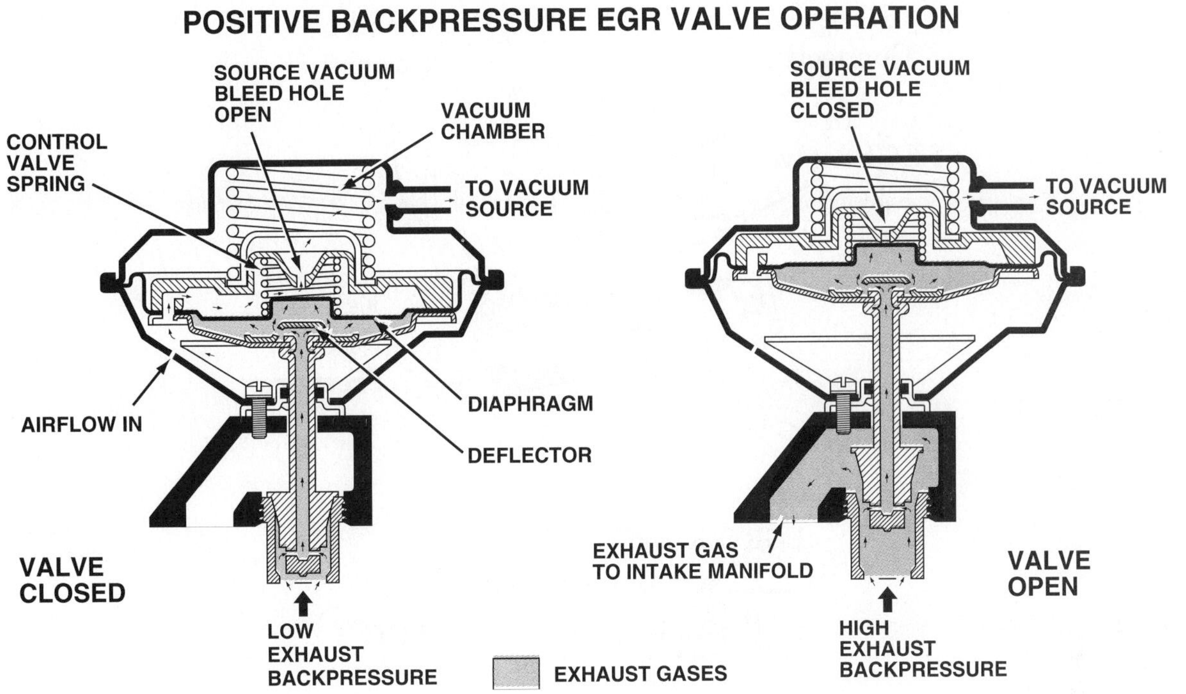

POSITIVE BACKPRESSURE EGR VALVE OPERATION

SOURCE VACUUM
BLEED HOLE
OPEN

VACUUM
CHAMBER

CONTROL
VALVE
SPRING

TO VACUUM
SOURCE

AIRFLOW IN

DIAPHRAGM

DEFLECTOR

**VALVE
CLOSED**

LOW
EXHAUST
BACKPRESSURE

EXHAUST GASES

SOURCE VACUUM
BLEED HOLE
CLOSED

TO VACUUM
SOURCE

EXHAUST GAS
TO INTAKE MANIFOLD

**VALVE
OPEN**

HIGH
EXHAUST
BACKPRESSURE

Figure 13-38. When exhaust backpressure is high enough to overcome control spring pressure and vacuum is applied, this integral backpressure EGR valve opens.

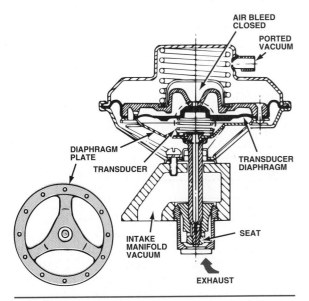

Figure 13-39. The integral negative backpressure EGR valve uses operating principles similar to the positive backpressure EGR.

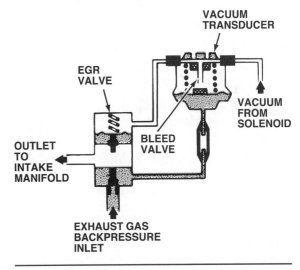

Figure 13-40. An external positive backpressure transducer and EGR valve schematic. (Courtesy of DaimlerChrysler Corporation)

computer determines when the solenoid is open and EGR flow is allowed based on sensor input.

Solenoid Vacuum Control

Many computer-controlled EGR systems have one or more solenoids controlling the vacuum

(EET). The engine computer controls the solenoid and the vacuum-operated transducer is controlled by exhaust system backpressure. When the exhaust backpressure is strong enough, the transducer opens its vacuum switch, but the engine

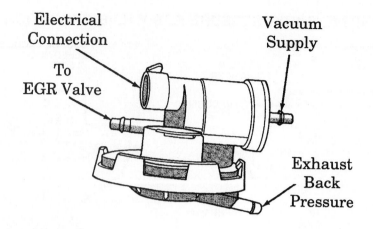

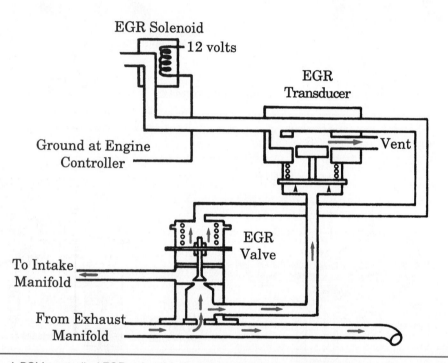

Figure 13-41. A PCM-controlled EGR solenoid. (Courtesy of DaimlerChrysler Corporation)

that is applied to the EGR valve. By using a solenoid, the computer can very precisely control the amount of vacuum applied to the EGR valve. This provides a more accurate control of the EGR valve and thus NO_xs.

Figure 13-41 shows an EGR system that utilizes an EGR solenoid, an EGR transducer, and an EGR valve. In this application, the solenoid is either on or off. When the engine controller determines that EGR valve is not required, the solenoid is de-energized and no vacuum is applied the EGR transducer. When the engine controller determines EGR is needed, it energizes the EGR solenoid and vacuum flows to the EGR transducer. The EGR transducer senses the ex-

haust backpressure. If the exhaust backpressure is not high enough, the vacuum will bleed off through the vent. If the exhaust backpressure is high enough, the vent will be closed and vacuum will be applied to the EGR valve.

In the above application, the EGR solenoid simply serves as a switch to apply vacuum to the EGR transducer. The amount that the EGR valve opens is controlled by the EGR transducer. In other applications, the EGR solenoid is used to control the amount of vacuum that reaches the EGR valve. This is accomplished by pulse-width modulating the signal to the EGR solenoid. When using pulse-width modulation, the amount of vacuum flow through the solenoid can be controlled very accurately.

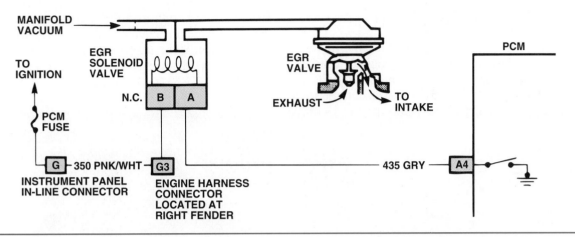

Figure 13-42. This EGR system uses the solenoid on-time to expose EGR supply vacuum to an air bleed to modulate the EGR valve. (Courtesy of General Motors Corporation)

The PCM-controlled EGR system used by GM, figure 13-42, contains a modulated solenoid and works much the same as the Ford system described earlier. In this design, the computer operates the solenoid using a duty cycle. This is similar to the operation of a carburetor mixture control solenoid but at a higher frequency. To regulate the amount of vacuum applied to the EGR valve, the computer varies the amount of the solenoid on time according to data from its sensors.

A temperature sensor responds to exhaust flow as it enters the intake through the EGR valve. If a failure occurs, the PCM will set a code in memory and turn on the malfunction indicator lamp (MIL).

Two variations of this system have been designed by Chevrolet and Buick, and installed on other GM models. The Chevrolet system uses the thermal sensor in the base of the EGR valve as discussed earlier. The Buick system includes a current-regulating module that maintains constant control voltage to the solenoid for more accurate EGR regulation.

EGR Valve Position Sensors

Late-model computer-controlled EGR systems use a sensor to indicate EGR operation. On Board Diagnostics Generation II (OBD II) EGR system monitors require an EGR sensor to do their job. Ford introduced the concept of an **EGR valve position sensor** with its EEC systems in the late 1970s, figure 13-43. A linear potentiometer on the top of the EGR valve stem indicates valve position for the computer. This is called an EGR valve position (EVP) sensor. Some later-model Ford EGR systems, however, use a feedback signal provided by an EGR exhaust backpressure sensor,

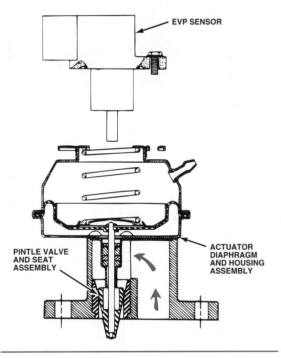

Figure 13-43. A linear potentiometer EGR valve position (EVP) sensor mounts on top of the EGR valve to report how far the valve is open.

which converts the exhaust backpressure to a voltage signal. This sensor is called a pressure feedback EGR (PFE) sensor.

The GM integrated electronic EGR valve uses a similar sensor figure 13-44. The top of the valve contains a vacuum regulator and EGR pintle-position sensor in one assembly sealed inside a nonremovable plastic cover, figure 13-45. The pintle-position sensor provides a voltage output to the PCM, which increases as the duty cycle increases, allowing the PCM to monitor valve operation.

Digital EGR Valves

GM introduced a completely electronic, digital EGR valve design on some 1990 engines. Unlike the previously mentioned vacuum-operated EGR valves, the digital EGR valve consists of three solenoids controlled by the PCM, figure 13-46.

Each solenoid controls a different size orfice in the base—small, medium, and large. The PCM controls each solenoid ground individually. It can produce any of seven different flow rates, using

Figure 13-44. An integrated EGR valve contains both a position sensor and a vacuum solenoid in a nonremovable cover.

Figure 13-46. The digital EGR valve relies on PCM modulated solenoids to control EGR flow.

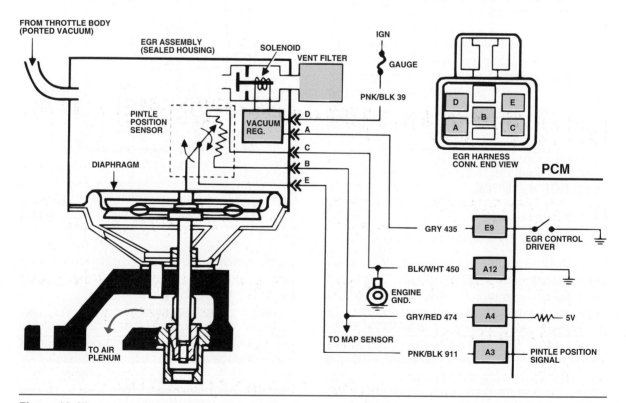

Figure 13-45. A schematic of the integrated EGR valve system showing internal and external sensor and actuator circuits. (Courtesy of General Motors Corporation)

Figure 13-47. The linear EGR valve is used on all GM vehicles since 1994.

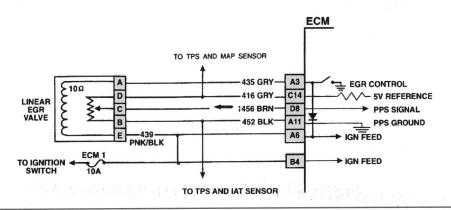

Figure 13-48. Linear EGR circuit. (Courtesy of General Motors Corporation)

the solenoids to open the three valves in different combinations. The digital EGR valve offers precise control, and using a swivel pintle design helps prevent carbon deposit problems.

Linear EGR Valve

In 1994 General Motors replaced all vacuum-operated EGR valves with the linear EGR valve, figure 13-47. The linear valve operates independently from manifold vacuum. The PCM controls the position of the EGR pintle based on inputs from the engine coolant temperature sensor, the mass air flow sensor, vehicle speed, and the throttle position sensor. Incorporated into the linear EGR valve is a pintle-position feedback sensor, which provides

EGR pintle position to the PCM. During operation the PCM adjusts the EGR pintle position based on the input of the pintle-position sensor. The linear EGR valve's response time is ten times faster than that of a vacuum-operated EGR valve.

Figure 13-48 shows the electrical schematic for the linear EGR valve. The EGR control pulse-width modulates the valves to achieve the desired position. The EGR valve position can range between 0 and 100 percent. The position of the EGR pintle is provided by the pintle-position sensor. A 5-volt reference is provided by the PCM to the pintle-position sensor. The pintle-position sensor signal is an analog voltage, which varies between 0 and 5 volts.

The linear EGR is generally active when the engine is warm and above idle. The PCM calculates the desired amount of EGR flow required for the present operating conditions. The pintle-position sensor provides diagnostic capabilities for the EGR valve.

On Board Diagnostics Generation II (OBD II) EGR Monitoring Strategies

In 1996, the U.S. EPA began requiring OBD II systems in all passenger cars and most light-duty trucks. These systems include emissions system monitors that alert the driver and the technician if an emissions system is malfunctioning. To be certain the EGR system is operating, the PCM runs a functional test of the system, when specific operating conditions exist. The OBD II system tests by opening and closing the EGR valve. The PCM monitors an EGR function sensor for a change in signal voltage. If the EGR system fails, a diagnostic trouble code (DTC) is set. If the system fails two consecutive times, the malfunction indicator light (MIL) is lit.

Chrysler monitors the difference in the exhaust oxygen sensor's voltage activity as the EGR valve opens and closes. Oxygen in the exhaust decreases when the EGR is open and increases when the EGR valve is closed. The PCM sets a DTC if the sensor signal does not change.

Depending on the vehicle application, Ford uses at least one of two types of sensors to evaluate exhaust-gas flow. The first type uses a temperature sensor mounted in the intake side of the EGR passageway. The PCM monitors the change in temperature when the EGR valve is open. When the EGR is open and exhaust is flowing, the sensor signal is changed by the heat of the exhaust. The PCM compares the change in the sensor's signal with the values in its look-up table.

The second type of Ford EGR monitor test sensor is called a delta pressure feedback EGR (DPFE) sensor. This sensor measures the pressure differential between two sides of a metered orifice positioned just below the EGR valve's exhaust side. Pressure between the orifice and the EGR valve decreases when the EGR opens because it becomes exposed to the lower pressure in the intake. The DPFE sensor recognizes this pressure drop, compares it to the relatively higher pressure on the exhaust side of the orifice, and signals the value of the pressure difference to the PCM. When the EGR valve is closed, the exhaust-gas pressure on both sides of the orifice is equal.

The OBD II EGR monitor for this second system runs when programmed operating conditions have been met. The monitor evaluates the pressure differential while the PCM commands the EGR valve to open. Like other systems, the monitor compares the measured value with the look-up table valve. If the pressure differential falls outside the acceptable value, a DTC sets.

GM uses the manifold absolute pressure (MAP) sensor as the EGR monitor on some applications. After meeting the enable criteria (operating condition requirements), the EGR monitor is run. The PCM monitors the MAP sensor while it commands the EGR valve to open. The MAP sensor signal should change in response to the sudden change in manifold pressure. If the signal value falls outside the acceptable value in the look-up table, a DTC sets. If the EGR fails on two consecutive trips, the PCM lights the MIL.

SUMMARY

Pressure in the engine cylinders forces combustion gases past the pistons. These gases, called blowby, settle in the engine crankcase, where they contaminate the lubricating oil and create harmful acids. Ventilation is necessary to remove the vapors from the crankcase. A positive crankcase ventilation (PCV) system recirculates crankcase vapors through the intake manifold to the cylinders, where they are burned. PCV systems use a metering valve or orifice to regulate airflow. A malfunctioning PCV system can cause driveability problems and create harmful acids and sludge that affect engine lubrication and result in premature wear.

Air injection is one of the oldest methods used to control HC and CO exhaust emissions. The injected air mixes with hot exhaust gas as it leaves the combustion chambers to further oxidize HC and CO emissions. All air-injection systems used on domestic cars operate in essentially the same way, regardless of manufacturer. Recent systems assist the converter by injecting air into the exhaust system to help warm up the catalyst. When a reduction catalyst is used, air injection is switched downstream once the engine is warm. Air pump injection systems have given way to pulse air systems on many later-model engines. Computers now control air-injection systems to provide a quicker response to changing engine requirements, but many fuel-injected engines no longer require air injection other than immediately after start-up.

Catalytic converters, which first appeared in 1975, promote a chemical reaction to change HC and CO emissions into harmless water and carbon dioxide. Two types of converters are used: oxidation and reduction. The oxidation type removes HC and CO emissions from the exhaust gases; reduction converters remove NO_x. Engines with feedback fuel systems or electronic engine controls use a three-way converter that combines oxidation and reduction functions. Three-way converters work best with an air-fuel ratio of 14.7 to 1. Converters use a monolithic or pellet-type substrate coated with platinum, palladium, or rhodium to provide a surface on which the oxidation or reduction reactions occur. Newer converters use cerium as a catalyst. Too much heat, the use of leaded fuel, or too much unburned fuel can damage any catalyst. Converters do not require periodic service, and are warranted for a 50,000-mile (80,000-kilometer) life span, as required by U.S. federal regulations.

Combustion chamber temperatures that exceed 2,500°F (1.370°C) cause a dramatic increase in NO_x formation, an air pollutant detrimental to health. In the presence of sunlight, hydrocarbons and NO_x cause photochemical smog. To reduce NO_x as much as possible, manufacturers use exhaust-gas recirculation (EGR) systems to meter a small amount of exhaust gas into the incoming air-fuel mixture. This dilutes the fuel charge and results in lower combustion chamber temperatures.

With the advent of sophisticated engine management systems, EGR system control has become a function of the PCM. The PCM controls when and how much EGR flow is allowed based on operating parameters of the engine.

Review Questions

Choose the letter that represents the best possible answer to the following questions:

1. In a PCV system, crankcase vapors are recycled to the:
   a. Exhaust system
   b. Road draft tube
   c. Oil-filler breather cap
   d. Intake manifold system

2. A separate flame arrester is used in a closed PCV system when inlet air is drawn from the:
   a. Clean side of the carburetor filter
   b. Dirty side of the carburetor filter
   c. The intake manifold
   d. The oil filler cap on a valve cover

3. Which is *not* part of a Type 4 PCV system?
   a. A PCV valve
   b. A vented oil filler cap
   c. An air inlet filter
   d. A manifold vacuum hose

4. The PCV valve operates in which of the following ways?
   a. Restricts airflow when intake manifold vacuum is high
   b. Increases airflow when intake manifold vacuum is low
   c. Acts as a check valve in case of carburetor backfire
   d. All of the above

5. The main reason for an air-injection system is to:
   a. Oxidize HC and CO exhaust emissions
   b. Reduce NO_x exhaust emissions
   c. Eliminate crankcase emissions
   d. Eliminate evaporative HC emissions

6. Oxidation of HC and CO emissions produces:
   a. HCO and CO_2
   b. H_2CO_3 and CO_2
   c. H_2O and CO_2
   d. HNO_3 and C_3PO

7. Two-vane air pumps have:
   a. An impeller-type fan for filter
   b. An integral wire mesh filter
   c. A hose to the clean side of the air cleaner
   d. A separate air filter

8. Air-injection nozzles are made of:
   a. Copper
   b. Stainless steel
   c. Aluminum
   d. Vanadium

9. The two types of air-injection backfire suppressor valves are:
   a. The check valve and the gulp valve
   b. The gulp valve and the diverter valve
   c. The diverter valve and the relief valve
   d. The diverter valve and the check valve

10. Photochemical smog is a result of:
    a. Sunlight + NO_x + HC
    b. Sunlight + NO_x + CO_2
    c. Sunlight + CO + HC
    d. Sunlight + NO_x + CO

11. NO_x forms in an engine under:
    a. High pressure and low temperature
    b. Low pressure and low temperature
    c. High temperature and high pressure
    d. All of the above

12. Which is *not* true of EGR valves?
    a. They are operated on venturi-vacuum systems.
    b. They are operated on ported vacuum.
    c. The may be mounted on the intake manifold.
    d. They are operated at wide-open throttle.

13. Technician A says air injection helps an exhaust catalyst reach its operating temperature faster.
    Technician B says a catalyst is most efficient when cooled by air injection. Who is right?
    a. A only
    b. B only
    c. Both A and B
    d. Neither A nor B

14. Catalytic converters:
    a. Increase the HC content in exhaust emissions
    b. Neither add nor remove the oxygen from exhaust emissions
    c. Improve oxidation of HC and NO
    d. Improve oxidation of HC and CO

15. The catalyst material in an oxidation catalytic converter is:
    a. Platinum or palladium
    b. Aluminum oxide
    c. Stainless steel
    d. Ceramic lead oxide

16. A catalyst:
    a. Slows a chemical reaction
    b. Heats a chemical reaction
    c. Increases, but is not consumed by, a chemical action
    d. Combines with the chemicals in the reaction

17. A reduction reaction:
    a. Adds oxygen to a compound
    b. Removes oxygen from a compound
    c. Reduces HC in exhaust gases
    d. Removes H_2O from exhaust gases

18. The outer shell of the oxidation catalyst is made of:
    a. Aluminum oxide pellets
    b. Platinum or palladium
    c. A honeycomb monolith
    d. Stainless steel

19. Which of the following is *not* true for cars with catalytic converters?
    a. Engines should not be cranked for more than 60 seconds.
    b. Spark plugs should always be disconnected to test ignition.
    c. Engine dieseling, surging, and stalling should be fixed immediately.
    d. Ignition should not be turned off while car is moving.

20. By pumping air into the exhaust manifold:
    a. Converter light-off occurs sooner
    b. HC and CO are oxidized
    c. Both a and b
    d. Neither a nor b

21. OBD II vehicles use a post-converter HO_2 sensor:
    a. As a failsafe in case the pre-converter sensor fails
    b. To determine if the converter fails to function properly
    c. To serve as a fuel trim device
    d. None of the above

22. In discussing an electronic-controlled EGR system, Technician A says a sensor informs the computer of how the system is functioning by a position signal. Technician B says a sensor informs the computer by a pressure sensor. Who is right?
    a. A only
    b. B only
    c. Both A and B
    d. Neither A nor B

23. A customer has just dropped off a 1996 Chrysler passenger car with the MIL on. One of the DTCs you retrieve indicates a malfunction in the EGR system. What sensor does the OBD II system rely on to test EGR flow?
    a. MAP sensor
    b. O_2S
    c. DPFE sensor
    d. EGR thermal sensor

24. On some GM OBD II-equipped vehicles, during the EGR system monitor test:
    a. The MAP sensor is used as a test sensor
    b. Preconditions must be met
    c. EGR monitoring begins as the PCM commands the EGR valve
    d. All of the above

25. Referring to the above question, Technician A says that the MAP sensor voltage should change when the EGR is activated. Technician B says that if the signal value does not meet those stored in the PCM's look-up table, a DTC sets. Who is correct?
    a. A only
    b. B only
    c. Both A and B
    d. Neither A nor B

14

The Ignition Primary Circuit and Components

OBJECTIVES

Upon completion and review of this chapter, you will be able to:

- Explain the process of mutual induction in the ignition coil.
- List the components in the primary ignition circuit.
- Describe the differences between electronic and distributorless ignition systems.
- Have knowledge of the different solid-state triggering devices.

KEY TERMS

breaker points	Hall Effect switch
capacitive-discharge ignition system	inductive-discharge ignition system
current limiting hump	magnetic pulse generator
distributor ignition (DI) system	magnetic saturation
dwell	mechanical ignition systems
dwell time	
electronic ignition (EI) systems	optical signal generator
firing line	self-induction
fixed dwell	variable dwell

INTRODUCTION

The primary ignition circuit is considered to be the heart of the ignition system. The secondary ignition circuit cannot function efficiently if the primary ignition circuit is damaged. This chapter explains components of the low-voltage primary ignition circuit and how the circuit operates. Breaker points opened and closed the low-voltage primary circuit until the 1970s when solid-state electronic switching devices took their place. Whether breaker points or electronic switches are used, the principles of producing high voltage by electromagnetic induction remain the same.

NEED FOR HIGH VOLTAGE

Energy is supplied to the automotive electrical system by the battery. The battery supplies about twelve volts, but the voltage required to ignite the air-fuel mixture ranges from about 5,000 to more than 40,000 volts, depending upon engine operating conditions.

High voltage is required to create an arc across the spark plug air gap. The required voltage level increases when the:

- Spark plug air gap increases
- Engine operating temperature increases
- Air-fuel mixture is lean
- Air-fuel mixture is at a greater pressure

Battery voltage must be greatly increased to meet the needs of the ignition system. Engine operating temperature increases the required voltage because resistance increases with greater temperature. A lean air-fuel mixture contains fewer volatile fuel particles, so resistance increases. More voltage is needed when the air-fuel mixture is at a greater pressure because resistance increases with an increase in pressure. The ignition system boosts voltage using electromagnetic induction.

HIGH VOLTAGE THROUGH INDUCTION

Any current-carrying conductor or coil is surrounded by a magnetic field. As current in the coil increases or decreases, the magnetic field expands or contracts. If a second coiled conductor is placed within this magnetic field, the expanding or contracting magnetic flux lines will cut the second coil, causing a voltage to be induced into the second coil. This transfer of energy between two unconnected conductors is called mutual induction.

Induction in the Ignition Coil

The ignition coil uses the principle of mutual induction to step up or transform low battery voltage to high ignition voltage. The ignition coil, figure 14-1, contains two windings of copper wire wrapped around a soft iron core. The primary winding is made of a hundred or so turns of heavy wire. It connects to the battery and carries current. The secondary winding is made of many thousand turns of fine wire. When current in the pri-

mary winding increases or decreases, a voltage is induced into the secondary winding, figure 14-1.

The ratio of the number of turns in the secondary winding to the number of turns in the primary winding is generally between 100:1 and 200:1. This ratio is the voltage multiplier. That is, any voltage induced in the secondary winding is 100 to 200 times the voltage present in the primary winding.

Several factors govern the induction of voltage. Only two of these factors are easily controllable in an ignition system. Induced voltage increases with:

- More magnetic flux lines
- More rapid movement of flux lines

More magnetic flux lines produce a stronger magnetic field because there is a greater current. The rapid movement of flux lines results in a faster collapse of the field because of the abrupt end to the current.

Voltage applied to the coil primary winding with breaker points is about nine to ten volts. However, at high speeds, voltage may rise to twelve volts or more. This voltage pushes from one to four amperes of current through the primary winding. Primary current causes a magnetic field buildup around the windings. Building up a complete magnetic field is called **magnetic saturation,** or coil saturation. When this current stops, the primary winding magnetic field collapses. A greater voltage is self-induced in the primary winding by the collapse of its own magnetic field. This **self-induction** creates from 250 to 400 volts in the primary winding. For example, in a typical electronic ignition system, when the switching device turns ON, the current in the primary ignition circuit increases to approximately 5.5 amps. As the magnetic field builds in the primary circuit, the magnetic flux lines cross over into the secondary windings of the coil and induce an even greater voltage charge. The amount of voltage in the primary circuit builds up to 240 volts. The secondary windings meanwhile are building 200 times the primary voltage, or 48,000 volts. This is enough voltage to ignite the air-fuel mixture under most operating conditions.

This collapsing magnetic field can be observed on an oscilloscope as a primary waveform pattern and is often referred to as the **firing line.** The voltage fluctuations vary by manufacturer, but most ignition systems range in amplitude between 250 to 400 volts. The secondary windings within the ignition coil are much thinner and have between 100 to 200 times the amount of windings. The

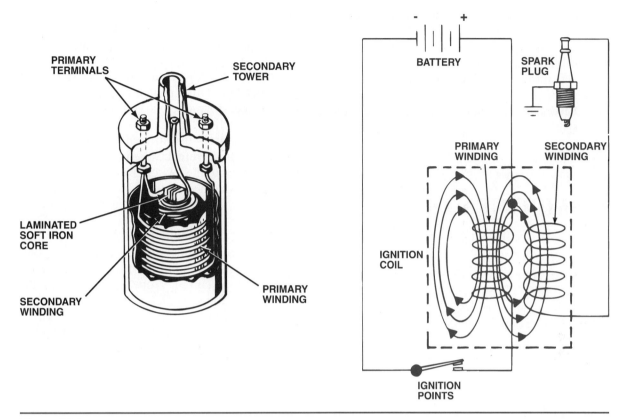

Figure 14-1. The ignition coil produces high-voltage current in the secondary winding when current is cut off in the primary winding.

voltage induced by the collapsing magnetic field is in the kilovolt (1000 X) range.

This kind of ignition system, based on the induction of a high voltage in a coil, is called an **inductive-discharge ignition system.** An inductive-discharge system, using a battery as the source of low-voltage current, has been the standard automotive ignition system for about eighty years.

Most ignition systems work on the inductive discharge principle except for one type, the **capacitive-discharge ignition system.** This high-performance electronic ignition system is explained in detail later in this chapter. Basically, in the capacitive-discharge system, the primary current voltage charges a capacitor inside the ignition module. The current does not travel through the coil during this period but instead, charges the storage capacitor. The stored voltage is released by the switching device at the proper time.

The ignition system transforms low battery voltage into high ignition voltage through induction. A look at the ignition system circuitry reveals how the system works.

BASIC CIRCUITS AND CURRENT

The ignition system consists of two interconnected circuits:

- The low-voltage primary circuit
- The high-voltage secondary circuit

When the ignition switch is turned on, battery current travels:

- Through the ignition switch or the primary resistor if present
- To and through the coil primary winding
- Through a switching device
- To ground at the negative and the grounded terminal of the battery

Low-voltage current in the coil primary winding creates a magnetic field. When the switching device interrupts this current:

- A high-voltage surge is induced in the coil secondary winding.

- Secondary current is routed from the coil to the distributor.
- Secondary current passes through the distributor cap, rotor, across the rotor air gap, and through another ignition cable to the spark plug.
- Secondary current creates an arc across the spark plug gap, as it travels to ground.

PRIMARY CIRCUIT COMPONENTS

The primary circuit contains the:

- Battery
- Ignition switch
- Pickup coil in the distributor
- Coil primary winding
- Ignition control module

Battery

The battery supplies low-voltage current to the ignition primary circuit. Battery current is available to the system when the ignition switch is in the START or the RUN position.

Ignition Switch

The ignition switch controls low-voltage current through the primary circuit. The switch allows current to the coil when it is in the START or the RUN position. Other switch positions route current to accessory circuits and lock the steering wheel in position.

Manufacturers use differing ignition switch circuitry. The differences lie in how battery current is routed to the switch. Regardless of variations, full system voltage is always present at the switch because it is connected directly to the battery.

General Motors automobiles draw ignition current from a terminal on the starter motor solenoid, figure 14-2. Ford Motor Company systems draw ignition current from a terminal on the starter relay, figure 14-3. In Chrysler Corporation vehicles, ignition current comes through a wiring splice installed between the battery and the alternator, figure 14-4. Imported and domestic automobiles may use different connections. Ignition current is drawn through a wiring splice between the battery and the starter relay to simplify the circuitry on some systems, figure 14-5.

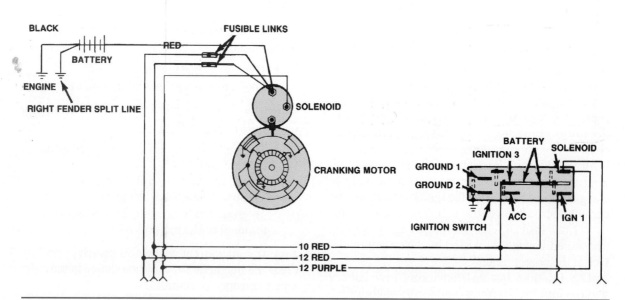

Figure 14-2.   General Motors products draw current from a terminal on the starter solenoid. (Courtesy of General Motors Corporation)

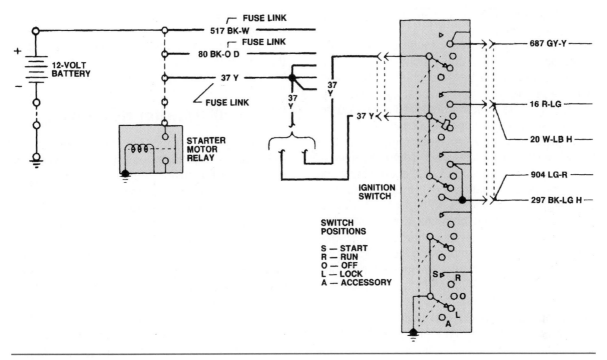

Figure 14-3. Ford products draw ignition current from a terminal on the starter relay. (Courtesy of General Motors Corporation)

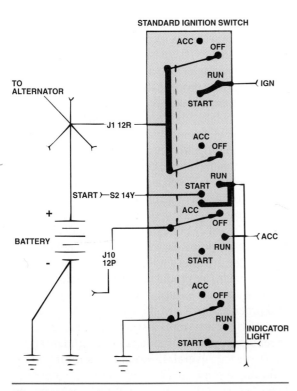

Figure 14-4. Chrysler products draw ignition current from a wiring splice between the battery and the alternator.

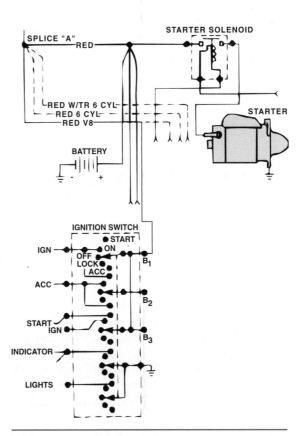

Figure 14-5. The ignition current is drawn from a wiring splice between the battery and the starter relay.

■ **Charles Franklin Kettering (1876–1958)**

Charles F. Kettering was a leading inventor and automotive engineer. After graduating from Ohio State University, he became chief of the inventions department at the National Cash Register Company. While there he designed a motor used in the first electrically operated cash register.

In 1909, he helped form a company called Dayton Engineering Laboratories Company, later to be known as Delco. In 1917, he became the president and general manager of the General Motors Research Corporation.

Kettering invented both the automobile self-starter and the inductive-discharge battery ignition system. The accompanying illustration is an early sketch by Kettering of his design for an ignition system. He was involved in the invention and perfection of high-octane gasoline; improvements for engines, especially diesel engines; electric refrigeration; and much more. He also helped establish the Sloan-Kettering Institute for Cancer Research in New York City.

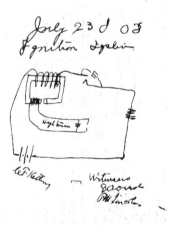

The Kettering inductive-discharge battery ignition system was invented by Charles F. Kettering in 1908. He used a set of contact points as a mechanical switch to open and close the circuit to the primary winding of the ignition coil. These contacts are opened by the rotation of a cam on the distributor shaft. They are called **breaker points** because they continually break the primary circuit.

The Kettering ignition system was quickly adopted as the standard for the automotive industry and used virtually unchanged for over sixty years. By the early 1970s, solid-state electronic components began to replace breaker points as switching devices for the primary circuit. The electronic or breakerless ignitions on late-model vehicles still use the inductive-discharge principles to produce a high-voltage spark. However, the primary circuits are controlled by electronic, rather than mechanical, switching devices.

Types of Ignition Systems

Technology has advanced over the years producing a variety of primary ignition circuit switching devices. Although different in design, they all function to control the primary circuit current.

The magnetic field of the coil primary winding must completely collapse in order to induce a high voltage in the secondary winding. For the field to collapse, current through the primary winding must rapidly stop. Current must then start and stop again to induce the next high-voltage discharge. The primary circuit needs a switching device that rapidly breaks and completes the circuit. For more than sixty years, **mechanical ignition systems** used breaker points as a switching device. Solid-state electronic devices have replaced breaker points on modern ignition systems. All non-points-type ignition systems are referred to as **electronic ignition (EI) systems.** Updated terminology lists a new category called the **distributor ignition (DI) system.** Any ignition system that incorporates a distributor, whether the system is mechanical or electronic, falls into this category. It is easier to remember the design of the switching device and then place the ignition system into the proper subcategory. Ignition system categories include the distributorless and the distributor types while subcategories include the mechanical and the electronic ignition systems. After closer examination, ignition systems are often referred to by their manufacturer-specific name: DuraSpark I, II or II, HEI, TFI, EDIS, etc. Refer to Chapter 16 for more information on ignition systems.

Electronic ignition systems are more efficient than the breaker points ignition systems because there are no points to wear, become damaged, or deteriorate by heat and corrosion. A hotter secondary spark is produced by electronic ignition systems, which in turn results in more complete and precise burning of the air-fuel mixture. Electronic ignition systems also help engine designers meet the tougher emission control regulations.

There are several different types of electronic ignition systems depending upon the methods used to trigger the coil and to direct the high voltage to the correct spark plug at the correct time.

The capacitive-discharge (CD) ignition system works similar to the inductive-discharge system. In

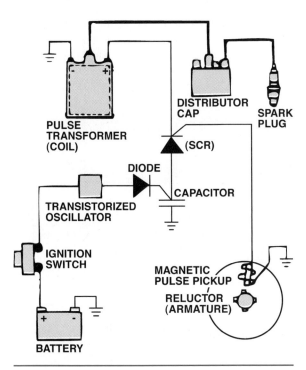

Figure 14-6. A schematic of a typical capacitive-discharge ignition system.

the capacitive-discharge system, the primary current charges a capacitor inside the ignition module. The dwell, or charge time, for the inductive-discharge systems corresponds to the capacitor charge time in the CD ignition system. In the CD system, the current does not travel through the coil during this period but, instead, travels to the storage capacitor. A charging circuit within the module powers a transformer that increases battery voltage to charge the capacitor to approximately 300 to 400 volts. The module is equipped with a silicon-controlled rectifier (SCR) that prevents capacitor discharge. When the switching device sends the trigger signal to the gate of the SCR, it closes the capacitor discharge circuit and discharges the voltage to the coil primary windings, figure 14-6. The abrupt change in coil primary voltage and current induces the secondary voltage in the coil.

In an inductive-discharge system, primary current travels through the coil during the dwell period and is turned off to induce the high secondary voltage. In a capacitive-discharge system, the primary voltage and current charge the capacitor. The capacitor is switched on to induce the high secondary voltage. These CD ignition systems charge the capacitor and discharge the coil faster than inductive-discharge systems. The CD ignition system produces a higher and faster firing voltage. The CD ignition system also starts

better in cold and wet weather and improves high-speed performance.

Current Limiting Devices and Systems

To prevent damage to ignition components, a current limiting system is built into the primary circuit. Ballast resistors work well in points ignition systems to limit current. With the arrival of modern electronic ignition systems, specialized electronic components were designed to control excess primary circuit current. The key to achieving precise primary coil saturation is by controlling the charge time, or dwell, of the coil by limiting the current in the primary ignition circuit. After saturation is achieved, excess current in the primary windings of the ignition coil results in damage to the coil and points or ignition module.

The current limiting functions of modern electronic ignitions are accomplished by several different designs. One type of current limiting system incorporates resistive coil windings and specialized transistors, which switch to resistor mode after excess current is detected, figure 14-7. Another type of current limiting system is equipped with an ignition module that monitors the maximum current during the previous coil saturation. If the proper coil saturation time is achieved, the module trims the dwell to limit current, thereby reducing the amount of electrical energy used by the ignition system. If maximum current is not achieved during the coil saturation time, the module increases the dwell time.

In order to understand current limiting systems, it is important to understand dwell and the ability of the ignition system to control the amount of current in the primary circuit. Dwell is the period of time when the switching device is closed and the current is traveling in the primary winding of the coil. Solid-state or electronic ignition systems require a dwell period for the same purpose, the charge time is controlled by an ignition module or computer. There are two basic designs for controlling dwell in the primary circuit: **fixed dwell** and **variable dwell**.

All points systems operate on fixed dwell. As the distributor rotates and the points close, the amount of time closed is determined by the degree settings. The time period the points open and close is identical for each cylinder and each rotation, providing the lobes on the distributor shaft are not worn. **Dwell** measured in distributor degrees remains constant at all speeds but the **dwell time** shortens at high speed and lengthens at slow speeds. Remember how the ballast resistor works in the points system. At low

■ Dwell and Engine Speed

As engine speed increases, the distributor cam rotates faster. Dwell angle is unchanged, but the time it takes the cam to rotate through this angle is decreased. That is, the amount of time that the points are closed is reduced as engine speed increases.

It takes a specific amount of time, about 0.010 second, for the primary current to reach full strength. At a 625-rpm idle speed, the 33-degree dwell angle of an 8-cylinder engine lasts about 0.0165 second. Primary current easily reaches full strength before the points open to interrupt it.

As engine speed increases to 1,000 rpm, the 33-degree dwell period decreases to 0.0099 second. Primary current is interrupted just before it reaches its maximum strength; therefore, available secondary voltage is decreased slightly.

At 2,000 rpm, the dwell period is reduced to 0.0045 second. Primary current is interrupted well before it reaches its maximum strength. Available secondary voltage is reduced considerably. If the engine and ignition system are not in excellent condition, the required voltage level may be greater than the available voltage level. If this occurs, the engine misfires.

The electronic ignition systems reduced the resistance in the primary circuit to allow faster coil saturation. It can now be done in about 0.004 second. Together with a 6-cylinder engine being more common, this has allowed a higher available voltage at higher engine rpm because of the higher coil saturation.

If a waste spark system, or one coil for two cylinders, is used the effective frequency of each coil firing is reduced by one-third of engine rpm, with the 6-cylinder engine. At 5,000 rpm, the time interval between each coil firing is 0.012 second. If each cylinder has an individual coil, the time interval doubles.

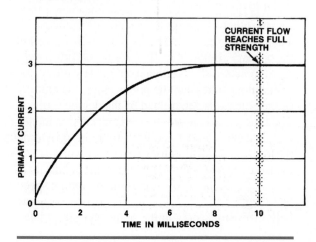

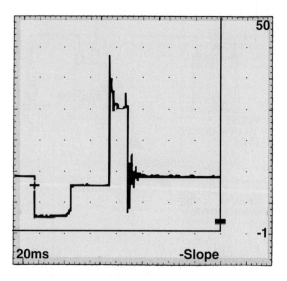

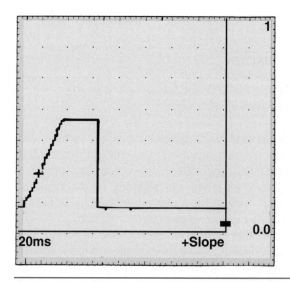

Figure 14-7. Some early electronic ignitions are equipped with a current limiting transistor.

speeds, current travels through the circuit for relatively long periods of time. As the current heats the resistor, its resistance increases, dropping the applied voltage at the coil. At higher speeds, the breaker points open more often and current travels for shorter periods of time. As the ballast resistor cools, its resistance drops. Higher voltage is applied to the coil but the shorter current duration results in about the same magnetic saturation of the coil. Dwell is fixed, but dwell time varies with engine speed. Because points are a mechanical component and could only be designed with fixed dwell, ballast resistors protect the primary circuit during engine rpm variations. Figure 14-8 shows the primary circuit oscilloscope pattern of a typical fixed-dwell electronic ignition. All fixed-dwell systems use a ballast resistor to limit current and voltage. Some

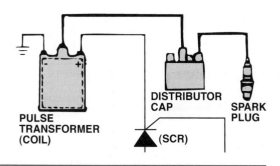

Figure 14-8. Dwell measured in degrees of distributor rotation remains relatively constant at all engine speeds in a fixed-dwell system.

early electronic ignition systems also incorporate the fixed-dwell system and ballast resistor.

■ Dual-Point, Dual-Coil, Dual-Plug Ignition

Some early Nash automobiles used a "Twin Ignition" system. This system had two ignition coils, two sets of spark plugs and cables, a distributor with 16-plug terminals and two coil wire terminals, a rotor with offset tips, two sets of breaker points, and two condensers.

The inline engines were designed with spark plugs on both sides of the cylinders. The overhead valves were located directly over each cylinder in a vertical position.

The breaker points were synchronized by using a dual-bulb test lamp. With the ignition switch on, the distributor cam was turned to just break a stationary set of points, thereby lighting one bulb. A movable set of points was then adjusted to break contact, and light the second bulb, at the same instant. This adjustment assured that both sparks would occur in a cylinder at the same time. The dual-coil, dual-plug system has also been used to reduce exhaust emissions on some Nissan engines.

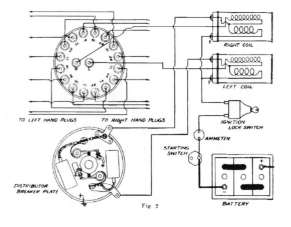

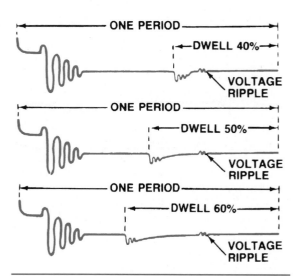

Figure 14-9. The dwell period increases with engine speed in the variable-dwell Delco HEI and some Motorcraft ignition systems.

Variable-dwell ignition systems do not use ballast resistors so the ignition coil and module receive full battery voltage. These systems do not conduct through the primary system until they get a signal to turn on. A module circuit senses primary current to the coil and limits the current when the magnetic field is saturated. The dwell time in milliseconds is relatively constant, but dwell in degrees increases with engine RPM.

The other method of limiting current through the coil is to limit the "on-time." This method can be used when the PCM controls all dwell and ignition timing. Although the dwell increases with RPM, the dwell time in milliseconds stays constant. Variable-dwell ignition systems have a higher available voltage capacity, lower primary resistance, and a hotter spark to burn leaner air-fuel mixtures designed for low emissions automobiles. Figure 14-9 shows the oscilloscope pattern of a typical variable-dwell electronic ignition system. Refer to Chapter 16 of this *Classroom Manual* for more information on ignition systems.

Dual trace digital and analog oscilloscopes are useful tools for diagnosing primary circuit malfunctions. Because the primary circuit operates on relatively small voltages, more accurate analysis of the primary circuit can be achieved over the standard automotive oscilloscopes. Current limiting can be monitored on the scope on all types of ignition systems. The rule to follow is: When viewing voltage on a circuit that is ground controlled, as voltage goes down, current

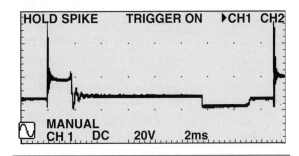

Figure 14-10. Abrupt voltage changes from the current limiting device cause a current limiting hump in the trace on electronic ignition systems.

increases and as voltage goes up, current decreases. Figure 14-10 shows the exact point in the dwell when the current limiting device in the ignition system switches ON and voltage increases. This portion of the waveform on an electronic ignition system is called the current limiting hump. The **current limiting hump** appears at different locations for different ignition systems. Some early electronic ignition systems are not equipped with sophisticated circuitry and therefore do not display the hump in the primary ignition trace. By learning the typical waveform patterns of properly operating ignition systems, ignition system degradation can be easily detected. Once the control of the ignition timing has been taken away from the ignition system, the primary turn-off point is stable and does not vary. At this point, any variation between the cylinders indicates an engine mechanical failure such as worn distributor bushings, distributor drive gears, loose timing chains, etc.

Solid-State Switching Devices

Breaker-point ignitions make it difficult for an engine to meet modern exhaust emission control standards. Such standards not only require maximum system performance, they also require consistent performance. Because breaker points wear during normal operation, ignition system settings and performance change.

Since solid-state switching devices do not wear, ignition system performance remains consistent and exhaust emissions are effectively maintained. All manufacturers now use electronic ignition systems.

The most basic electronic ignition is the transistor-switching device. Early electronic ignition systems use breaker points as the trigger de-

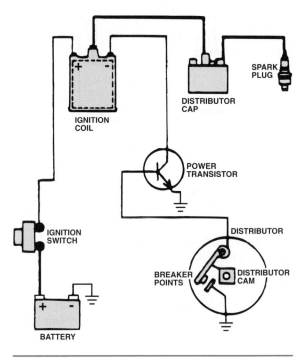

Figure 14-11. Schematic of a typical transistorized ignition system.

vice and a power transistor as the switching device. Although the points trigger the primary circuit, it is the transistor that actually opens and closes the primary circuit. The transistor acts as a load-carrying switch. A small amount of current is used to control a large amount of current, similar to a relay. Therefore, this type of ignition is considered electronic. The points enjoy extended life because only a small amount of current is applied to them, see figure 14-11. This system is much more efficient than mechanical ignition systems and the trigger device does require periodic replacement.

Electronic ignition systems quickly progressed to more sophisticated designs. Points were no longer used to trigger the transistor. Instead, a solid-state control module is responsible for switching the primary current on and off. A big design change with electronic ignition systems is that no primary current passes through the distributor. All primary current passes through the module by way of the transistors, resistors, and capacitors. The ignition coil positive terminal (+) is connected to the ignition switch, same as breaker point systems, but the negative (−) side of the coil is connected to the ignition module and then to ground. The transistors switch the primary current from the ground side.

The module must be signaled when to turn the current off. Of the four devices commonly used to do this, magnetic pulse generators and Hall Effect switches are the most common. Chrysler and some Japanese manufacturers, such as Isuzu, use optical signal generators.

A simple and very common ignition electronic switching device is the magnetic pulse generator. The **magnetic pulse generator** is installed in the distributor housing where the breaker points used to be. The pulse generator, figure 14-12, consists of a trigger wheel (reluctor) and a pickup coil. The pickup coil consists of an iron core wrapped with fine wire in a coil at one end and attached to a permanent magnet at the other end. The center of the coil is called the pole piece.

The trigger wheel is made of steel with a low reluctance, which cannot be permanently magnetized. Therefore, it provides a low-resistance path for magnetic flux lines. The trigger wheel has as many teeth as the engine has cylinders.

As the trigger wheel rotates, its teeth come near the pole piece. Flux lines from the pole piece concentrate in the low-reluctance trigger wheel, increasing the magnetic field strength and inducing a positive voltage signal in the pickup coil. When the teeth and the pole piece are aligned, the magnetic lines of force from the permanent magnet are concentrated and the voltage drops to zero.

The reluctor continues and passes the pole piece, but now the voltage induced is negative, creating an ac voltage signal, figure 14-13. The pickup coil is connected to the electronic control module, which senses this AC voltage, converts the signal to DC voltage and switches the primary current off, figure 14-14. Each time a trigger wheel tooth comes near the pole piece, the control module is signaled to switch off the primary current. Solid-state circuitry in the module determines when the primary current is turned on again. Figure 14-15 shows the typical construction of a

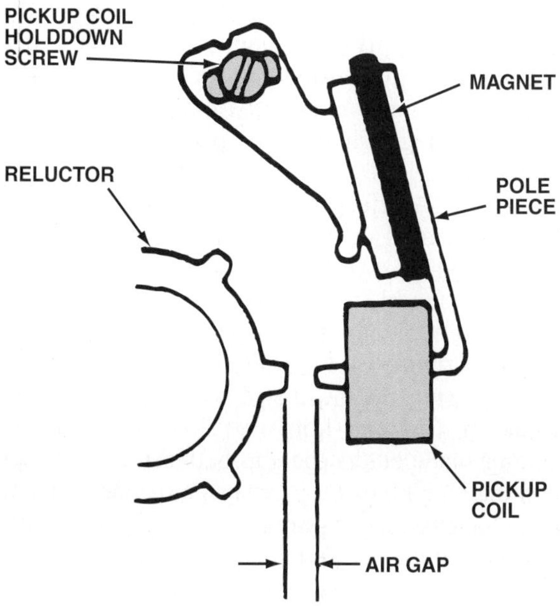

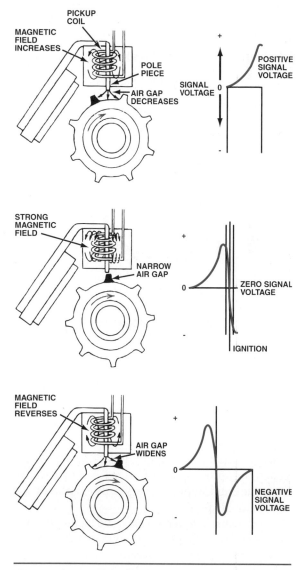

Figure 14-13. The pickup coil generates an AC voltage signal as the reluctor moves closer and then away from the magnetic field.

PICKUP COIL HOLDDOWN SCREW

MAGNET

RELUCTOR

POLE PIECE

PICKUP COIL

AIR GAP

Figure 14-12. A typical magnetic pulse generator.

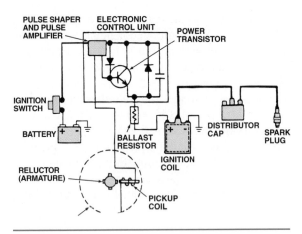

Figure 14-14. Schematic of a typical magnetic pulse generator ignition system.

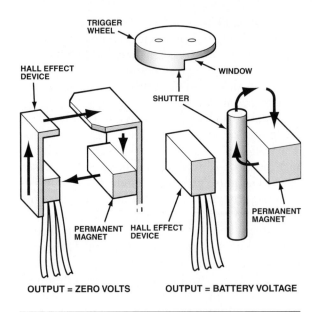

Figure 14-16. Hall Effect ignition systems use a reference voltage to power the Hall device.

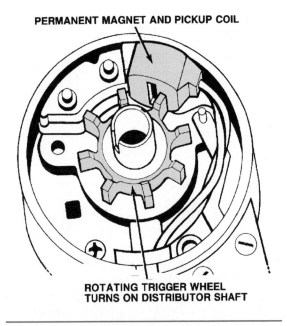

PERMANENT MAGNET AND PICKUP COIL

ROTATING TRIGGER WHEEL TURNS ON DISTRIBUTOR SHAFT

Figure 14-15. A magnetic pulse generator installed in the distributor housing. (Courtesy of Ford Motor Company)

pulse generator for an 8-cylinder engine. These generators produce an AC signal voltage whose frequency and amplitude vary in direct proportion to rotational speed.

The terms "trigger wheel" and "pickup coil" describe the magnetic pulse generator. Various manufacturers have different names for these components. Other terms for the pickup coil are pole piece, magnetic pickup, or stator. The trigger

wheel is often called the reluctor, armature, timer core, or signal rotor.

A **Hall Effect switch** also uses a stationary sensor and rotating trigger wheel, figure 14-16. Unlike the magnetic pulse generator, it requires a small input voltage in order to generate an output or signal voltage. Hall Effect is the ability to generate a small voltage signal in semiconductor material by passing current through it in one direction and applying a magnetic field to it at a right angle to its surface. If the input current is held steady and magnetic field fluctuates, the output voltage changes in proportion to field strength.

Most Hall Effect switches in distributors have a Hall element or device, a permanent magnet, and a ring of metal blades similar to a trigger wheel. Some blades are designed to hang down. These are typically found in Bosch and Chrysler systems, figure 14-17, while others may be on a separate ring on the distributor shaft, typically found in GM and Ford distributors. When the shutter blade enters the gap between the magnet and the Hall element, it creates a magnetic shunt that changes the field strength through the Hall element, thereby creating an analog voltage signal. The Hall element contains a logic gate that converts the analog signal into a digital voltage sig-

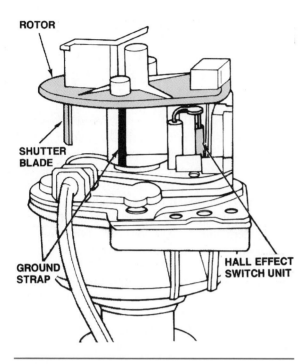

ROTOR

SHUTTER BLADE

GROUND STRAP

HALL EFFECT SWITCH UNIT

Figure 14-17. Shutter blades rotating through the Hall Effect switch air gap bypass the magnetic field around the pickup and drop the output voltage to zero.

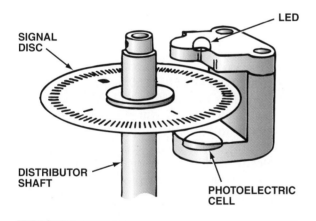

SIGNAL DISC

LED

DISTRIBUTOR SHAFT

PHOTOELECTRIC CELL

Figure 14-18. This Nissan optical signal generator works by interrupting a beam of light passing from the LEDs to photodiodes.

nal. This digital signal triggers the switching transistor. The transistor transmits a digital square waveform at varying frequency to the powertrain control module (PCM).

The Hall Effect switch requires an extra connection for input voltage; however, its output voltage does not depend on the speed of the rotating

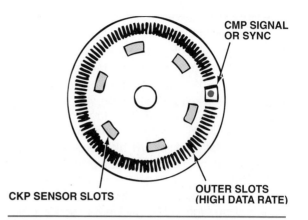

CMP SIGNAL OR SYNC

CKP SENSOR SLOTS

OUTER SLOTS (HIGH DATA RATE)

Figure 14-19. Each row of slots in this Chrysler optical distributor disc acts as a separate sensor, creating signals used to control fuel injection, ignition timing, and idle speed.

trigger wheel. Therefore, its main advantage over the magnetic pulse generator is that it generates a full-strength output voltage signal even at slow cranking speeds. This allows precise switching signals to the ignition primary circuit and accurate and fine adjustments of the air-fuel mixture.

The **optical signal generator** uses the principle of light beam interruption to generate voltage signals. Many optical signal distributors contain a pair of light-emitting diodes (LED) and photo diodes installed opposite each other, figure 14-18. A disc containing two sets of chemically etched slots is installed between the LEDs and photo diodes. Driven by the camshaft, the disc acts as a timing member and revolves at half engine speed. As each slot interrupts the light beam, an alternating voltage is created in each photo diode. A hybrid integrated circuit converts the alternating voltage into on-off pulses sent to the PCM.

The high-data-rate slots, or outer set, are spaced at intervals of two degrees of crankshaft rotation, figure 14-19. This row of slots is used for timing engine speeds up to 1,200 rpm. Certain slots in this set are missing, indicating the crankshaft position of the number one cylinder to the PCM. The low-data-rate slots, or inner set, consist of six slots correlated to the crankshaft top-dead-center angle of each cylinder. The PCM uses this signal for triggering the fuel-injection system and for ignition timing at speeds above 1,200 rpm. This way, the optical signal generator acts both as the crankshaft position (CKP) sensor and engine speed (RPM) sensor, as well as a switching device.

■ Dual Breaker Points

In the early days of the automobile, short dwell time was an obstacle to increasing the rpm limit and power of high-performance street engines. Racing engines used magnetos, which, by design, increase spark strength with rpm. Street engines, because of cost and starting requirement, were stuck with point-coil ignitions. At high engine speeds, above 4,500 or 5,000 rpm, a typical 33-degree dwell angle on a V-8 engine did not allow enough coil saturation time. The result was insufficient voltage, making high-speed misfires a familiar problem. Reducing the point gap to increase dwell time drastically cut point life because the narrow gap caused point arcing and burning at low speeds.

By approximately the 1930s, manufacturers built the first production dual-point distributors that eliminated this problem. The two point sets had staggered openings and closings. The dwell on each set of points might be only 27 to 31 degrees; however, the combined dwell could be 36 to 40 degrees, which provided adequate coil saturation time and voltage for strong ignition at high speeds.

The point sets were connected parallel, so the primary circuit was complete, or closed, whenever either or both point sets were closed. However, both sets of points had to be open to break the primary circuit and fire the spark plug. The first set of points to open did not break the circuit because the other set still provided a current path through the primary circuit. The second set to open was the "opening" set, since both sets were then open simultaneously for a short period. Because the first set to open was also the first to close, it was the "closing" set, and vital coil saturation time could begin. Dual-point distributors survived into the 1970s, but were not needed with the universal adoption soild-state switching devices and electronic ignition.

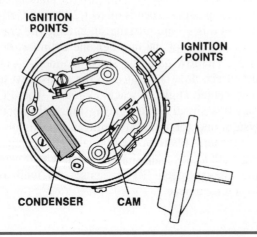

IGNITION POINTS

IGNITION POINTS

CONDENSER   CAM

MONITORING IGNITION PRIMARY CIRCUIT VOLTAGES

An ignition system voltage trace, both primary and secondary, is divided into three sections: firing, intermediate, and dwell. Deviations from a normal pattern indicate a problem. In addition, most scopes display ignition traces in three different patterns. Each pattern is best used to isolate and identify particular kinds of malfunctions. The three basic patterns are superimposed pattern, parade pattern, and stacked, or raster, pattern.

In a superimposed pattern, voltage traces for all cylinders are displayed one on top of another to form a single pattern. This display provides a quick overall view of ignition system operation and also reveals certain major problems. The parade pattern displays voltage traces for all cylinders one after another across the screen from left to right in firing order sequence. A parade display is useful for diagnosing problems in the secondary circuit. A raster pattern shows the voltage traces for all cylinders stacked one above another in firing order sequence. This display allows you to compare the time periods of the three sections of a voltage trace.

All ignition waveforms contain a vast amount of information about circuit activity, mechanical condition, combustion efficiency, and fuel mixture. Electronic ignition system waveform patterns vary with manufacturer and ignition types; therefore, it is important to be able to recognize the normal waveform patterns and know how they relate to the particular system. The differences between the scope trace of breaker point and electronic ignition systems occur in the intermediate and the dwell sections, figure 14-20. The most noticeable difference is that no condenser oscillations are present in the intermediate section of the waveform trace on electronic ignition systems. Also, the beginning of the dwell section in the breaker point system is the points closing, while on electronic ignition systems it is the current on signal from the switching device. Another difference is that the current limiting devices on electronic ignition systems often display an abrupt rise in dwell voltage called a current limiting hump. These and other waveform characteristics are used to diagnose many primary circuit malfunctions.

With ignition scope experience and knowledge of waveform data, primary ignition traces are categorized according to the particular features of the ignition system. The GM HEI ignition system

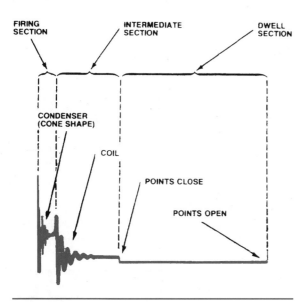

Figure 14-20. The primary superimposed pattern traces all the cylinders, one upon the other.

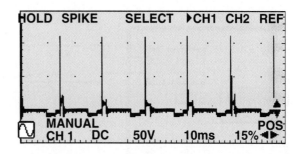

Figure 14-21. The firing line on Ford electronic distributorless ignition system (EDIS) should be approximately 342 volts for the primary circuit.

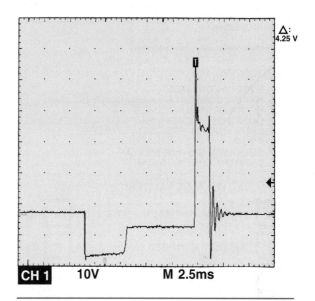

Figure 14-22. Ford thick film integrated (TFI) ignition systems use low-resistance coils that allow current acceleration and produce a distinct current limiting hump in the dwell trace.

is capable of producing 30,000 volts of secondary voltage at the spark plug. The primary firing line measures approximately 250 volts, figure 14-21. As the secondary firing voltage goes up or down, it performs the same in the primary firing voltage. Ford TFI systems display a dwell line with 4.25 milliseconds between the primary turn-on and the current limiting hump, figure 14-22. The low-resistance coil of a TFI system produces current acceleration to about 6 amps before the current limiting restricts the current. Early electronic ignition systems display features not particular to the newer, more efficient ignition systems.

Observing Switching Devices and Their Waveforms

The ignition scope has its place in diagnosing many primary ignition system malfunctions. Because the primary circuit deals with low voltage values, the lab scope is the best tool for diagnosing primary ignition system problems. A lab scope has greater resolution, which provides more clarity than an ignition scope. Many digital scopes are capable of storing waveform patterns for future review and comparison. Lab scopes are convenient because they observe other activity on the ignition primary circuit. The various primary circuit-switching devices also display voltage fluctuations during their prescribed trigger cycle that can be observed and diagnosed.

Magnetic pulse generators provide an AC voltage signal easily observed using a lab scope. The GM HEI ignition system magnetic pickup coil applies this ac symmetrical sine wave signal voltage to the "P" and "N" terminals on the HEI module, figure 14-23. The solid-state circuitry inside the ignition module converts the AC signal to a DC square wave that peaks at 5.0 volts. These AC and DC voltage fluctuations are observed separately. While the engine is cranking, the AC voltage signal from the pickup coil is accessed from terminal "P," figure 14-24. The square-wave signal HEI reference signal, is accessed from terminal "R." A dual-trace scope simultaneously displays the primary circuit voltage and the reference signal voltage, figure 14-25.

A Hall Effect switching device requires a three-wire circuit: power input, signal output,

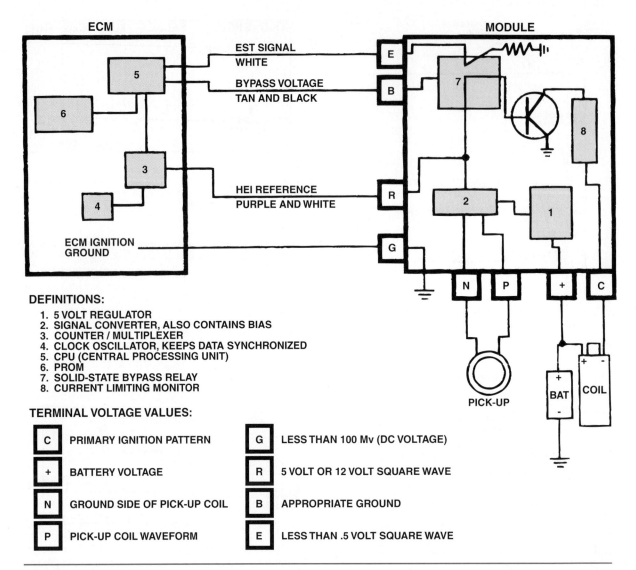

Figure 14-23. Schematic of the GM HEI magnetic pulse generator ignition system.

and ground. The Hall element receives an input voltage from the ignition switch or PCM, figure 14-26. As the shutter blade opens and closes, so does the input signal ground circuit. This opening and closing transmits a digital square-wave signal to the PCM. Many ignition systems are designed with the Hall Effect square waves peaking at 5.0 volts. Others may use 7.0, 9.0, or some other voltage.

More complex Hall Effect ignition systems require more details. The Hall Effect TFI ignition system on many Ford vehicles uses a camshaft position (CMP) sensor, which is a Hall Effect device inside the distributor. The CMP sensor produces a digital profile ignition pickup (PIP) signal. The shutter blade passes over a smaller vane, which identifies the number one cylinder for fuel injector firing. This signal is called the Signature PIP. The PIP signal is sent from the CMP sensor to the PCM and the ignition control module (ICM). The computer uses this signal to calculate the spark angle data for the ignition control module to control ignition coil firing. The ICM acts as a switch to ground in the primary circuit. The falling edge of the SPOUT signal controls the battery voltage applied to the primary circuit. The rising edge of the SPOUT signal controls the actual switch opening, figure 14-27. The PCM uses the inductive voltage spike when the primary field collapses and converts it into an ignition diagnostic monitor signal (IDM), figure 14-28. This

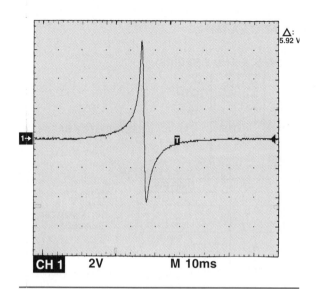

Figure 14-24. A lab scope displays an AC sine wave during cranking on HEI Terminal "P."

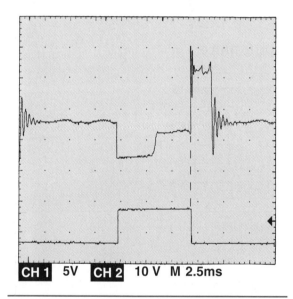

Figure 14-25. The HEI primary circuit voltage trace is displayed on the top and the square-wave reference signal is displayed on the bottom of the scope screen.

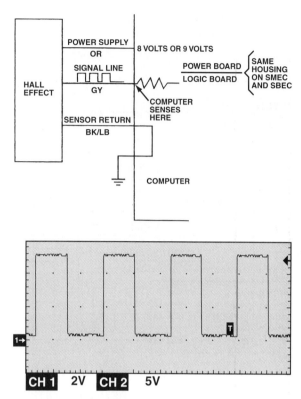

Figure 14-26. The lab scope displays the Hall Effect output signal transmitted to the PCM.

diagnostic monitor signal is used by the computer to observe the primary circuit. Because the ICM mounts on the exterior of the distributor, the various voltage waveforms are observed on a scope.

Optical ignition systems create distinct voltage signals. The Chrysler Optical ignition system on the 3.0L engine uses two photocells and two LEDs with solid-state circuitry to create two 5.0-volt signals. The inner set, or low data, of slots signals TDC for each cylinder while the outer slots (high data) monitor every two degrees of crankshaft rotation, figure 14-29. When observing the waveforms, if the low data rate or the high data rate signals fail on the high section, then the LED is most likely malfunctioning. If the signals fail on the low section, then the PCM is most likely not delivering the 5.0-volt reference signal. If the low data signal is lost, the engine normally starts. If the high data signal is lost, the engine starts, but on default settings from the PCM, figure 14-30.

■ Useful Distributor Tool

When timing an engine and rotating the distributor to advance or retard the spark, sometimes the distributor does not want to budge. This may occur even after the holddown bolt has been loosened. If it does, try using an oil filter wrench around the distributor. It may give the grip and leverage needed to free the distributor.

DISTRIBUTOR IGNITION SCHEMATIC EXAMPLE

DISTRIBUTOR IGNITION SCHEMATIC (5.8./7.5. E AND F SERIES)

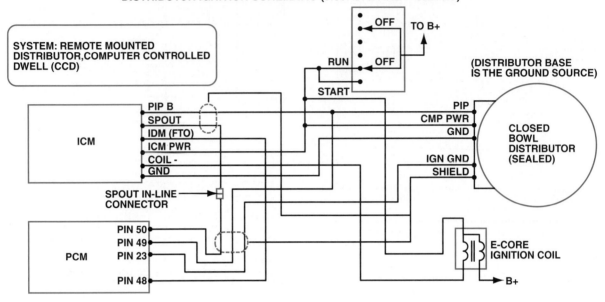

Figure 14-27. Schematic of the Ford TFI CCD ignition system on late-model trucks. (Courtesy of Ford Motor Company)

ICM WAVEFORM EXAMPLES (DISTRIBUTOR IGNITION)

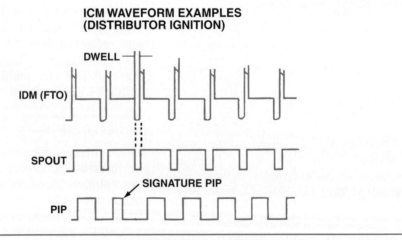

Figure 14-28. The IDM, SPOUT, and PIP signals are monitored to check power input, power output, and ground.

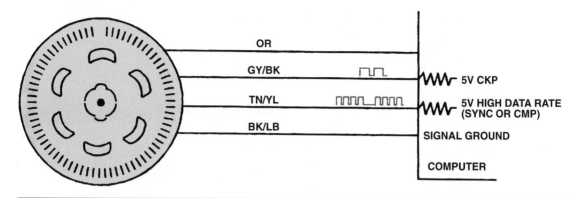

Figure 14-29. In a Chrysler Optical Ignition system, the low data rate square wave looks different than the high data rate square wave.

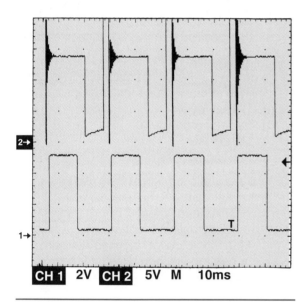

Figure 14-30. The primary circuit voltage trace of a Chrysler Photo Optical Ignition system is displayed on the top while the CKP square wave is displayed on the bottom of the scope screen.

SUMMARY

Through electromagnetic induction, the ignition system transforms the low voltage of the battery into the high voltage required to fire the spark plugs. Induction occurs in the ignition coil where current travels through the primary winding to build up a magnetic field. When the field rapidly collapses, high voltage is induced in the coil secondary winding. All domestic original-equipment ignitions are the battery-powered, inductive-discharge type.

The ignition system is divided into two circuits: primary and secondary. The primary circuit contains the battery, the ignition switch, the coil primary winding, a switching device, and signal devices to determine crankshaft position.

For over sixty years, mechanical breaker points were used as the primary circuit-switching device. Solid-state electronic components replaced breaker points as the switching device in the 1970s. The two most common solid-state switching devices are the magnetic pulse generator and the Hall Effect switch.

The breaker points are a mechanical switch that opens and closes the primary circuit. The time period when the points are closed is called the dwell angle. The dwell angle varies inversely with the gap between the points when they are open. As the gap decreases, the dwell increases. The ignition condenser is a capacitor that absorbs primary voltage when the points open. This prevents arcing across the points and premature burning.

Three types of common ignition electronic signal devices are the magnetic pulse generator, the Hall Effect switch, and the optical signal generator. The magnetic pulse generator creates an ac voltage signal as the reluctor continually passes the pole piece, changing the induced voltage to negative. The Hall Effect switch uses a stationary sensor and rotating trigger wheel just like the magnetic pulse generator. Its main advantage over the magnetic pulse generator is that it can generate a full-strength output voltage signal even at slow speeds. The optical signal generator acts both as the crankshaft position CKP sensor and TDC sensor, as well as a switching device. It uses light beam interruption to generate voltage signals.

Review Questions

Choose the letter that represents the best possible answer to the following questions:

1. The voltage required to ignite the air-fuel mixture ranges from:
   a. 5 to 25 volts
   b. 50 to 250 volts
   c. 500 to 2,500 volts
   d. 5,000 to 40,000 volts

2. Which of the following does *not* require higher voltage levels to cause an arc across the spark plug gap?
   a. Increased spark plug gap
   b. Increased engine operating temperature
   c. Increased fuel in air-fuel mixture
   d. Increased pressure of air-fuel mixture

3. In an Inductive Discharge ignition, the coil transforms low voltage from the primary circuit to high voltage for the secondary circuit through:
   a. Magnetic induction
   b. Capacitive discharge
   c. Series resistance
   d. Parallel capacitance

4. Voltage induced in the secondary winding of the ignition coil is how many times greater than the self-induced primary voltage?
   a. 1 to 2
   b. 10 to 20
   c. 100 to 200
   d. 1,000 to 2,000

5. The two circuits of the ignition system are:
   a. The "Start" and "Run" circuits
   b. The point circuit and the coil circuit
   c. The primary circuit and the secondary circuit
   d. The insulated circuit and the ground circuit

6. Which of the following components is part of both the primary and the secondary circuits?
   a. Ignition switch
   b. Distributor rotor
   c. Switching device
   d. Coil

7. Which of the following is *not* contained in the primary circuit of an ignition system?
   a. Battery
   b. Spark plugs
   c. Ignition switch
   d. Coil primary winding

8. Which of the following is true of the coil primary windings?
   a. They consist of 100 to 150 turns of very fine wire.
   b. The turns are insulated by a coat of enamel.
   c. The negative terminal is connected directly to the battery.
   d. The positive terminal is connected to the switching device and to ground.

9. In order to collapse the magnetic field of the coil, the primary circuit requires a:
   a. Ballast resistor
   b. Switching device
   c. Condenser
   d. Starting bypass circuit

10. In a electronic ignition system, the distributor cam are replaced by:
    a. RFI filter capacitors
    b. Auxiliary ballast resistors
    c. Magnetic pickup triggering devices
    d. Integrated coil and distributor cap assemblies

11. All of the following are electronic ignition switching devices e*xcept:*
    a. Pickup coil
    b. Photo optical LED
    c. Capacitor
    d. Hall Effect switch

12. A Hall Effect switch requires _____ to generate an output voltage signal.
    a. A reference voltage
    b. A crankshaft position sensor
    c. Pickup coil
    d. A shunt transistor

13. Dwell is measured in distributor degrees while dwell time is measured in:
    a. Kilovolts
    b. Milliseconds
    c. Millimeters
    d. Ohms

14. The distributer rotates at _____ the speed of the crankshaft.
    a. One-half
    b. The same speed as
    c. Twice
    d. One-quarter

15. A current limiting hump displays:
    a. Voltage changes
    b. Time increments
    c. Firing line
    d. Current changes

16. As voltage increases, current _____.
    a. Induces
    b. Decreases
    c. Lowers
    d. Stays the same

15

The Ignition Secondary Circuit and Components

OBJECTIVES

Upon completion and review of this chapter, you will be able to:

- Describe the operation of the secondary ignition system.
- List the secondary ignition components.
- List conditions that cause high resistance in the secondary circuit.
- Explain the various design features of spark plugs.

KEY TERMS

available voltage
bakelite
burn time
extended-core
 spark plug
firing voltage
heat range
ionize
no-load oscillation
reach

required voltage
resistor-type
 spark plugs
spark voltage
television-radio-
 suppression (TVRS)
 cables
voltage decay
voltage reserve

INTRODUCTION

The general operation of the ignition primary circuit and some of the system components have already been covered. This chapter examines the secondary circuit and its components.

The secondary circuit must conduct surges of high voltage. To do this, it has large conductors and terminals, and heavy-duty insulation. The secondary circuit in a distributor ignition system, figure 15-1, consists of the:

- Coil secondary winding
- Distributor cap and rotor
- Ignition cables
- Spark plugs

The secondary circuit in a distributorless ignition system (DIS), figure 15-2, consists of the:

- Coil packs
- Ignition cables, where applicable
- Spark plugs

The secondary circuit in a coil over plug DIS consists of the same components as a DIS system except that each cylinder has its own separate coil and there are usually no ignition wires to the

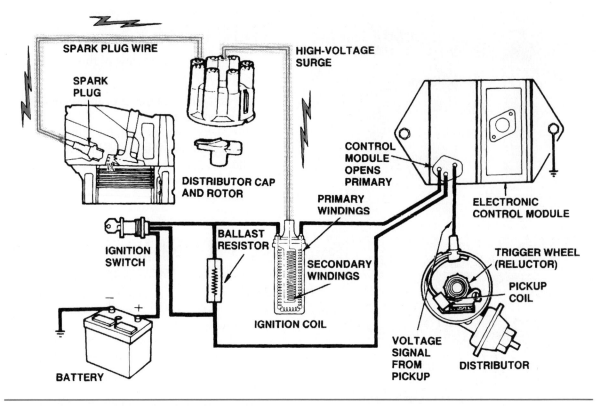

Figure 15-1. Typical early Chrysler electronic ignition system with a distributor, cap, and rotor.

spark plugs, figure 15-3. Also, the coil over plug system fires each cylinder sequentially rather than using the waste spark method. Refer to Chapter 16 of this *Classroom Manual* for additional information.

IGNITION COILS

The ignition coil steps up voltage in the same way as a transformer. When the magnetic field of the coil primary winding collapses, it induces a high voltage in the secondary winding.

Coil Secondary Winding and Primary-to-Secondary Connections

Two windings of copper wire compose the ignition coil. The primary winding of heavy wires consists of 100 to 150 turns; the secondary winding is 15,000 to 30,000 turns of a fine wire. The ratio of secondary turns to primary turns is usually between 100 and 200. To increase the strength of

the magnetic field, the windings are wrapped around a laminated core of soft iron, figure 15-4.

The coil must be protected from the underhood environment to maintain its efficiency. Four coil designs are used:

- Oil-filled coil
- Laminated E-core coil
- DIS coil packs
- Coil over plug assemblies

Oil-filled Coil

In the oil-filled coil, the primary winding is wrapped around the secondary winding, which is wrapped around the iron core. The coil windings are insulated by layers of paper and the entire case is filled with oil for greater insulation. The top of the coil is molded from an insulating material such as Bakelite. Metal inserts for the winding terminals are installed in the cap. Primary and secondary terminals are generally marked with a + and −, figure 15-5. Leads are attached with nuts and washers on some coils; others use push-on connectors. The entire unit is sealed to keep out dirt and moisture.

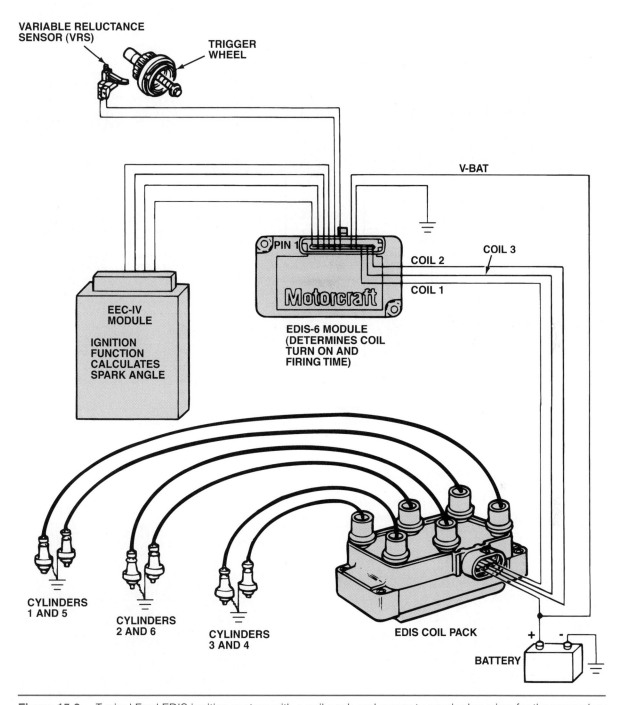

Figure 15-2. Typical Ford EDIS ignition system with a coil pack and separate spark plug wires for the secondary system. (Courtesy of Ford Motor Company)

Laminated E-Core Coil

Unlike the oil-filled coil, the E-core coil uses an iron core laminated around the windings and potted in plastic, much like a small transformer, figure 15-6. The coil is so named because of the "E" shape of the laminations making up its core. Because the laminations provide a closed magnetic path, the E-core coil has a higher energy transfer. The secondary connection looks much like a spark plug terminal. Primary leads are housed in a single snap-on connector that attaches to the coil's blade-type terminals. The E-core coil has very low primary resistance and is used without a ballast resistor in Ford TFI and some GM HEI ignitions.

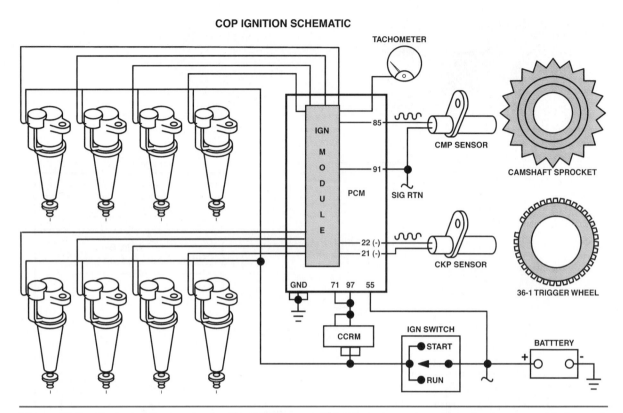

Figure 15-3. Typical coil over plug ignition system on a V-8 with a separate coil/spark plug assembly for each cylinder.

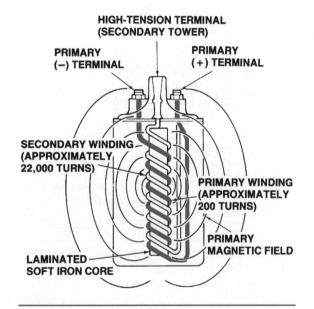

Figure 15-4. The laminated iron core within the coil strengthens the coil's magnetic field.

Figure 15-5. Many oil-filled coils are marked with the polarity of the primary leads.

DIS Coil Packs

Distributorless ignitions use two or more coils in a single housing called a coil pack, figure 15-7. Because the E-core coil has a primary and sec- ondary winding on the same core, it uses a common terminal. Both ends of the E-core coil's primary winding connect to the primary ignition circuit; the open end of its secondary winding connects to the center tower of the coil, where the distributor high-tension lead connects.

Figure 15-6. The E-core coil is used without a ballast resistor.

Figure 15-7. Typical DIS coil packs used on cylinder engines.

Coil packs are significantly different, using a closed magnetic core with one primary winding for each two high-voltage outputs. The secondary circuit of the coil pack is wired in series. Each coil in the coil pack directly provides secondary voltage for two of the spark plugs, which are in series with the coil secondary winding. Coil pack current is limited by transistors called output drivers, which are located in the ignition module attached to the bottom of the pack. The output drivers open and close the ground path of the coil primary circuit. Other module internal circuits control timing and sequencing of the output drivers.

Coil Over Plug Coil Assemblies

Each cylinder has its own separate coil and there are no ignition wires to the spark plugs on all coil over plug ignition systems. Each cylinder is fired only on its compression stroke.

Coil Voltage

A coil must supply the correct amount of voltage for any system. Since this amount of voltage varies, depending on engine and operating conditions, the voltage available at the coil is generally greater than what is required by the system. If it is less, the engine may not run.

Available Voltage

The ignition coil supplies much more secondary voltage than the average engine requires. The peak voltage that a coil produces is called its **available voltage.**

Three important coil design factors determine available voltage level:

- Secondary-to-primary turns ratio
- Primary voltage level
- Primary circuit resistance

The turns ratio is a multiplier that creates high secondary voltage output. The primary voltage level that is applied to a coil is determined by the ignition circuit design and condition. Anything affecting circuit resistance such as incorrect parts or loose or corroded connections, affects this voltage level. Generally, a primary circuit voltage loss of one volt may decrease available voltage by 10,000 volts.

If there were no spark plug in the secondary circuit, the coil secondary voltage would have no place to discharge quickly. That is, the circuit would remain open so there is no path to ground for the current to follow. As a result, the voltage oscillates in the secondary circuit and dissipates as heat. The voltage is completely gone in just a few milliseconds. Figure 15-8 shows the trace of this no-load, open-circuit voltage. This is called secondary voltage **no-load oscillation.** The first peak of the voltage trace represents the maximum available voltage from that particular coil. Available voltage is usually between 20,000 and 50,000 volts.

Required Voltage

When there is a spark plug in the secondary circuit, the coil voltage creates an arc across the plug air gap to complete the circuit. Figure 15-9 compares a typical no-load oscillation to a typical secondary

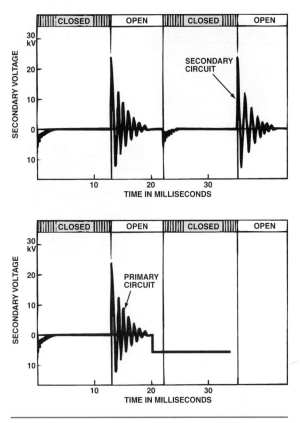

Figure 15-8. A secondary circuit no-load voltage trace.

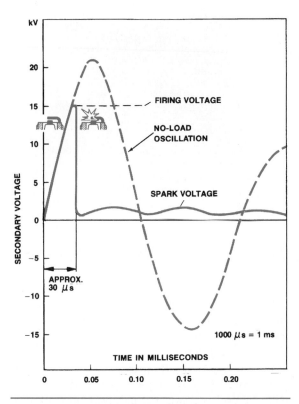

Figure 15-9. The dashed line shows no-load voltage; the solid line shows the voltage trace of firing voltage and spark voltage.

firing voltage oscillation. At about 15,000 volts, the spark plug air gap ionizes and becomes conductive. This is the ionization voltage level, also called the **firing voltage,** or **required voltage.**

As soon as a spark has formed, less voltage is required to maintain the arc across the air gap. This reduces the energy demands of the spark causing the secondary voltage to drop to the much lower spark voltage level. This is the inductive portion of the spark. **Spark voltage** is usually about one-fourth of the firing voltage level.

Figure 15-10 shows the entire trace of the spark. The spark duration or **burn time** of the trace indicates the amount of resistance and efficiency of the spark voltage. Burn time on most ignition systems is between 1.6 and 1.8 milliseconds. The traces shown are similar to the secondary circuit traces seen on an oscilloscope screen. High secondary resistance causes the spark voltage to increase and a quick burn time, or short spark duration. This occurs as the secondary system overcomes the resistance by reducing current and increasing voltage. This section of the waveform trace indicates secondary efficiency. When the secondary voltage falls below the inductive air-gap voltage level, the spark can no longer be maintained. The spark gap

becomes nonconductive. The remaining secondary voltage oscillates in the secondary circuit, dissipating as heat. This is called secondary **voltage decay.** At this time, the primary circuit closes and the cycle repeats, figure 15-11.

It is a good idea to memorize the three most typical firing voltages and their respective air-gap settings. The firing voltage on the spark plug of 0.035 inch is about 6 to 8 kV; 0.045-inch air gap is 8 to 10 kV; and 0.060 air gap is 10 to 12 kV. There are many more plug air-gap settings but it is important to know the three most popular spark plug gap settings and their kilovolt requirements to compare to other systems. The scope pattern for the secondary circuit quickly shows the firing voltage, figure 15-12.

Secondary ignition component failure due to wear is a common problem that results in high circuit resistance. Whether caused by a damaged plug wire, worn distributor cap, or excessive spark plug gap, this additional resistance disrupts current and alters the scope trace. Higher voltage, which is required to overcome this type of resistance, reduces current. The energy in the coil dissipates more rapidly than normal. As a result, high circuit resistance produces a high firing spike followed by high, short spark line.

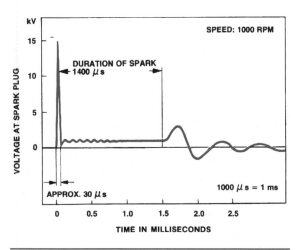

Figure 15-10. The voltage trace of an entire secondary ignition pulse.

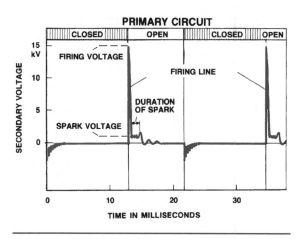

Figure 15-11. As the primary circuit opens and closes, the ignition cycle repeats.

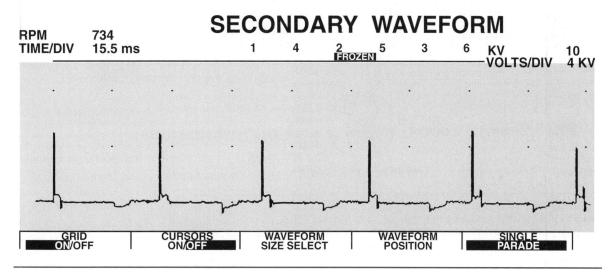

Figure 15-12. Normal secondary firing voltages for the secondary system with an 0.045-inch air gap.

Firing voltage for the DIS exhibits a slightly different waveform pattern than distributor ignition systems because of the design of the "waste spark" secondary system. Distributorless ignition systems pair two cylinders, called companion cylinders, and fire both spark plugs at the same time. One cylinder fires on the compression stroke, while the other cylinder fires on the exhaust stroke. The spark plugs are wired in series. One fires in the conventional method, from center electrode to ground electrode, while the other fires from ground electrode to center electrode. Every other cylinder has reverse polarity, figure 15-13. The cylinder on the compression stroke requires more voltage to fire the air-fuel mixture than the other cylinder. The exhaust waste spark, or stroke pattern, shows less voltage because there is much less

resistance across the spark plug gap of the waste spark, figure 15-14. Also, the DIS fires once per engine revolution as opposed to every other revolution in a distributor engine. As a result, DIS plugs wear twice as fast as distributor ignition plugs.

Secondary waveform patterns should be viewed first to gather the most complete list of ignition and engine driveability problems. The secondary system clearly indicates the operating resistance—or the systems that are expending too much resistance. The secondary waveform patterns include data for resistance in three areas: the electrical resistance (air gaps and secondary circuit), the cylinder compression resistance (mechanical), and the combustion resistance (air-fuel ratio performance). The resistance expended by faulty secondary ignition wires, excessive spark plug and rotor air gaps, weak compression, incorrect valve timing, or poor air-fuel mixing may be detected on the secondary waveform pattern.

Some other conditions of high resistance that cause required voltage levels to increase are:

- Eroded electrodes in the distributor cap, rotor, or spark plug

- Damaged ignition cables
- Reversed plug polarity
- High compression pressures
- Lean air-fuel mixture that is more difficult to ionize

Voltage Reserve

The physical condition of the automotive engine and ignition system affects both available and required voltage levels. Figure 15-15 shows available and required voltage levels in a particular ignition system under various operating conditions. **Voltage reserve** is the amount of coil voltage available in excess of the voltage required.

Under certain poor circuit conditions, there may be no voltage reserve. At these times, some spark plugs do not fire and the engine runs poorly or not at all. Ignition systems must be properly maintained to ensure that there is always some voltage reserve. Typically, an ignition system should have a voltage reserve of about 60 percent of available voltage under most operating conditions.

Coil Installations

Ignition coils are usually mounted with a bracket on a fender panel in the engine compartment or on the engine, figure 15-16.

Some ignition coils have an unusual design and location. The Delco-Remy high energy ignition (HEI) electronic ignition system has a coil

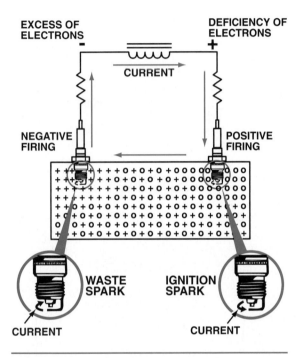

Figure 15-13. Although the current in the secondary windings of the DIS coil does not reverse, one spark plug fires with normal polarity, while the other spark plug fires with reverse polarity.

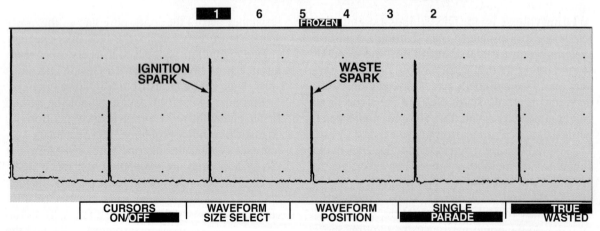

Figure 15-14. The cylinder with the greatest resistance (compression) exhibits high firing voltage while the cylinder with the least resistance (waste) exhibits the lower firing voltage.

mounted in the distributor. The coil output terminal is connected directly to the center electrode of the distributor cap. The connections to the primary winding are made through a multi-plug connector.

Many ignition systems manufactured in Asia have the coil inside the distributor along with the pickup coil, rotor, and module. Toyota mounts the ignition coil inside the distributor, figure 15-17, but positions the module (igniter) on the bulkhead or shock tower for cooling purposes.

Distributorless ignitions use an assembly containing one or more separate ignition coils and an electronic ignition module, figure 15-18. Control circuits in the module discharge each coil separately in sequence, with each coil serving two cylinders 360 degrees apart in the firing order. Chapter 16 of this *Classroom Manual* covers this type of system in more detail.

In any system, the connections to the primary winding must be made correctly. If spark plug po-

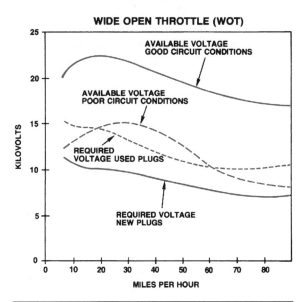

Figure 15-15. Available and required voltage levels under different system conditions.

Figure 15-16. A typical distributorless ignition system coil pack may be mounted on the engine in a location that provides easy access.

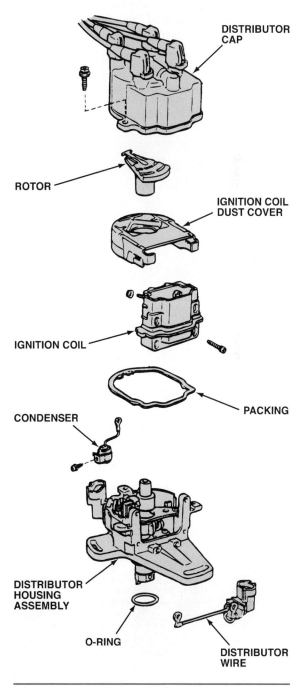

Figure 15-17. Toyota mounts the ignition coil inside the distributor.

larity is reversed, greater voltage is required to fire the plug. Plug polarity is established by the ignition coil connections.

One end of the coil secondary winding is connected to the primary winding, figure 15-19, so the secondary circuit is grounded through the ignition primary circuit. When the coil terminals are properly connected to the battery, the grounded end of the secondary circuit is electrically positive. The other end of the secondary circuit, which is the center electrode of the spark plug, is electrically negative. Whether the secondary winding

Figure 15-18. The Buick C3I distributorless ignition uses a single coil pack with three coils, each of which serves two cylinders 360 degrees apart in the firing order.

is grounded to the primary + or [−] terminal depends on whether the windings are wound clockwise or counterclockwise.

The secondary circuit must have negative polarity, or positive ground, for two main reasons. First, electrons flow more easily from negative to positive than they do in the opposite direction. Second, high temperatures of the spark plug center electrode increase the rate of electron movement, or current. The center electrode is much hotter than the side electrode because it cannot transfer heat to the cylinder head as easily. The electrons move quickly and easily to the side electrode when the air-fuel mixture is ignited. Although the secondary operates with negative polarity, it is a positive ground circuit.

If the coil connections are reversed, figure 15-20, spark plug polarity is reversed. The grounded end of the secondary circuit is electrically negative. The plug center electrode is electrically positive, and the side electrode is negative. When plug polarity is reversed, 20 to 40 percent more secondary voltage is required to fire the spark plug.

Coil terminals are usually marked BAT or +, and DIST or [−]. To establish the correct plug polarity with a negative-ground electrical system, the + terminal must be connected to the positive terminal of the battery through the ignition switch, starter relay, and other circuitry. The [−] coil terminal must be connected to the ignition control module.

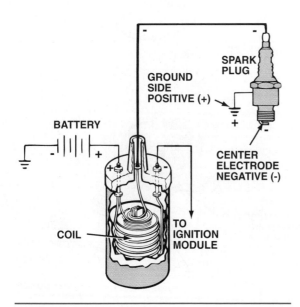

Figure 15-19. When coil connections are made properly, the spark plug center electrode is electrically negative.

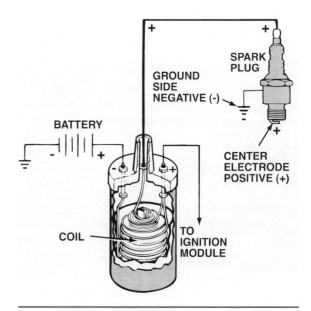

Figure 15-20. When coil connections are reversed, spark plug polarity is reversed.

DISTRIBUTOR CAP AND ROTOR

The distributor cap and rotor, figure 15-21, receive high-voltage current from the coil secondary winding. Current enters the distributor cap through the central terminal called the coil tower. The rotor carries the current from the coil tower to the spark plug electrodes in the rim of the cap. The rotor mounts on the distributor shaft and rotates with the shaft so the rotor electrode moves from one spark plug electrode to another in the cap to follow the designated firing order.

Distributor Rotor

A rotor is made of silicone plastic, **bakelite,** or a similar synthetic material that is a very good insulator. A metal electrode on top of the rotor conducts current from the carbon terminal of the coil tower.

The rotor is keyed to the distributor shaft to maintain its correct relationship with the shaft and the spark plug electrodes in the cap. The key may be a flat section or a slot in the top of the shaft. Delco-Remy V-6 and V-8, shown at the left in

figure 15-22, are keyed in place by two locators and secured by two screws. On the right is shown a plug-on or push-on distributor rotor. Most other rotors are pressed onto the shaft by hand. The rotor in a Chrysler optically triggered distributor is retained by a horizontal capscrew.

Rotors used with Hall Effect switches often have the shutter blades attached, figure 15-23, serving a dual purpose: In addition to distributing the secondary current, the rotor blades bypass the Hall Effect magnetic field and create the signal for the primary circuit to fire.

Rotor Air Gap

An air gap of a few thousandths of an inch, or a few hundredths of a millimeter, exists between the tip of the rotor electrode and the spark plug electrode of the cap. If they actually touch, both

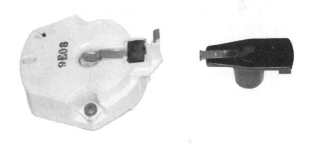

Figure 15-22. Typical distributor rotors.

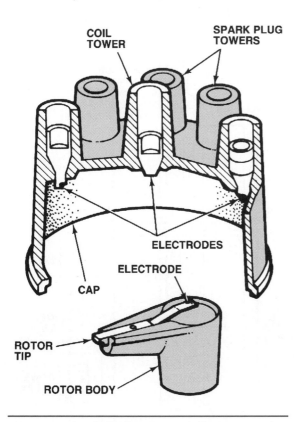

Figure 15-21. A distributor rotor and a cutaway view of the distributor cap.

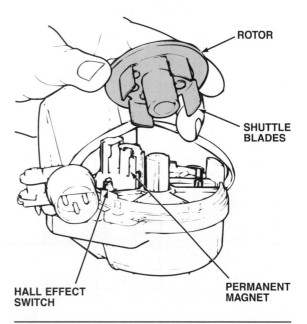

Figure 15-23. A Hall Effect triggering device attached to the rotor. (Courtesy of DaimlerChrysler Corporation)

would wear very quickly. Because the gap cannot be measured when the distributor is assembled, it is usually described in terms of the voltage required to create an arc across the electrodes. Only about 3,000 volts are required to create an arc across some air gaps, but others require as much as 9,000 volts. As the rotor completes the secondary circuit and the plug fires, the rotor air gap adds resistance to the circuit. This raises the plug firing voltage, suppresses secondary current, and reduces RFI.

Distributor Cap

The distributor cap is also made of silicone plastic, Bakelite, or a similar material that resists chemical attack and protects other distributor parts. Metal electrodes in the spark plug towers and a carbon insert in the coil tower provide electrical connections with the ignition cables and the rotor electrode, figure 15-24. The cap is keyed to the distributor housing and is held on by two or four spring-loaded clips or by screws.

■ Dual Plug DIS

In 1989, Ford introduced the dual plug distributorless ignition system (DP-DIS) on the 2.3L Ranger. This system was added to the 2.3L EFI Mustang engine in 1991. The dual plug system is made up of a crankshaft position sensor mounted near the crankshaft, two 4-tower DIS coil packs, and a DIS ignition module. Each coil fires two spark plugs at the same time. This system is also a typical waste spark system. The dual plug DIS features simultaneous firing voltages to the same cylinder to provide more efficient combustion and cleaner emissions. The coil pack and plugs on the right side operate continuously while the coil pack and plugs on the left side are controlled by the computer and are switched on or off according to engine temperatures, engine load, and air-fuel mixture demands.

2.3L RANGER DP-DIS SCHEMATIC

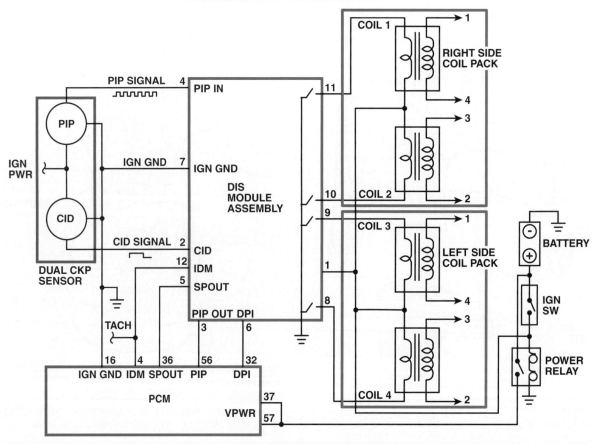

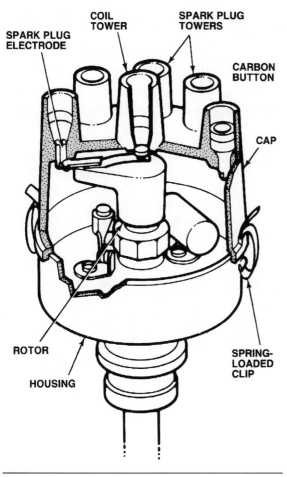

Figure 15-24. The distributor cap and rotor assembled with the distributor housing.

Delco-Remy HEI distributor caps with an integral coil, figure 15-25, are secured by four spring-loaded clips. When removing this cap, ensure that all four clips are disengaged and clear of the housing. Then lift the cap straight up to avoid bending the carbon button in the cap and the spring that connects it to the coil. If the button and spring are distorted, arcing can occur that will burn the cap and rotor.

Positive-engagement spark plug cables are used with some Chrysler and Ford 4-cylinder ignition systems. There are no electrodes in distributor caps used with these cables. A terminal electrode attached to the distributor-cap end of the cable locks inside the cap to form the distributor contact terminal. The secondary terminal of the cable is pressed into the cap, figure 15-26.

Ford distributors used with some electronic engine control (EEC) systems have caps and rotors with the terminals on two levels to prevent secondary voltage arcing. Spark plug cables are not connected to the caps in firing order sequence, but the caps are numbered with the engine cylinder numbers, figure 15-27. The caps have two sets of numbers, one set for 5.0-liter standard engines, and the other for 5.7-liter and 302-cid high-performance engines. Cylinder numbers must be carefully checked when changing spark plug cables.

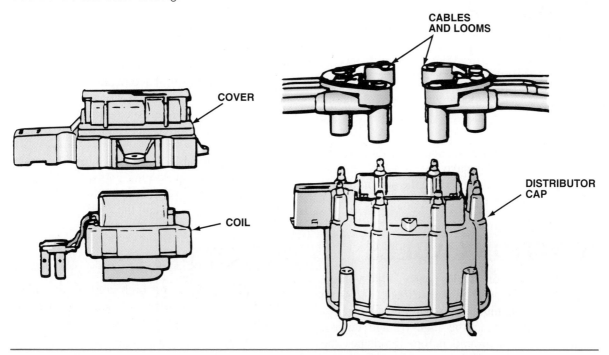

Figure 15-25. Many Delco-Remy HEI electronic ignition systems used on V-6 and V-8 engines have a coil mounted in the distributor cap.

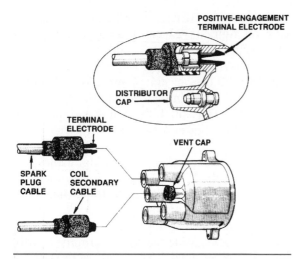

Figure 15-26. Chrysler 4-cylinder distributors have used positive locking terminal electrodes as part of the ignition cable since 1980. (Courtesy of DaimlerChrysler Corporation)

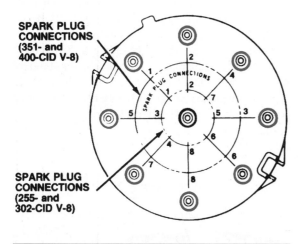

Figure 15-27. Spark plug cable installation order for Ford V-8 EEC systems.(Courtesy of Ford Motor Company)

Distributor caps used on some late-model Ford and Chrysler vehicles have a vent to prevent the buildup of moisture and reduce the accumulation of ozone inside the cap.

IGNITION CABLES

Secondary ignition cables carry high-voltage current from the coil to the distributor and from the distributor to the spark plugs. They use heavy insulation to prevent the high-voltage current from jumping to ground before it reaches the spark plugs. Ford, GM, and some other elec-

tronic ignitions use an 8-mm cable; all others use a 7-mm cable.

Conductor Types

Spark plug cables originally used a solid steel or copper wire conductor. Cables manufactured with these conductors were found to cause radio and television interference. While this type of cable is still made for special applications such as racing, most spark plug cables have been made of high-resistance, nonmetallic conductors for the past 30 years. Several nonmetallic conductors may be used, such as carbon, and linen or fiberglass strands impregnated with graphite. The nonmetallic conductor acts as a resistor in the secondary circuit and reduces RFI and spark plug wear due to high current. Such cables are often called **television-radio-suppression (TVRS) cables,** or just suppression cables.

When replacing spark plug cables on vehicles with computer-controlled systems, ensure that the resistance of the new cables is within the original equipment specifications to avoid possible electromagnetic interference with the operation of the computer.

Terminals and Boots

Secondary ignition cable terminals are designed to make a strong contact with the coil and distributor electrodes. They are, however, subject to corrosion and arcing if not firmly seated and protected from the elements.

Positive-engagement spark plug cable terminals, figure 15-28, lock in place inside the distributor cap and cannot accidentally come loose. They can only be removed when the cap is off the distributor. The terminal electrode is then compressed with pliers and the wire is pushed out of the cap.

The ignition cables must have special connectors, often called spark plug boots, figure 15-29. The boots provide a tight and well-insulated contact between the cable and the spark plug.

SPARK PLUGS

Spark plugs allow the high-voltage secondary current to arc across a small air gap. The three basic parts of a spark plug, figure 15-30, are:

- A ceramic core, or insulator, that insulates the center electrode and acts as a heat conductor.

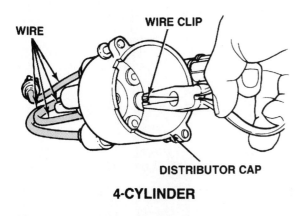

4-CYLINDER

Figure 15-28. Positive locking terminal electrodes are removed by compressing the wire clips with pliers and removing the wire from the cap.

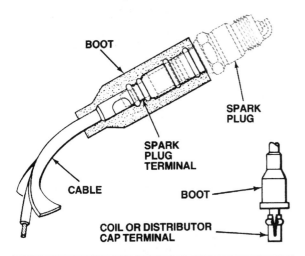

Figure 15-29. Ignition cables, terminals, and boots work together to carry the high-voltage secondary current.

- Two electrodes, one insulated in the core and the other grounded on the shell.
- A metal shell that holds the insulator and electrodes in a gas-tight assembly and has threads to hold the plug in the engine.

The metal shell grounds the side electrode against the engine. The other electrode is encased in the ceramic insulator. A spark plug boot and cable are attached to the top of the plug. High-voltage current travels through the center of the plug and arcs from the tip of the insulated electrode to the side electrode and ground. This spark ignites the air-fuel mixture in the combustion chamber to produce power.

The burning gases in the engine can corrode and wear the spark plug electrodes. Electrodes are

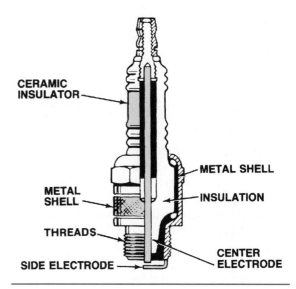

Figure 15-30. A cutaway view of the spark plug.

made of metals that resist this attack. Most electrodes are made of high-nickel alloy steel, but platinum and silver alloys are also used.

Spark Plug Firing Action

The arc of current across a spark plug air gap provides two types of discharge:

- Capacitive
- Inductive.

When a high-voltage surge is first delivered to the spark plug center electrode, the air-fuel mixture in the air gap cannot conduct an arc. The spark plug acts as a capacitor, with the center electrode storing a negative charge and the grounded side electrode storing a positive charge. The air gap between the electrodes acts as a dielectric insulator. This is the opposite of the normal negative-ground polarity, and results from the polarity of the coil secondary winding.

Secondary voltage increases, and the charges in the spark plug strengthen until the difference in potential between the electrodes is great enough to **ionize** the spark plug air gap. That is, the air-fuel mixture in the gap is changed from a nonconductor to a conductor by the positive and negative charges of the two electrodes. The dielectric resistance of the air gap breaks down and current travels between the electrodes. The voltage level at this instant is called ionization voltage. The current across the spark plug air gap at the instant of ionization is the capacitive portion of the spark. It flows from negative to positive and uses the energy stored in the plug itself when the plug was acting as a capacitor, before ionization. This is the portion of the spark that starts the combustion process within the engine.

The ionization voltage level is usually less than the total voltage produced in the coil secondary winding. The remainder of the secondary voltage (voltage not needed to force ionization) is dissipated as current across the spark plug air gap. This is the inductive portion of the spark discharge, which causes the visible flash or arc at the plug. It contributes nothing to the combustion of the air-fuel mixture, but is the cause of electrical interference and severe electrode erosion. High-resistance cables and spark plugs suppress this inductive portion of the spark discharge and reduce wear.

SPARK PLUG CONSTRUCTION

Spark Plug Design Features

Spark plugs are made in a variety of sizes and types to fit different engines. The most important differences among plugs are:

- Reach
- Heat range
- Thread and seat
- Air gap

These are illustrated in figure 15-31.

Reach
The **reach** of a spark plug is the length of the shell from the seat to the bottom of the shell, including both threaded and unthreaded portions. If an incorrect plug is installed and the reach is too short, the electrode will be in a pocket and the spark will not ignite the air-fuel mixture very well, figure 15-32.

If the spark plug reach is too long, the exposed plug threads could get hot enough to ignite the air-fuel mixture at the wrong time. It may be difficult to remove the plug due to carbon deposits on the plug threads. Engine damage can also result from interference between moving parts and the exposed plug threads.

Heat Range
The **heat range** of a spark plug determines its ability to dissipate heat from the firing end. The length

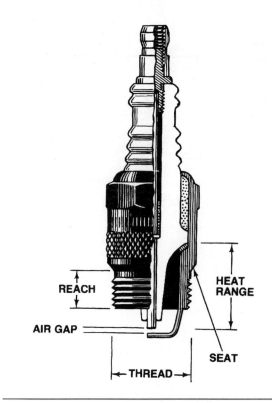

Figure 15-31. The design features of a spark plug.

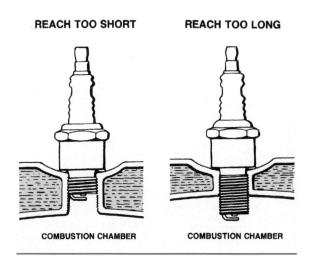

Figure 15-32. Spark plug reach.

of the lower insulator and conductivity of the center electrode are design features that primarily control the rate of heat transfer, figure 15-33. A "cold" spark plug has a short insulator tip that provides a short path for heat to travel, and permits the heat to rapidly dissipate to maintain a lower firing tip temperature. A "hot" spark plug has a long insulator tip that creates a longer path for heat to travel. This slower heat transfer maintains a higher firing tip temperature.

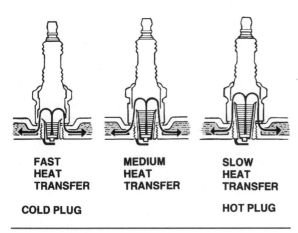

FAST HEAT TRANSFER   **MEDIUM HEAT TRANSFER**   **SLOW HEAT TRANSFER**

COLD PLUG   **HOT PLUG**

Figure 15-33. Spark plug heat range.

Engine manufacturers choose a spark plug with the appropriate heat range required for the normal or expected service for which the engine was designed. Proper heat range is an extremely important factor because the firing end of the spark plug must run hot enough to burn away fouling deposits at idle, but must also remain cool enough at highway speeds to avoid preignition. It is also an important factor in the amount of emissions an engine will produce.

Current spark plug designations use an alphanumeric system that identifies, among other factors, the heat range of a particular plug. Spark plug manufacturers are gradually redesigning and redesignating their plugs. For example, a typical AC-Delco spark plug carries the alphanumeric

■ **Making Tracks**

You often read instructions to inspect ignition parts for carbon tracks. Although you may have heard about or seen carbon tracks, have you ever thought about what they are and what causes them?

Carbon tracks are deposits or defects on distributor rotors, caps, spark plugs, and cables that create a short-circuit path to ground for secondary high voltage. They also cause crossfiring, in which the high voltage jumps from the distributor rotor to the wrong terminal in the cap.

The problems caused by carbon tracks are all pretty similar, but the causes for these defects are rather complex. A distributor cap or rotor may develop a hairline crack because of rough handling, a manufacturing defect, or some other problem. Under certain conditions, moisture collects in the crack and creates a lower-resistance path for high voltage. High-voltage arcing to ground in a distributor ionizes molecules and forms air-conductive deposits along its path. If any dirt or grease is in the short-circuit path, the combination of high voltage and its accompanying current causes carbon deposits to form around the crack. Thus, a carbon track develops.

Carbon tracks also form without a crack in a cap or rotor. High voltage ionizes air and oil molecules in the distributor and causes deposits to form. The deposits have high resistance, but if they are the least bit conductive, secondary voltage can arc to them. Over a period of time, the deposits build up and create a short circuit.

Outside a distributor, similar carbon tracks form on spark plug insulators and ignition cables due to grease deposits and weak points in damaged cable insulation.

Carbon tracks inside a distributor cap are often tricky to diagnose. Sometimes an engine runs smoothly at idle but misfires at high speed. As the distributor advance mechanisms operate, the rotor moves farther away from the cap terminals as the coil discharges. The high-voltage current must cross an increasing air gap. If a nearby carbon track provides lower resistance, the voltage will jump to ground and the engine misfires.

A typical carbon track has about the same, or a little less, resistance as a TVRS ignition cable. That is quite conductive enough to cause a misfire or a no-start problem. The accompanying photo shows a classic set of carbon tracks inside a distributor cap.

designation R45LTS6; the new all-numeric code for a similar AC spark plug of the same length and gap is 41-600. This makes it more difficult for those who attempt to correct driveability problems by installing a hotter or colder spark plug than the manufacturer specifies. Eventually, it no longer will be possible for vehicle owners to affect emissions by their choice of spark plugs.

Thread and Seat

Most automotive spark plugs are made with one of two thread diameters: 14 or 18 millimeters, figure 15-34. All 18-mm plugs have tapered seats that match similar tapered seats in the cylinder head. The taper seals the plug to the engine without the use of a gasket. The 14-mm plugs are made either with a flat seat that requires a gasket or with a tapered seat that does not. The gasket-type, 14-mm plugs are still quite common, but the 14-mm tapered-seat plugs are now used in most late-model engines. A third thread size is 10 millimeters; 10-mm spark plugs are generally used on motorcycles, but some automotive engines also use them; specifically, Jaguar's V-12 uses them as well.

The steel shell of a spark plug is hex-shaped so a wrench fits it. The 14-mm, tapered-seat plugs have shells with a 5/8-inch hex; 14-mm gasketed and 18-mm tapered-seat plugs have shells with a 13/16-inch hex. A 14-mm gasket plug with a 5/8-inch hex is used for special applications, such as a deep recessed with a small diameter opening.

Air Gap

The correct spark plug air gap is important to engine performance and plug life. A gap that is too narrow causes a rough idle and a change in the exhaust emissions. A gap that is too wide requires higher voltage to jump it; if the required voltage is greater than the available ignition voltage, misfiring results.

Special-Purpose Spark Plugs

Specifications for all spark plugs include the design characteristics just described. In addition, many plugs have other special features to fit particular requirements.

Resistor-Type Spark Plugs

This type of plug contains a resistor in the center electrode, figure 15-35. The resistor generally has a value of 7,500 to 15,000 ohms and is used to reduce RFI. **Resistor-type spark plugs** can be used in place of nonresistor plugs of the same size, heat range, and gap without affecting engine performance.

Extended-Tip Spark Plugs

Sometimes called an **extended-core spark plug,** this design uses a center electrode and insulator that extend farther into the combustion chamber, figure 15-36. The extended-tip operates hotter under slow-speed driving conditions to burn off

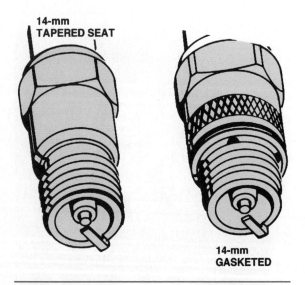

Figure 15-34. Spark plug thread and seat types.

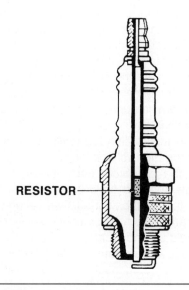

Figure 15-35. A resistor-type spark plug.

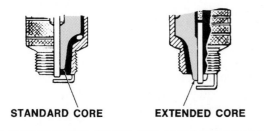

STANDARD CORE   **EXTENDED CORE**

Figure 15-36. A comparison of a standard and an extended-core spark plug.

■ Spark Plug Design

Many people have tried to redesign the spark plug. Not all of the "new" designs have worked out. For example, a plug manufactured before World War I had an insulated handle at the top. By pulling this handle up, an auxiliary gap was opened, presumably to create a hotter spark and stop oil fouling. A window in the side of the plug showed whether the gap was open or closed.

Another "revolutionary" type of plug had a screw connector that allowed the inner core assembly to be removed and cleaned quickly.

Still another design had threads and electrodes at each end of the plug. The plug could be removed, the terminal cap installed on the other end, and then reinstalled upside down. All of the photos were provided by Champion Spark Plug Company.

combustion deposits and cooler at high speed to prevent spark plug overheating.

Wide-Gap Spark Plugs

The electronic ignition systems on some late-model engines require spark plug gaps in the 0.045- to 0.080-inch (1.0- to 2.0-mm) range. Plugs for such systems are made with a wider gap than other plugs. This wide gap is indicated in the plug part number. Do not try to open the gap of a narrow-gap plug to create the wide gap required by such ignitions.

Copper-Core Spark Plugs

Many plug manufacturers are making plugs with a copper segment inside the center electrode. The copper provides faster heat transfer from the electrode to the insulator and then to the cylinder head and engine coolant. Copper-core plugs are also extended-tip plugs. This combination results in a more stable heat range over a greater range of engine temperatures and greater resistance to fouling and misfire.

Platinum-Tip Spark Plugs

Platinum-tip plugs are used in some late-model engines to increase firing efficiency. The platinum center electrode increases electrical conductivity, which helps prevent misfiring with lean mixtures and high temperatures. Since platinum is very resistant to corrosion and wear from combustion chamber gases and heat, recommended plug life is twice that of other plugs.

Long-Reach, Short-thread Spark Plugs

Some late-model GM engines, Ford 4-cylinder engines, and Ford 5.0-liter V-8 engines use 14-mm, tapered-seat plugs with a 3/4-inch reach but have threads only for a little over half of their length, figure 15-37. The plug part number includes a suffix that indicates the special thread design, although a fully threaded plug can be substituted if necessary.

Advanced Combustion Igniters

This extended-tip, copper-core, platinum-tipped spark plug was introduced by GM in 1991. It combines all the attributes of the individual plug designs described earlier and uses a nickel-plated shell for corrosion protection. This combination

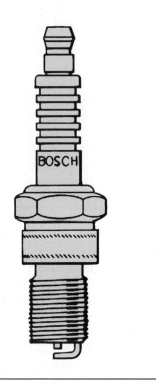

Figure 15-37. A long-reach, short-thread spark plug.

delivers a plug life in excess of 100,000 miles (160,000 km). No longer called a spark plug, the GM advanced combustion igniter (ACI) has a smooth ceramic insulator with no cooling ribs. The insulator is coated with a baked-on boot release compound that prevents the spark plug wire boot from sticking and causing wire damage during removal.

SUMMARY

The ignition secondary circuit generates the high voltage and distributes it to the engine to ignite the combustion charge. This circuit contains the coil secondary winding, the distributor cap and rotor, the ignition cables, and the spark plugs.

The ignition coil produces the high voltage necessary to ionize the spark plug gap through electromagnetic induction. Low-voltage current in the primary winding induces high voltage in the secondary winding. A coil must be installed with the same primary polarity as the battery to maintain proper secondary polarity at the spark plugs.

Available voltage is the amount of voltage the coil is capable of producing. Required voltage is the voltage necessary to ionize and fire the spark plugs under any given operating condition. Voltage reserve is the difference between available voltage and required voltage. A well-tuned ignition system should have a 60-percent voltage reserve.

The spark plugs allow the high voltage to arc across an air gap and ignite the air-fuel mixture in the combustion chamber. Important design features of a spark plug are its reach, heat range, thread and seat size, and the air gap. Other special features of spark plugs are the use of resistors, extended tips, wide gaps, and copper cores. For efficient spark plug firing, ignition polarity must be established so that the center electrode of the plug is negative and the ground electrode is positive.

Review Questions

Choose the letter that represents the best possible answer to the following questions:

1. Which of the following is used as insulation to protect the windings of coils?
   a. Steel
   b. Wood
   c. Iron
   d. Plastic

2. Which of the following statements is true about ignition coils?
   a. Easily repaired
   b. Adjustments made by setscrews
   c. Requires periodic adjustment
   d. Can be replaced

3. When the coil terminals are properly connected to the battery, the grounded end of the secondary circuit is electrically:
   a. Positive
   b. Negative
   c. Neutral
   d. Ionized

4. A loss of one volt in the primary circuit may decrease available secondary voltage by:
   a. 10 volts
   b. 100 volts
   c. 1,000 volts
   d. 10,000 volts

5. Firing voltage is usually about _____ as high as spark voltage.
   a. One-fourth
   b. One-half
   c. Four times
   d. Two times

6. The voltage delivered by the coil is:
   a. Its full voltage capacity under all operating conditions
   b. Approximately half of its full voltage capacity at all times
   c. Only the voltage necessary to fire the plugs under any given operating condition
   d. Its full voltage capacity only while starting

7. The voltage reserve is the:
   a. Voltage required from the coil to fire a plug
   b. Maximum secondary voltage capacity of the coil
   c. Primary circuit voltage at the battery side of the ballast resistor
   d. Difference between the required voltage and the available voltage of the secondary circuit

8. A well-tuned ignition system should have a voltage reserve of about _____ of available voltage, under most operating conditions.
   a. 30 percent
   b. 60 percent
   c. 100 percent
   d. 150 percent

9. Bakelite is a synthetic material used in distributors because of its good:
   a. Permeability
   b. Conductance
   c. Insulation
   d. Capacitance

10. Which of the following are basic parts of a spark plug?
    a. Plastic core
    b. Paper insulator
    c. Fiberglass shell
    d. Two electrodes

11. Which of the following is *not* an important design feature among types of spark plugs?
    a. Reach
    b. Heat range
    c. Polarity
    d. Air gap

12. In the illustration below, the dimension arrows indicate the:

    a. Heat range
    b. Resistor portion of the electrode
    c. Extended core length
    d. Reach

13. All spark plugs have:
    a. A resistor
    b. An extended core
    c. A ceramic insulator
    d. A series gap

16

Electronic Ignition Systems

OBJECTIVES

Upon completion and review of this chapter, you will be able to:

- Describe the advantages of electronic ignition systems.
- Describe ignition transistor operation.
- Identify various electronic ignition systems.
- List the different primary triggering devices used in electronic ignition systems.
- Identify a waveform from a Hall Effect and magnetic-type primary trigger.

KEY TERMS

closed-loop dwell control	Hall Effect
coil charge time	Hall Effect switch
current limiting hump	magnetic pulse generator
current ramping	module
dwell	optical signal generator
dwell time	variable-dwell
fixed-dwell	waste spark

INTRODUCTION

During the 1960s, electronic ignition systems were used only on a few high-performance engines. Electronic systems were not standard on domestic automobiles until the early 1970s; foreign manufacturers followed a few years later with their versions. Within less than one decade, electronic systems completely replaced breaker-point ignitions. Breaker-point and electronic ignition systems do the same thing, except that a breaker-point system switches the primary circuit mechanically while an electronic system does this by means of a transistor. The major driving force behind the rapid transition to electronic systems was their ability to help meet stringent emission-control standards.

Although electronic systems are more expensive to produce than breaker-point systems, their advantages greatly outweigh the drawback of increased cost:

- Greater available voltage, especially at high engine speeds
- Reliable system performance at all engine speeds

- Potential for more responsive and variable advance curves
- Decreased maintenance
- Engine operation and exhaust emissions are more accurately controlled by solid-state circuitry

BASIC ELECTRONIC IGNITION SYSTEMS

Electronic ignition systems differ from breaker-point systems in the devices used to control primary circuit current. Breaker-point systems use mechanical breaker points to open and close the primary circuit. The points are operated by a cam on the distributor shaft. Most solid-state systems use an electronic switch in the form of a sealed module to control primary circuit current. This module contains one or more transistors, integrated circuits, or other solid-state control components. A triggering device in the distributor functions with the control module.

As we learned in Chapter 14, this triggering device is usually a magnetic pulse generator or a Hall Effect switch. Electronic ignitions also can be triggered by a set of breaker points, as will be explained later in this chapter.

Electronic ignitions can be classified in terms of their primary circuit operation. There are two general types:

- Inductive discharge
- Capacitive discharge.

The difference between these types lies in how they use primary current to produce a high-voltage secondary current.

Inductive Discharge

The inductive discharge system uses battery voltage to create current through the coil primary winding. When a signal is sent by the triggering device, primary current is interrupted. This sudden *decrease* in primary current collapses the coil's magnetic field and induces a high-voltage surge in the secondary circuit.

In an electronic ignition, primary current passes through the ignition control **module,** not through the distributor breaker points. Most modules contain one or more large power transistors that switch the primary current. A power or a

switching transistor can transmit as much as 10 amperes of current—far more than a set of breaker points can. The power transistor is controlled by a driver transistor that receives voltage signals from the distributor signal generator.

Original equipment electronic ignitions use inductive discharge to provide about 30,000 volts of available voltage and sustain a spark for about 1.8 milliseconds (a millisecond is one-thousandth of a second, or 0.001 second).

Breaker Points

Inductive discharge is used in both electronic and breaker-point systems; only the device used to open the circuit differs. The earliest transistorized ignitions used breaker points as a mechanical switch to control voltage applied to a power transistor. Battery voltage causes current to travel through the control module to the breaker points, figure 16-1. This current biases the transistor base so that current can also travel through the coil primary winding. When the points open, current stops. The power transistor no longer conducts current to the coil primary winding, and an ignition spark is produced.

While full primary current travels through the transistor, the points carry less than one ampere of current. This minimized the pitting and burning problems encountered with breaker points. However, they were still subject to mechanical wear and bounced at high speeds. Electronic ignitions

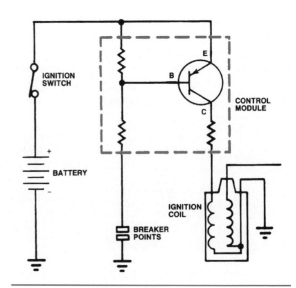

Figure 16-1. A typical breaker-point-triggered electronic ignition system.

replaced the points with a solid-state signal generator containing no moving parts.

Capacitive Discharge

The capacitive-discharge system uses battery voltage to charge a large capacitor in the control module. The capacitor charging time, while the module is on, corresponds to dwell in the inductive-discharge system. Current travels to the storage capacitor instead of to the coil during this period. The module charging circuit uses a transformer to increase the voltage in the capacitor to as high as 400 volts. The module also contains a silicon-controlled rectifier (SCR), which functions as an open switch to prevent the capacitor from discharging as it charges.

When the triggering device signals the module, the SCR closes the capacitor discharge circuit allowing the capacitor to discharge through the coil primary winding. This sudden *increase* in primary current expands the coil's magnetic field and induces a high-voltage surge in the secondary circuit.

In an inductive-discharge system, primary current travels through the coil during the dwell period and then is turned off to induce the high secondary voltage. In a capacitive-discharge system, the primary voltage and current charge the capacitor. The capacitor is then switched on, to induce the high secondary voltage. CD ignition systems charge the capacitor and discharge the coil faster than inductive-discharge systems. The CD ignition system can produce a higher (more kilovolts) and faster firing voltage. The CD ignition system also starts better in cold and wet weather and improves high-speed performance.

Capacitive-discharge systems are available in the automotive aftermarket, but no major domestic manufacturer installs them as original equipment. A few imported cars such as Audi and Mercedes-Benz have used these ignition systems in the past. They provide a greater available voltage than inductive-discharge systems, but can sustain a spark for only about 200 microseconds. The spark is much more intense than that produced by an inductive-discharge system and fires plugs that are in very poor condition. Under certain engine operating conditions, however, a longer spark time is required or the air-fuel mixture does not burn completely.

Figure 16-2 shows the time and voltage characteristics of a breaker-point ignition system, a capacitive-discharge system, and a solid-state inductive system.

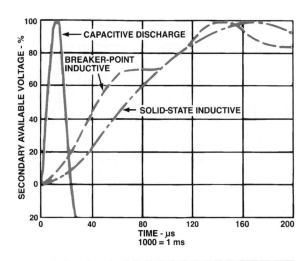

Figure 16-2. A major difference in ignition system performance is the time at which maximum secondary voltage is developed.

CONTROL MODULES AND PRIMARY CIRCUITRY

Although various manufacturers use different circuitry within their ignition system control modules, the basic function of all modules is the same. The following paragraphs explain how these electronic ignition modules work. Because the electronic components within the module are delicate and complex, the modules are sealed during assembly. If any individual component within the module fails, the entire unit is replaced rather than repaired. Our explanation of module circuitry and operation is brief, because the technician never needs to service module components.

Primary Circuit Control

Technology has advanced over the years, producing a variety of primary ignition circuit triggering devices and increasingly sophisticated ignition modules to switch primary ignition current on and off. Although different in design, all triggering devices function to control the timing of primary current switching. There are several types of electronic ignition systems in use today. They are loosely defined by the triggering devices they use. Photoelectric ignitions use a light emitting diode (LED) and a photoelectric cell to

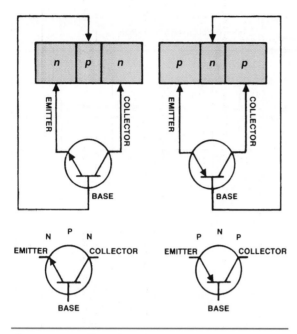

Figure 16-3. Transistor construction and symbols.

produce a DC square wave as the triggering signal. Magnetic pulse generators use an induction coil and a rotating reluctor to produce an AC signal that is sent to the ignition module. Hall Effect ignitions produce a DC triggering signal through use of a Hall Effect device and a rotating shutter wheel.

Primary circuit control is made possible in an electronic ignition by the ability of transistors to control a high current in response to a very small current. The transistor is like a solid-state relay. It has a base of one type of semiconductor material and an emitter and collector of the other type of material, figure 16-3. A certain amount of current must travel through the base-to-emitter or base-to-collector circuit before any current travels through the emitter-to-collector circuit. In this way, a small amount of current controls the travel of a large amount of current.

Ignition Modules

The ignition module will contain a transistor or a series of transistors that are controlled by the primary triggering device. The transistor responds to a signal from the primary triggering device, either a magnetic, an optical, or a Hall Effect sensor. In turn, the transistor controls the primary current flow in the primary side of the ignition coil. When

the transistor is turned on by the primary trigger, it turns on and allows current to flow in the primary windings of the coil. When the primary triggering device's signal turns off, the transistor turns off and primary coil current stops flowing. When this happens, the changing magnetic field in the primary windings induces voltage in the coil's secondary windings. This voltage is then delivered to the spark plug.

Ignition modules take various forms; however, they all work on these principles. Some older modules were large units that mounted remotely from the distributor, figure 16-4. Other units were much smaller and mounted in or on the distributor, figure 16-5.

TRIGGERING DEVICES AND IGNITION TIMING

All triggering devices produce a pulsating voltage that signals the generation of an ignition spark. Three triggering devices are commonly used in a solid-state ignition:

- Magnetic pulse generator
- Hall Effect switch
- Optical (light detection) signal generator

Magnetic Pulse Generator

A simple and very common ignition electronic switching device is the magnetic pulse generator system. Most manufacturers use the rotation of the distributor shaft to time the voltage pulses. The **magnetic pulse generator** is installed in the distributor housing where the breaker points used to be. The pulse generator, figure 16-6, consists of a trigger wheel (reluctor) and a pickup coil. The pickup coil consists of an iron core wrapped with fine wire, in a coil at one end and attached to a permanent magnet at the other end. The center of the coil is called the pole piece.

The trigger wheel is made of steel with a low reluctance that cannot be permanently magnetized. Therefore, it provides a low-resistance path for magnetic flux lines. The trigger wheel usually has as many teeth as the engine has cylinders.

As the trigger wheel rotates, its teeth come near the pole piece. Flux lines from the pole piece concentrate in the low-reluctance trigger wheel, in-

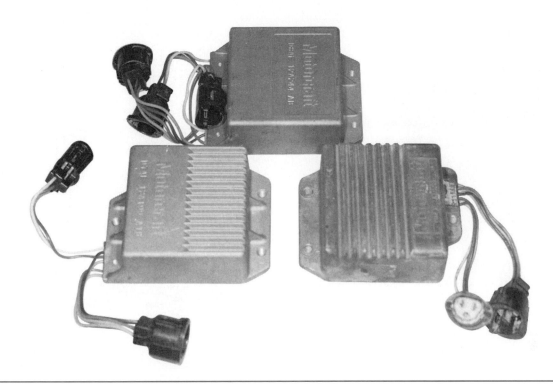

Figure 16-4. Typical external ignition modules from Ford.

Figure 16-5. The ignition module may be installed inside the distributor, as in the HEI system.

creasing the magnetic field strength and inducing a positive voltage signal in the pickup coil. When the teeth and the pole piece are aligned, the magnetic lines of force from the permanent magnet are concentrated and the voltage drops to zero. The point at which the trigger wheel tooth aligns with the pole piece is the ignition timing point for one cylinder and corresponds to the point where breaker points open in the older breaker-point systems. The reluctor continues and passes the pole piece but now the voltage induced is negative, creating an alternating current (AC) voltage signal, figure 16-7. The pickup coil is connected to the electronic control module, which senses this AC voltage, converts the signal to a direct current (DC) voltage, and switches the primary current on and off. Each time a trigger wheel tooth comes near the pole piece, the control module is signaled to switch off the primary current. Solid-state circuitry in the module determines when the primary current will be turned on again.

We have used the terms *trigger wheel* and *pickup coil* in describing the magnetic pulse generator. Various manufacturers have different names for these components. Other terms for the pickup coil are pole piece, magnetic pickup, or stator. The trigger wheel is often called the reluctor, armature, timer core, or signal rotor.

Figure 16-6. This Delco-Remy magnetic pulse generator has a pole piece for each cylinder (arrows), but operates in the same manner as single pole piece units.

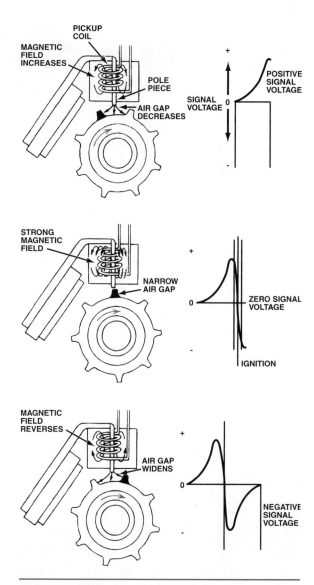

Figure 16-7. The pickup coil generates an AC voltage signal as the trigger wheel moves close to the pole piece.

The magnetic pulse generator produces an AC voltage signal that can be observed using a lab scope. Although the ignition module converts the AC signal to DC, the ignition system can be disabled (coil wire grounded) and the AC voltage signal can be observed. Typically, most magnetic pulse generator ignition systems produce between 4 to 5 volts peak-to-peak voltage signals while cranking, figure 16-8. This GM high-energy ignition (HEI) system displays approximately 2.24 volts positive peak, which gives approximately 4.48 volts peak-to-peak volts. An AC voltage that fluctuates less than 1 volt peak-to-peak causes stalling and driveability problems. Severe HEI problems such as weak or cracked pickup coil magnets, incorrect air gaps, and so on, may cause

engine performance problems during all ranges. Figure 16-9 shows an AC voltage signal from a defective HEI system with a peak-to-peak voltage value of only 776 millivolts (mV). With the engine running and a dual trace lab scope installed, the converted AC to DC voltage signal can be observed in conjunction with the firing voltages. Figure 16-10 shows that the rising edge of the converted AC signal (D/REF) lines up directly with the primary switching on point, while the falling edge lines up directly with the primary switching off point where the coil fires the plug. This perfect alignment of voltage signals happens only at idle because as the rpm increases, the HEI internal

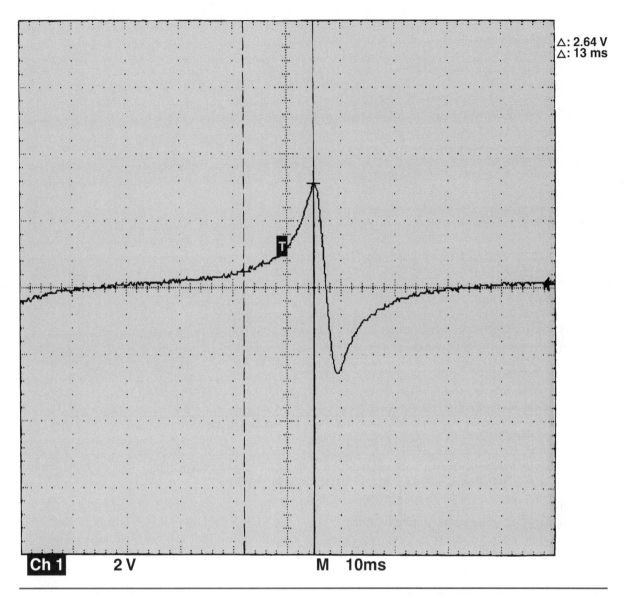

△: 2.64 V
△: 13 ms

Ch 1   2 V   M   10ms

Figure 16-8. The AC sine wave can be observed while the engine is cranking (GM HEI ignition system shown).

dwell circuit varies the turn-on signal sooner to compensate for advance timing changes. The HEI magnetic pulse generator can be observed at many different stages to decisively conclude the working condition of the ignition system.

As mentioned in Chapter 15, GM, Ford, and other manufacturers have used magnetic pulse generators that rely on engine crankshaft rotation (instead of distributor shaft rotation) to produce a signal voltage.

Ford's EEC-I system takes its crankshaft rotation signal from the flywheel end of the crankshaft. As we will see later in this chapter, Buick's C3-I system uses a Hall Effect crankshaft sensor.

Hall Effect Switch

The Hall Effect switch is the most recent switching technology in electronic ignitions. A **Hall Effect switch** also uses a stationary sensor and rotating trigger wheel (shutter), figure 16-11. Unlike the magnetic pulse generator, the Hall Effect switch requires a small input voltage in order to generate an output or signal voltage. **Hall Effect** is the ability to generate a voltage signal in semiconductor material (gallium arsenate crystal) by passing current through it in one direction and applying a magnetic field to it at a right angle to its surface. If the input current is held steady and

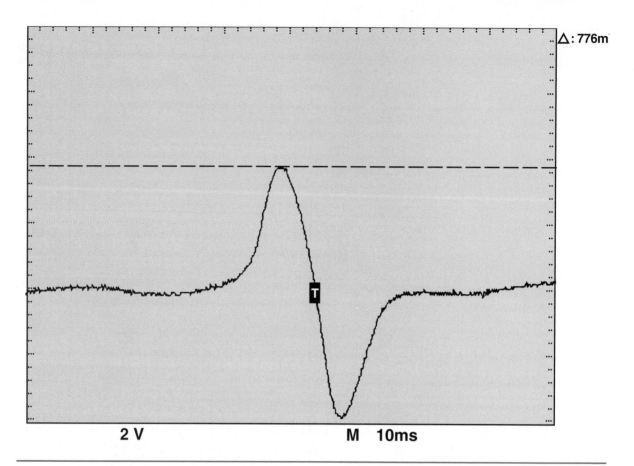

2 V   **M   10ms**

Figure 16-9.   A defective pickup coil will often display a low peak-to-peak voltage range—this example shows a limited 776 millivolt range peak-to-peak value (GM HEI ignition system shown).

magnetic field fluctuates, an output voltage is produced that changes in proportion to field strength, figure 16-12.

Most Hall Effect switches in distributors have a Hall element or device, a permanent magnet, and a rotating ring of metal blades (shutters) similar to a trigger wheel (another method uses a stationary sensor with a rotating magnet). Some blades are designed to hang down, typically found in Bosch and Chrysler systems, while others may be on a separate ring on the distributor shaft, typically found in GM and Ford Hall Effect distributors. When the shutter blade enters the gap between the magnet and the Hall element, it creates a magnetic shunt that changes the field strength through the Hall element, thereby creating an analog voltage signal. The Hall element contains a logic gate that converts the analog signal into a digital voltage signal, which triggers the switching transistor. The transistor transmits a digital square waveform at varying frequency to the ignition module or onboard computer.

A Hall Effect switch is a complex electronic circuit. Figure 16-13 shows the relationship of the Hall Effect switch and ignition operation. An important point to understand is that ignition occurs when the Hall shutter *leaves* the gap between the Hall semiconductor element and the magnet. Also, notice that primary current "builds up" during the coil saturation period after primary voltage is switched on. This waveform is referred to as **current ramping.** The pattern will often display itself in the shape of a ramp.

Hall Effect switches require extra connections for input voltage; however, their output voltage does not depend on the speed of the rotating trigger wheel. Magnetic pulse generators depend upon induction to create the signal voltage. The strength of an induced voltage varies if the magnetic lines move more quickly or slowly. The Hall Effect is not induction, and the speed of the magnetic lines has no effect on the signal voltage. This constant-strength signal voltage offers more reliable ignition system performance throughout a wide range of

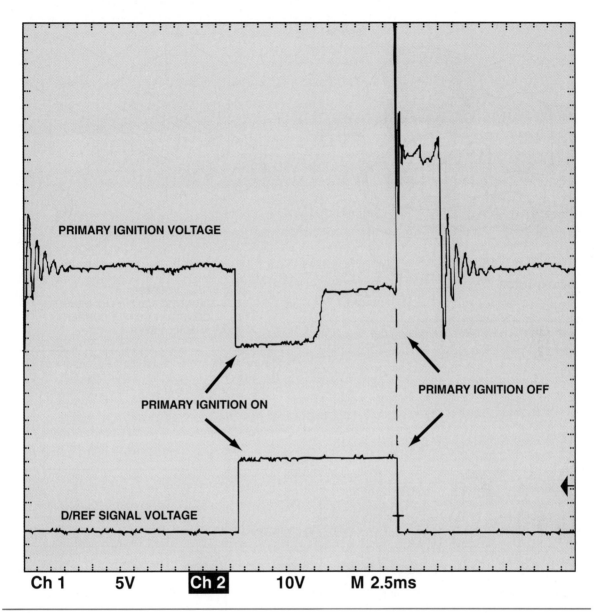

PRIMARY IGNITION VOLTAGE

PRIMARY IGNITION ON

PRIMARY IGNITION OFF

D/REF SIGNAL VOLTAGE

Ch 1   5V   Ch 2   10V   M 2.5ms

Figure 16-10. The rising edge of the magnetic pulse generator (turn-on) signal aligns with the primary current switching (turn-on) signal.

engine speeds. Moreover, a Hall Effect switch provides a uniform digital voltage pulse regardless of rotation speed. This makes a Hall Effect switch ideal as a digital engine sensor for fuel-injection timing and other functions besides ignition control. Therefore, its main advantage over the magnetic pulse generators is that it can generate a full-strength output voltage signal even at slow cranking speeds. This allows precise switching signals to the ignition primary circuit and accurate adjustments of the air-fuel mixture.

Lab scopes easily display the digital voltage signals from the Hall Effect switching devices. Hall Effect switches typically display square waves with a rising edge (leading edge), a flat top, the falling edge, and finally the flat bottom, figure 16-14. Square wave amplitude is the distance between the high level and the low level. Square waves with time durations on the top and bottom portion of the waveform are symmetrical. Most electronic ignition Hall Effect switching devices are three-wire; ground (−), operating voltage (reference), and output signal. The output signal is the square wave digital signal that has been processed by the signal processing circuit and converted from AC to a DC digital signal, figure 16-15. Some ignition systems

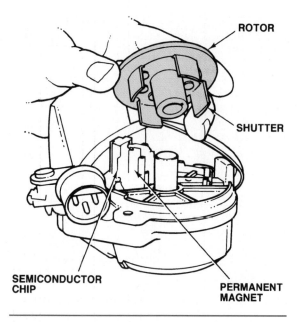

Figure 16-11. A Hall Effect triggering device.

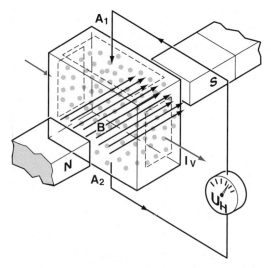

U$_H$ = HALL VOLTAGE
B = MAGNETIC FIELD (FLUX DENSITY)
I$_V$ = CONSTANT SUPPLY CURRENT
A$_1$,A$_2$ = HALL LAYER

Figure 16-12. The magnetic field generates a voltage signal as the electrons from the supply current move (generate) toward the Hall negative layer.

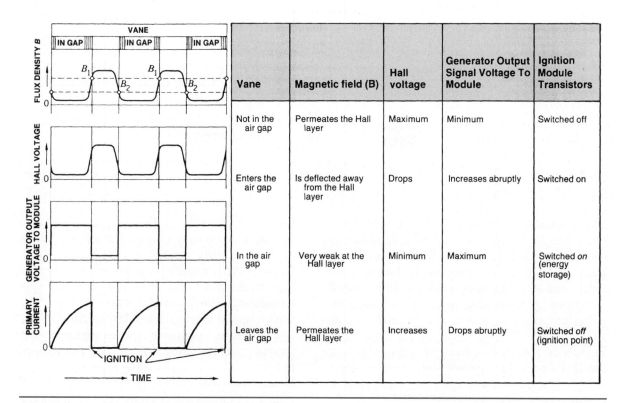

Vane	Magnetic field (B)	Hall voltage	Generator Output Signal Voltage To Module	Ignition Module Transistors
Not in the air gap	Permeates the Hall layer	Maximum	Minimum	Switched off
Enters the air gap	Is deflected away from the Hall layer	Drops	Increases abruptly	Switched on
In the air gap	Very weak at the Hall layer	Minimum	Maximum	Switched on (energy storage)
Leaves the air gap	Permeates the Hall layer	Increases	Drops abruptly	Switched off (ignition point)

Figure 16-13. The relationship of Hall Effect signal voltage and ignition discharge.

374

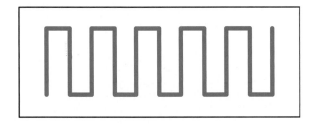

Figure 16-14. A typical Hall Effect square wave output signal.

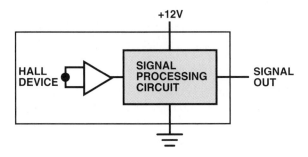

Figure 16-15. A schematic of a typical 3-wire Hall Effect sensor.

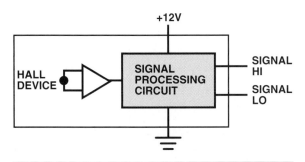

Figure 16-16. A schematic of a typical 4-wire Hall Effect sensor.

are equipped with 4-wire Hall Effect switching devices that are designed with two output signal voltage wires: signal HI and signal LO, figure 16-16. Signal HI is used for the lab scope hookup and the display of the square waveform for the circuit. Signal LO is a bias voltage signal of approximately 0.7 volt for the signal processing circuit.

Chrysler 2.2L and 2.5L engines equipped with the TBI system use 3-wire Hall Effect switching devices. Early Chrysler ignition systems require two separate signals from the distributor, one for the sync signal for TDC number 1 and the other for the primary ignition turn-on signal. Later systems required only the primary ignition turn-on signal from the Hall Effect sensor, figure 16-17. This uniform 5-volt square wave toggles from the GY wire.

Optical (Light Detection) Signal Generator

The **optical signal generator,** often called a photo optical signal generator, uses the principle of light beam interruption to generate voltage signals. Many optical signal distributors contain a pair of light-emitting diodes (LED) and photo diodes installed opposite each other, figure 16-18. A disc containing two sets of chemically etched slots is installed between the LEDs and photo diodes. Driven by the camshaft, the disc acts as a timing member and revolves at half engine speed. As each slot aligns with the light beam, an alternating voltage is created in each photo diode. A hybrid integrated circuit converts the alternating voltage into on-off pulses sent to the engine controller. Because the sensors are self-contained within the distributor, a protective inner cover is used to separate them from the high-tension distributor part of the distributor housing. The cover prevents actuation errors caused by electrical noise.

The high-data-rate slots, or outer set, are spaced at intervals of two degrees of crankshaft rotation, figure 16-19. This row of slots is used for timing engine speeds up to 1,200 rpm. Certain slots in this set are missing, indicating the crankshaft position of number 1 cylinder to the engine controller. For sequential fuel injection (SFI), the low-data-rate slots, or inner set, consists of six slots correlated to the crankshaft top-dead-center angle of each cylinder. The engine computer uses this signal for triggering the fuel injectors and for ignition timing at speeds above 1,200 rpm. In this way, the optical signal generator acts both as the crankshaft position sensor and TDC sensor, as well as a triggering device.

Until Chrysler introduced one on some 1987 models, no domestic manufacturer had used an optical signal generator as original equipment. The Isuzu I-TEK and other Japanese ignition systems, however, rely on optical signal generators, which operate in essentially the same way as the Chrysler version.

The optical signal generator provides a more reliable signal voltage at much lower engine speeds than a magnetic pulse unit. However, periodic LED and photocell cleaning may be required. Several aftermarket, or add-on, ignition systems have been produced using this switching device.

Optical ignition systems create distinct voltage signals that can be observed using a lab

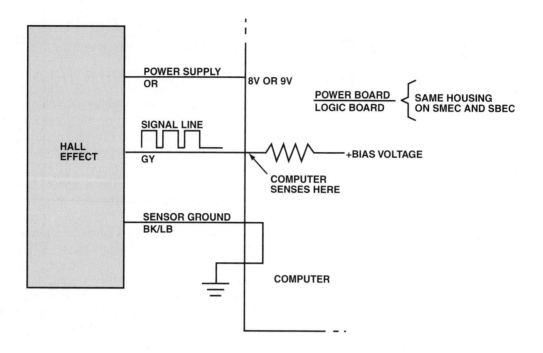

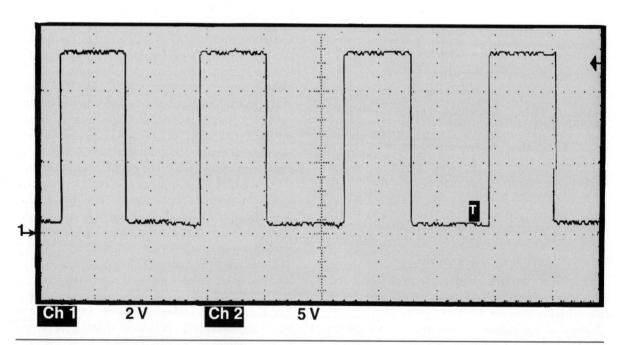

Figure 16-17. The converted ac sine wave can be observed on the GY wire as a square wave switching (toggle) signal.

scope. The Chrysler optical ignition system on the 3.0L engine uses two photocells and two LEDs with solid-state circuitry to create two 5.0-volt signals. The inner sets of slots (low data) signal TDC for each cylinder while the outer slots (high data) monitor every two degrees of crankshaft rotation. Observing the waveforms, if the low-data-rate or the high-data-rate signals fail on the high section, then the LED is most likely malfunctioning. If the signals fail on the low section, then the computer is most likely not delivering the 5.0-volt reference signal. If the

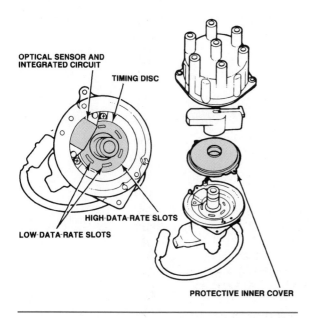

Figure 16-18. Components of the Chrysler optical distributor. The slotted timing disc interrupts the light beam in the optical sensor to produce on-off signals used to control fuel injection, idle speed, and ignition timing. (Courtesy of DaimlerChrysler Corporation)

low-data signal is lost, the engine will not start. If the high-data signal is lost, the engine will run, but only on default settings from the computer, figures 16-20 and 16–21.

ELECTRONIC IGNITION DWELL, TIMING, AND ADVANCE

Dwell is the period of time when the breaker points are closed and current is traveling in the primary winding of the coil. Although electronic ignitions require a dwell period for the same purpose, it is controlled by a timing or current sensing circuit in the ignition module, rather than by a signal from the distributor.

The initial timing adjustment for most basic electronic ignitions is similar to that for breaker-point ignition systems. Timing is set by rotating

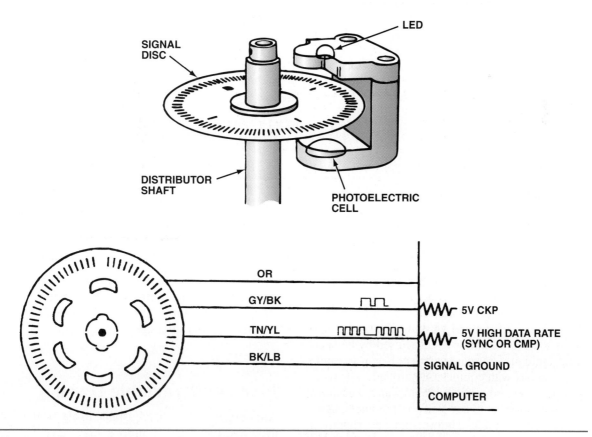

Figure 16-19. The high-data-rate slots coincide with ignition timing while the low-data-rate slots coincide with the TDC mark for each cylinder (3.0L V-6 engine shown).

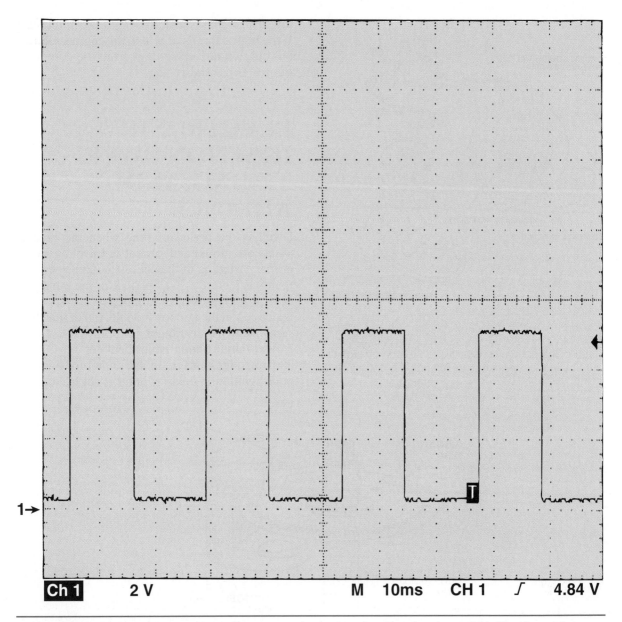

Figure 16-20. Typical known good waveform pattern for the low-data-rate signal on a 3.0L Chrysler engine.

the distributor housing with the engine idling at normal operating temperature.

The first generation of electronic ignitions used the same centrifugal and vacuum advance mechanisms to advance timing as those used by breaker-point ignitions. As manufacturers began to equip their engines with computer-controlled systems in the late 1970s, however, timing advance became a function of the computer. Since a computer can receive, process, and send information very rapidly, it can change ignition timing with far more efficiency and accuracy than any mechanical device.

The distributor in an electronic ignition used with a computer engine control system contains a triggering device for basic timing. Because the computer actually controls ignition timing, however, the distributor's primary function is to distribute secondary voltage.

Fixed and Variable Dwell

Although dwell is not adjustable in electronic distributors, electronic ignitions may have one of two kinds of dwell control:

- Fixed dwell
- Variable dwell

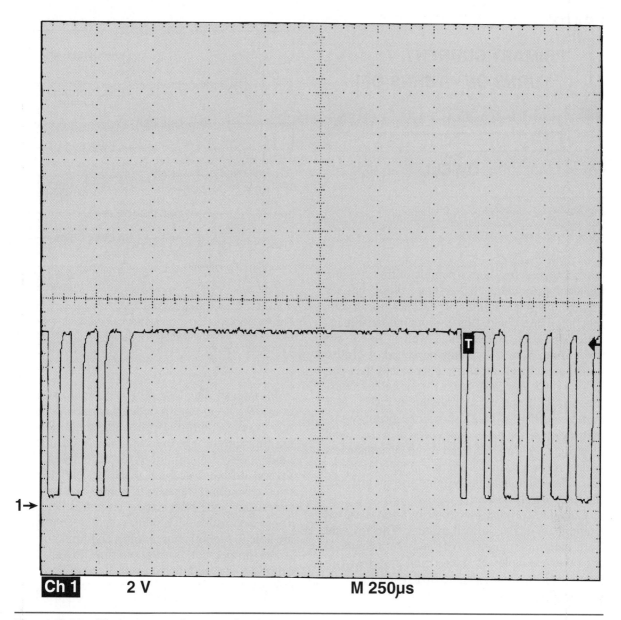

| Ch 1 | 2 V | | M 250μs |

Figure 16-21. The long pause between signals is the result of the missing slot, which indicates TDC number 1 to the computer (high-data-rate signal on a 3.0L Chrysler engine).

Fixed Dwell

A ballast resistor is placed in the primary circuit of a **fixed-dwell** electronic ignition to limit current and voltage. Dwell is the length of time the switching transistor sends current to the primary coil windings. It begins once the secondary voltage and current have fallen below predetermined levels. The ballast resistor functions just as it does in a breaker-point system to control primary voltage and current. Dwell, measured in distributor degrees of rotation, remains constant at all engine speeds.

All points systems operate on fixed dwell. As the distributor rotates and the points close, the amount of time they remain closed is determined by the dwell settings. Each time period the points open and close is identical for each cylinder and each rotation, provided the lobes on the distributor shaft are not worn. **Dwell,** measured in distributor degrees of rotation, remains constant at all speeds, but the **dwell time** shortens at high speed and lengthens at slow speeds. Remember how the ballast resistor works in the points system. At low speeds, current travels through the circuit for relatively long periods of time. As the current heats the resistor, its resistance increases, dropping the applied voltage at the coil. At higher speeds, the breaker points open more often and

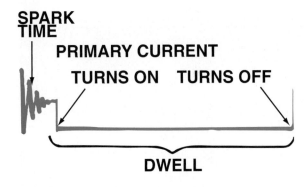

Figure 16-22. Dwell, measured in degrees of distributor rotation, remains relatively constant at all engine speeds in a fixed-dwell system.

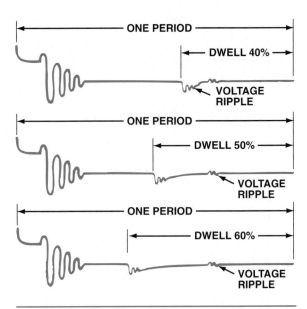

Figure 16-23. The dwell period increases with engine speed in the variable-dwell Delco HEI and some Motorcraft systems.

current travels for shorter periods of time. As the ballast resistor cools, its resistance drops. Higher voltage is applied to the coil but the shorter current travel duration results in about the same magnetic saturation of the coil. Dwell is fixed, but dwell time varies with engine speed. Because points are a mechanical component and they can be designed only with fixed dwell, ballast resistors protect the primary circuit during engine rpm variations. Figure 16-22 shows the primary circuit oscilloscope pattern of a typical fixed-dwell electronic ignition. All fixed-dwell systems use a ballast resistor to limit current and voltage. Some early electronic ignition systems also incorporate the fixed-dwell system.

Chrysler's original electronic ignition is a good example of a fixed-dwell system. The original Ford electronic ignition (EI) and Dura-Spark II are other examples of fixed-dwell systems with ballast resistors, as are some Bosch and Japanese electronic ignitions.

Variable dwell

Variable-dwell ignition systems do not use ballast resistors and the ignition coil and module receive full battery voltage. A module circuit senses primary current to the coil and reduces the current when the magnetic field is saturated. Unlike the fixed-dwell system, dwell varies, but the dwell *time* remains relatively constant. Dwell time in variable-dwell ignition systems is often referred to as **coil charge time.** In general, variable-dwell ignition systems have a higher available voltage capacity, lower primary resistance, and a hotter spark to burn leaner air-fuel mixtures designed for low-emissions automobiles. Figure 16-23 shows the primary circuit oscilloscope pattern of a typical variable-dwell electronic ignition.

In order to diagnose primary circuit malfunctions, it is important to understand how the ignition system controls dwell while simutaneously limiting current. The current limiting functions of modern electronic ignitions are accomplished by several different designs. Each design incorporates specialized transistor circuitry that switches to resistor mode ofter excess current is detected. These sophisticated modules and/or computers are able to set coil charge times and limit current. If the proper coil saturation time (dwell) was achieved, the module trims the dwell to limit current, thereby reducing the amount of electrical energy used by the ignition system. If maximum current was not achieved during the coil saturation time, the module increases the dwell time.

Dual trace digital and analog oscilloscopes are useful tools for diagnosing primary circuit malfunctions. Because the primary circuit operates on relatively small voltages, more accurate analysis of the primary circuit can be achieved over the standard automotive oscilloscopes. Current limiting can be monitored on the scope on all types of ignition systems. The basic rule here is—when viewing voltage on a circuit that is ground controlled, as voltage goes down, current increases, and as voltage goes up, current decreases. A waveform pattern can show the exact point in the dwell when the current limiting device in the ignition module switches on and voltage increases. Figure 16-24 displays a typical primary

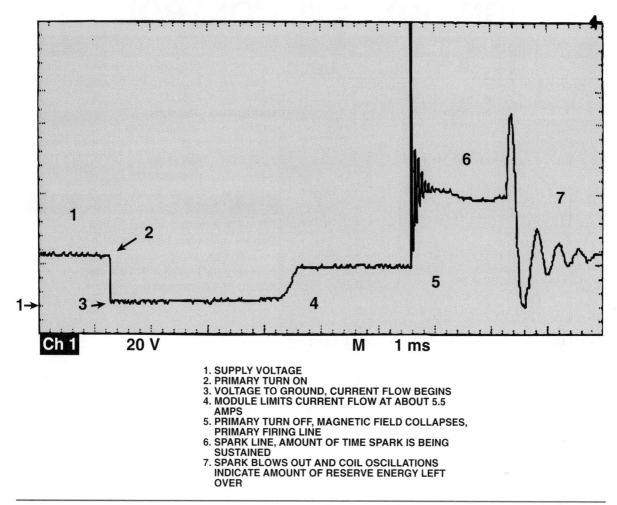

Figure 16-24. A typical operating primary current waveform.

1. SUPPLY VOLTAGE
2. PRIMARY TURN ON
3. VOLTAGE TO GROUND, CURRENT FLOW BEGINS
4. MODULE LIMITS CURRENT FLOW AT ABOUT 5.5 AMPS
5. PRIMARY TURN OFF, MAGNETIC FIELD COLLAPSES, PRIMARY FIRING LINE
6. SPARK LINE, AMOUNT OF TIME SPARK IS BEING SUSTAINED
7. SPARK BLOWS OUT AND COIL OSCILLATIONS INDICATE AMOUNT OF RESERVE ENERGY LEFT OVER

ignition firing sequence. Primary current is switched on (step 3) and then the ignition module detects that the coil is saturating (step 4) and limits current travel just before turning off primary circuit voltage (number 5). This portion of the waveform on electronic ignition systems is called the **current limiting hump.** The current limiting hump will appear at different locations for different manufacturers' ignition systems. By learning the typical waveform patterns of properly operating ignition systems, ignition system degradation can be easily monitored. Once the control of the ignition timing has been taken away from the ignition system (distributor vacuum, computer timing disconnects, and so on), the primary turn-off point will be stable and will not vary. At this point, any variation between the cylinders may indicate an engine mechanical failure (worn distributor bushings, distributor drive gears, loose timing chains, and so on).

Technicians frequently use voltage waveform patterns to diagnose problems in modern electronic circuits. In addition to the latest digital oscilloscopes and computer programs, technicians are using current waveform patterns to test modern ignition systems for circuit operation and component performance. Voltage is electrical "pressure" and current is electrical "volume." When testing modern fuel systems with electrical fuel pumps, the fuel system often exhibits proper fuel pressure but falls short of fuel volume. The fuel pump, under load, cannot supply enough fuel! In electrical systems, voltage specifications will frequently read properly, but once the system is under load, current fluctuations will quickly pinpoint defective circuit and component operations.

The primary current waveform, after it is recorded by a digital storage oscilloscope, will slope upward much like a ramp, figure 16-25.

IGNITION COIL CONTROL

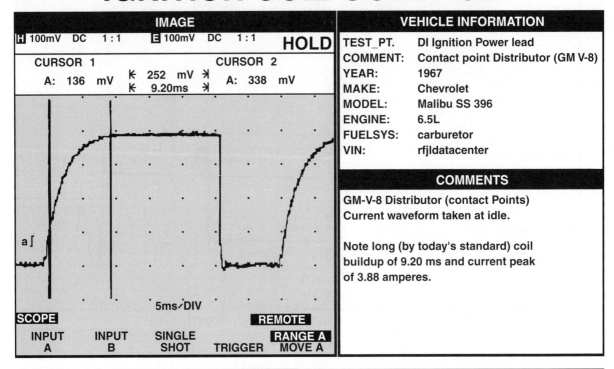

IMAGE		VEHICLE INFORMATION	

Figure 16-25. Current ramp for the primary ignition on a points distributor.

This type of waveform diagnostics is commonly called current ramping. Once the coil primary circuit switches on, the coil charge time or primary charge time slopes upward as current builds. The current ramp will flatten at the top during primary coil saturation. Quickly, the coil primary current is switched off (falling edge) and the dwell period ends. The secondary system fires the plug (not indicated in the illustration) and the primary current remains off, indicated by a flat bottom line. The coil primary will then be switched back on and the cycle repeats. The sloped current waveforms resemble ramps.

The two most important specifications that must be observed on a typical primary current ramp is the coil charge time, specified in milliseconds, and the amount of current in the circuit at the limiting hump (peak amperage rating). Most electronic distributor ignition systems will limit the current between 6.0 and 6.5 amps. Distributorless ignition systems limit the current slightly higher, between 8.0 and 10 amps, because of the different saturation times of the hot coil (forward polarity) and cold coil (reverse polarity) in the waste spark system. The following list is a general guideline for the coil charge times on most domestic electronic ignition systems.

Ford electronic distributor ignition systems
  3.6 milliseconds

GM electronic distributor ignition systems
  3.6 milliseconds

GM electronic distributor ignition systems w/OBD II
  2.5 milliseconds

GM electronic distributorless ignition systems
  2.6 milliseconds

Chrysler electronic distributor ignition systems w/external module
  8.6 milliseconds

Chrysler electronic distributor ignition systems w/computer
  3.8 milliseconds

Figure 16-25 displays the primary ignition current ramp for a Chevrolet V-8 with points ignition. Notice that the coil charge time is 9.20 milliseconds and the peak amperage is only 3.88 amps. Points ignition systems became obsolete because of the long coil charge times and the inability to switch primary currents above 4 amps. Figure 16-26 shows a normally operating TFI ignition on a 1992 Ford Taurus with a coil charge time of 3.6 milliseconds and a current limit at

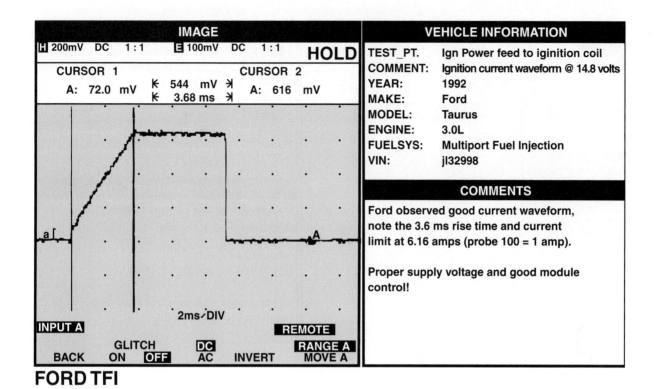

IMAGE		VEHICLE INFORMATION	

H 200mV DC 1:1   E 100mV DC 1:1   **HOLD**

CURSOR 1		CURSOR 2
A: 72.0 mV	⊬ 544 mV ⊬	A: 616 mV
	⊬ 3.68 ms ⊬	

TEST_PT.	Ign Power feed to iginition coil
COMMENT:	Ignition current waveform @ 14.8 volts
YEAR:	1992
MAKE:	Ford
MODEL:	Taurus
ENGINE:	3.0L
FUELSYS:	Multiport Fuel Injection
VIN:	jl32998

COMMENTS

Ford observed good current waveform, note the 3.6 ms rise time and current limit at 6.16 amps (probe 100 = 1 amp).

Proper supply voltage and good module control!

2ms/DIV

INPUT A   REMOTE

BACK   GLITCH ON **OFF**   **DC** AC   INVERT   **RANGE A** MOVE A

FORD TFI

Figure 16-26. Current ramp for the primary ignition on a Ford distributor ignition (TFI). Coil charge time is 3.6 milliseconds with the current limited at 6.16 amps.

6.16 amps. Figure 16-27 shows this same ignition system with a defective relay. The coil charge time is only 3.92 milliseconds while the current limits at 4.88 amps. Low supply voltage to the ignition coil resulted in high resistance and less current.

In general, ignition coils used with variable dwell systems have a higher available voltage capability, lower primary resistance, and a higher turns ratio than the coils used with fixed-dwell systems.

All variations of the GM Delco-Remy High Energy Ignition (HEI) are variable dwell systems, as well as Ford Dura-Spark I and thick-film integrated (TFI) ignition systems. Other examples are most Chrysler Hall Effect ignitions, some Bosch, Marelli, and several Japanese electronic ignitions.

Basic Timing and Advance Control

We learned in Chapter 15 that ignition takes place at a point just before or just after top dead center is reached during the compression stroke. This time is measured in degrees of crankshaft rotation and is established in distributor-type ignitions by the mechanical coupling between the crankshaft and distributor. Basic or initial ignition timing is usually set at idle speed.

As engine speed increases, however, ignition must take place earlier. This is necessary to ensure that maximum compression pressure from combustion develops as the piston starts downward on its power stoke, figure 16-28. This change in ignition timing is called spark advance and is controlled in basic electronic ignitions (those not integrated with electronic engine controls) by the same mechanical devices used with a breaker-point ignition:

- Centrifugal advance weights
- Vacuum advance diaphragm

The centrifugal advance mechanism responds to changes in engine speed and moves the position of the trigger wheel relative to the distributor shaft. The vacuum advance mechanism responds to changes in engine load and moves the position of the pickup coil. These changes in position alter the time, relative to crankshaft position, at which the primary circuit is opened.

When electronic ignitions are integrated with electronic control systems, the computer monitors

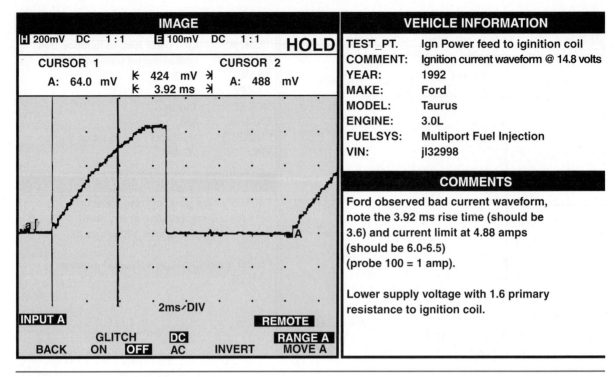

Figure 16-27. Abnormal current ramp for the primary ignition on a Ford TFI ignition system. Coil charge time is 3.92 milliseconds with a 4.88 amps output.

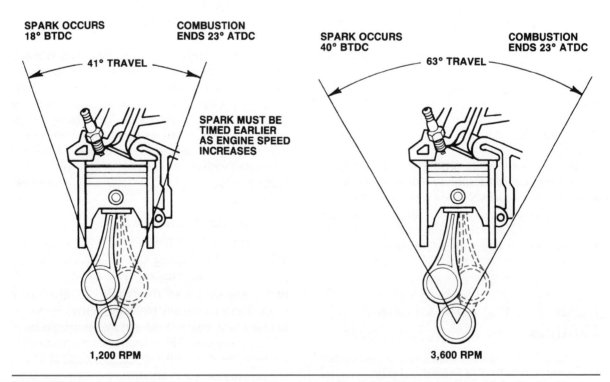

Figure 16-28. As rpm increases, the spark must be advanced to deliver the maximum spark for the best torque. (Courtesy of Ford Motor Company)

engine speed and load changes, engine temperature, manifold pressure (vacuum), air travel, exhaust oxygen content, and other factors. The computer then changes ignition timing to produce the most efficient combustion. Electronic engine controls are used on late-model vehicles.

ORIGINAL-EQUIPMENT ELECTRONIC DISTRIBUTOR IGNITIONS

This section contains brief descriptions of the major distributor electronic ignitions used by domestic and foreign manufacturers. All are inductive-discharge systems, but their triggering devices and module circuits vary somewhat from manufacturer to manufacturer.

Because these systems have been under constant development, they have been modified from year to year and model to model. The descriptions given in this section summarize the basic changes resulting from this ongoing development. Whenever an electronic ignition requires service, you should always refer to the manufacturer's shop manual or an appropriate repair manual to determine the exact specifications and whether the system has any unique features you should know about.

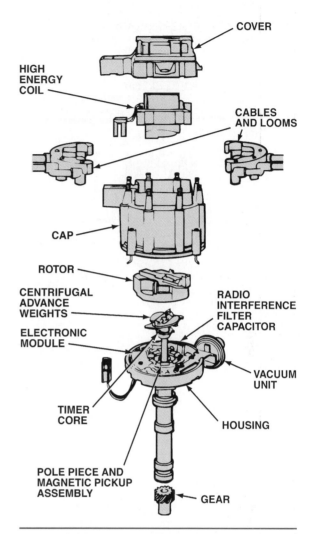

Figure 16-29. A basic HEI distributor.

Delco-Remy

High-Energy Ignition (HEI)
The Delco-Remy high-energy ignition (HEI) system, figure 16-29, was introduced on some 1974 GM V-8 engines and became standard equipment on all GM engines in 1975. The HEI system was developed from an earlier Delco-Remy Unitized ignition used on a limited number of 1972–74 engines. The HEI and Unitized ignitions have all of the ignition components built into the distributor.

The HEI system was the first domestic original-equipment manufacturer (OEM) electronic ignition to use a variable-dwell primary circuit and no ballast resistor. The control module lengthens the dwell period as engine rpm increases to maintain uniform primary current and coil saturation throughout all engine speed ranges.

The HEI module is installed on the breaker plate inside the distributor, and the basic ignition module has four terminals: two connected to the primary circuit and two attached to the pickup coil of a magnetic pulse ignition pickup. The back of the HEI module is coated with a silicone dielectric compound before installation.

The HEI pickup looks different from those of other manufacturers, but it operates in the same basic manner. The rotating trigger wheel of the HEI distributor is called the "timer core," while the coil is attached to a fixed ring-shaped magnet called the "pole piece." The pole piece and trigger wheel have as many equally spaced teeth as the engine has cylinders, except on uneven-firing V-6 engines, which have three teeth on the trigger wheel and six unevenly spaced teeth on the pole pieces.

The most common early HEI system has an integral (built-in) coil mounted in the distributor cap,

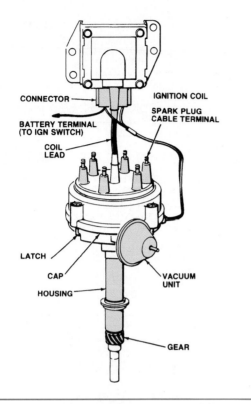

Figure 16-30. The HEI distributor used with some in-line and V-6 GM engines.

CONNECTOR

IGNITION COIL

SPARK PLUG
CABLE TERMINAL

BATTERY TERMINAL
(TO IGN SWITCH)

COIL
LEAD

LATCH

CAP

HOUSING

VACUUM
UNIT

GEAR

figure 16-29. Some 4- and 6-cylinder engines have HEI systems with a separate coil, figure 16-30, that provides additional distributor clearance on the engine. Both designs operate in the same way and have similar wiring connections. All early HEI systems use 8-mm spark plug cables with silicone insulation to minimize crossfiring caused by the high secondary voltage capability of the HEI coil. Some engines use wide-gap spark plugs to take advantage of the system's high-voltage capability.

In addition to distributing the spark to the appropriate cylinder, the HEI rotor serves as a fuse to protect the module. If an open circuit occurs in the ignition secondary and voltage rises above a certain level, the center of the rotor will burn through and allow the spark to travel to ground rather than arcing through and destroying the module. Early black rotors have a dielectric strength of approximately 70,000 volts; later white ones are designed to ground the secondary circuit at around 100,000 volts.

The newer design rotor may be used in place of the earlier one, but the cap and rotor must be replaced as a matched set. The air gap between the rotor tip and cap electrodes was 0.090 inch (2.29 mm) on the earlier parts, but the later design has an air gap of 0.125 inch (3.18 mm) to better suppress radiofrequency interference (RFI) that could interfere with the engine control computers. Caps and rotors from the early and late designs should never be mixed.

In addition to this basic HEI system, GM has used six other HEI versions with electronic engine controls and for certain spark timing requirements. The six systems are HEI with:

- Electronic spark selection (ESS)
- Electronic spark control (ESC)
- Electronic module retard (EMR)
- Electronic spark timing (EST)
- EST and ESC
- EST and a Hall Effect switch

The HEI system with electronic spark control (ESC) is a detonation-control system used on 1980 and later turbocharged and high-compression engines. The system consists of a knock sensor, a controller unit, and a special five-terminal HEI module.

When detonation occurs, the knock sensor sends a signal to the controller, which then instructs the module to retard the ignition timing by a small amount. If the knock sensor continues to detect detonation, the controller instructs the module to further retard timing by another small increment. This sensor-controller-module cycle goes on continuously, and the controller's instructions to the module are updated many times a second until the detonation is eliminated. As soon as that happens, the process reverses itself and timing is advanced in small steps as long as detonation does not reoccur. Once the detonation-producing conditions have been eliminated, the timing will return to normal within 20 seconds.

The HEI system with electronic module retard (EMR) is a simple 10-degree timing retard system used for cold starts and was also introduced in 1980. The retard circuitry is contained within the special five-terminal HEI module and is activated by a simple vacuum switch on most models. On cars with the computer-controlled catalytic converter (C-4) system, the module is controlled by the C-4 computer.

The HEI system with electronic spark timing (EST) was introduced on 1981 engines with computer command control (CCC), except those the minimum-function CCC systems (Chevette, Pontiac T1000, and Acadian). The seven-terminal module converts the pickup coil signal into a crankshaft position signal used by the powertrain control module (PCM) to advance or retard ignition timing for optimum spark timing. HEI-EST distributors have no centrifugal or vacuum advance units.

The HEI system with EST and ESC was also introduced in 1981 and combines the electronic spark control of EST with the detonation sensor of ESC. It is used primarily with turbocharged engines.

The HEI system with EST and a Hall Effect switch combines the basic magnetic pulse generator of the HEI distributor with a Hall Effect switch and is used with CCC engine control systems. The pickup coil sends timing signals to the HEI module during cranking. Once the engine starts, the Hall Effect switch overrides the pickup coil and sends crankshaft position signals to the computer for electronic control of timing.

Chrysler

Electronic Ignition

In 1971, Chrysler became the first domestic manufacturer to introduce a basic electronic ignition on some models, figure 16-31. The system became standard on all Chrysler Motors cars in 1973. The same basic system, with very few modifications, was used on most 6- and 8-cylinder carbureted Chrysler engines.

The Chrysler electronic ignition system, figure 16-32, is a fixed-dwell design using a magnetic pulse distributor, a remote-mounted electronic control module, and a unit-type ballast resistor. The control module is mounted on the firewall or inner fender panel and has an exposed switching transistor that controls primary current. *Don't touch the transistor when the ignition is on, because enough voltage is present to give you a shock.* The distributor housing, cap, rotor, and advance mechanisms are all similar to breaker-point components, as are the ignition coil and 7-mm spark plug cables.

All early Chrysler electronic ignition systems have a single magnetic pickup in the distributor. However, the distributors used with some electronic lean burn (ELB) and electronic spark control (ESC) systems use a dual-pickup distributor containing a start pickup and a run pickup. The run pickup is positioned to advance the ignition trigger signal compared to the start pickup. Under normal engine operation, the module uses the signal from the run pickup. When the ignition switch is in the start position, however, the retarded trigger signal from the start pickup is used to ensure faster starts.

Distributors with early ELB systems have only a centrifugal advance mechanism. Those for later ESC systems have neither centrifugal nor vacuum

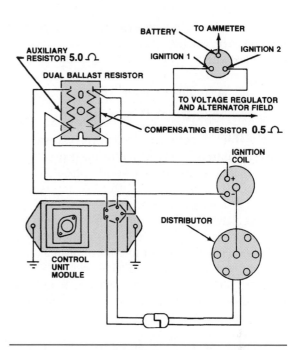

Figure 16-31. A primary circuit diagram of Chrysler's electronic ignition system. (Courtesy of DaimlerChrysler Corporation)

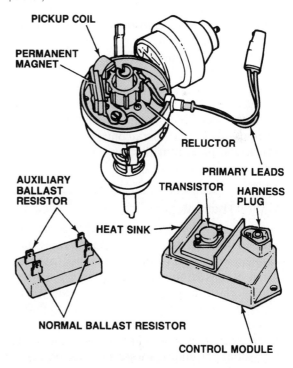

Figure 16-32. The basic Chrysler electronic ignition system.

advance mechanisms; all spark advance is controlled by the computer.

The air gap between the reluctor and pickup coil, or coils, is adjustable, but has no effect on the dwell period, which is determined by the control module. The air gap must be set to a specific clearance with a nonmetallic feeler gauge when a new pickup unit is installed. Air-gap specifications vary according to model year.

Hall Effect Electronic Ignition System

Chrysler introduced a different electronic ignition in 1978 on its first 4-cylinder, front-wheel-drive (FWD) cars. A Hall Effect switch is used instead of a magnetic pulse generator. The original fixed-dwell ignition on 1978 4-cylinder engines was used with an analog computer and had a 0.5-ohm ballast resistor to control primary current and voltage. The 1978 ELB and 1979 ESC system distributors had both centrifugal and vacuum advance mechanisms and were similar in operation to the 6- and 8-cylinder versions described above.

A changeover to a digital spark-control computer in 1981 resulted in the electronic spark advance (ESA) system. This meant several changes to the system and its operation. Since the computer took over spark control timing, the distributor had no advance mechanisms. The system uses no ballast resistor, and dwell is variable; that is, it increases as engine speed increases. In 1984, the ESA system was incorporated into the electronic fuel injection (EFI) spark control system used on fuel-injected engines. A logic module and a power module replaced the spark control computer, but the ignition portion of this system works essentially the same as in the ESA system.

■ Mind Those Magnets

When you are servicing a distributor from an electronic ignition system, be sure that no metal particles or iron filings get inside. The magnetic pickup coils and pole pieces in the breakerless distributors used by many manufacturers will attract metal debris that can foul up ignition performance. Use a clean, soft-bristled brush or low-pressure compressed air to clean the inside of a distributor and keep scrap metal off of the pickup coil.

Thick-Film Integrated (TFI) Ignition

The TFI-I ignition system, figure 16-33, was introduced in 1982 on the 1.6-liter Escort engine. It differs in many respects from the earlier Ford sys-

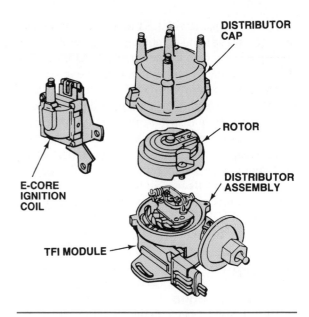

Figure 16-33. Ford TFI ignition system components. (Courtesy of Ford Motor Company)

tems already discussed. Instead of a remote-mounted control module, the TFI-I system uses an integrated circuit control module attached to the outside of the distributor housing. The module connects directly to the distributor stator.

Inside the distributor is the familiar magnetic pulse generator, but TFI-I is a variable-dwell system that operates without a ballast resistor. The conventional oil-filled coil is replaced with a special low-resistance E-core part. The distributor cap, rotor, and spark plug cables, however, are similar to those used with Dura-Spark systems.

There are two different versions of the TFI-I control module. Early production parts were made of blue plastic and are called "non-push-start" modules. They contain protection circuitry that will shut off voltage to the coil if an ignition trigger signal is not detected for a period of 10 to 15 seconds. When shutdown occurs, the ignition switch must be turned off and back on again before the engine will start.

All later modules are made of gray plastic and contain revised circuitry that still turns off voltage to the coil after a preset period, but switches the power back on as soon as a trigger signal is detected.

A revised TFI ignition system appeared in 1983 as part of Ford's fifth generation of electronic engine controls (EEC-IV). Called TFI-IV, this design uses a gray ignition module that appears similar to the TFI-I module, but has a six-wire connector instead of the TFI-I's three-wire connector. The additional three wires connect the module to the

ECA. The TFI-IV module is used with a universal distributor that contains a Hall Effect switch. The TFI ignitions replaced the Dura-Spark ignitions on virtually all Ford vehicles until OBD II.

The TFI module attaches to the distributor with several screws and depends on good contact with the distributor housing for its cooling. Whenever a module is replaced, silicone dielectric compound must be applied to its back to improve heat conductivity and prevent premature module burnout. The module should never be used as a handle to turn the distributor when setting the initial timing. This kind of careless handling can cause it to warp and fail soon after.

Early TFI-IV ignition systems were independent of the EEC-IV computer and are referred to as non-CCD TFI-IV (non-computer-controlled dwell). The TFI-IV ignition systems share similarities that make checking the TFI ignition easy. Here is a list of the non-CCD TFI-IV ignition system features.

1. The module is powered from the ignition switch at pin number 4, figure 16-34. This voltage is then applied to the internal module circuit and used as a reference voltage signal for the profile ignition pickup (PIP) sensor at pin number 6. The 12-volt signal is toggled by a Hall Effect sensor to control the primary ignition circuit during cranking. The voltage for the Hall Effect sensor is supplied on pin number 3 during running mode. The computer receives the PIP signal at terminal number 56 and turns on the fuel pump relay and injectors. Pin number 2 controls the primary circuit and sets the dwell with the on/off (rising/falling) square waveforms.

2. Once the engine is running, power is switched to the ignition module from pin number 4 to pin number 3, allowing the SPOUT signal to control spark timing. A dual trace lab scope can monitor the SPOUT signal from pin number 5 and the PIP (rpm signal) from pin number 6, see figure 16-35. The square waveforms on the non-CCD TFI-IV ignition systems for the PIP and SPOUT contain timing bites on the falling edge on the PIP and on the rising edge on the SPOUT. These are typical for this system and indicate that the computer is controlling spark advance but NOT primary current dwell.

3. The computer observes the firing events directly off the coil wire and corrects ignition timing. Because voltage is high, a 22K-ohm resistor buffers the signal. In the event the resistor burns out, the computer will set a Code 18, which is also preceded by a lack of ignition timing advance. The IDM signal can be observed on pin number 4. A dual trace lab scope can distinguish the primary current waveform and the PIP signal simultaneously, figure 16-36.

The TFI-IV non-CCD system uses the computer to monitor the module and the firing events but the ignition module is responsible for controlling the primary circuit switching.

Later in 1989, Ford introduced the TFI-IV CCD (computer-controlled dwell) ignition systems. These differ from the non-CCD TFI-IV systems in that the computer controls the internal dwell circuitry of the module. In figure 16-37, the 22K-ohm resistor and IDM circuit are removed and pin number 4 incorporates the filtered tach output (FTO). The FTO signal represents the coil charge time and is directly monitored by the computer. The SPOUT signal controls the primary circuit switching, thereby firing the ignition coil. Here is a list of the major differences between the two TFI systems.

1. The SPOUT signal on pin number 5 lines up perfectly with the primary current waveform turn-on point and turn-off point (firing event), figure 16-38. A dual trace lab scope can monitor the SPOUT signal from pin number 5 and the primary current from pin number 2. The SPOUT signal is monitored by the computer and then used to control the dwell of the primary current.

2. The FTO signal represents the coil charge time and is directly monitored by the computer on pin number 4. The FTO signal aligns with the SPOUT signal because the computer uses both signals to control the dwell and timing, figure 16-39.

ORIGINAL-EQUIPMENT DISTRIBUTORLESS IGNITIONS

This section contains brief descriptions of the distributorless electronic ignitions used by domestic manufacturers. Whenever a distributorless ignition requires service, you should always refer to

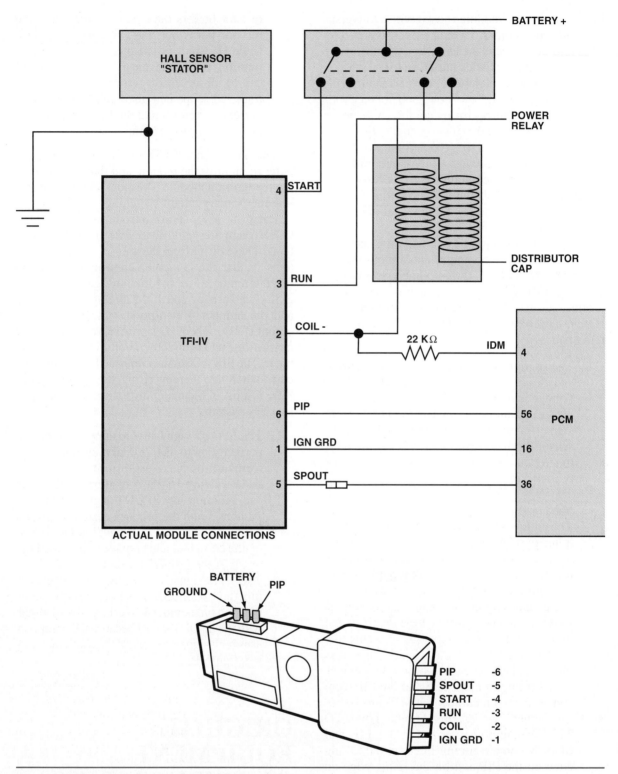

Figure 16-34. Schematic of a Ford TFI-IV non-CCD ignition system. (Courtesy of Ford Motor Company)

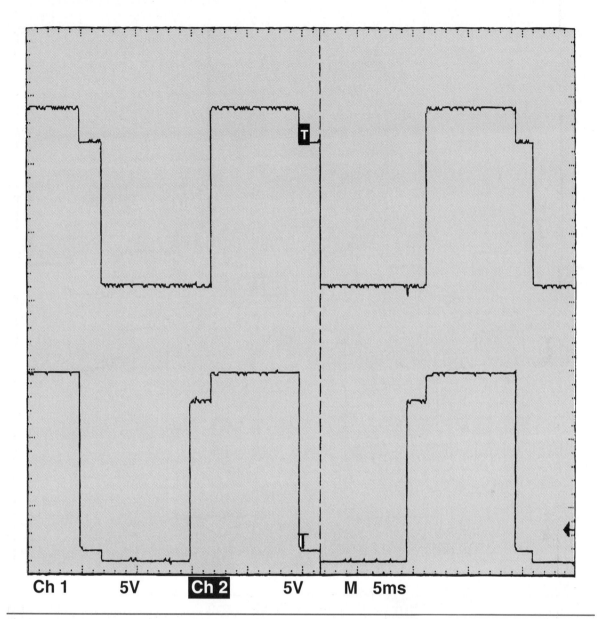

Ch 1   5V   Ch 2   5V   M   5ms

Figure 16-35. At idle, the PIP waveform (top trace) accessed on pin number 6 will align with the SPOUT signal (bottom trace) accessed on pin number 5 (TFI-IV non-CCD ignition system).

the manufacturer's shop manual or an appropriate repair manual to determine the exact specifications and whether the system has any unique features you should know.

Principles of Distributorless Ignitions

The term *distributorless ignition system* (DIS) refers to any ignition system without a distributor. The DIS system fires the spark plugs with a mul-

tiple coil pack containing two or three separate ignition coils (according to the number of engine cylinders) and an ignition control module. Control circuits in the module discharge each coil separately in sequence, with each coil serving two cylinders 360 degrees apart in the firing order. Each coil fires two plugs simultaneously in what is called a **waste spark** method. One spark goes to a cylinder near TDC on the compression stroke, while the other fires the plug in a cylinder near TDC of the exhaust stroke. The plug in the cylinder on the exhaust stroke requires very little volt-

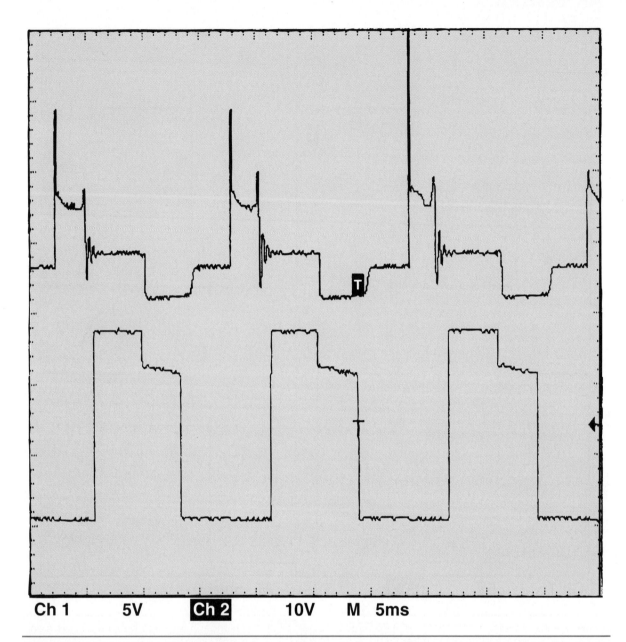

| Ch 1 | 5V | Ch 2 | 10V | M | 5ms |

Figure 16-36. The PIP (rpm) signal does not align with the primary current turn-on signal because the TFI-IV non-CCD ignition module controls the dwell, not the computer.

age (about four kV) to fire and has no effect on engine operation.

As we learned in Chapter 15, the ignition coil secondary windings in a distributor ignition generally are wound to give the spark plug center electrode positive polarity and the side electrode negative polarity. When the spark plug fires, electrons travel from the coil secondary windings to the center electrode, across the plug gap to the side electrode where they return to the coil secondary windings through the engine block. A spark plug fired in this manner is said to have forward polarity, figure 16-40. If the electrons travel to the side electrode and across the plug gap to the center electrode, the plug is said to have reverse polarity.

In a distributorless ignition, each pair of spark plugs is connected to one coil. In this system, one plug is always fired with forward polarity and the other is always fired with reverse polarity, figure 16-40. Because firing a spark plug

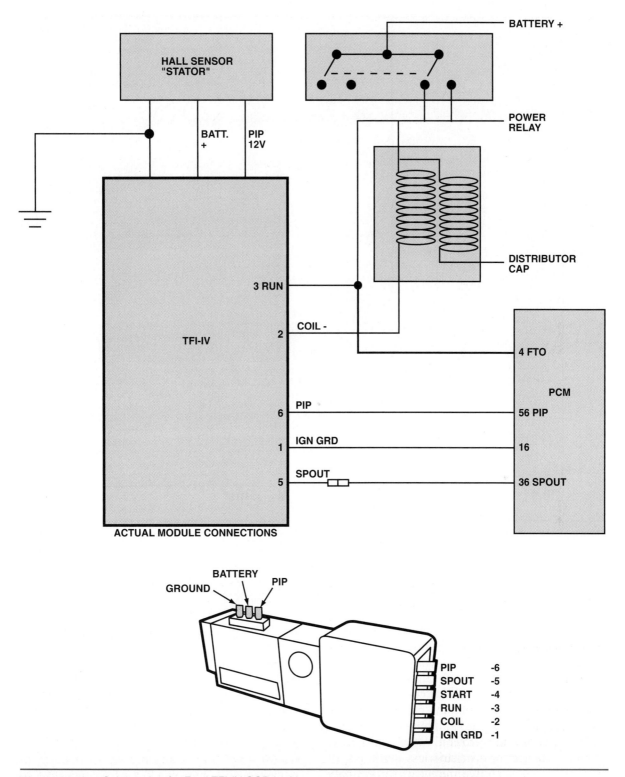

Figure 16-37. Schematic of a Ford TFI-IV CCD ignition system. (Courtesy of Ford Motor Company)

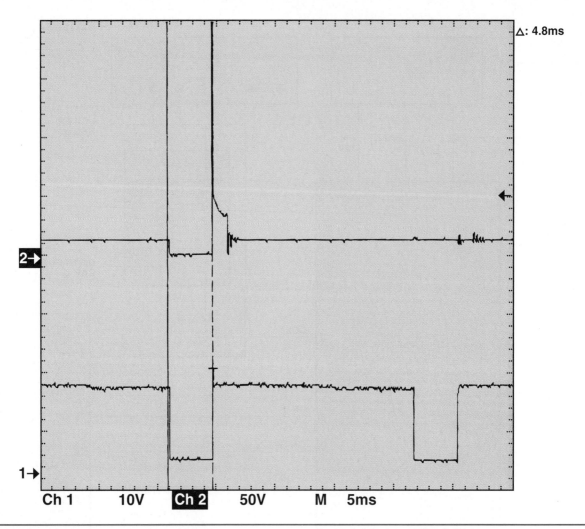

Figure 16-38. Observe that the SPOUT signal (bottom trace) aligns with the primary current turn-on signal (falling edge) and with the spark plug firing (rising edge).

with reverse polarity takes about 30 percent more energy, a misfire could result if the DIS coils did not have a different saturation time and primary current travel than a conventional coil. This provides more than 40 kilovolts of available energy—as much as 20 percent more than a conventional coil. Refer to the current ramping waveform patterns at the beginning of this chapter for additional information.

Firing voltage for distributorless ignition systems exhibits a slightly different waveform pattern than distributor ignition systems because of the design of the "waste spark" secondary system. Distributorless ignition systems pair two cylinders (companion cylinders) and fire both spark plugs at the same time. One cylinder will be on

the compression stroke while the other cylinder will be on the exhaust stroke. The spark plugs are wired in series. One fires in the conventional method, from center electrode to ground electrode, while the other fires from ground electrode to center electrode. Every other cylinder has reverse polarity. The cylinder on the compression stroke will require more voltage to fire the air-fuel mixture than the other cylinder (exhaust or waste). The firing event fires two spark plugs simultaneously. One plug fires from negative to positive polarity while the other fires from positive to negative polarity. The exhaust stroke pattern will show less voltage because there is much less resistance across the spark plug gap of the waste spark, figure 16-41. Also, the distributor-

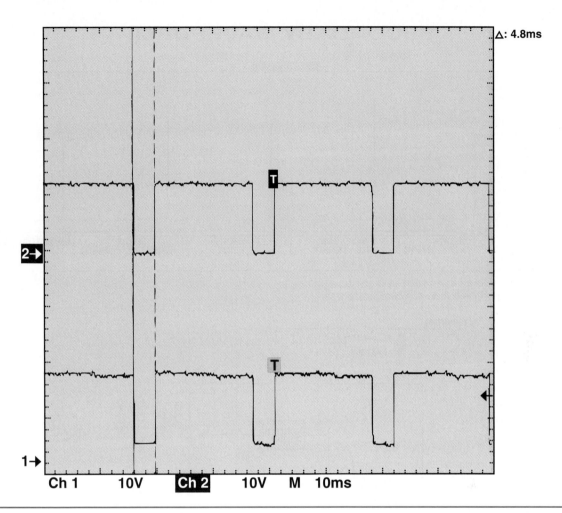

△: 4.8ms

Ch 1   10V   Ch 2   10V   M   10ms

Figure 16-39. Observe that the SPOUT and FTO signals align on the TFI-IV CCD ignition.

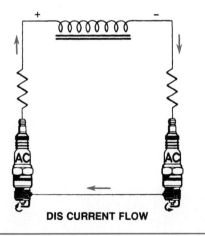

DIS CURRENT FLOW

Figure 16-40. In a DIS system, one spark plug in each pair always fires with forward polarity. The other plug fires with reverse polarity, with current traveling from the side to the center electrode. (Courtesy of General Motors Corporation)

less ignition system fires once per engine revolution as opposed to every other revolution in a distributor-equipped engine.

The ignition control module determines and maintains the coil firing order. When it orders the coil to fire, one spark plug fires forward and the other fires backward. The voltage drop across each plug is determined by firing polarity and cylinder pressure. The ignition module controls primary current travel and limits dwell time. The low resistance of the primary coil winding, combined with an applied voltage of 14 volts, results in a theoretical current travel greater than 14 amperes, helping to decrease the coil's saturation time. Such a high current travel, however, will damage the system components unless it can be limited to a range of 8.5 to 10 amperes. Limiting of the circuit current to the safe range is done by a control circuit inside the module, figure 16-42.

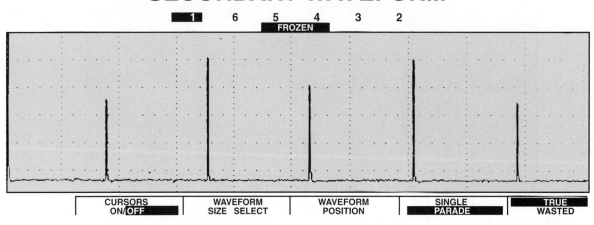

Figure 16-41. On a DIS ignition system, the secondary waveform pattern indicates less resistance across the spark plug gap on the exhaust stroke (waste spark).

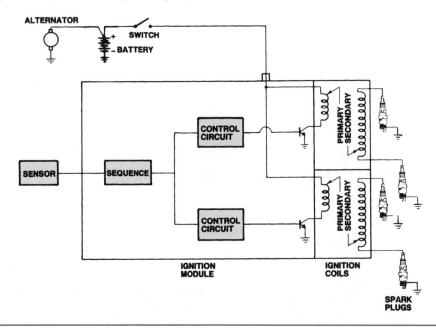

Figure 16-42. A functional schematic of the DIS module, showing the current-limiting control circuitry. (Courtesy of General Motors Corporation)

Some modules use a type of **closed-loop dwell control,** figure 16-43. In this system, the module continuously monitors coil buildup for maximum current. If maximum current was reached during the previous buildup, the module shortens dwell time to lower the wattage used by the system. If minimum current was not reached during the previous buildup, the module lengthens dwell time to permit full saturation of the coil. When current limiting takes place before coil discharge, the module decreases dwell time for the next cycle.

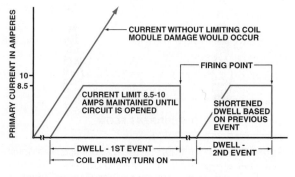

Figure 16-43. Closed-loop dwell allows full saturation of the ignition coil by increasing or decreasing dwell time. (Courtesy of General Motors Corporation)

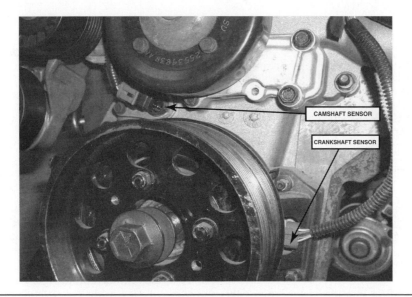

Figure 16-44. This photo shows the location of the camshaft and crankshaft Hall Effect sensors used on GM's 3.8-liter and 3800 engines.

Delco-Remy

Computer-Controlled Coil Ignition (C³I)

This distributorless ignition was introduced in 1984 and is used on Buick-built V-6 engines. There is no distributor with signal generator, rotor, and cap in the C³I system. The ignition trigger signal is provided by Hall Effect crankshaft and camshaft position sensors. The crankshaft sensor provides a signal that indicates basic timing, crankshaft position, and engine speed. The camshaft sensor, figure 16-44, provides a firing order signal. On 3800 and 3.8-liter non-turbo engines, the camshaft sensor is mounted in the timing cover. With 3.8-liter turbocharged engines, the camshaft sensor replaces the normal distributor and is mounted in the distributor location to drive the oil pump. On 3300 and 3.0-liter engines, a synchronization ("sync") signal sensor is combined with the crankshaft sensor at the front of the engine. Regardless of their appearance or location, all camshaft sensors have the same purpose. They are used for sequential fuel injection, misfire detection, and may be used to sync the coil firing sequence. A sync signal that is on the crankshaft is used only to start the coil firing sequence.

The crankshaft and camshaft sensors use Hall Effect switches with revolving interrupter rings to synchronize and fire the coils at the required time. The ignition module sends a reference voltage through a semiconductor wafer in the Hall

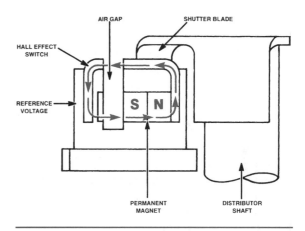

Figure 16-45. When the shutter blade is not blocking the magnet a voltage is produced. (Courtesy of Daimler-Chrysler Corporation)

Effect switch. A permanent magnet mounted in line with the semiconductor wafer induces a voltage across the semiconductor, figure 16-45. As a metal blade on the interrupter ring passes between the permanent magnet and semiconductor wafer, the magnetic field is broken and Hall Effect voltage drops, figure 16-46. The 3800 and 3.8-liter non-turbo engines reverse the process. On these engines, the permanent magnet is mounted on the camshaft sprocket and the Hall Effect switch is part of the timing cover sensor. As the camshaft sprocket revolves, it turns the Hall Effect switch on and off.

The camshaft sensor serves only to establish the initial ignition firing sequence during engine

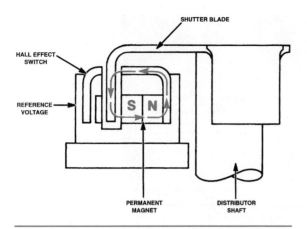

Figure 16-46. When the shutter blade blocks the magnet no voltage is produced. Courtesy of DaimlerChrysler Corporation)

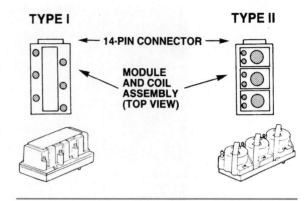

Figure 16-47. Identifying the Type I and Type II coil and ignition module assemblies used by Delco. Courtesy of General Motors Corporation)

cranking. The ignition module synchronizes the initial cam sensor signal with one of the crankshaft sensor signals during cranking and remembers the crankshaft sensor sequence as long as the ignition remains on.

The operation of the C^3I system is very similar to that of the HEI with EST system described earlier. During starting, the ignition module controls both ignition timing and spark distribution. When the engine reaches a programmed speed between 200 and 400 rpm, the PCM overrides the ignition module and assumes control of timing based on signals from the crankshaft sensor and other engine sensors. In case of PCM failure, the ignition module assumes timing control and operates the ignition with a fixed advance of 10 degrees BTDC.

Since introducing the C^3I system, GM has used three different variations:

- Type 1: All three coils are molded into a single housing with a smooth exterior surface, figure 16-47. Three spark plug cable terminals are provided on each side of the housing. If one coil malfunctions, the entire coil pack must be replaced.
- Type 1 Fast Start: The coil pack can be interchanged with a Type 1, but ignition module circuitry differs and connector plugs are not compatible.
- Type 2: Similar to Type 1, but the coils can be replaced individually.

The Type 1 Fast Start system measures crankshaft sensor signals more precisely, resulting in a faster startup. A dual crankshaft sensor is located beside the harmonic balance/crankshaft pulley on the front of the engine. The harmonic balancer has

two sets of interrupter rings. The outside ring consists of 18 evenly spaced interrupter blades that deliver 18 pulses every crankshaft revolution. These pulses are called the 18X signal. The inside ring consists of three blades with gaps of 10, 20, and 30 degrees spaced at 100, 90, and 110 degrees apart respectively. The inside ring pulses are called the 3X signal, figure 16-48.

Variations in the 3X signal allow the ignition module to synchronize the correct coil without need of the camshaft signal. Since the module can determine the correct coil within 120 degrees of crankshaft rotation, it starts firing on the first coil identified. The 18X pulse acts as a "clock pulse" to measure the length of each 3X pulse. The 18X pulse changes once during the 3X 10-degree gap, twice during the 20-degree gap, and three times during the 3X 30-degree gap. Once the module determines which 3X pulse it is reading, it can energize the correct coil.

Direct Ignition System (DIS)

The basic operation of the DIS system used on many Chevrolet and Pontiac 4-cylinder and V-6 engines is quite similar to that of the C^3I system, except for the method used to sense crankshaft position. Instead of Hall Effect switches located on the front of the engine, this DIS system uses a magnetic sensor installed in the side of the engine block. When used with the 2.5-liter engine, the sensor is installed on the back of the module. With other 4-cylinder and V-6 engines, the sensor is installed in the block below the module and is connected externally, figure 16-49.

A notched wheel or reluctor is cast into the crankshaft. The crankshaft reluctor on both 4-cylinder and V-6 engines is machined with seven

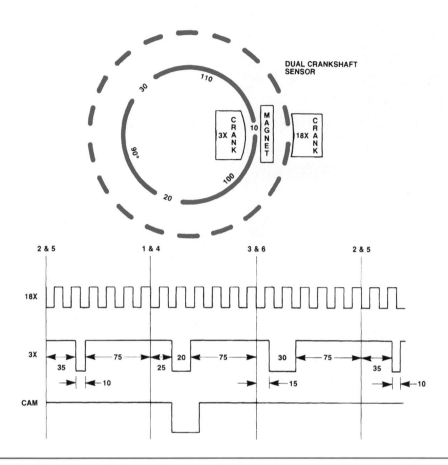

Figure 16-48. Dual Hall Effect switches and two sets of interrupter rings on the harmonic balancer are used in the "fast start" system to measure crankshaft sensor signals with more accuracy. (Courtesy of General Motors Corporation)

Figure 16-49. On GM 2.5-liter engines the crankshaft sensor connected directly to the coil packs.

notches or slots and serves as the field interrupter. The sensor head consists of a permanent magnet with a wire winding and is positioned a specified distance from the reluctor. As the crankshaft ro-

tates, the reluctor notches interrupt the sensor's magnetic field, causing a small AC voltage to be induced in the sensor's wire winding, figure 16-50. Because the sensor is installed in a fixed position in the engine block and the reluctor is an integral part of the crankshaft, there is no timing adjustment possible or required with this system.

Six reluctor notches are evenly spaced around the reluctor surface at 60-degree intervals; the seventh notch is spaced 10 degrees from one of the six notches. The signal from the seventh notch is used by the module to synchronize coil firing sequence to the crankshaft position. While reluctor configuration is the same for 4-cylinder and V-6 engines, coil firing order and determination of crankshaft position are calculated by the ignition module differently.

In the V-6 engine system, figure 16-51, the synchronization notch tells the module to ignore the first notch and establish base timing for cylinders 2 and 5 with the second notch. The module ignores the third notch, and relies on the fourth notch to set base timing of cylinders 3 and 6. The

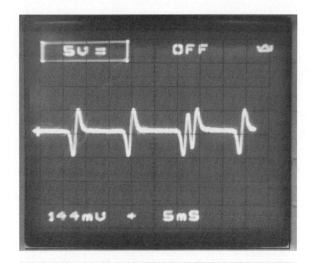

Figure 16-50. This waveform shows the AC voltage produced by the magnetic sensor as the reluctor teeth rotate past the sensor. Each notch on the reluctor wheel produces an AC voltage.

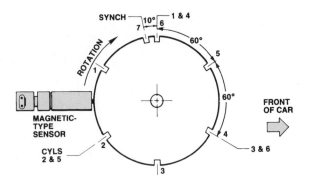

Figure 16-51. The V-6 ignition module recognizes notches 2, 4, and 6 as the signal to fire cylinders 2 and 5, 3 and 6, and 1 and 4, in that order. (Courtesy of General Motors Corporation)

fifth notch also is ignored in favor of the sixth notch for cylinders 1 and 4. After the seventh or synchronization notch passes, the entire sequence is repeated. As a result, the firing order of the first crankshaft revolution is 2-5, 3-6, and 1-4.

With 4-cylinder engines, the module starts the firing sequence on the seventh notch. If engine speed is below a predetermined value, the module fires each coil at a specified interval based only on engine speed. The synchronization notch tells the module to ignore the first notch and use the second notch to establish 10 degrees BTDC timing for cylinders 2 and 3. The module ignores the third and fourth notches, but uses the fifth notch to establish an equivalent timing setting for cylinders 1 and 4. In this way, the 2/3 coil is fired during startup.

The reference pulse in both systems is pulled low by the notch ahead of the one that is used to fire the cylinder, returning to its high state when the cylinder firing notch passes. The change in reference voltage is sent to the PCM for use in electronic spark timing (EST) and fuel injection.

Integrated Direct Ignition (IDI)
This system is used only on the Oldsmobile-built Quad 4 engine and differs from other 4-cylinder DIS systems primarily in the configuration of the system components, figure 16-52. The coil pack and module are contained in a unit that connects directly to the spark plugs. This eliminates the

use of spark plug cables, but the entire housing must be removed when changing spark plugs. However, the coils and housing can be replaced individually.

Chrysler

Direct Ignition System (DIS)
Chrysler uses a direct ignition system, figure 16-53, on its 3.3-liter and 3.5-liter engines. Crankshaft timing is determined by a magneto-resistive sensor installed in the transaxle bellhousing. The single-board engine controller (SBEC) sends an 8-volt reference signal to the sensor. The transaxle drive plate contains three groups of four slots. Each group of slots is positioned 20 degrees apart and provides a signal for two spark plugs. Transaxle drive plate rotation makes and breaks the sensor's magnetic field, causing sensor output voltage to the SBEC engine controller to vary between zero and five volts. The SBEC uses this voltage signal to determine engine speed and calculate both timing advance and the required fuel delivery, as displayed in figure 16-54.

Also a magneto-resistive type, the camshaft sensor is installed in the timing chain case cover and functions in the same way as the crankshaft sensor. The camshaft timing gear contains a series of slots that give a one–two–three–one–two–blank sequence. The SBEC uses the signals, with the CKP signal, to determine coil firing and fuel injector sequence, figure 16-55. The proper ignition coil can be fired within one crankshaft revolution. The injectors can be pulsed at the same time.

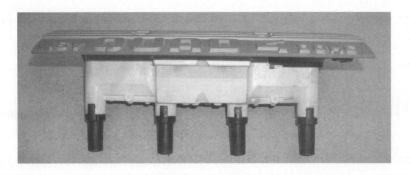

Figure 16-52. The integrated direct ignition system contains the coils, module, and spark plug boots in one assembly.

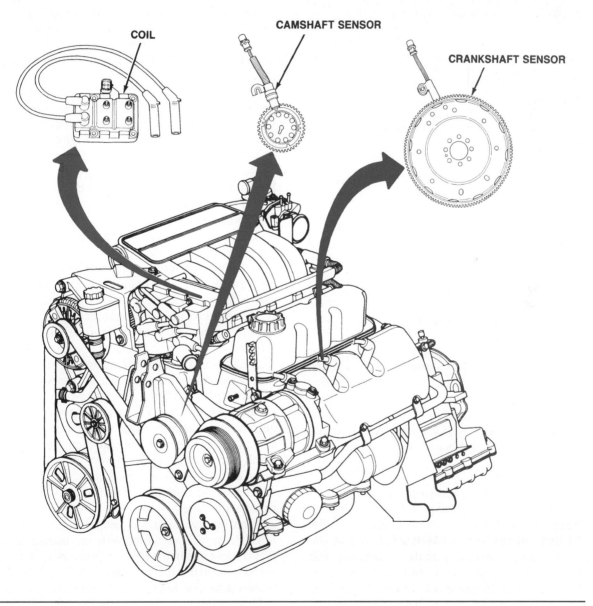

Figure 16-53. Major components of the Chrysler DIS system. (Courtesy of DaimlerChrysler Corporation)

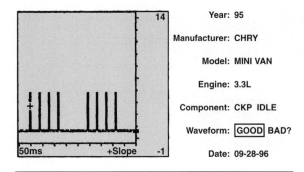

Figure 16-54. The CKP at idle. Transaxle drive plate rotation makes and breaks the magneto-resistive sensor's magnetic field, varying sensor output voltage to the SBEC. (Courtesy of Snap-on Vantage®)

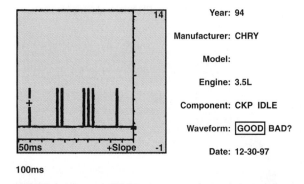

Figure 16-55. CMP sensor's signals, along with CKP signal, help the SBEC determine coil firing and fuel injector sequence. (Courtesy of Snap-on Vantage®)

Ford (Motorcraft)

All Ford distributorless ignitions operate on the same basic principles as those used by GM and Chrysler. Each coil fires two spark plugs, with one spark igniting the mixture at the top of the compression stroke and the other spark wasted at the top of the exhaust stroke. One plug of each pair has positive polarity; the other has negative polarity.

4-2 Distributorless Ignition System

Sometimes called the dual plug DIS, this unusual distributorless ignition was introduced on 1989 2.3-liter, 4-cylinder truck engines, figure 16-56. Each cylinder uses two sprak plugs, with one plug installed on each side of the combustion chamber. Those plugs on the right side of the engine form the primary system, and are responsible for engine operation at all times. The plugs on the left side of the engine form the secondary system and

are switched on and off by the EEC-IV computer, according to engine speed and load requirements.

Only the primary plugs fire when the engine is cranking. Once the engine is running, the EEC-IV computer commands the DIS module through the dual plug inhibit (DPI) circuit to switch from single-to dual-plug operation. The EEC-IV computer also is responsible for ignition timing and dwell.

The 4-2 DIS uses two four-tower DIS coil packs, figure 16-57, with a single remote DIS module, figure 16-58. A dual Hall Effect crankshaft sensor bracket-mounted near the crankshaft damper, figure 16-59, completes the system. The right coil fires the primary plugs during normal operating conditions; the left coil fires the secondary plugs as directed by the EEC-IV computer and the DIS module.

The crankshaft sensor works on the same principles as other Hall Effect switches covered previously, and is very similar in operation to the dual sensor used by GM in its C^3I ignition. A pair of rotating vane cups on the crankshaft damper produces a profile ignition pickup (PIP) signal for base timing data, and a cylinder identification (CID) signal used by the DIS module to determine which coil to fire.

The EEC-IV module sends a spark output (SPOUT) signal to the DIS module. The leading edge triggers the coil, and the trailing edge controls dwell time. This feature is called computer-controlled dwell (CCD). A buffered tach signal called ignition diagnostic monitor (IDM) supplies ignition system diagnostic information used for self-test.

A CID sensor or circuit failure will not result in a no-start condition, as the DIS module randomly selects and fires one of the two coils under such circumstances. The result of the module's guess may be an engine that is hard to start. However, turning the key off and then back on to crank the engine again allows the module to make another guess. After a few tries, the module will make the right choice and select the correct firing sequence.

If an ignition failure results in the loss of the SPOUT signal, a failure-effects management (FEM) program in the EEC-IV computer memory prevents total driveability loss. The EEC-IV computer opens the SPOUT line, allowing the DIS module to fire the coils directly from the PIP output. This results in a fixed spark angle of 10 degrees BTDC and fixed dwell.

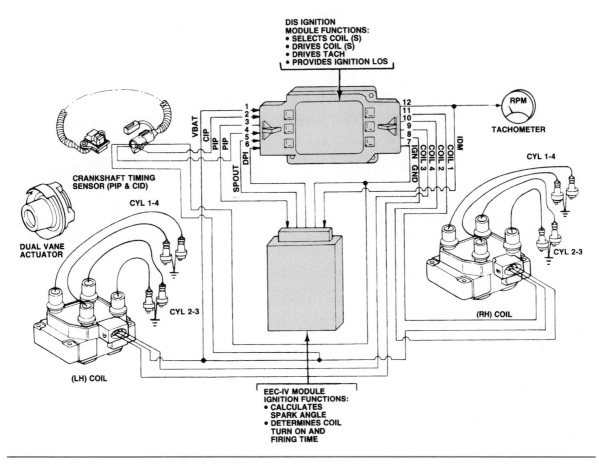

Figure 16-56. A diagram of the Ford 4-2 DIS system. (Courtesy of Ford Motor Company)

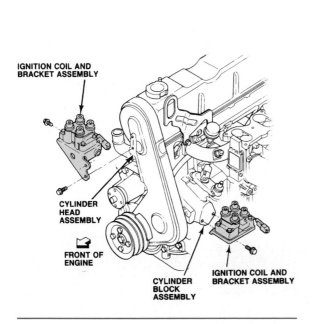

Figure 16-57. 4-2 DIS coil pack locations. (Courtesy of Ford Motor Company)

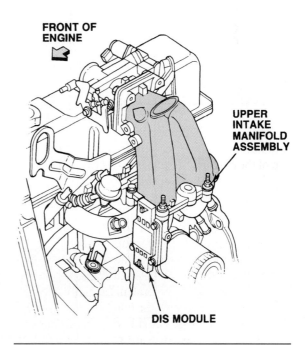

Figure 16-58. 4-2 DIS ignition module location. (Courtesy of Ford Motor Company)

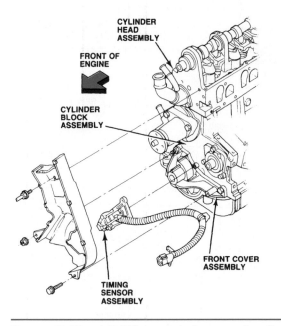

Figure 16-59. 4-2 DIS crankshaft sensor location. (Courtesy of Ford Motor Company)

V-6 distributorless Ignition System (DIS)

This DIS system also was introduced in 1989 on the 3.0-liter SHO V-6 engine and the 3.8-liter super-charged V-6 engine. The system functions in essentially the same way as the 4-2 DIS discussed above, but its components differ, figure 16-60:

- A single six-tower coil pack contains three coils with an individual tach wire for each coil; the coil pack is serviced as an assembly.
- A single set of spark plugs is used, with one plug per cylinder.
- The 3.0-liter cylinder identification (CID) sensor is installed on the end of one camshaft.
- The 3.8-liter CID sensor is installed in the engine block where the distributor would normally be located.
- A single Hall Effect switch is used as the crankshaft sensor.

When the engine is cranked, the DIS module looks for a change in the CID signal, from high

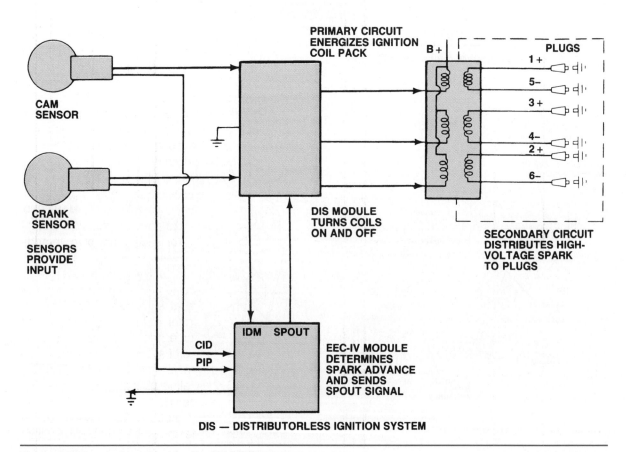

Figure 16-60. A diagram of the Ford V-6 DIS. (Courtesy of Ford Motor Company)

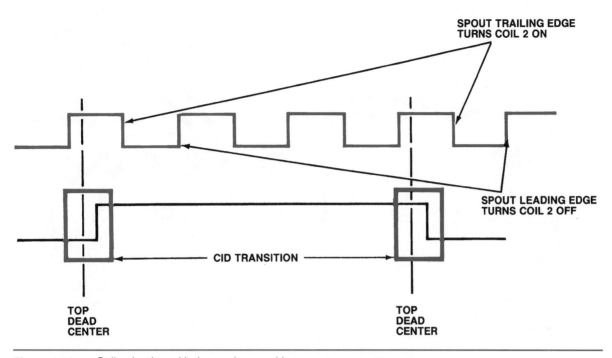

Figure 16-61.   Coil selection with the engine cranking. (Courtesy of Ford Motor Company)

to low or low to high. As soon as the module sees the leading or trailing edge of the CID signal, it prepares to fire coil 2 of the coil pack. Once the change in the CID singal occurs, the module looks for the trailing edge of SPOUT to turn on coil 2. When the module sees the next leading edge of SPOUT, it turns the primary current off to coil 2, which fires coil 2, figure 16-61. The DIS module always fires the coils in one coil pack in a given order. At engine start-up, the coil firing sequence is always 2, 3, 1. Because the coils continue to fire in the same order, they are synchronized with compression and remain synchronized as long as the engine is running, even if the DIS module loses the CID signal. SPOUT tells the DIS module when to fire the next coil as long as the engine is running. If the SPOUT circuit opens, the PIP signal is used by the module to fire the coils.

Electronic Direct Ignition System (EDIS)

A second-generation DIS system, EDIS functions faster and with greater accuracy than the V-6 DIS system just described. Because the system has been used on 4-cylinder, V-6, and V-8 engines since its introduction in 1990, the number of coils and coil packs differ according to engine application. Other EDIS components also differ in appearance and location, but the system functions

■ Individual-Coil DIS

The distributorless ignition systems (DIS) we have detailed in this chapter commonly employ half as many coils as they have spark plugs to fire. These systems fire each pair of spark plugs simultaneously: one with forward polarity and the other with reverse polarity. Depending on piston position, one spark is a waste spark. These systems eliminate moving parts and reduce the associated friction and wear-and-tear. They are also economical to produce and quite adequate for their intended use. However, there is a better, though more expensive, way.

Many manufacturers have models with distributorless ignition coil for each spark plug. In addition to the advantages of waste-spark DIS, individual-coil DIS fire all plugs individually with forward polarity for a hotter spark, which provides better performance at high rpm or under load. These systems further reduce the length of the secondary cable or eliminate it altogether, decreasing the chance of secondary arcing or leakage, improving the long-term reliability of the system, and eliminating the chance of cross-fire. Furthermore, shortening or eliminating the spark plug cable greatly reduces the amount of radiofrequency interference (RFI) emitted by the ignition system.

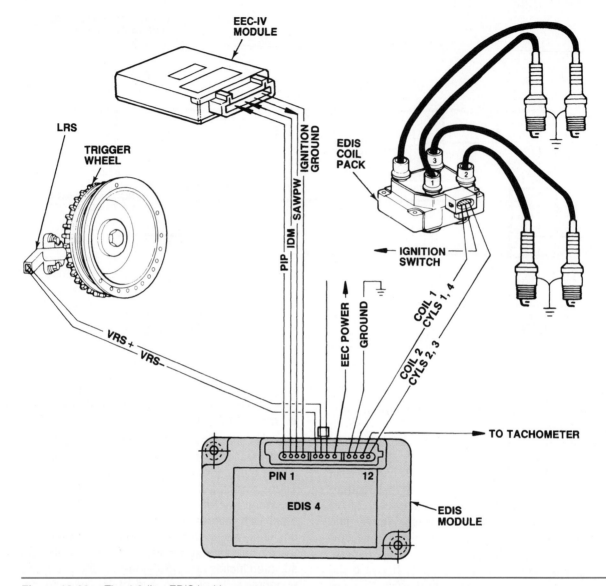

Figure 16-62.   The 1.9-liter EDIS ignition system components. (Courtesy of Ford Motor Company)

essentially the same regardless of engine application. Figure 16-62 shows the EDIS components used on the 1.9-liter SEFI engine. There are other differences between EDIS and DIS:

- EDIS does not use a cylinder identification (CMP) sensor. The variable reluctance sensor is a CKP sensor with a sync signal for ignition coil firing sequence.
- EDIS crankshaft position signals are more sophisticated and complex than those used on DIS systems.
- EDIS uses a spark angle word (SAW) signal in place of the DIS SPOUT signal. The SAW signal also is more complex than a SPOUT signal.

- The EDIS module is smarter, faster, has more "thinking" ability and decision-making responsibilities, and better diagnostic ability than the earlier DIS module.

The variable reluctance sensor (VRS) is a type of crankshaft position sensor (CKP); it has a magnetic transducer containing a pole piece wrapped with fine wire, figure 16-63. If the transducer is exposed to a change in flux lines, a differential voltage will be induced across the windings. Thus, when a ferromagnetic toothed timing wheel on the crankshaft rotates in the presence of the VRS, the passing teeth cause the VRS reluctance to change, resulting in a varying analog voltage signal. The timing wheel has 35 teeth and a blank

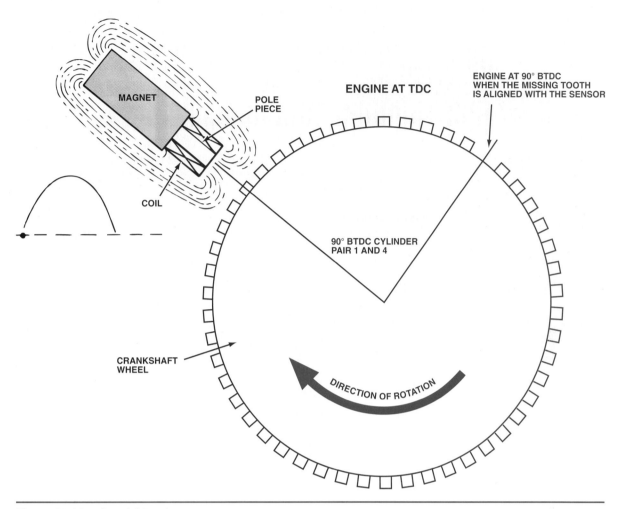

Figure 16-63. A variable reluctance sensor. (Courtesy of Ford Motor Company)

spot spaced at 10-degree increments. The blank spot for the thirty-sixth tooth serves as a fixed reference point for No. 1 piston travel identification.

During cranking, the EEC-IV module refers to a predetermined fuel control strategy stored in its memory to supply the necessary fuel for starting the engine. At the same time, the EDIS module looks for any significant change in the VRS signal. When it recognizes the missing tooth, figure 16-64, the EDIS module has a reference point and is synchronized to fire the proper coil. While the engine continues to crank, the EDIS module fires the coil at 10 degrees BTDC (base timing). When the coil for piston No. 1 is fired, a profile ignition pickup (PIP) signal is sent from the EDIS module to the EEC-IV computer. The PIP signal provides the EEC-IV computer with crankshaft position and engine speed data. Because the PIP signal is synthesized by the EDIS module, it takes the form of a digital square wave with a 50 percent duty cycle, figure 16-65. When the EEC-IV computer

recognizes the PIP signal, it enables fuel and spark functions. It also processes the PIP signal with other information to determine the spark angle word (SAW) signal. The SAW signal replaces the former SPOUT signal on the TFI and DIS ignition systems as the signal for an ignition coil firing event (primary current turn-off). This signal is sent to the EDIS module, which uses it to determine whether ignition timing should be advanced or retarded.

Integrated Electronic Ignition (EI) System
Later Ford models equipped with the EEC-V OBD II system use a coil-over-plug distributorless ignition system. This system equips each cylinder with a separate coil/module assembly mounted directly over the spark plug. The EI ignition system also uses a crankshaft position sensor (CKP) and a wiring harness to connect each coil/module assembly to the computer. This system also eliminates the need for spark plug wires

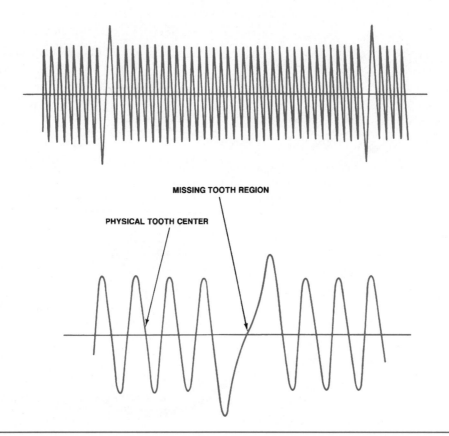

Figure 16-64. The VRS analog signal wave. (Courtesy of Ford Motor Company)

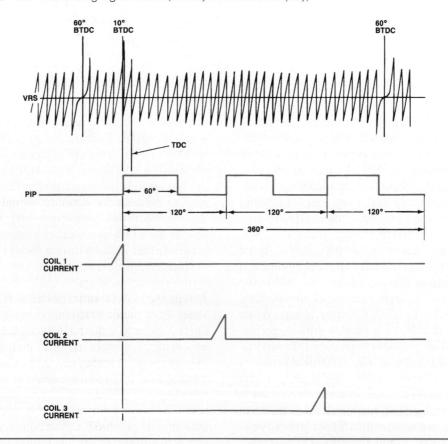

Figure 16-65. Comparison of analog CKP sensor signals with digital PIP signals created by the EDIS module. (Courtesy of Ford Motor Company)

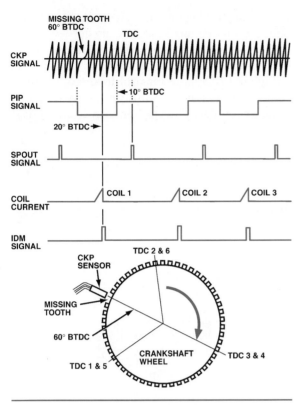

Figure 16-66. Ignition waveforms for the Ford integrated electronic ignition (EI) system (coil-over-plug). (Courtesy of Ford Motor Company)

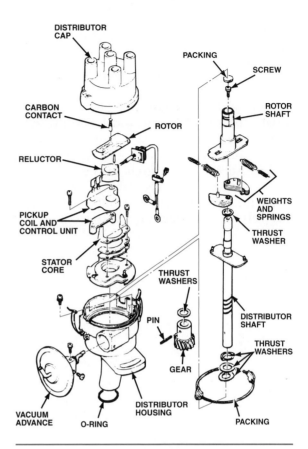

Figure 16-67. Exploded view of a Hitachi distributor installed on early model Subaru engines.

but requires an input from the camshaft position sensor (CMP). Here is a list of the integrated electronic ignition system features:

1. Much like the EDIS system, the CKP sensor is used to detect crankshaft position and engine speed by sensing the missing tooth on a pulsed wheel mounted onto the crankshaft. The CMP sensor is used to identify top dead center (TDC) of the cylinder number 1 on compression stroke, figure 16-66.

2. On most distributorless ignition systems, each coil will fire its companion coil (waste spark) simultaneously on the exhaust stroke. Integrated electronic ignition (EI) systems fire only one coil/module assembly for each compression stroke.

3. The PCM controls primary current to each coil/module assembly. Kickback voltage occurs when the voltage spike collapses. The PCM monitors the firing event using the ignition diagnostic monitor (IDM) circuit.

4. The EI systems require slightly less current limit amperage ratings than distributorless ignition systems. EI systems limit between 6.5 to 7.0 amps while distributorless ignition systems limit between 8.0 to 10.0 amps.

Imported Car Electronic Ignitions

Imported vehicles appeared on the automotive scene with electronic ignitions about the same time or shortly after the domestic systems we have discussed. All major imported cars now use an inductive-discharge electronic ignition. In most of these systems, timing signals are sent to the ignition module from a magnetic pulse generator in the distributor, although Bosch makes a Hall Effect distributor used by Volkswagen.

Hitachi electronic ignition systems are installed onto many Japanese import vehicles. The Hitachi system for Subaru varies with year and models. The 1977 through 1986 Subaru models use a different version for the turbocharged engines. The turbocharged models use a four-terminal igniter (module) instead of the two-terminal igniter. These Subaru Hitachi ignition systems consists of an external igniter (module) on early versions or an internal pickup coil/control unit (module) on later versions, figure 16-67, a conventional coil, a centrifugal and vacuum advance, an external ballast resistor (early) and a 4-tooth reluctor. There are some versions that mount the igniter onto the coil.

By the middle 1980s and later, Hitachi also equipped many Subarus with a photoelectric ignition system.

Nippondenso electronic ignition systems are installed on many domestic models built overseas. Nippondenso ignition systems are also installed onto Japanese models such as the Lexus, Subaru, Toyota, Suzuki, and Mitsubishi. Later Nippondenso ignition systems equip the igniter (module) inside the distributor along with the ignition coil and pickup coil assembly, figure 16-68.

The electronic ignition used on the Mitsubishi-built 2.6-liter engine in some Chrysler vehicles is typical of most Japanese designs, figure 16-69. The distributor contains a magnetic pickup with an integral IC igniter (control module), although some Toyota ignitions use an external igniter mounted on the coil. The variable dwell ignition contains no ballast resistor. Full battery voltage is supplied to both the coil and igniter whenever the ignition switch is in the start or the run position.

The early Toyota had a distributor with two or three magnetic sensors. The CKP (Ne) could be either a 4- or 24-tooth sensor inside the distributor.

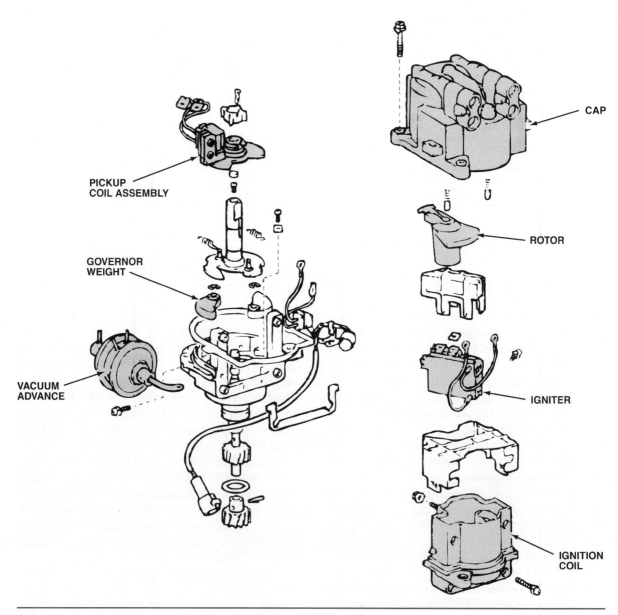

Figure 16-68.   Exploded view of a Nippondenso distributor with the integral ignition coil.

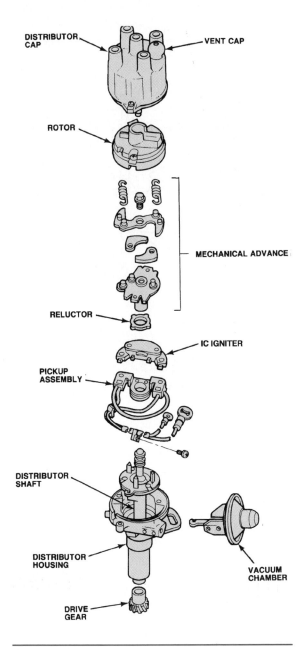

DISTRIBUTOR CAP

VENT CAP

ROTOR

MECHANICAL ADVANCE

RELUCTOR

IC IGNITER

PICKUP ASSEMBLY

DISTRIBUTOR SHAFT

DISTRIBUTOR HOUSING

VACUUM CHAMBER

DRIVE GEAR

Figure 16-69. The Mitsubishi ignition system used on 2.6-liter engines is typical of the inductive-discharge ignitions used on Japanese vehicles.

The CMP (G) signal was a 1, 2- or 2-tooth signal to sync the Ne signal for injection and ignition firing, figure 16-70. Some systems used two (G) signals. These signals were sent to the PCM, which sent a square wave signal (IGt) to the ICM (igniter) to trigger the coil firing. The ICM monitored the inductive spike on the coil negative. If the plug fired, it sent a confirmation (IGf) signal to the PCM. The PCM required this signal for fuel injection.

With OBD II, the CKP (Ne) sensor was moved to the crankshaft for more accurate misfire detection.

The Honda distributor contains the coil, ICM, and three magnetic sensors. The CMP sensor has 1 tooth, the TDC sensor has 4 teeth, and the CKP has 24 teeth. The CKP signal is used for engine speed and crankshaft position.

The TDC signal is TDC for each cylinder. It is used for starting and as a backup for the CKP or CMP. The CMP signal is used to sync the CP and sequential fuel injection.

OBD II systems moved the CKP and TDC sensors to the crankshaft.

These signals are sent to the PCM, which controls the ICM for ignition timing. Many of the later systems use an individual coil for each cylinder. There are two basic types: one ICM for all coils and individual ICMs built into each coil.

An example of the latter is the Isuzu 3.5-liter engine. The CKP is a 58X (60-2) digital signal taken from the crankshaft. The CMP sensor is on the end of the camshaft. These signals are sent to the PCM, which controls timing and dwell with the ICM control signal to each coil, figure 16-71.

When cranking, the 58X signal is used to start ignition and fuel injection. Both companion coils are fired for a compression and waste spark effect. The fuel injectors are pulsed in groups. After the CMP signal is received, the PCM fires the coils only on compression, and the injectors are pulsed sequentially. This allows a faster start, with ignition and injection before the CMP signal is received.

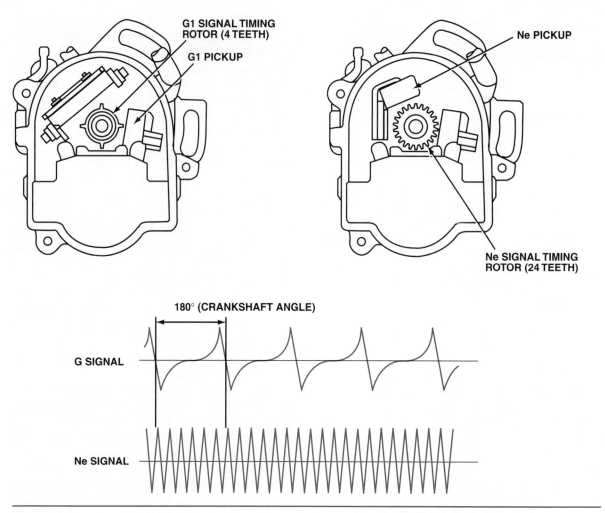

Figure 16-70. Early Toyota distributors used 4- or 24-tooth sensors. Different rotors generated different signals.

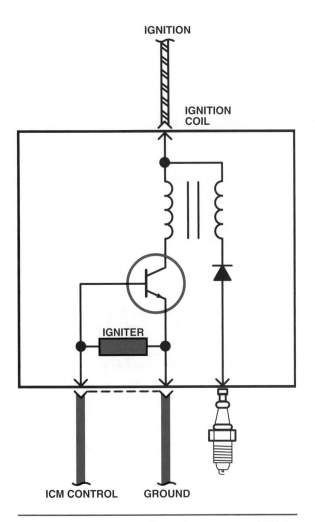

IGNITION

IGNITION COIL

IGNITER

ICM CONTROL **GROUND**

Figure 16-71. Isuzu 3.5L engine uses one ICM for each coil. The PCM controls timing and dwell. (Courtesy of Isuzu Motors America)

SUMMARY

Electronic ignition systems came into widespread use on both domestic and import vehicles in the early 1970s. Today, they are standard on all domestically built cars and light trucks, as well as imported cars and light trucks. Electronic ignitions provide greater available voltage and reliable performance for longer periods under varying operating conditions, with decreased maintenance.

Electronic ignitions are either the inductive-discharge type or the capacitive-discharge type. Original-equipment systems are the inductive-discharge type. The primary circuits of electronic ignitions can be triggered by breaker points, a magnetic pulse generator, a Hall Effect switch, a metal detector, or an optical triggering device signal. Original-equipment systems generally use a magnetic pulse generator or a Hall Effect switch in the distributor to trigger the primary circuit.

The primary circuit is switched on and off by transistors in the ignition control module. The electronic ignition module also controls the length of the ignition dwell period. Electronic ignitions with fixed dwell use a ballast resistor; those with variable dwell have no ballast resistor.

Basic electronic ignition systems use the same centrifugal and vacuum advance devices as breaker-point ignitions. Those electronic ignitions used with electronic engine control systems have no advance mechanisms. The systems computer controls spark advance.

Electronic ignitions have undergone major change and improvement since their introduction in the mid-1970s. Distributor-type ignitions are being replaced by distributorless ignitions, with increasing responsibility for their operation assigned to the electronic engine control system.

Review Questions

Choose the letter that represents the best possible answer to the following questions:

1. Which of the following is true of electronic ignition systems?
   a. Provide less available voltage than breaker-point systems
   b. Provide greater available voltage than breaker-point systems
   c. Give less reliable system performance at all engine speeds
   d. Require more maintenance than breaker-point systems

2. Which of the following is true of inductive discharge systems?
   a. Used as original equipment by most car manufacturers
   b. A common aftermarket installation
   c. Provide about 50,000 volts of available voltage
   d. Sustain a spark for about 200 microseconds

3. Capacitive-discharge systems provide _____ secondary voltage when compared to inductive-discharge systems.
   a. Exactly the same
   b. About the same
   c. Greater
   d. Less

4. Transistors have:
   a. A base of one type of material and an emitter and collector of another
   b. An emitter of one type of material and a base and collector of another
   c. A collector of one type of material and a base and emitter of another
   d. A collector, base, and emitter all of the same material

5. The earliest electronic ignition systems used which of the following triggering devices?
   a. Metal detectors
   b. Breaker points
   c. Magnetic pulse generators
   d. Light detectors

6. The most common type of original-equipment triggering devices are:
   a. Breaker points
   b. Light detectors
   c. Metal detectors
   d. Magnetic pulse generators

7. The accompanying illustration shows a:
   a. Magnetic pulse generator
   b. Metal detector
   c. Light detector
   d. Breaker-point assembly

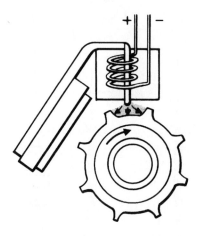

8. The dwell period of a solid-state system can be measured with:
   a. A voltmeter
   b. An ammeter
   c. An ohmmeter
   d. An oscilloscope

9. In the Delco high-energy ignition (HEI) system, the dwell period is controlled by the:
   a. Pole piece
   b. Timer core
   c. RFI filter capacitor
   d. Ignition control module (ICM)

10. The Delco HEI system uses:
    a. A single-ballast resistor
    b. A dual-ballast resistor
    c. No ballast resistor
    d. Calibrated resistance wire

11. The rotating component attached to the distributor shaft in the Chrysler electronic ignition is called the:
    a. Armature
    b. Reluctor
    c. Timer core
    d. Trigger wheel

12. The rotation component attached to the distributor shaft in the Ford electronic ignition is called the:
    a. Armature
    b. Reluctor
    c. Time core
    d. Trigger wheel

13. A Hall Effect switch requires _____ to generate an output voltage.
    a. A magnetic pulse generator
    b. A varactor diode
    c. A 45-degree magnetic field
    d. An input voltage

14. Thick-film integrated (TFI) ignition modules are:
    a. Mounted inside the distributor
    b. Mounted on the side of the distributor
    c. Mounted on the vehicle fenderwell
    d. Used as a handle to rotate the distributor

15. A Hall Effect switch is used with which Chrysler electronic ignitions?
    a. 4-cylinder
    b. Inline-six
    c. V-6
    d. V-8

16. When a distributorless ignition coil fires its two sparks plugs, one spark ignites the mixture at the _____ of the compression stroke and the other spark is wasted at the _____ of the exhaust stroke.
    a. Top, top
    b. Top, bottom
    c. Bottom, bottom
    d. Bottom, top

17. When a DIS module fires the spark plug in a cylinder on its exhaust stroke, there is _____ effect on engine operation.
    a. No
    b. A minor
    c. A stabilizing
    d. A detrimental

18. The Ford dual plug DIS system is used on:
    a. 3.0-liter SHO V-6 engines
    b. 3.8-liter supercharged V-6 engines
    c. 2.3-liter truck engines
    d. 4-cylinder, V-6, and V-8 engines

Glossary of Technical Terms

Actuator: An electrical or mechanical device that receives an output signal from a computer and does something in response to that signal.

Adaptive Memory: A feature of computer memory that allows the microprocessor to adjust its memory for computing engine settings, based on both sudden and gradual changes in engine operating conditions. Adaptive memory accounts for engine and systems wear or degradation.

Adsorption: A chemical action by which liquids or vapors gather on the surface of a material. In a vapor storage canister, chemical properties force fuel vapors to attach themselves (adsorb) to the surface of charcoal granules.

AIR: Air Injection Reactor. The secondary air system that supplies fresh air to the exhaust manifold, exhaust pipe, and/or converter to assist in oxidizing exhaust emissions.

Air Bleed: A small opening in the fuel inlet passage used to help break up liquid fuel for better atomization.

Air Charge Temperature (ACT) Sensor: A thermistor used to measure intake air temperature or air-fuel mixture temperature. Now known as an intake air temperature sensor (IAT).

Air-Fuel Ratio: The ratio of air to gasoline in the air-fuel mixture that enters an engine.

Air-Fuel Ratio Sensor: An oxygen sensor that produces a positive or negative current based on the air-fuel ratio. When no current is produced, the air-fuel ratio is 14.7:1. A negative current indicates a rich exhaust stream and a positive current indicates a lean exhaust stream.

Airhorn: The topmost section of a carburetor, above the venturi, and usually housing the choke plate.

Air Injection: A way of reducing exhaust emissions by injecting air into each of the exhaust system. It mixes with the hot exhaust system. It mixes with the hot exhaust and oxidizes the HC and CO to form H_2O and CO_2.

Alternating Current, AC: Current that flows alternatively from B+ to B−, then flows from B− to B+, and so on. Alternating current continually changes direction back and forth.

Ampere (also Amp, Amperage, Intensity, or I): The unit for measuring the rate of electrical current flow through a conductor, calculated in billions of electrons per second.

Analog: A voltage signal or processing action that is continuously variable relative to the operation being measured or controlled.

Analog-to-Digital (ad): An electronic conversion process for changing analog voltage signals to digital voltage signals.

Aneroid Bellows: One or a series of sealed accordion-shaped membranes exposed to barometric (atmospheric) pressure. Pressure changes cause the bellows to expand and contract, or flex. On a carburetor, this flexing actuates a valve or metering rods to change fuel or air flow.

Antiknock Value: The characteristic of gasoline that helps prevent detonation.

Armature: The movable part in a relay. The revolving part in a generator or motor.

ASM: Acceleration Simulation Mode. A loaded mode test performed on a dynamometer at a fixed load and speed. The load on the vehicle is based on the vehicle's weight.

Atmospheric Pressure: The pressure that the earth's atmosphere exerts on objects. At sea level, this pressure is 14.7 psi (101 kPa) at 32°F (0°C).

Atomization: Breaking a liquid down into small particles or a fine mist.

Backfire: The accidental combustion of gases in an engine's intake or exhaust system.

Backpressure: The resistance, caused by turbulence and friction, that is created as a gas or liquid is forced through a restrictive passage.

Baffle: A plate or obstruction that restricts the flow of air or liquids. The baffle in a fuel tank keeps the fuel from sloshing as the vehicle moves.

Barometric Manifold Absolute Pressure (BMAP) Sensor: A MAP and BARO sensor combined into one housing.

Barometric Pressure (BARO) Sensor: A sensor, similar in operation to the MAP sensor, that reads barometric (ambient) air pressure.

Bifurcated: Separated into two parts. A bifurcated exhaust manifold has four primary runners that converge into two secondary runners; these converge into a single outlet into the exhaust system.

Bimetal Temperature Sensor: A device made of two strips of metal welded together. When heated, one side expands more than the other, causing the sensor to bend.

Binary: A mathematical system consisting of only two digits (0 and 1) that allows a digital computer to read and process input voltage signals.

Block-Learn: The long-term effects of integrator corrections. As such, block-learn complements adaptive memory. If, for example, the computer continually must overcompensate fuel metering to maintain the stoichiometric ratio, it "learns" the correction and adapts its memory to make the correction factor part of its basic program. As of OBD II, known as long-term fuel trim.

Blowby: The leakage of combustion gases and unburned fuel past an engine's piston rings.

Blowoff Valve: A spring-loaded valve that opens when boost pressure overcomes the spring tension to vent excess pressure.

Boost: A measure of the amount of air pressure above atmospheric that a supercharger or turbocharger can deliver. Boost remains constant regardless of altitude.

Bore: The diameter of an engine cylinder; to enlarge or finish the surface of a drilled hole.

Bottom Dead Center (BDC): The exact bottom of a piston stroke.

Breaker Points: The metal contact points that act as an electrical switch in a distributor to open and close the ignition primary circuit.

Camshaft Overlap: The period, measured in degrees of crankshaft rotation, during which both the intake and exhaust valves are open. It occurs at the end of the exhaust stroke and the beginning of the intake stroke.

CAN: Controller Area Network. A high-speed communication protocol, 125K bps to 1M bps and used for real-time control.

Capacitive-Discharge Ignition: A method of igniting the air-fuel mixture in an engine cylinder through the storage of a voltage potential within a capacitor.

CARB: California Air Resources Board. Responsible for developing and enforcing I/M programs and repair standards in California.

Carbon Monoxide (CO): An odorless, colorless, tasteless, poisonous gas. A major pollutant given off by an internal combustion engine.

Catalyst: A substance that causes a chemical reaction, without being changed by the reaction.

Catalytic Converter: A device installed in an exhaust system that converts pollutants to harmless by-products through a catalytic chemical reaction.

Catalytic Cracking: An oil refining process that uses a catalyst to break down (crack) the larger components of crude oil. The gasoline produced by this method usually has a lower sulfur content than gasoline produced by thermal cracking.

Central Processing Unit (CPU): The processing and calculating portion of a computer.

Check Valve: A valve that permits flow in only one direction.

Circuit: A path through which an electric current is designed to flow.

Clearance Volume: The combustion chamber volume with the piston at TDC.

Closed Loop: An operation mode whereby output of a device is monitored and fed back to affect the next operating cycle of the component. In the automobile, the oxygen sensor continuously detects oxygen levels in the exhaust and provides "feedback" for fuel mixture adjustments at the carburetor, the adjustments are made, the results are sampled again and fed back; hence, the system is operating in closed loop. A thermostatically controlled home heating system is a classic example of a closed-loop system.

CO: Carbon Monoxide. An odorless, poisonous gas formed whenever a hydrocarbon is burned and there is insufficient oxygen.

CO_2: Carbon dioxide. A gas that is produced as the result of the combustion process. CO_2 is not considered a pollutant; however, it does lead to global warming.

Compression Ratio: A ratio of the total cylinder volume when the piston is at BDC to the volume of the combustion chamber when the piston is at TDC.

Conductor: A material that allows easy flow of electrons. Good conductors include silver, gold, copper, aluminum, and so on.

Continuous Injection System (CIS): A fuel-injection system in which fuel is injected constantly whenever the engine is running, as opposed to intermittently. Bosch K-Jetronic and KE-Jetronic are typical examples of CIS.

Conventional Theory of Current Flow: The theory that electrons flow from a positive power source (B+) to negative (B−). Also called positive current flow theory.

Current Limiting Hump: The portion of the waveform that indicates the exact point in the dwell when the current limiting device switches on. The device decreases current, so voltage level increases.

Delivery System: The fuel pump, strainer, fuel lines, valves, pressure regulator, and so on, used to deliver fuel to the carburetor or fuel injector(s).

Diagnostic Executive: The program within the PCM that coordinates the OBD II self-monitoring system. It manages the comprehensive component and emissions monitors, DTC and MIL operation, freeze-frame data, and scan tool interface.

Diagnostic Trouble Code (DTC): A type of OBD II test result that indicates a faulty circuit or system using an alphanumeric code.

Dieseling: A condition in which either a localized source of heat such as carbon deposits, or extreme heat buildup exists in the combustion chamber at the time the engine ignition is turned off. Dieseling occurs when heat ignites the fuel without a spark, and the engine continues to run, even though the ignition has been switched off.

Digital: A two-level voltage signal or processing function that is either on/off or high/low.

Digital-to-Analog (da): An electronic conversion process for changing digital voltage signals to analog voltage signals.

Direct Current, DC: Current that flows from positive (B+) to negative (B−) only (conventional theory of current flow).

Displacement: A measurement of engine volume. It is calculated by multiplying the piston displacement of one cylinder by the number of cylinders. The total engine displacement is the volume displaced by all the pistons.

Distributor Ignition System (DI): An ignition system, either mechanical or electronic, that incorporates a distributor in the design.

Diverter Valve: A valve used in an air-injection system to prevent backfire. During deceleration, it "dumps" the air from the air pump into the atmosphere. Also called a dump valve.

DLC: The SAE abbreviation for the diagnostic link connector required on all OBD II complaint vehicles.

DPFE Sensor: A device that converts exhaust backpressure into an electrical signal to report EGR flow to an engine control computer.

Duty Cycle: Duty cycle is a specific number of on/off cycles in a given length of time, for example, ten times per second.

Duty-Cycle Solenoid: A solenoid that cycles on and off at regular intervals in a specific amount of time, for example, ten times per second, and is used to adjust its performance based on outside information. The time the solenoid is energized is referred to as "on-time" and is usually measured as a percentage of the duty cycle or as dwell.

Dwell: Period of time when the ignition switching device is closed. During dwell, the primary circuit is complete so there is current in the primary winding of the coil.

Dwell Time: The duration or actual time that the dwell period, in distributor degrees, cycles.

E-85: The designation for vehicles equipped to run on different blends of ethanol up to 85 percent.

Eccentric: Off center. A shaft lobe that has a center different from that of the shaft.

EGR Valve Position Sensor: A device that senses EGR valve pintle position and reports it to an engine control computer.

Electron Theory of Current Flow: The theory that electrons flow from negative (B−) to positive (B+).

Electronic Fuel Injection (EFI): A computer-controlled fuel-injection system that precisely controls mixture for all operating conditions and at all speed ranges.

Electronic Ignition (EI): Any type of ignition that uses electronic components to switch the primary current on and off.

Engine Coolant Temperature (ECT) Sensor: An engine coolant temperature sensor, usually a negative temperature coefficient (NTC) type thermistor, that reports coolant temperature to the PCM.

Engine Mapping: Vehicle operation simulation procedure used to tailor the onboard computer program to a specific engine/powertrain combination. This program is stored in a PROM or calibration assembly. Engine mapping allows engineers to program the ECM with optimal specifications for ignition timing, fuel trim, EGR, and other parameters at specific engine load-and-rpm combinations.

Ethanol: Ethyl alcohol distilled from grain or sugar cane.

Evaporative Emission Control (EVAP): A way of controlling HC emissions by collecting fuel vapors from the fuel tank and carburetor fuel bowl vents and directing them through an engine's intake system.

Exhaust Gas Recirculation (EGR): A way of reducing NO_x formation by directing a metered amount of exhaust gas through the engine intake, reducing the temperature in the combustion chamber.

External Combustion Engine: An engine, such as a steam engine, in which fuel is burned outside the engine.

Firing Line: The section of the ignition waveform trace reflecting the collapsing magnetic field. The primary firing line is usually between 250 to 400 volts.

Firing Order: The sequence by cylinder number in which combustion occurs in the cylinders of an engine.

Fixed Dwell: The ignition dwell begins when the switching device turns on and remains constant, specified in distributor degrees, through all speeds.

Flags: Special OBD II messages that indicate the completion of an emission monitor test. If the Diagnostic Executive has not set all of the flags, the technician may have to drive the vehicle through the I/M Readiness Drive Cycle.

Flat Spot: The brief hesitation or stumble of an engine caused by a momentary overly lean air-fuel mixture.

Float Valve: A valve that is controlled by a hollow ball floating in a liquid, such as in the fuel bowl of a carburetor.

Four-Stroke Engine: The Otto-cycle engine. An engine in which a piston must complete four strokes to make up one operating cycle. The strokes are: intake, compression, power, and exhaust.

Freeze Frame Record: The part of the Diagnostic Executive that stores various vehicle operating data when setting an emissions-related DTC and lighting the MIL.

FTP: Federal Test Procedure. The test procedure used to certify all new vehicles.

Fuel Composition Sensor: A sensor that measures the percentage of ethanol in the fuel as well as the fuel temperature. Fuel composition sensors are used on E-85 vehicles.

Fuel Injection: A means of delivering fuel into the intake manifold or directly into the combustion chamber under pressure, as opposed to relying on vacuum as with a carburetor.

Galvanic Battery: A direct current voltage source, generated by the chemical action of an electrolyte.

Gasohol: Also referred to as an oxygenated fuel, a blend of ethanol and unleaded gasoline, usually at a 1 to 9 ratio.

Ground Path: In the automobile, the metal chassis and frame serve as paths by which current from more than one circuit may flow back to the battery.

Gulp Valve: A valve used in an air-injection system to prevent backfire. During deceleration it also directs air from the air pump to the intake manifold where the air leans out the rich air-fuel mixture.

Hall Effect Switch: A signal-generating switch that develops a transverse voltage across a current-carrying semiconductor when subjected to a magnetic field. A Hall device transmits a digital DC signal that varies in frequency.

HC: The chemical symbol for hydrocarbons, which is unburned fuel in the exhaust stream.

Hertz: Frequency, measured in cycles per second. If an actuator operates at 10 hertz, it operates at 10 cycles per second.

High-Side Driver: The term used to refer to a transistor that switches B+ to a device.

High-Speed Surge: A subtle increase and decrease of vehicle speed that occurs independent of throttle movement.

HO$_2$S: SAE abbreviation for heated oxygen sensors.

Hydrocarbon (HC): A chemical compound made up of hydrogen and carbon. A major pollutant given off by an internal combustion engine. Gasoline is a hydrocarbon compound.

IAT: The SAE abbreviation for the intake air temperature sensor, which is used to measure the temperature of the incoming airstream.

Idle Air Control (IAC) Motor: The idle air control motor is a solenoid or stepper motor that allows variable amounts of bypass air around the throttle to control idle speed under various engine operating conditions.

Idle Speed Control (ISC) Motor: The idle speed control motor is a permanent magnet motor with a threaded rod that extends or retracts to advance or relax the throttle position linkage as required. An ISC is a mechanical version of an IAC.

Ignition Interval (Firing Interval): The number of degrees of crankshaft rotation between ignition sparks. Sometimes called firing interval.

I/M: The abbreviation for Inspection/Maintenance programs, which are emission inspection programs in nonattainment areas.

Impeller: A rotor or rotor blade used to force a gas or liquid in a certain direction under pressure.

Inductive-Discharge Ignition: A method of igniting the air-fuel mixture in an engine cylinder through the induction of high voltage in the secondary winding of the coil.

Inert Gas: A gas that will not aid the combustion process.

Inertia: The tendency of an object at rest to remain at rest, and of an object in motion to remain in motion.

Infinite Resistance: Infinite resistance indicates that an open circuit exists.

Initialization: The initial setting or calibrating of a computer-controlled device such as a stepper motor when the key is turned on.

Injection Pump: A pump used on diesel engines to deliver fuel under high pressure at precisely timed intervals to the fuel injectors.

Input Conditioning: The process of amplifying or converting a voltage signal into a form usable by the computer's central processing unit.

Integrator: The ability of the computer to make short-term—minute-by-minute— corrections in fuel metering. As of OBD II, known as short-term fuel trim.

Intercooler: An air-to-air or air-to-liquid heat exchanger used to lower the temperature of the air-fuel mixture by removing heat from the intake air charge.

Internal Combustion Engine: An engine, such as a gasoline or diesel engine, in which fuel is burned inside the engine.

Keep-Alive Memory (KAM): A form of long-term RAM used mostly with adaptive strategies, requiring a separate power supply circuit to maintain voltage when the ignition is off.

Lightoff: The temperature at which a catalytic converter starts to function properly (at 50 percent efficiency) to oxidize exhaust emissions; close to 500°F (260°C).

Liquid-Vapor Separator Valve: A valve in some EVAP systems that separates liquid fuel from fuel vapors.

Low-Side Driver: The term used to refer to a transistor that switches the ground side of a circuit.

M85: A clean-burning alternative fuel with an octane rating of 100 made by blending 85 percent methanol with 15 percent unleaded gasoline. When burned, it produces fewer hydrocarbons than does pure gasoline.

Magnetic Pulse Generator: A signal generating switch that creates a voltage pulse as magnetic flux changes around a pickup coil. The device transmits an analog AC signal voltage that varies in frequency and amplitude.

Magnetic Saturation: The condition when a magnetic field reaches full strength and maximum flux density.

Malfunction Indicator Lamp (MIL): A part of the OBD system that lights to alert the driver of a faulty automotive system circuit.

Manifold Absolute Pressure (MAP) Sensor: A manifold absolute pressure sensor is a device that sends a signal to the PCM indicating the air pressure, either below or above atmospheric, present in the intake manifold.

Manifold Vacuum: Low pressure in an engine's intake manifold, located below the carburetor throttle valve.

Mechanical Ignition System: Any type of ignition system that uses a mechanical set of points to switch the primary current on and off.

Methanol: A clear, tasteless, and highly toxic form of alcohol made from natural gas, coal, or wood. Contains about 60 percent as much energy as gasoline.

Micron: A unit of length equal to one-millionth of a meter or one one-thousandth of a millimeter.

Multipoint (Port) Injection: A fuel-injection system in which individual injectors are installed in the intake manifold at a point close to the intake valve. Air passing through the manifold mixes with the injector spray just as the intake valve opens.

Multiplexing: Simultaneous communications between various modules using a common data line.

Negative Temperature Coefficient (NTC): Negative temperature coefficient refers to the action of resistance decreasing as temperature increases.

Noble Metals: Metals, such as platinum and palladium, that resist oxidation.

Normally Aspirated: An engine that uses atmospheric pressure and the normal vacuum created by the downward movement of the pistons to draw in its air-fuel charge. Not supercharged or turbocharged.

Octane Rating: The measurement of the antiknock value of a gasoline.

Ohm: The unit for measuring electrical resistance.

Onboard Diagnostic (OBD) System: A type of automotive diagnostic system mandated by the California Air Resources Board (CARB) and the U.S. EPA. Although the system's abilities vary depending on the specific version, generally, OBD seeks to have the vehicle serviced sooner, and to improve the technician's ability to repair emission-related problems.

On-time: The time, measured in milliseconds, that an actuator is "turned on." (See also Pulse Width.)

Open Circuit: A circuit that is open due to a break in the circuit (either intentional or unintentional) that prevents current from flowing, even though potential may exist.

Open Loop: An operational mode in which the engine control powertrain control module adjusts the system using other sensor information and does not respond to feedback signals from the oxygen sensor.

Orifice: A small, often calibrated, opening, in a tube, pipe, or valve.

OSC: Oxygen Storage Capacity. The ability of a TWC to store oxygen momentarily before releasing. A converter's measure of effectiveness can be determined by monitoring its OSC.

Oxidation: The combining of an element with oxygen in a chemical process that often produces extreme heat as a by-product.

Oxides of Nitrogen (NO_x): Chemical compounds of nitrogen given off by an internal combustion engine. They combine with hydrocarbons to produce ozone, a primary component of smog.

Ozone: A gas with a penetrating odor, and a primary component of smog. Ground-level ozone forms when HCs and NO_x in certain proportions, react in the presence of sunlight. Ozone irritates the eyes, damages the lungs, and aggravates respiratory problems.

Parallel Circuit: A circuit with more than one path for the current to follow, each path having one load device. Most home heating and automotive circuits are parallel circuits.

Particulate Matter (PM10): Microscopic particles of materials, such as lead and carbon, that are given off by an internal combustion engine as pollution.

PCV: Positive Crankcase Ventilation. A means of scavenging and either venting (older cars) or burning blowby gases from the crankcase before they can cause engine damage or pollution.

Percolation: A condition caused by heat expanding the fuel in a fuel line. The fuel pushes the carburetor inlet needle valve open and fills up the fuel bowl even when more fuel is not needed. Also, the presence of too much fuel in the intake manifold.

Photochemical Smog: A combination of pollutants that forms harmful chemical compounds when acted upon by sunlight.

Piezoelectric: Voltage caused by physical pressure applied to the faces of certain crystals.

Piezoresistive: A sensor whose resistance varies in relation to pressure or force applied to it. A piezoresistive sensor receives a constant reference voltage and returns a variable signal in relation to its varying resistance.

Pintle Valve: A valve shaped much like a tapered pin. In an EGR valve, the pintle is normally closed and is attached to a diaphragm. When vacuum is applied, the diaphragm lifts the pintle from its seat and allows exhaust gas to flow to the engine intake.

Plenum: A chamber that stabilizes the air-fuel mixture and allows it to rise to a pressure slightly above atmospheric pressure.

Poppet Valve: A valve that plugs and unplugs its opening by linear movement.

Port Fuel Injection: A fuel-injection system in which individual injectors are installed in the intake manifold at a point close to the intake valve. Air passing through the manifold mixes with the injector spray just as the intake valve opens.

Ported Vacuum: A vacuum source located just above the closed throttle valve in a throttle body. Ported vacuum increases as the throttle valve is opened.

Positive Crankcase Ventilation (PCV): A way of controlling engine emissions by directing crankcase vapors (blowby) back through an engine's intake system.

Positive Temperature Coefficient (PTC) Resistor: A thermistor whose resistance decreases as the temperature decreases.

Powertrain Control Module (PCM): The computer that controls a number of functions, which include the engine and transmission. The PCM is essentially an ECM and TCM combined into one unit that may also control related suspension and traction control functions.

Preignition: An unwanted, early ignition of the air-fuel mixture.

Pressure Differential: The difference between the atmospheric pressure outside the engine and the area of low pressure inside the engine created by the downward piston intake stroke.

Pressure Drop: A reduction of pressure between two points.

Programmable Read-Only Memory (PROM): An integrated circuit chip installed in a computer that contains appropriate operating instructions and database information for a particular application.

Pulsating: Expanding and contracting rhythmically.

Pulse Width: The portion of one duty cycle that the circuit is "on," typically measured in milliseconds.

Pulse-Width Modulation: The variation in voltage of an automotive electronic signal to a solenoid, regulating the current flow and consequently the work performed.

Purge Valve: A vacuum-operated or electronically controlled solenoid valve used to draw fuel vapors from a vapor storage canister.

Random-Access Memory (RAM): Temporary short-term or long-term computer memory that can be read and changed, but is lost whenever power to the computer is shut off.

Read-Only Memory (ROM): The permanent part of a computer's memory storage function. ROM can be read but not changed, and is retained when power is shut off to the computer.

Reciprocating Engine: Also called piston engine. An engine in which the pistons move up and down or back and forth as a result of combustion of an air-fuel mixture at one end of the piston cylinder.

Rectified: Alternating current that has been converted to direct current is said to have been rectified. For example, the AC generator's (alternator) output is rectified to DC before it leaves the generator.

Reduction: A chemical process in which oxygen is taken away from a compound.

Reed Valve: A one-way check valve. A reed, or flap, opens to admit a fluid or gas under pressure from one direction, while closing to deny movement from the opposite direction.

Reference Voltage: A constant voltage signal (below battery voltage) applied to a sensor by the computer. The sensor alters the voltage according to engine operating conditions and returns it as a variable input signal to the computer that adjusts system operation accordingly.

Remote Injector Driver: The integrated circuit chip that controls fuel injector operation. Normally located in the PCM housing, the chip is installed in a separate housing to protect the PCM from excessive heat generated by the high current draw of the low-impedance injectors.

Residual or Rest Pressure: The pressure held in the fuel system after the engine is shut off, which prevents vapor lock and assists in rapid restarting of the engine.

Resistance: Opposition to electrical current flow.

Returnless: A fuel-injection system that does not have a return line from the fuel rail on the engine.

Rotary Valve: A valve that rotates to cover and uncover the intake port of a two-stroke engine at the proper time. Rotary valves are usually flat discs driven by the crankshaft. Rotary valves are more complex than reed valves, but are effective in broadening a two-stroke engine's power band.

Runners: The passages or branches of an intake manifold that connect the manifold's plenum chamber to the engine's inlet ports.

Scavenging: A slight suction caused by a vacuum drop through a well-designed header system. Scavenging helps pull exhaust gases out of an engine cylinder.

Sealed Housing for Evaporative Determine (SHED): An EPA-required test that must be performed by the manufacturer on all new model vehicles. This test measures the evaporative emissions from the vehicle.

Self-induction: A self-induced voltage potential formed by the building or induction of current through a conductor, such as the primary windings in the coil.

Series Circuit: A circuit with only one path for the current, but with two or more loads.

Series-Parallel Circuit: A circuit that contains series and parallel branches, with at least two loads in parallel and in series. An instrument panel light circuit is a series-parallel circuit.

Sintered: Bonded together with pressure and heat, forming a porous material, such as the metal disk used in some vacuum delay valves.

Siphoning: The flow of a liquid as a result of a pressure differential, from high to low, without the aid of a mechanical pump.

Spark Timing: A way of controlling exhaust emissions by controlling ignition timing. Vacuum advance is delayed or shut off at low and medium speeds, reducing NO_x and HC emissions.

Speed-Density: A term describing a fuel-injection system using primarily engine rpm (speed), intake air temperature, and MAP (density) to calculate the amount of fuel injected.

State Implementation Plan (SIP): The plan that each state in nonattainment areas must submit to the EPA. This plan specifies the type of I/M program that will be implemented.

Stepper Motor: A direct current motor that moves in incremental steps from de-energized to fully energized. It is often used to control throttle linkage position.

Stoichiometric Ratio: An ideal air-fuel mixture for combustion, in which all oxygen and all fuel will burn completely.

Stroke: One complete top-to-bottom or bottom-to-top movement of an engine piston.

Substrate: The layer, or honeycomb, of aluminum oxide upon which the catalyst in a catalytic converter is deposited.

Sulfur Oxides (SO_x): Sulfur given off by processing and burning gasoline and other fossil fuels. As it decomposes, sulfur dioxide combines with water to form sulfuric acid, or "acid rain."

Supercharging: Use of an air pump to deliver an air-fuel charge to the engine cylinders at a pressure greater than atmospheric pressure.

Temperature Inversion: A weather pattern in which a layer, or "lid," of warm air keeps a layer of cooler air trapped beneath it.

Tetraethyl Lead (TEL): A gasoline additive that helps prevent detonation.

Thermal Cracking: A common oil refining process that uses heat to break down (crack) the larger components of crude oil. The gasoline produced by this method usually has a higher sulfur content than gasoline produced by catalytic cracking.

Thermistor (Thermal Resistor): A resistor especially built to change its resistance as the temperature changes.

Thermostatic: Referring to a device that automatically responds to temperature changes in order to activate a switch.

Throttle-Body Fuel Injection (TBI): A fuel-injection system in which one or two injectors are installed in a carburetor-like throttle body mounted on a conventional intake manifold. Fuel is sprayed at a constant pressure above the throttle plate to mix with the incoming air charge.

Top Dead Center (TDC): The exact top of a piston stroke. Also a specification used when tuning an engine.

Transducer: A device that changes one form of energy into another.

Turbo Lag: The time interval required for a turbocharger to overcome inertia and spin up to speed and produce useful boost.

Turbocharger: A supercharging device that uses exhaust gases to turn a turbine that forces extra air-fuel mixture into the cylinders.

TWC: A two-stage, three-way converter, having a reduction catalyst followed by the oxidation catalyst in a single housing. Air may be supplied between the two sections of the converter by the air pump.

Two-Stroke Engine: An engine in which a piston makes two strokes to complete one operating cycle.

Vacuum: Low pressure within an engine created by a downward piston intake stroke with the intake valve open.

Vacuum Lock: A stoppage of fuel flow caused by insufficient air intake to the fuel tank.

Vane Airflow (VAF) Sensor: A vane airflow sensor is a device that measures incoming air volume by using an air vane or flap that rotates against spring pressure causing a laser calibrated potentiometer to register the amount of rotation. A second "dampening chamber" helps to steady the vane as it rotates.

Vapor Lock: A condition in which bubbles in a vehicle's fuel system stop or restrict fuel flow. High underhood temperatures sometimes cause fuel to boil within fuel lines.

Vaporization: Changing a liquid, such as gasoline, into a gas (vapor) by evaporation or boiling.

Variable Dwell: The ignition dwell period varies in distributor degrees at different engine speeds, but remains relatively constant in duration.

Venturi: A restriction in an airflow, such as in a carburetor, that speeds the airflow and creates a vacuum.

Venturi Vacuum: Low pressure in the venturi of a carburetor, caused by fast airflow through the venturi.

Volatile Organic Compounds (VOCs): Those parts of a fuel's composition that when released can easily vaporize and contribute to photochemical smog.

Volatility: The ease with which a liquid changes into a gas or vapor. Gasoline is more volatile than water because it evaporates more quickly.

Volt: The unit for measuring the amount of electrical force.

Voltage Drop: The difference in electrical force or potential when measured across two sides of a load device (resistance) in a live circuit.

Voltage, Potential, Electromotive Force (EMF, or E): The force or pressure that pushes electrons (current) through a circuit causing current flow.

Voltage Spike: A sudden, sharp, and often extreme rise in voltage.

Volumetric Efficiency: The actual volume of air-fuel mixture an engine draws in compared to the theoretical maximum it could draw in, written as a percentage.

Wastegate: A diaphragm-actuated bypass valve used to limit turbocharger boost pressure by limiting the speed of the exhaust turbine.

Water Injection: A method of lowering the air-fuel mixture temperature by injecting a fine spray of water that cools the intake charge as it evaporates.

Water Jackets: Passages in the head and block that allow coolant to circulate through the engine.

Appendix A — OBD II Diagnostic Trouble Codes

P0100 Mass or Volume Airflow Circuit Problem
P0101 Mass or Volume Airflow Circuit Range or Performance Problem
P0102 Mass or Volume Airflow Circuit Low Input
P0103 Mass or Volume Airflow Circuit High Input
P0105 Manifold Absolute Pressure or Barometric Pressure Circuit Problem
P0106 Manifold Absolute Pressure or Barometric Pressure Circuit Range or Performance Problem
P0107 Manifold Absolute Pressure or Barometric Pressure Circuit Low Input
P0108 Manifold Absolute Pressure or Barometric Pressure Circuit High Input
P0110 Intake Air Temperature Circuit Problem
P0111 Intake Air Temperature Circuit Range or Performance Problem
P0112 Intake Air Temperature Circuit Low Input
P0113 Intake Air Temperature Circuit High Input
P0115 Engine Coolant Temperature Problem
P0116 Engine Coolant Temperature Circuit Range or Performance Problem
P0117 Engine Coolant Temperature Circuit Low Input
P0118 Engine Coolant Temperature Circuit High Input
P0120 Throttle Position Circuit Problem
P0121 Throttle Position Circuit Range or Performance Problem
P0122 Throttle Position Circuit Low Input
P0123 Throttle Position Circuit High Input
P0125 Excessive Time to Enter Closed-Loop Fuel Control
P0130 O2 Sensor Circuit Problem (Bank 1 Sensor 1)
P0131 O2 Sensor Circuit Low Voltage (Bank 1 Sensor 1)
P0132 O2 Sensor Circuit High Voltage (Bank 1 Sensor 1)
P0133 O2 Sensor Circuit Slow Response (Bank 1 Sensor 1)
P0134 O2 Sensor Circuit No Activity Detected (Bank 1 Sensor 1)
P0135 O2 Sensor Circuit Heater Problem (Bank 1 Sensor 1)
P0136 O2 Sensor Circuit Problem (Bank 1 Sensor 2)
P0137 O2 Sensor Circuit Low Voltage (Bank 1 Sensor 2)
P0138 O2 Sensor Circuit High Voltage (Bank 1 Sensor 2)
P0139 O2 Sensor Circuit Slow Response (Bank 1 Sensor 2)
P0140 O2 Sensor Circuit No Activity Detected (Bank 1 Sensor 2)
P0141 O2 Sensor Circuit Heater Problem (Bank 1 Sensor 2)
P0142 O2 Sensor Circuit Problem (Bank 1 Sensor 3)
P0143 O2 Sensor Circuit Low Voltage (Bank 1 Sensor 3)
P0144 O2 Sensor Circuit High Voltage (Bank 1 Sensor 3)
P0145 O2 Sensor Circuit Slow Response (Bank 1 Sensor 3)
P0146 O2 Sensor Circuit No Activity Detected (Bank 1 Sensor 3)
P0147 O2 Sensor Circuit Heater Problem (Bank 1 Sensor 3)
P0150 O2 Sensor Circuit Problem (Bank 2 Sensor 1)
P0151 O2 Sensor Circuit Low Voltage (Bank 2 Sensor 1)
P0152 O2 Sensor Circuit High Voltage (Bank 2 Sensor 1)
P0153 O2 Sensor Circuit Slow Response (Bank 2 Sensor 1)
P0154 O2 Sensor Circuit No Activity Detected (Bank 2 Sensor 1)
P0155 O2 Sensor Circuit Heater Problem (Bank 2 Sensor 1)
P0156 O2 Sensor Circuit Problem (Bank 2 Sensor 2)
P0157 O2 Sensor Circuit Low Voltage (Bank 2 Sensor 2)
P0158 O2 Sensor Circuit High Voltage (Bank 2 Sensor 2)
P0159 O2 Sensor Circuit Slow Response (Bank 2 Sensor 2)
P0160 O2 Sensor Circuit No Activity Detected (Bank 2 Sensor 2)
P0161 O2 Sensor Circuit Heater Problem (Bank 2 Sensor 2)
P0162 O2 Sensor Circuit Problem (Bank 2 Sensor 3)
P0163 O2 Sensor Circuit Low Voltage (Bank 2 Sensor 3)
P0164 O2 Sensor Circuit High Voltage (Bank 2 Sensor 3)
P0165 O2 Sensor Circuit Slow Response (Bank 2 Sensor 3)
P0166 O2 Sensor Circuit No Activity Detected (Bank 2 Sensor 3)
P0167 O2 Sensor Circuit Heater Problem (Bank 2 Sensor 3)
P0170 Fuel Trim Problem (Bank 1)
P0171 System Too Lean (Bank 1)
P0172 System Too Rich (Bank 1)
P0173 Fuel Trim Problem (Bank 2)
P0174 System Too Lean (Bank 2)
P0175 System Too Rich (Bank 2)
P0176 Fuel Composition Sensor Circuit Problem
P0177 Fuel Composition Sensor Circuit Range or Performance
P0178 Fuel Composition Sensor Circuit Low Input
P0179 Fuel Composition Sensor Circuit High Input
P0180 Fuel Temperature Sensor Circuit Problem
P0181 Fuel Temperature Sensor Circuit Range or Performance
P0182 Fuel Temperature Sensor Circuit Low Input
P0183 Fuel Temperature Sensor Circuit High Input

FUEL AND AIR METERING

P0201 Injector Circuit Problem Cylinder 1
P0202 Injector Circuit Problem Cylinder 2
P0203 Injector Circuit Problem Cylinder 3
P0204 Injector Circuit Problem Cylinder 4
P0205 Injector Circuit Problem Cylinder 5
P0206 Injector Circuit Problem Cylinder 6
P0207 Injector Circuit Problem Cylinder 7
P0208 Injector Circuit Problem Cylinder 8
P0209 Injector Circuit Problem Cylinder 9
P0210 Injector Circuit Problem Cylinder 10
P0211 Injector Circuit Problem Cylinder 11
P0212 Injector Circuit Problem Cylinder 12
P0213 Cold-Start Injector 1 Problem
P0214 Cold-Start Injector 2 Problem

IGNITION SYSTEM OR MISFIRE

P0300 Random Misfire Detected
P0301 Cylinder 1 Misfire Detected
P0302 Cylinder 2 Misfire Detected
P0303 Cylinder 3 Misfire Detected
P0304 Cylinder 4 Misfire Detected
P0305 Cylinder 5 Misfire Detected
P0306 Cylinder 6 Misfire Detected
P0307 Cylinder 7 Misfire Detected
P0308 Cylinder 8 Misfire Detected
P0309 Cylinder 9 Misfire Detected
P0310 Cylinder 10 Misfire Detected
P0311 Cylinder 11 Misfire Detected
P0312 Cylinder 12 Misfire Detected
P0320 Ignition or Distributor Engine Speed Input Circuit Problem
P0321 Ignition or Distributor Engine Speed Input Circuit Range or Performance
P0322 Ignition or Distributor Engine Speed Input Circuit No Signal
P0325 Knock Sensor 1 Circuit Problem
P0326 Knock Sensor 1 Circuit Range or Performance
P0327 Knock Sensor 1 Circuit Low Input
P0328 Knock Sensor 1 Circuit High Input
P0330 Knock Sensor 2 Circuit Problem
P0331 Knock Sensor 2 Circuit Range or Performance
P0332 Knock Sensor 2 Circuit Low Input
P0333 Knock Sensor 2 Circuit High Input
P0335 Crankshaft Position Sensor Circuit Problem
P0336 Crankshaft Position Sensor Circuit Range or Performance
P0337 Crankshaft Position Sensor Circuit Low Input
P0338 Crankshaft Position Sensor Circuit High Input

AUXILIARY EMISSION CONTROLS

P0400 Exhaust Gas Recirculation Flow Problem
P0401 Exhaust Gas Recirculation Flow Insufficient Detected
P0402 Exhaust Gas Recirculation Flow Excessive Detected
P0405 Air Conditioner Refrigerant Charge Loss
P0410 Secondary Air-Injection System Problem
P0411 Secondary Air-Injection System Insufficient Flow Detected
P0412 Secondary Air-Injection System Switching Valve or Circuit Problem
P0413 Secondary Air-Injection System Switching Valve or Circuit Open
P0414 Secondary Air-Injection System Switching Valve or Circuit Shorted
P0420 Catalyst System Efficiency Below Threshold (Bank 1)
P0421 Warm-up Catalyst Efficiency Below Threshold (Bank 1)
P0422 Main Catalyst Efficiency Below Threshold (Bank 1)
P0423 Heated Catalyst Efficiency Below Threshold (Bank 1)
P0424 Heated Catalyst Temperature Below Threshold (Bank 1)
P0430 Catalyst System Efficiency Below Threshold (Bank 2)
P0431 Warm-up Catalyst Efficiency Below Threshold (Bank 2)
P0432 Main Catalyst Efficiency Below Threshold (Bank 2)
P0433 Heated Catalyst Efficiency Below Threshold (Bank 2)
P0434 Heated Catalyst Temperature Below Threshold (Bank 2)
P0440 Evaporative Emission Control System Problem
P0441 Evaporative Emission Control System Insufficient Purge Flow
P0442 Evaporative Emission Control System Leak Detected
P0443 Evaporative Emission Control System Purge Control Valve Circuit Problem
P0444 Evaporative Emission Control System Purge Control Valve Circuit Open
P0445 Evaporative Emission Control System Purge Control Valve Circuit Shorted
P0446 Evaporative Emission Control System Vent Control Problem
P0447 Evaporative Emission Control System Vent Control Open
P0448 Evaporative Emission Control System Vent Control Shorted
P0450 Evaporative Emission Control System Pressure Sensor Problem
P0451 Evaporative Emission Control System Pressure Sensor Range or Performance
P0452 Evaporative Emission Control System Pressure Sensor Low Input
P0453 Evaporative Emission Control System Pressure Sensor High Input

VEHICLE SPEED AND IDLE CONTROL SYSTEM

P0500   Vehicle Speed Sensor Problem
P0501   Vehicle Speed Sensor Range or Performance
P0502   Vehicle Speed Sensor Low Input
P0505   Idle Control System Problem
P0506   Idle Control System RPM Lower than Expected
P0507   Idle Control System RPM Higher than Expected
P0510   Closed-Throttle Position Switch Problem

COMPUTER AND OUTPUT CIRCUITS

P0600   Serial Communication Link Problem
P0605   Internal Control Module (Module Identification Defined by J1979)

TRANSMISSION

P0703   Brake Switch Input Problem
P0705   Transmission Range Sensor Circuit Problem (PRNDL Input)
P0706   Transmission Range Sensor Circuit Range or Performance
P0707   Transmission Range Sensor Circuit Low Input
P0708   Transmission Range Sensor Circuit High Input
P0710   Transmission Fluid Temperature Sensor Circuit Problem
P0711   Transmission Fluid Temperature Sensor Circuit Range or Performance
P0712   Transmission Fluid Temperature Sensor Circuit Low Input
P0713   Transmission Fluid Temperature Sensor Circuit High Input
P0715   Input or Turbine Speed Sensor Circuit Problem
P0716   Input or Turbine Speed Sensor Circuit Range or Performance

P0717   Input or Turbine Speed Sensor Circuit No Signal
P0720   Output Speed Sensor Circuit Problem
P0721   Output Speed Sensor Circuit Range or Performance
P0722   Output Speed Sensor Circuit No Signal
P0725   Engine Speed Input Circuit Problem
P0726   Engine Speed Input Circuit Range or Performance
P0727   Engine Speed Input Circuit No Signal
P0730   Incorrect Gear Ratio
P0731   Gear 1 Incorrect Ratio
P0732   Gear 2 Incorrect Ratio
P0733   Gear 3 Incorrect Ratio
P0734   Gear 4 Incorrect Ratio
P0735   Gear 5 Incorrect Ratio
P0736   Reverse Incorrect Ratio
P0740   Torque Converter Clutch System Problem
P0741   Torque Converter Clutch System Performance or Stuck Off
P0742   Torque Converter Clutch System Stuck On
P0743   Torque Converter Clutch System Electrical
P0745   Pressure Control Solenoid Problem
P0746   Pressure Control Solenoid Performance or Stuck Off
P0747   Pressure Control Solenoid Stuck On
P0748   Pressure Control Solenoid Electrical
P0750   Shift Solenoid A Problem
P0751   Shift Solenoid A Performance or Stuck Off
P0752   Shift Solenoid A Stuck On
P0753   Shift Solenoid A Electrical
P0755   Shift Solenoid B Problem
P0756   Shift Solenoid B Performance or Stuck Off
P0757   Shift Solenoid B Stuck On
P0758   Shift Solenoid B Electrical
P0760   Shift Solenoid C Problem
P0761   Shift Solenoid C Performance or Stuck Off
P0762   Shift Solenoid C Stuck On
P0763   Shift Solenoid C Electrical
P0765   Shift Solenoid D Problem
P0766   Shift Solenoid D Performance or Stuck Off
P0767   Shift Solenoid D Stuck On
P0768   Shift Solenoid D Electrical
P0770   Shift Solenoid E Problem
P0771   Shift Solenoid E Performance or Stuck Off
P0772   Shift Solenoid E Stuck On
P0773   Shift Solenoid E Electrical

Appendix B — Evaporative System Pressure Conversion Chart

PSI	in Hg	in H$_2$O
14.7	29.93	407.19
1.0	2.036	27.7
0.9	108	24.93
0.8	1.63	22.16
0.7	1.43	19.39
0.6	1.22	16.62
0.5	1.018	13.85
0.4	0.814	11.08
0.3	0.611	8.31
0.2	0.407	5.54
0.1	0.204	2.77
0.09	0.183	2.49
0.08	0.163	2.22
0.07	0.143	1.94
0.06	0.122	1.66
0.05	0.102	1.385

1 psi = 28 inches of water

1/4 psi = 7 inches of water

Appendix C — Major Elements of Operating I/M Programs

State	Network Type	Test Type	Cutpoints	Visual Checks	
Alaska **basic**	Test & Repair (85% Test Only credit)	2 speed Idle 96+ OBD	220/0.5	Catalyst Air pump EGR PCV Evap disable	
Arizona (Phoenix) enhanced	Test Only	81–95: IM147 <81: Loaded Idle 96+: OBD	0.8/12/2 220/1.2	Catalyst Air pump PCV Gas cap	
Arizona (Tucson) basic	Test Only	<96: Loaded Idle 96+: OBD	220/1.2	Catalyst Air pump	
California basic	Hybrid	2 speed Idle	130/1.0	Catalyst Air pump EGR Fuel inlet Evap disable	
California enhanced	Hybrid	ASM2 96+ OBD	5015 85/0.5/992 2525 37/0.47/852	Catalyst Air pump EGR PCV Evap disable	
Colorado (Denver & Boulder) enhanced	Test Only	82+: IM240 <82: 2 speed Idle	2/20/4 220/1.2	O2 sensor Catalyst Air pump Fuel inlet Gas cap	
Colorado (Colo Springs Aspen, Greely Ft Collins) basic	Test & Repair (50% Test Only credit)	81+: 2 speed Idle <81: Idle	220/1.2	O2 sensor Catalyst Air pump Fuel inlet	
Connecticut enhanced	Test Only	81+: ASM2525	101/0.57/786	Catalyst Fuel inlet Gas cap	
Delaware enhanced	Test Only	2 speed Idle 96+ OBD	220/1.2	Catalyst Fuel inlet	
Georgia enhanced	Hybrid (100% Test Only credit)	ASM2 96+ OBD	2525 78/0.42/566 5015 78/0.44/625	Catalyst	
Idaho basic	Test Only (20% Test & Repair)	2 speed Idle 96+ OBD	220/1.2	Catalyst Air pump	
Illinois enhanced	Test Only	81+: IM240 <80: Idle	0.6/10/NA 300/3.0	Catalyst Fuel inlet Gas cap	

Evap Tests	Tech Training	Frequency	Vehicle Types	Model Years	Start Date	OBD Testing
None		Biennial	LDGVs LDGTs HDGVs	Anchorage 1968+ Fairbanks 1975+ <2 exempt	Jul-85	pass/fail: 7/01
Pressure Gas Cap		Annual 1967–80 Biennial 1981+	LDGVs LDGTs HDGVs MC	1967+ <4 exempt	Jan-95	pass/fail: 1/02
Gas Cap		Annual	LDGVs LDGTs HDGVs MC	1967+ <4 exempt	Jan-95	pass/fail: 1/02
Gas Cap Functional EGR	100% TTC	Biennial	LDGVs LDGTs HDGVs	1974+ <4 exempt	1984	MIL fail pass/fail: 5/02
Gas Cap Functional EGR	100% TTC	Biennial	LDGVs LDGTs HDGVs	1974+ <4 exempt	Jul-98	MIL fail pass/fail: 5/02
Gas Cap		82+: Biennial <82: Annual	LDGVs LDGTs HDGVs	All except <4 exempt	Jan-95	MIL fail only
Gas Cap		82+: Biennial <82: Annual	LDGVs LDGTs HDGVs	All except <4 exempt	Jan-95	MIL fail only
Gas Cap		81+: Biennial <81: Annual	LDGVs LDGTs	up to 25 years old	Jan-98	advisory: pass/fail: 1/05
Pressure Gas Cap		Biennial	LDGVs LDGTs	1968+ <5 exempt	Jan-94	advisory: pass/fail: 7/02
Gas Cap		Annual	LDGVs LDGTs	<3 exempt	Oct-98	advisory: 10/01 pass/fail: 5/02
None		Annual <1 Exempt	LDGVs LDGTs HDGVs	1965+	Jan-84	advisory: pass/fail: 3/02
Gas Cap		Biennial	LDGVs LDGTs HDGVs	1968+ <4 exempt	Feb-99	advisory: scan pass/fail: 7/02 OBD used as clean screen

State	Network Type	Test Type	Cutpoints	Visual Checks	
Indiana enhanced	Test Only	81–95: IM93 <81: Idle 96+: OBD	0.6/10/NA 220/1.2	Catalyst	
Kentucky basic	Test Only	Loaded Idle- Louis Idle - N. Kentucky	220/1.2	Catalyst Air pump Evap disable	
Louisiana low enhanced	Test & Repair	no tailpipe test	NA	Catalyst Air pump EGR PCV Fuel inlet Evap disable	
Maine SIP measure low enhanced	Test & Repair	no tailpipe test	NA	Catalyst	
Maryland enhanced	Test Only	77–83: Idle 84+: IM240	220/1.2 0.7/15/1.8	Catalyst Gas cap	
Massachusetts enhanced	Test & Repair (100% Test Only credit)	MA31	0.8/15/2.0	Gas cap	
Missouri enhanced	Test Only	81+: IM240 <80: Idle	0.6/10/1.5 220/1.2	Catalyst Gas cap	
Nevada low enhanced	Test & Repair (50% Test Only credit)	2 speed Idle	220/1.2	Catalyst Air pump EGR Fuel inlet Gas cap	
New Hampshire SIP measure low enhanced	Test & Repair	no tailpipe test	NA	Catalyst Air pump PCV Gas cap	
New Jersey low enhanced	Hybrid (80% Test Only credit)	81+: ASM5015 <80: Idle	78/0.44/625 100/0.5	Catalyst Fuel inlet	
New Mexico basic	Test & Repair (50% Test Only credit)	2 speed Idle	220/1.2	Catalyst Air pump	
New York enhanced	Test & Repair (88/84/86%) TO credit for HC/CO/NOx	81+: NY test <81: Idle	0.6/10/1.5 220/1.2	Catalyst Air pump EGR PCV Evap disable	
North Carolina basic	Test & Repair (50% Test Only credit)	Idle OBD	220/1.2	Catalyst Air pump EGR PCV Fuel inlet Evap disable	
Ohio enhanced	Test Only	ASM2525	76/0.42/566	Catalyst Gas cap	
Oregon basic	Test Only	<80: 2 speed Idle 81–95:BAR31 96+: OBD	220/1.0 1.2/30/3.00	Catalyst Air pump EGR PCV Evap disable	

Evap Tests	Tech Training	Frequency	Vehicle Types	Model Years	Start Date	OBD Testing
Gas Cap		Biennial	LDGVs LDGTs	1976+ <4 exempt	Jan-97	pass/fail: 12/01
Pressure Gas Cap		Annual: Louisville Biennial: N. KY	LDGVs LDGTs HDGVs	1968+	Jan-94	advisory: pass/fail: 7/02
Gas Cap		Annual	LDGVs LDGTs HDGVs	1980+	Jan-00	advisory: pass/fail: 6/03?
Gas Cap		Annual	LDGVs LDGTs	1974+	Jan-99	pass/fail: 1/01
Gas Cap		Biennial	LDGVs LDGTs HDGVs	<2 exempt 1977+	Oct-97	advisory: MIL fail pass/fail: 7/02
Gas Cap	100% TTC	Biennial	LDGVs LDGTs HDGVs	1984+ <2 exempt	Oct-99	advisory: pass/fail: 7/03 LEP clean screen
Gas Cap		Biennial	LDGVs LDGTs	1971+ <2 exempt	Apr-00	advisory: pass/fail: 1/05
None	100% TTC	Annual	LDGVs LDGTs HDGVs	1968+ <2 exempt	1995	advisory: pass/fail: 4/02
None		Annual	LDGVs LDGTs HDGVs	1980+	Jan-99	advisory: pass/fail: 1/04
Gas Cap	50% TTC	Biennial	LDGVs LDGTs HDGVs	1968+ <4 exempt	Dec-99	advisory: pass/fail: 6/03?
		Biennial	LDGVs LDGTs HDGVs	1975+ <2 exempt	Mar-89	advisory: pass/fail: 1/03
Gas Cap	50% TTC	Annual	LDGVs LDGTs HDGVs	2–25 years old	Nov-98	advisory: Upstate pass/fail: 6/03 (NYC) 6/04 (upstate)
None		Annual	LDGVs LDGTs HDGVs	1–25 yrs old	Jan-90	advisory: 7/01 pass/fail: 5/02
Gas Cap		Biennial	LDGVs LDGTs	2–25 years old	Jan-96	advisory: pass/fail: 1/03
None		Biennial	LDGVs LDGTs HDGVs	1975+ <2 exempt	Fall 1997	advisory: pass/fail: 12/00

State	Network Type	Test Type	Cutpoints	Visual Checks	
Pennsylvania enhanced	Test & Repair (100% Test Only credit)	ASM5015 in Philadelphia (75–80 get idle) 2 speed Idle in Pittsburgh	105/0.59/869 220/1.2	Catalyst Air pump Fuel inlet EGR PCV Evap disable	
Rhode Island enhanced	Test & Repair	R12000	1.84/16.5/2.42	None	
Tennessee basic	Test Only	Idle	220/1.2	Catalyst Gas cap Fuel inlet	
Texas enhanced	Test & Repair (100% Test Only credit)	ASM2 96+ OBDII	2525 76/0.42/566 5015 78/0.44/625	Catalyst Air pump EGR PCV Evap disable	
Utah (Weber and Utah Counties) basic	Test & Repair (100% Test Only credit)	<96: 2 speed Idle 96+: OBD	220/1.2	Catalyst Air pump EGR PCV Fuel inlet Evap disable	
Utah (Salt Lake County) basic	Test & Repair (100% Test Only credit)	<96: ASM2 96+: OBD	5015 97/0.51/1004 2525 87/0.46/914	Catalyst Air pump EGR PCV Fuel inlet Evap disable	
Utah (Davis county) basic	Test & Repair	<96: Idle 96+: OBD	220/1.2	Catalyst Air pump EGR PCV Fuel inlet Evap disable	
Vermont low enhanced SIP measure	Test & Repair (50% Test Only credit)	<96: no tailpipe test 96+: OBD	NA	Catalyst Gas cap	
Virginia enhanced	Test & Repair (94% Test Only credit)	81+: ASM2 <80: 2 speed Idle	5015 105/0.59/869 2525 101/0.57/786 220/1.2	Catalyst Air pump EGR PCV Evap canister Gas cap	
Washington basic	Test Only	2 speed Idle in Puget Sound ASM2525 in Spok./Vancouver	220/1.2 180/1.3/NA		
Washington DC enhanced	Test Only	85–95: IM240 <85: Idle 96+: OBD	0.8/15/2 220/1.2	Catalyst	
Wisconsin enhanced	Test Only	<96: IM240 96+: OBD	0.6/10.0/1.5	Catalyst Air pump EGR PCV Fuel inlet Evap disable	

Evap Tests	Tech Training	Frequency	Vehicle Types	Model Years	Start Date	OBD Testing
Gas Cap	100% TTC	Annual	LDGVs LDGTs	1975+	Nov-97	advisory: pass/fail: 9/03
Gas Cap		Biennial	LDGVs LDGTs	2–25 years old	Jun-99	advisory: pass/fail: 1/03
None		Annual	LDGVs LDGTs HDGVs	1975+	Jan-91	advisory: 10/01 pass/fail: 4/02
Gas Cap	100% TTC	Annual	LDGVs LDGTs HDGVs	2–24 yrs	Nov-97	advisory: pass/fail: 5/02
None	100% TTC	Annual	LDGVs LDGTs HDGVs	1968+	Jan-97	advisory: pass/fail: 1/97
Gas Cap	100% TTC	Annual	LDGVs LDGTs HDGVs	1968+	Jan-97	advisory: pass/fail: 1/02
None	100% TTC	Annual	LDGVs LDGTs HDGVs	1968+	Jan-97	advisory: pass/fail: 1/02
None		Annual	LDGVs LDGTs	1968+	Jan-97	pass/fail: 1/01
Gas Cap	100% TTC	Biennial	LDGVs LDGTs HDGVs	2–24 years old	May-98	advisory: pass/fail: 9/02
Gas Cap		Biennial	LDGVs LDGTs HDGVs	5–25 years old	Jan-93	advisory: pass/fail: 7/02
Gas Cap	100% TTC	Biennial	LDGVs LDGTs HDGVs	1968+ <4 exempt	Apr-99	pass/fail: 2/02
Gas Cap		Biennial	LDGVs LDGTs HDGVs	1968+	Jan-95	pass/fail: 7/01

Appendix D — Vehicle Manufacturer's Service Information

This is a list of manufacturers' websites where service information is available for emission-related repairs. In most cases there is a fee required for access; however, some information may be provided free. Subscriptions may be available on a daily, monthly, or yearly basis.

Aston Martin	www.astonmartintechinfo.com
Audi	http://www.audi.ddsltd.com
BMW	www.bmwtechinfo.com
Chrysler/Dodge/Eagle/Jeep	www.techauthority.com
Ferrari	www.ferrariusa.com
Ford/Lincoln/Mercury	www.motorcraft.com
General Motors	www.gmtechinfo.com
Honda/Acura	www.ServiceExpress.Honda.com
Hyundai	http://www.hmaservice.com
Isuzu	www.isuzutechinfo.com
Jaguar	www.jaguartechinfo.com
Kia	www.kiatechinfo.com
Land Rover	www.landrovertechinfo.com
Maserati	www.maseratiusa.com
Mazda	www.mazdatechinfo.com
Mercedes-Benz	www.startekinfo.com
Mini	www.minitechinfo.com
Mitsubishi	www.mitsubishitechinfo.com
Nissan/Infiniti	www.nissantechinfo.com
Porsche	www.porsche.com
Saab	www.saabtechinfo.com
Subaru	www.subaru.com
Suzuki	www.suzukitechinfo.com
Toyota/Lexus	http://techinfo.toyota.com
Volkswagen	http://www.vw.ddsltd.com
Volvo	http://www.volvoira.com

Appendix E — Automotive Fuel and Emissions Related Websites

http://www.fedworld.gov/cleanair/index.htm
FedWorld Emission Related Information

http://www.obdii.com/links.html
OBD II Information page and links

http://www.arb.ca.gov/homepage.htm
California Air Resources Board

http://www.epa.gov/OMSWWW/
EPA website

http://www.lindertech.com/
Linder Technical Services

http://www.iatn.net/
International Automotive Technician's Network

http://www.autoshop101.com/
Kevin Sullivan's Autoshop 101

http://www.nhtsa.dot.gov/
National Highway Traffic Safety Administration

http://arbis.arb.ca.gov/msei/on-road/ldv.htm
Emission repair information

http://www.fuelinjection.com/

Appendix F — Links to State Emission Programs

Alaska
http://www.state.ak.us/dmv/reg/imtest.htm
http://www.state.ak.us/local/akpages/ENV.
CONSERV/title18/aac52ndx.htm
Arizona
http://www.adeq.state.az.us/environ/air/vei/
index.html
California
http://www.arb.ca.gov/
Colorado
http://www.aircarecolorado.com
Connecticut
http://www.ctemissions.com
Delaware
http://www.dmv.de.gov/Vehicle_Services/Other/
ve_oth_inspections.html
District of Columbia
http://dmv.washingtondc.gov/serv/inspections.
shtm
Georgia
http://www.cleanairforce.com
Illinois
http://www.epa.state.il.us/air/vim/index.html
Indiana
http://www.state.in.us/bmv/platesandtitles/
cleanAir.html
Kentucky
http://www.nr.state.ky.us/nrepc/dep/daq/pubinfo/
NKyemissions.htm
Maryland
http://www.mde.state.md.us/arma/Veip/veiphome.
html
Massachusetts
http://vehicletest.state.ma.us/home.html
Missouri
http://www.gatewaycleanair.com/

New Jersey
http://www.state.nj.us/mvc/cleanair/index.html
Nevada
http://nevadadmv.state.nv.us/emission.htm
New Mexico
http://www.cabq.gov/aircare
New York
http://www.nydmv.state.ny.us/vehsafe.htm
North Carolina
http://www.dmv.dot.state.nc.us/enforcement/
emissionsinspections/
Ohio
http://www.epa.state.oh.us/dapc/mobile.html
Oregon
http://www.deq.state.or.us/aq/vip/
Pennsylvania
http://www.drivecleanpa.state.pa.us/drivecleanpa/
Rhode Island
http://www.riinspection.com/
Tennessee
http://www.tennessee.gov/safety/
titlingandregistering.html#emissions
Texas
http://www.txdps.state.tx.us/vi/
Utah
http://dmv.utah.gov/registerrequirements.
html#emission
Virginia
http://www.deq.state.va.us/mobile/
Washington
http://www.ecy.wa.gov/programs/air/CARS/
Automotive_Pages.htm
Wisconsin
http://www.dot.state.wi.us/dmv/im.html

Index

AC, 87
Acceleration simulation mode (ASM), 285
Actuator, 108
Adaptive memory, 168
Adsorption, 81
AIR, 295–296
    pump, 296
    system monitor, 191
Air-conditioning sensors, 176
Air control system, 215
Airflow, 51–56
Airflow requirements, 48–49
Air-fuel mixture compression, 238
Air-fuel mixture control, 201–202
Air-fuel ratios, 49–51, 164, 281–282
Air-fuel ratio sensor, 142
Air injection, 295–300
    and catalytic converters, 299–300
Air-injection reactor. *See* AIR
Air-management valves, 298–299
Air pollution, 2
    and the automobile, 4–5
    legislation and regulatory agencies,
      5–11
    major pollutants, 2–4
    smog-climate reaction with, 5
Air pressure, 48
Air Quality Act, 5, 7
Airtex, 87
Alcohol additives, 61–62
Alternating current (AC), 102
Amperage, 103
Ampere, 103
Analog, 110
Analog inputs, 129–131
Analog-to-digital (AD), 111–112
Anti-backfire valves, 297–298
Antiknock value, 60
Armature, 86
Atmospheric pressure, 51
Atomization, 55
Audi, 34
Automotive emissions control, 11–14
Available voltage, 347

Backfire, 297
Baffle, 68
Bakelite, 353
Battery, 324
Baud rate, 114–115

Bi-fuel vehicles, 269–271
Binary, 111
Block learn, 169
BMW, 34
Body computer module (BCM), 171
    concept, 194
Body control module (BCM), 120
Boost, 238
Boost pressure control, 251–253
Bore, 19
Bosch, R., 118
Bottom dead center (BDC), 29, 31
Breaker points, 366–367
Buick, 306
Burn time, 37–38, 348
Bus link, 192

Cadillac, 201
California Air Resources Board (CARB),
    5, 116–117, 185, 190
California Pilot Program, 9
Camshaft position (CMP), 175
CAN protocol, 118
Capacitive-discharge (CD), 367
Capacitive-discharge ignition system, 323,
    326–327
Carbon dioxide. *See* CO2
Carbon monoxide. *See* CO
Carbon tracks, 359
Carburetor, 53–54
Carnot, N. L. S., 56
Carter, 87
Catalyst, 300
Catalytic converter monitor, 187
Catalytic converters, 266, 300–305
    and air injection, 299–300
    converter design, 301–302
    light-off, 301
    longevity of, 304–305
    oxidation and reduction reactions, 300
Catalytic cracking, 60
Cecil, W., 32
Central processing unit (CPU), 112
Check valve, 70
Chemical impurities, 59–60
Chevrolet, 34, 57, 201
Chrysler Corporation, 87, 92, 174,
    181–182, 201, 202, 225, 229, 250,
    262, 306, 307, 308, 310, 324, 331,
    332, 337, 375, 376, 387–388,

    400–401, 410. *See also*
      DaimlerChrysler
Circuit components, 104–106
Circuit protection devices, 104–105
Circuits, 103, 106–107
Clean Air Act, 5, 6, 7, 62
Clearance volume, 31
Clock pulses, 114
Closed-loop control, 122
Closed-loop dwell control, 396
Closed-loop operating mode, 164
Closed PCV systems, 292–294
CO, 2, 164, 275, 278, 300
    control, 8
CO_2, 275, 279–280
Coil charge time, 380
Coil packs, 346–347
Combustion process, 276–277
Compressed natural gas (CNG), 269–271
Compression ratio, 31
Computer, parts of, 112–115
Computer control, 107–112
    basic functions, 108–110
Computer integration, 166
Computer memory, 112–114
Computers, types of, 120
Conductors, 103, 106
Continuous injection system (CIS),
    204–205
Control devices, 105
Control system development, 171, 194–195
Conventional theory of current flow, 102
Coolant vacuum switch cold open
    (CVSCO), 298
Corporate Average Fuel Economy (CAFE)
    Standards, 9–11
Crankcase, 22
Crankcase ventilation, 292–295
Crankshaft, 22–23
Crankshaft position (CKP), 37, 175
Critical outputs, 152
Current, 102
Current limiting hump, 330, 381
Current ramping, 372
Curtis-Wright, 41
Cylinder block, 22

DaimlerChrysler, 267. *See also* Chrysler
    Corporation
Data lines, 116–120

Data link connector, 116, 191–192
Dayton Engineering Laboratories
    Company. *See* Delco
Delco, 173, 326
Delco-Remy, 350, 355, 385, 397–400
Delivery system, 67–68
Department of Transportation (DOT), 9–10
de Rivaz, I., 30
Detonation, 60
Detonation sensors, 176–177
Diagnostic Executive, 186
Diagnostic trouble codes (DTCs), 185,
    192–193
Diesel, R., 39
Diesel engine, 39–41
Digital, 110–111
Digital inputs, 131–132
Digital outputs, 152
Digital-to-analog (DA), 111
Direct current (DC), 102
Displacement, 29
Distributor cap, 354–356
Distributor drive, 37
Distributor ignition (DI) system, 326
Distributor ignitions, 385–389
Distributorless ignitions, 389–413
    principles of, 391–396
Distributor rotor, 353–354
Diverter valve, 297–298
(DPFE) sensor, 316
Draft tube ventilation, 292
Dual-point, dual-coil, dual-plug
    ignition, 329
Duty cycle, 202
Dwell, 327, 377–383
    and engine speed, 328
    fixed, 379–380
    variable, 380–383
Dwell time, 327, 379–380

ECT sensor, 133–134, 152
E-85 vehicles, 267–269
EGR valve position sensor, 313
Electric fuel pumps, 88–93
Electromagnetic interference (EMI),
    118, 133
Electronically erasable programmable
    read-only memory (EEPROM), 167
Electronic brake control module
    (EBCM), 120
Electronic control module (ECM), 118
Electronic control systems, 102
    with auxiliary sensors and actuators,
        216–220
Electronic engine controls, 51
Electronic fuel injection (EFI), 54–55,
    200, 203
Electronic ignition (EI) systems, 326
    basic, 366–367
    control modules and primary circuitry,
        367–368
    distributor ignitions, 385–389
    distributorless ignitions, 389–413
    dwell, timing, and advance, 377–385
    imported car ignitions, 409–413

triggering devices and ignition timing,
    368–377
Electronic injection, 205–206
Electronic injectors, 212–215
Electronic spark timing (EST), 153
Electronic throttle control (ETC)
    system, 153
Electron theory of current flow, 102
Enabling criteria, 186
Energy Policy Act of 1992, 269
Engine air-fuel requirements, 50
Engine compression, 237–238
Engine control systems, history of,
    180–185
Engine coolant temperature (ECT),
    219–220
Engine coolant temperature sensor. *See*
    ECT sensor
Engine cooling system, 31–33
Engine lubrication system, 33
Engine mapping, 166–167
Engine operation
    compression and combustion, 18–19
    cylinder arrangement, 27–28
    engine cooling system, 31–33
    engine displacement and compression
        ratio, 28–31
    engine-ignition synchronization, 37–38
    engine lubrication system, 33
    the four-stroke cycle, 19–20
    the ignition system, 33–37
    initial timing, 38–39
    major engine components, 20–27
    other engine types, 39–42
    vacuum, 19
Engine size, 28
Engine speed sensors, 175
Environmental Protection Agency (EPA),
    5, 10
Erasable programmable read-only
    memories (EPROMs), 167
ETBE, 64
Ethanol, 61, 63
Ethyl Tertiary Butyl Ether. *See* ETBE
Evaporative emission control (EVAP)
    systems, 80–85, 158–159, 266
Evaporative systems monitor. *See* EVAP
    system monitor
EVAP system monitor, 189–190
Exhaust check valves, 298
Exhaust gas recirculation, 166
Exhaust gas recirculation (EGR) system,
    157–158, 173, 175–176, 179,
    190–191, 298, 305
    computer-controlled, 310–316
    history, 306–310
    system operation, 306
Exhaust oxygen sensor monitor, 187–188
Exhaust scavenging, 242
Extended-core spark plug, 360–361
External combustion engine, 18

Fail records, 193–194
Federal Energy Act of 1975, 9
Federal test procedure. *See* FTP

Ferrari, 34
Fiber optics, 114
Filler cap, 264
Filler tubes, 69
Firing line, 322
Firing order, 35
Firing voltage, 348
Fixed dwell, 327, 378, 379–380
Flag, 193
Flame arresters, 264
Flexible lines, 73–74
Float valve, 70
Flywheel, 23–24
Ford, H., 56
Ford Motor Company, 87, 88, 92, 95, 137,
    182–184, 224–225, 227–228, 262,
    307, 324, 345, 354, 355, 382,
    402–409
    emission control systems, 267
    flexible fuel system, 266–267
Four-stroke engine, 20
Freeze-frame record, 193
FTP, 283–284
Fuel compensation sensor, 267
Fuel composition, 58–64
Fuel control system operating modes,
    121–123
Fuel delivery system, 207–215
Fuel filters, 93–96, 208–209
    types of, 95
Fuel injection, 200
    advantages of, 201
    common subsystems and components,
        207–220
    and filters, 96
    operating requirements, 200–201
    specific systems, 220–229
    trends, 229–232
    types of systems, 202–207
Fuel-injection fittings, 76
Fuel injectors, 152
Fuel line layout, 77–79
Fuel line mounting, 74
Fuel lines, 71–76, 209
Fuel-metering actuators, 178–179
Fuel pickup tube, 69
Fuel pump, 155–156, 207–208, 264–265
    speed controller, 265
Fuel ratio control, 165
Fuel system monitor, 188–189
Fuel tank, 68–71, 264
Full-function control systems, 170

Gasohol, 61–62
Gasoline additives, 60–61
General Motors (GM), 10, 34–35, 36, 40,
    41, 69, 76, 87, 90, 92, 95, 102, 153,
    171, 184–185, 194, 218, 220–224,
    226–227, 267, 315, 334, 335, 345,
    382, 383
    VFV emissions control systems, 266
    VFV fuel system, 261–266
General Motors Research Corporation, 326
Ground path, 103
Gulp valve, 297

Hall Effect, 371, 402, 409
Hall Effect switch, 331–333, 336, 371–375, 397
HC, 275, 277, 300
Heat range, 358
High-energy ignition (HEI), 350, 355, 385–387, 398
High-side drivers, 149–151
High-speed surge, 308
High voltage
    need for, 321–322
    through induction, 322–323
Hitachi, 173, 409, 410
Honda, 267, 269
HO_S, 140
Hydrocarbons. *See* HC
Hydrocarbons (HC), 2, 164

Idle air control (IAC) motor, 217
Idle speed control (ISC), 154
Idle speed control (ISC) motor, 218
Ignition cables, 356
Ignition coils, 344–352
    coil packs, 346–347
    installations, 350–352
    laminated e-core, 345
    oil-filled, 344
    voltage, 347–350
Ignition control (IC), 152–153
Ignition modules, 368
Ignition reference signal, 137–138
Ignition switch, 324–325
Ignition system, 33–37
    basic circuits and current, 323–324
    cables, 356
    high voltage through induction, 322–323
    need for high voltage, 321–322
    primary circuit components, 324–334
    spark plugs, 356–362
    types of, 326–334
    voltage trace, 334–339
Ignition timing, 166
Ignition timing sensors, 175
I/M, 276, 282–288
    types of tests, 284
Impeller, 88
I/M readiness test, 193
I/M 240 test, 285–286
Inductive-discharge ignition system, 323, 366
Inert, 306
Inertia, 23
Infinite resistance, 104
Injector patterns, 230–231
Inline filters, 94
Inline fuel filter, 264
Input circuits, 114
Input conditioning, 108
Input devices, electrical operation of, 128–133
Inspection/maintenance programs. *See* I/M
Intake air temperature (IAT), 142–143, 152, 174
Intake manifold, 56–57

Intake mixture cooling, 253
Intake system, 51–57, 58–59
Integrator, 169
Intercooler, 240
Internal combustion engine, 17–18
Ionize, 357
Isuzu, 331, 375

Jaguar, 34

Keep-alive memory (KAM), 114
Kettering, C. F., 326

Laminated e-core ignition coils, 345
Lenoir, J. J. E., 56
Lexus, 410
Light-off, 301
Liquid-vapor separator, 80
Load devices, 105–106
Loaded mode test, 284–285
Low-pressure fittings and clamps, 74–75
Low-side drivers, 149

MAF sensor, 136–137, 152, 172–173
Magnetic pulse generator, 331, 368–371
Magnetic saturation, 322
Malfunction indicator lamp (MIL), 156–157, 185, 192–193
Manifold absolute pressure sensor. *See* MAP sensor
MAP sensor, 135–137, 152, 167, 174
Mass air flow sensor. *See* MAF sensor
Mazda, 250, 267
Mechanical fuel injection, 203
Mechanical fuel pumps, 86–88
Mechanical ignition systems, 326
Mechanical-injection nozzles, 212
M85, 257
    blended fuels, 259–260
Mercedes, 34
Methanol, 61, 257
    and automotive engines, 260
    blended fuels, 259–260
    symptoms of exposure, 260
Methyl Tertiary Butyl Ether. *See* MTBE
Microns, 96
Misfire monitor, 188
Mitsubishi, 171, 229, 250, 410
Modified-loop operating mode, 164
Module, 366
MTBE, 63–64
Multiplexing, 117–118, 195
Multiport (port) injection, 200

Natural vacuum leak detection (NVLD), 190
Negative temperature coefficient (NTC) resistors, 133
Network configurations, 118
Nissan, 225, 252
Noble metals, 300
No-load oscillation, 347
Normally aspirated, 238
NOx, 275, 276, 278–279, 300
    formation, 305–306

NSU, 41
Nylon lines, 76

O^2, 275, 280–281, 300
OBD II testing, 286–287
Octane rating, 60
Ohm, 104
Oil-filled ignition coils, 344
Onboard diagnostic (OBD) systems, 11, 185–195
Onboard diagnostics (OBD), 84–85
One-speed and two-speed idle test, 284
On-time (pulse width), 202
Open circuit, 103
Open-loop control, 121–122
Open-loop operating mode, 163–164
Optical signal generator, 333, 375–377
Orifice, 82
Orifice-controlled systems, 294
O^2S, 139–140, 152
OSC, 304
Otto, N., 20, 22
Otto and Langen, 56
Otto-cycle engine, 20
Output circuits, 114
Oxidation, 300
Oxides of nitrogen (NOx), 3–4, 278–279
Oxygen. *See* O2
Oxygenates, 63–64
Oxygen sensors, 138–143
Oxygen storage capacity. *See* OSC
Ozone, control, 8

Papin, D., 56
Parallel circuit, 106
Particulate matter (PM10), 4
PCV, 292–295
    closed PCV systems, 292–294
    orifice-controlled systems, 294
    separator systems, 295
    system efficiency, 295
    system service, 299
Percolation, 87
Photochemical smog, 2
Pintle valve, 307
Pontiac, 201
Poppet valves, 25
Port fuel injection, 203–204
Positive crankcase ventilation systems. *See* PCV
Positive temperature coefficient (PTC) resistors, 133
Post-catalytic converter oxygen sensors, 177
Potential, 103
Power, 104
Power sources, 104
Powertrain control module (PCM), 118, 120, 128, 129, 131, 136, 137, 138, 142, 152, 153, 155, 156, 157, 158, 166, 169
Pre-catalytic converter oxygen sensors, 177
Pressure differential, 48
Pressure drop, 82
Pressure regulator, 209–212
Primary circuit control, 367–368

Programmable read-only memory (PROM), 113
Propane, as alternative fuel, 262
Pulsating, 88
Pulse width modulation (PWM), 151–152, 157
Pump
    operation, 85–86
    types, 86–93
Purge solenoids, 158–159

Random-access memory (RAM), 113–114
Random roadside testing, 287–288
Reach, 358
Read-only memory (ROM), 113
Reciprocating engine, 20
Rectified, 102
Reduction, 300
Reformulated fuels, 62–63
Remote injector driver, 261–262
Remote sensing, 287
Required voltage, 348
Residual or rest pressure, 88
Resistance, 104
Resistor-type spark plugs, 360
Returnless, 78
Rigid lines, 72–73
Robert Bosch Company, 93, 102, 180–181, 201, 203, 204, 210, 212, 213, 214, 218, 221, 224, 332, 409
Rochester Products, 213, 214
Rollover leakage protection, 70–71
Rotary engine, 41–42

SAE, classifications of networking protocols, 118
Safety cells, 75
Sealed housing for evaporative determination (SHED), 283
Secondary air. See AIR
Self-induction, 322
Sensor outputs, 113
Sensors
    critical inputs, 133
    important characteristics of, 132–133

Separator systems, 295
Serial data, 156
Series-parallel circuit, 106
Silicon-controlled rectifier (SCR), 327
Society of Automotive Engineers. See SAE
Solenoid operation, 178
Solenoid vacuum control, 311–313
Solid-state switching devices, 330–333
Spark plugs, 356–362
    construction, 358–362
    firing action, 357–358
Spark voltage, 348
Speed-density, 224
State implementation plan (SIP), 282
Stoichiometric air-fuel ratio, 49
Stoichiometric ratio, 165
Stroke, 19
Subaru, 409, 410
Substrate, 301
Sulfur oxides (SOx), 4
Supercharging, 238–245
Suzuki, 410
Switched input, 128–129
System actuators, 178–180

TAME, 64
Tank venting requirements, 70
Temperature inversion, 5
Tertiary Amyl Methyl Ether. See TAME
Thermal cracking, 60
Three-way converters. See TWC
Throttle, 215–216
Throttle body injection (TBI), 54–55, 200, 203–204
Throttle body injection (TBI) systems, 226
Throttle position sensors (TPS), 134–135, 152, 219
Throttle position (TP) switch, 174
Timing marks, 39
Titania exhaust gas oxygen sensors, 141
Top dead center (TDC), 29, 31
Toyo Kogyo, 41
Toyota, 225, 269, 410
Transmission control module (TCM), 120, 166

Trevithick, R., 56
Trip, 186
Turbochargers, 246–250
    controls, 251–253
    design and operation, 246–249
    installation, 249–250
    size and response time, 249
Turbo lag, 249
TWC, 302–304
Two-stroke engine, 20

Universal asynchronous receiver transmitter (UART), 118

Vacuum lock, 70
Valves, 25–27
Vane-type airflow sensor (VAF), 171–172
Vaporization, 55–56
Vapor lock, 61, 72, 85
Vapor purging, 82–83
Vapor storage, 80–81
Variable dwell, 327, 380–383
Variable fuel sensor, 263–264
Vehicles, variable or flexible fuel, 258–259
Vehicle speed sensor (VSS), 142–143, 175
Visual tampering checks, 284
Volatility, 55, 59
Volkswagen, 40, 41, 69, 409
Voltage, 103–104
Voltage drop, 103
Voltage reserve, 350
Voltage spike, 105
Volumetric efficiency, 48–49
V-type engines, 36

Wankel, F., 41
Wastegate, 250
Waste spark, 391
Water injection, 253
Water jackets, 31
Watt, J., 56